T0388435

Energy from Waste

Energy from Waste

Production and Storage

Edited by
Ram K. Gupta and Tuan Anh Nguyen

CRC Press is an imprint of the
Taylor & Francis Group, an **informa** business

First edition published 2022
by CRC Press
6000 Broken Sound Parkway NW, Suite 300, Boca Raton, FL 33487-2742

and by CRC Press
2 Park Square, Milton Park, Abingdon, Oxon, OX14 4RN

CRC Press is an imprint of Taylor & Francis Group, LLC

Library of Congress Cataloging-in-Publication Data
Names: Gupta, Ram K., editor. | Nguyen, Tuan Anh (Chemist), editor.
Title: Energy from waste : production and storage / edited by Ram K. Gupta
and Tuan Anh Nguyen.
Other titles: Energy from waste (CRC Press)
Description: First edition. | Boca Raton, FL : CRC Press, 2022. | Includes
bibliographical references and index. |
Summary: "Conversion of waste into value-added products such as energy transforms a potential environmental problem into a sustainable solution. This book focuses on the conversion of waste from various sources for use in energy production and storage applications. It provides state-of-the-art methods for developing advanced materials and chemicals for energy applications using waste, and discusses the various treatment processes and technologies. This work provides new direction to scientists, researchers, and students in materials and chemical engineering and related subjects seeking sustainable solutions to energy production and waste management"- Provided by publisher.
Identifiers: LCCN 2021047610 (print) | LCCN 2021047611 (ebook) |
ISBN 9781032013596 (hbk) | ISBN 9781032013732 (pbk) | ISBN 9781003178354 (ebk)
Subjects: LCSH: Waste products as fuel. | Refuse as fuel. | Biomass energy.
Classification: LCC TP360 .E568 2022 (print) | LCC TP360 (ebook) |
DDC 662/.88—dc23/eng/20211209
LC record available at https://lccn.loc.gov/2021047610
LC ebook record available at https://lccn.loc.gov/2021047611

ISBN: 978-1-032-01359-6 (hbk)
ISBN: 978-1-032-01373-2 (pbk)
ISBN: 978-1-003-17835-4 (ebk)

DOI: 10.1201/9781003178354

Typeset in Times
by codeMantra

Contents

PART 4 Waste for Advanced Energy Devices

Preface

The development of agriculture, science, and technology plays a major role in raising living standards; however, rapid urbanization, increasing population, and industrialization also play a significant role in producing a large quantity of wastes globally. Over two billion tons of waste is produced every year, and most of them are dumped untreated, which causes serious environmental and health issues. Waste management is an economic approach that plays a crucial role in preserving the environment and improving living standards. Waste management can be integrated alongside energy-generating technologies as a solution to handling waste as well as producing chemicals and energy. Various methods such as gasification, pyrolysis, anaerobic digestion, fermentation, esterification, and fuel cells can be used to convert waste into chemicals and bioenergy. Various types of energy materials such as syngas, bio-oil, biochar, bioelectricity, biogas, bioethanol, biohydrogen, and biodiesel can be easily generated using wastes. The recent development utilized biowastes for the production of high-performance electrode materials for fuel cells, batteries, and supercapacitors.

The main purpose of our book is to provide current state-of-the-art development in the field of waste management, particularly for generating chemicals and energy. Various processes and technologies for converting wastes into energy are covered, which will be attractive to academia as well as industries. Several new approaches for the use of wastes for green energy production and storage are provided. The book covers an introduction, techniques, the importance of waste management for a sustainable future, and the use of waste in producing many chemicals and energy devices. This book provides fundamentals as well as advanced concepts to the readers for developing new ideas and approaches to utilizing wastes for energy applications.

Ram K. Gupta, Associate Professor
Department of Chemistry
Kansas Polymer Research Center
Pittsburg State University
Pittsburg, Kansas, United States

Tuan Anh Nguyen, Professor
Institute for Tropical Technology
Vietnam Academy of Science and Technology
Hanoi, Vietnam

Preface

Editors

Dr. Ram K. Gupta is an Associate Professor at Pittsburg State University. Dr. Gupta's research focuses on conducting polymers and composites, green energy production and storage using biowastes and nanomaterials, optoelectronic and photovoltaic devices, organic–inorganic heterojunctions for sensors, bio-based polymers, flame-retardant polymers, biocompatible nanofibers for tissue regeneration, scaffold and antibacterial applications, corrosion-inhibiting coatings, and biodegradable metallic implants. Dr. Gupta has published over 235 peer-reviewed articles, made over 300 national, international, and regional presentations, chaired many sessions at national/international meetings, edited many books, and written several book chapters. He has received more than $2.5 million for research and educational activities from many funding agencies. He serves as Editor-in-Chief, Associate Editor, and editorial board member of numerous journals.

Tuan Anh Nguyen earned his BSc in Physics from Hanoi University in 1992 and his Ph.D. in Chemistry from Paris Diderot University (France) in 2003. He was a Visiting Scientist at Seoul National University (South Korea, 2004) and the University of Wollongong (Australia, 2005). He then worked as a Postdoctoral Research Associate & Research Scientist at Montana State University (the USA), 2006–2009. In 2012, he was appointed as Head of the Microanalysis Department at the Institute for Tropical Technology (Vietnam Academy of Science and Technology). He has managed four Ph.D. theses as thesis director, and three are in progress. He is Editor-In-Chief of "Kenkyu Journal of Nanotechnology & Nanoscience" and Founding Co-Editor-In-Chief of "Current Nanotoxicity and Prevention". He is the author of 4 Vietnamese books and Editor of 32 Elsevier books in the Micro & Nano Technologies Series.

List of Contributors

Rose Fadzilah Abdullah
Institute of Advanced Technology
Universiti Putra Malaysia
Serdang, Malaysia

Mohamed Aboughaly
Faculty of Energy Systems and Nuclear Science
University of Ontario Institute of Technology
Oshawa, Ontario, Canada

Jeffin James Abraham
Center for Advanced Materials
Qatar University
Doha, Qatar

Benjamin Agyei-Tuffour
Department of Materials Science and Engineering
University of Ghana
Legon-Accra, Ghana

Stefania Akromah
Department of Materials Engineering
College of Engineering, Kwame Nkrumah
University of Science and Technology
Kumasi, Ghana

Jazib Ali
Electronic Engineering Department
University of Rome Tor Vergata
Rome, Italy

Amornchai Arpornwichanop
Center of Excellence in Process and Energy Systems Engineering
Department of Chemical Engineering, Faculty of Engineering
Chulalongkorn University
Bangkok, Thailand
and
Bio-Circular-Green-economy Technology & Engineering Center
Department of Chemical Engineering, Faculty of Engineering
Chulalongkorn University
Bangkok, Thailand

Christian Randell A. Arro
Department of Chemistry and Earth Sciences
College of Arts and Sciences
Qatar University
Doha, Qatar

Ghulam Abbas Ashraf
Department of Physics
Zhejiang Normal University
Zhejiang, China

Gita Bagheri
Department of Chemical Engineering
Islamic Azad University
Shahryar, Iran

Dona Susan Baji
Centre for Nanosciences and Molecular Medicine
Amrita Vishwa Vidyapeetham
Kochi, India

Souhardya Bera
Department of Chemical Engineering
University of Calcutta
Kolkata, India

Muhammad Bilal
School of Life Science and Food Engineering
Huaiyin Institute of Technology
Huaian, China

Stephanie C. Bolyard
Florida State University
Environmental Research and Education Foundation
Raleigh, North Carolina

Dan J.L. Brett
Electrochemical Innovation Lab (EIL)
Department of Chemical Engineering
University College London (UCL)
London, United Kingdom
and
The Faraday Institution
Harwell Campus
Didcot, United Kingdom

Ayşe Çelik Bedeloğlu
Polymer Materials Engineering Department
Bursa Technical University
Yıldırım, Turkey

T.S Chandra
Department of Biotechnology
Indian Institute of Technology Madras
Chennai, India

Yingna Chang
Al-ion Battery Research Center
Department of Electrical Engineering and Automation
Shandong University of Science and Technology
Shandong, China

Nidhi Chauhan
Amity Institute of Nanotechnology
Amity University Uttar Pradesh
Noida, India

Jonghyun Choi
Department of Chemistry
Kansas Polymer Research Center
Pittsburg State University
Pittsburg, Kansas

Elisangela Pacheco da Silva
Department of Chemistry
UEM – State University of Maringa
Maringá, Brazil

David Dodoo-Arhin
Department of Materials Science and Engineering
University of Ghana
Legon-Accra, Ghana

Prashant Dubey
Advanced Carbon Products and Metrology Department
CSIR-National Physical Laboratory (CSIR-NPL)
New Delhi, India

Ali A. El-Samak
Department of Civil Engineering and Architectural Engineering
Qatar University
Doha, Qatar

Elizângela Hafemann Fragal
IMP
Univ Lyon
Villeurbanne, France

Vanessa Hafemann Fragal
Department of Chemistry
UEM – State University of Maringa
Maringá, Brazil

Hossam A. Gabbar
Faculty of Energy Systems and Nuclear Science
University of Ontario Institute of Technology
Oshawa, Ontario, Canada

Luiz Fernando Gorup
LIEC – Laboratory Interdisciplinar de Eletroquímica e Cerâmica
Department of Chemistry
University of São Carlos
São Carlos SP, Brazil
and
School of Chemistry and Food Science
Federal University of Rio Grande
Rio Grande, Brazil
and
Materials Engineering
Federal University of Pelotas
Pelotas, Brazil

Ram K. Gupta
Department of Chemistry
Kansas Polymer Research Center
Pittsburg State University
Pittsburg, Kansas

Jennifer Hack
Electrochemical Innovation Lab (EIL)
Department of Chemical Engineering
University College London (UCL)
London, United Kingdom

Essia Hannachi
Department of Nuclear Medicine Research
Institute for Research and Medical Consultations (IRMC)
Imam Abdulrahman Bin Faisal University
Dammam, Saudi Arabia

Alaa H. Hawari
Department of Civil Engineering and Architectural Engineering
Qatar University
Doha, Qatar

Balkis Hazmi
Institute of Advanced Technology
Universiti Putra Malaysia
Serdang, Malaysia

Guanjie He
Electrochemical Innovation Lab (EIL)
Department of Chemical Engineering
University College London (UCL)
London, United Kingdom
and
School of Chemistry
University of Lincoln
Lincoln, United Kingdom

Fabeena Jahan
School of Chemical Sciences
Kannur University
Payyanur, India

Utkarsh Jain
Amity Institute of Nanotechnology
Amity University Uttar Pradesh
Noida, India

Mahfuz Kabir
Bangladesh Institute of International and Strategic Studies (BIISS)
Dhaka, Bangladesh

Zobaidul Kabir
School of Environmental and Life Sciences
The University of Newcastle
New South Wales, Australia

Ramji Kalidoss
Department of Biomedical Engineering
Bharath Institute of Higher Education and Research
Selaiyur, India

Ashwinder Kaur
CSIR-Central Scientific Instruments Organization (CSIR-CSIO)
Chandigarh, India
and
Department of Physics
Punjabi University
Patiala, India

Fahmeeda Kausar
School of Chemistry & Chemical Engineering
Shanghai Jiao Tong University
Shanghai, China

Radhakrishnan Kothalam
Department of Chemistry
College of Engineering and Technology
SRM Institute of Science and Technology
Kattankulathur, India

Anuj Kumar
Nano-Technology Research Laboratory
Department of Chemistry
GLA University
Mathura, India

Ashish Kumar
Department of Chemistry
Faculty of Technology and Science
Lovely Professional University
Phagwara, India

Hao Li
Institutes of Physical Science and Information Technology
Anhui University
Hefei, China

Zongge Li
Al-ion Battery Research Center
Department of Electrical Engineering and Automation
Shandong University of Science and Technology
Shandong, China

Yiyang Liu
Electrochemical Innovation Lab (EIL)
Department of Chemical Engineering
University College London (UCL)
London, United Kingdom

Quanwei Ma
Institutes of Physical Science and Information Technology
Anhui University
Hefei, China

Sindhu Thalappan Manikkoth
School of Chemical Sciences
Kannur University
Payyanur, India

B.C. Meikap
Department of Chemical Engineering
Indian Institute of Technology (IIT) Kharagpur
Kharagpur, India
and
Department of Chemical Engineering
School of Chemical Engineering
University of Kwazulu-Natal (UKZN)
Durban, South Africa

Kwadwo Mensah-Darkwa
Department of Materials Engineering
College of Engineering
University of Science and Technology
Kumasi, Ghana

Asmita Mishra
Department of Chemical Engineering
Indian Institute of Technology (IIT) Kharagpur
Kharagpur, West Bengal

Sunita Mishra
CSIR-Central Scientific Instruments Organization (CSIR-CSIO)
Chandigarh, India

Wan Nur Aini Wan Mokhtar
Department of Chemical Sciences
Faculty of Science
Universiti Kebangsaan Malaysia
Bangi, Malaysia

Vanessa Bongalhardo Mortola
School of Chemistry and Food Science
Federal University of Rio Grande
Rio Grande, Brazil

Isha Mudahar
Department of Basic and Applied Sciences
Punjabi University
Patiala, India

Tasnim Munshi
School of Chemistry
University of Lincoln
Lincoln, United Kingdom

Anjali V. Nair
Centre for Nanosciences and Molecular Medicine
Amrita Vishwa Vidyapeetham
Kochi, India

Shantikumar Nair
Centre for Nanosciences and Molecular Medicine
Amrita Vishwa Vidyapeetham
Kochi, India

Nathada Ngamsidhiphongsa
Center of Excellence in Process and Energy Systems Engineering
Department of Chemical Engineering, Faculty of Engineering
Chulalongkorn University
Bangkok, Thailand

C. Nithya
Department of Chemistry
PSGR Krishnammal College for Women
Coimbatore, India

Alfred Nkhama
Department of Physics
Kansas Polymer Research Center
Pittsburg State University
Pittsburg, Kansas

Deepthi Panoth
School of Chemical Sciences
Kannur University
Payyanur, India

Anjali Paravannoor
School of Chemical Sciences
Kannur University
Payyanur

Yaneeporn Patcharavorachot
Department of Chemical Engineering
School of Engineering
King Mongkut's Institute of Technology Ladkrabang
Bangkok, Thailand

Michelly Cristina Galdioli Pellá
Department of Chemistry
UEM – State University of Maringa
Maringá, Brazil

Aditi Podder
Brown and Caldwell
Houston, Texas

Deepalekshmi Ponnamma
Center for Advanced Materials
Qatar University
Doha, Qatar

Phuet Prasertcharoensuk
Center of Excellence in Process and Energy Systems Engineering
Department of Chemical Engineering, Faculty of Engineering
Chulalongkorn University
Bangkok, Thailand

Cristian Tessmer Radmann
School of Chemistry and Food Science
Federal University of Rio Grande
Rio Grande, Brazil

M. Ashiqur Rahman
School of Engineering
Design and Built Environment
Western Sydney University
Sydney, Australia

Mofijur Rahman
Department of Mechanical Engineering
University of Technology Sydney (UTS)
New South Wales, Australia

Ranjusha Rajagopalan
College of Chemistry and Chemical Engineering
Central South University
Changsha, P.R. China

Umer Rashid
Institute of Advanced Technology
Universiti Putra Malaysia
Serdang, Malaysia

Debra R. Reinhart
Civil, Environmental and Construction Engineering Department
University of Central Florida
Orlando, Florida

Subhasis Roy
Department of Chemical Engineering
University of Calcutta
Kolkata, India

Dhamodaran Santhanagopalan
Centre for Nanosciences and Molecular Medicine
Amrita Vishwa Vidyapeetham
Kochi, India

Lip Huat Saw
Lee Kong Chian Faculty of Engineering and Science
UTAR
Kajang, Malaysia

Kirti Saxena
Amity Institute of Nanotechnology
Amity University Uttar Pradesh
Noida, India

Ian Scowen
School of Chemistry
University of Lincoln
Lincoln, United Kingdom

Thiago Sequinel
Faculty of Exact Sciences and Technology (FACET)
Federal University of Grande Dourados
Dourados, Brazil

Shadab Shahsavari
Department of Chemical Engineering
Islamic Azad University
Varamin, Iran

Shahin Shahsavari
Process Engineering Department
Faculty of Chemical Engineering
Tarbiat Modares University
Tehran, Iran

Shveta Sharma
Department of Chemistry
Faculty of Technology and Science
Lovely Professional University
Phagwara, India

Paul R. Shearing
Electrochemical Innovation Lab (EIL)
Department of Chemical Engineering
University College London (UCL)
London, United Kingdom
and
The Faraday Institution
Didcot, United Kingdom

Zahra Shokri
Pharmaceutical Incubator Center
Tehran University of Medical Sciences
Tehran, Iran

Vishal Shrivastav
CSIR-Central Scientific Instruments Organization (CSIR-CSIO)
Chandigarh, India

Hammad Siddiqi
Department of Chemical Engineering
Indian Institute of Technology (IIT) Kharagpur
Kharagpur, India

Rafael Silva
Department of Chemistry
UEM – State University of Maringa
Maringá, Brazil

Yassine Slimani
Department of Biophysics
Institute for Research and Medical Consultations (IRMC)
Imam Abdulrahman Bin Faisal University
Dammam, Saudi Arabia

Muhammad Rizwan Sulaiman
Department of Chemistry
Kansas Polymer Research Center
Pittsburg State University
Pittsburg, Kansas

Shashank Sundriyal
Advanced Carbon Products and Metrology Department
CSIR-National Physical Laboratory (CSIR-NPL)
New Delhi, India

Weng Cheong Tan
Lee Kong Chian Faculty of Engineering and Science
UTAR
Kajang, Malaysia

Abhinay Thakur
Department of Chemistry
Faculty of Technology and Science
Lovely Professional University
Phagwara, India

Bhim Sen Thapa
Department of Biological Environment
Kangwon National University
Gangwon-do, Republic of Korea
and
Department of Biotechnology
Indian Institute of Technology Madras
Chennai, India

Kunnambeth M. Thulasi
School of Chemical Sciences
Kannur University
Payyanur, India

Ömer Faruk Ünsal
Polymer Materials Engineering Department
Bursa Technical University
Yıldırım, Turkey

Ganesh Vattikondala
Department of Physics and Nanotechnology
College of Engineering and Technology
SRM Institute of Science and Technology
Kattankulathur, India

Baiju Kizhakkekilikoodayil Vijayan
School of Chemical Sciences
Kannur University
Payyanur, India

Haiyan Wang
College of Chemistry and Chemical Engineering
Central South University
Changsha, China

Rui Wang
Institutes of Physical Science and Information Technology
Anhui University
Hefei, China

Ming Chian Yew
Lee Kong Chian Faculty of Engineering and Science
UTAR
Kajang, Malaysia

Ming Kun Yew
Lee Kong Chian Faculty of Engineering and Science
UTAR
Kajang, Malaysia

Changzhou Yuan
School of Material Science & Engineering
University of Jinan
Jinan, P. R. China

Chaofeng Zhang
Institutes of Physical Science and Information Technology
Anhui University
Hefei, China

Guoxin Zhang
Al-ion Battery Research Center
Department of Electrical Engineering and Automation
Shandong University of Science and Technology
Shandong, China

Longhai Zhang
Institutes of Physical Science and Information Technology
Anhui University
Hefei, China

Tengfei Zhou
Institutes of Physical Science and Information Technology
Anhui University
Hefei, China

Part 1

Introduction

1 Biowastes for Energy
An Introduction

Alfred Nkhama, Muhammad Rizwan Sulaiman, Jonghyun Choi, and Ram K. Gupta
Pittsburg State University

CONTENTS

1.1 INTRODUCTION

Today's modern society is in dire need of energy since every technological advancement and day-to-day living are supported by energy. For decades, the main power supplies have primarily been based on limited and nonrenewable fossil fuels. These fossil fuels were earlier consumed excessively, which resulted in drastic consequences such as ozone layer depletion, environmental degradation, air pollution, global warming, and deforestation. With the continuous increase in the global population and increasing need for energy for daily activities, fossil fuels may not satisfy the energy demand in the future. Importantly, the current hydrocarbon resource

DOI: 10.1201/9781003178354-2

discoveries are getting smaller than those found in the past, which implies that they are being depleted. Therefore, the exploration of clean and sustainable energy, such as solar energy, hydropower, tidal energy, bioenergy, wind energy, and geothermal energy, is necessary for a sustainable future.

Various stakeholders, including international communities, governments, and individuals, were taking an interest in developing a sustainable future based on the techniques established over the past few decades to substitute hydrocarbon-derived fuels with renewable energy sources. Among various renewable energy resources, including solar energy, wind energy, tidal energy, and fuel cells, energy production using biowastes could be a cost-effective approach and provides a clean solution to ever-growing wastes. Biowastes derived from living organisms or having an organic origin are potential sources of chemicals and energy.

1.2 SOURCE AND SIGNIFICANCE OF BIOWASTES

Wastes such as crop residues, livestock effluents, sawdust, and sewage treatment products are biomass that can be decomposed under aerobic and anaerobic conditions to convert into value-added products [1,2]. Biowaste management is an economic approach that plays a crucial role in safeguarding the environment and enhancing living standards. Most of the biowastes are landfilled or burned to avoid trash accumulation. The landfill process requires human resources, energy, and a large dump area, while burning produces undesirable gases. Both processes severely affect the global environment and the health of living beings. To overcome these issues, waste management can be integrated alongside energy-generating technologies. Biowaste treatment with different biowaste-to-bioenergy (BtB) techniques is an executable way for treating biowaste as well as producing energy [3–5]. The worldwide market value of BtB technology is around $25 billion and is expected to reach $40 billion by 2023 [6]. Various biological techniques such as fermentation, esterification, anaerobic digestion, and electro-fuel cells are applied to convert waste into energy. Other physiochemical processes such as gasification, pyrolysis, landfills, and incineration are also commonly used. The classification of biowastes is based on the origin of waste and requires different strategies for pretreatment and transformation into bioenergy. The following are the classes of biowastes and processes to convert them into bioenergy.

1.2.1 Biowastes from Forest and Wood Processing Industries

Forest wastes can be categorized into two groups: (i) waste produced during wood reaping from the forest and (ii) waste generated during wood processing in industries such as plywood and timber. Unutilized forest wastes could result in forest fires that can lead to economic damage, soil degradation, environmental pollution, and wildlife deterioration. The use of forest biomass to generate energy can minimize forest fires and act as a suitable alternative for sustainable energy. Wood wastes are also generated via many other activities such as thinning of plantations, constructions, and furniture industries. Furniture industries mostly generate trim and sawdust. The biomass generated from different sources contains different amounts of cellulose, lignocellulose, and lignin, which impact energy production efficiency. The content of the components helps to decide the pretreatment steps in BtB. Biomass with little lignin is regarded as a good raw material for generating energy [5].

1.2.2 Biowaste from Food Processing

The food processing industry is one of the rapidly growing industries in the world. The continuously growing population requires even more supermarkets and restaurants to supply and satisfy the need for food. Breweries, edible oil production, bakeries, juice factories, meat processing, etc., are a few food industries that produce a large number of biowastes. The wastes generated from the food industries can be categorized into liquids or solids. Fruits that do not meet and comply with the set standards are considered solid wastes in fruit industries or grocery stores. The so-called liquid wastes are obtained after washing meat, fruits, and vegetables and are made of starch, sugar, and organic matter. Other sources of liquid waste are oils or grease and household liquids. Vegetable oil used in cooking food is discarded after certain uses. Food-derived wastes can be processed and transformed into useful products such as enzymes, biofuels, and nutraceuticals by biorefinery, thus helping reduce waste from the environment.

1.2.3 Biowaste from the Paper Industry

The paper and pulping industries are considered to be the third largest contributor to pollution [7]. The process of breaking bonds in the structure of wood is known as pulping. The pulping process in the paper industry is carried out by applying different mechanical and chemical processes [7]. The employed pulping method also determines the yield and quality of the pulp produced. The disposable papers such as napkins and newspapers are produced by mechanical processes. Special papers such as rayons and photographic films are produced by sulfite pulping and mostly consist of cellulose. Waste in solid form is mostly generated during pulping, de-inking, and wastewater treatment processes. It is estimated that 40–50 kg of sludge is produced during the production of one ton of paper [8]. The control and management of such wastes is a challenge. Anaerobic digestion or different technologies have been employed to efficiently deal with this challenge by transforming these wastes into energy.

1.2.4 Biowaste from Municipal Solid

The economic booming, population growth, and urbanization result in the rapid increase in municipal solid wastes. A person produces 100–400 kg of municipal solid wastes yearly. About 2.2 billion tons of municipal solid wastes are projected to be produced every year until 2025, which would be increased to 4.2 billion tons by 2050 [9]. Many challenges have been faced by the authorities and governments in developing nations in managing such wastes. Municipal solid wastes are mainly managed by four methods: landfills, thermal, recycling, and biological treatments.

1.2.5 Animal Waste

Animal wastes include animal dung, meat products from slaughterhouses, and wastes from the dairy/poultry industry. Skin, blood, meat, tendons, hair, feathers, bone, and internal organs are the main wastes generated in the meat industry. If these wastes are discarded untreated, they might convert into a medium favoring various pathogens' growth. Livestock producers also face a challenge in managing animal waste. Most of the wastes are kept in lagoons for future use as manure that generates flammable methane gas through a natural decomposition process, creating a safety hazard if not taken care of properly. The handling of animal waste relies on the solids content; those with 20%–25% solids can be handled easily. One approach that has acquired tremendous attention in recent years for animal waste handling is bioenergy production from animal waste via microbes. Animal excreta is composed of microbes that help in biogas generation and is directly applicable in biofuel production by BtB technologies.

1.3 PRETREATMENT OF BIOWASTE

As discussed previously, biowastes are derived from various sources including animal waste, agricultural yield processing, food processing, and municipal generated waste. Some wastes consist of complex structures and, therefore, need to be pretreated to be digestible and fermentable. The pretreatment methods depend on the composition of biowastes.

1.3.1 Pretreatment of Animal Fat Waste

The solid or semi-solid animal waste materials are first ground into small sizes and then subjected to high temperatures (115 °C–145 °C) to eliminate moisture and discharge fat [10]. Melting and supercritical CO_2 are the common methods used for recovering utilizable fat from animal wastes. During the melting of the fat via heating the animal waste, it can be separated from the indissoluble solids. Taher et al. applied supercritical CO_2 (3 mL/min, 55 °C, and 500 bar of pressure) to draw out 87.4% fats from animal wastes and converted the recovered fat into biodiesel via the transesterification process [11]. Animal fat wastes are good sources for producing biodiesel mostly using the transesterification process. Animal fat wastes contain proteins, water, fatty acids, and phosphoacylglycerols, which can influence the transesterification process and the quality of the produced biodiesel, and hence need to be pretreated before converting into biodiesel. Phosphoacylglycerols are removed by the degumming process, which involves applying 60% orthophosphoric acid to remove the gum through centrifugation [12]. Water contained in animal fat wastes is taken away by vacuum-drying before applying transesterification. Animal fat wastes contain a high amount of fatty acid that can lead to soap formation during the transesterification process. Fatty acids can be removed by acid-catalyzed esterification before base-catalyzed transesterification.

1.3.2 Lignocellulosic Waste Pretreatment

Lignocellulosic materials consist of cellulose, hemicellulose, and lignin. They also consist of small fractions of pectin, protein, ash, and extractives (including soluble nonstructural materials such as sugars, nitrogenous compounds, chlorophyll pigments, and waxes). The pretreatment goal is to break down supramolecular structures, which bind cellulose–lignin–hemicellulose matrix. The breakdown allows carbohydrate polymers to be accessible by different hydrolases such as acids or enzymes [5,13]. Various pretreatment methods such as chemical, physical, biological, and physicochemical methods are used to transform biomass into suitable raw materials that are fermentable to microbes. Mulakhudair et al. applied microbubbles with high ozone content formed by fluidic oscillation for pretreating lignocellulose [14]. The free radicals produced in the gas–liquid interfacial sites of microbubbles attack the disintegrated lignocellulosic material, causing it to become more susceptible and hydrolyzable. There are still no adequate details showing the correlation between the defiant nature of biomass and how its composition impacts enzymic hydrolysis. An immense study and effort are encouraged to utilize biomass as a useful feedstock for bioenergy generation.

1.3.3 Pretreatment of Waste Cooking Oil

Fatty acid content, dimer, and polymer production increases due to the accelerated hydrolysis process by heat and water during the frying process, which affects biodiesel yield and transesterification process [15]. These effects might be avoided through pretreatment techniques, which include vacuum distillation to remove water, filtration to get rid of particles, and adsorption for removing fatty acids [16]. Schnieder et al. used an adsorption method to decrease the acidity of waste cooking oil as low as 63% using rice husks [17]. A re-esterification process that included mixing metal catalysts with glycerol and heating the formed mixture at 200°C was used to remove fatty acids from waste cooking oil [18].

1.3.4 Removal of Inhibitory Compounds and Salts

Biowastes with high amounts of salts and heavy metals such as cadmium, lead, and mercury are not suitable for direct application in anaerobic digestion or composting because of their inhibitory tendency to microbes. Anwar et al. investigated the influence of salt content on anaerobic digestion and observed that the presence of salts reduces the rate of methane production [19]. These problems can be managed by searching and identifying salts and metals before fermenting waste materials. The famous approach applied to removing salts is electrodialysis, and metal removal is done by the adsorption method using activated carbon [20]. Methods such as oxidation and chemical precipitation are applicable, but they are not favored due to the generation of toxic side products [21].

1.4 BIOWASTE TO BIOENERGY

Various types of bioenergies can be produced by processing wastes via various conversion pathways. These bioenergies include syngas, bio-oil, biochar, bioelectricity, biogas, bioethanol, biohydrogen, and biodiesel (Figure 1.1) [22]. Organic components-enriched biowastes make them a potential candidate for bioenergy production. These different types of bioenergy have been produced by applying biological and physicochemical methods. The biological methods include anaerobic digestion, transesterification, fermentation, and microbial fuel cells. Similarly, physicochemical methods include incineration, gasification, landfill, and pyrolysis. In this section, various biological methods that are mostly applied to produce bioenergies are addressed.

1.4.1 Biodiesel from Biowaste

Biodiesel can be produced via transesterification using animal fat, used cooking oils, and microbial oil [23,24]. Biodiesel can be directly applied in powering engines in its pure form (B100) or as a mixture with petrodiesel to generate fuel [25]. Used cooking oils contain a large amount of triacylglycerol, which can be converted to biodiesel via different techniques such as chemical, enzymatic, and mechanical processes. Supercritical fluid, microwave-assisted, and accelerated solvent extraction are other methods that are commonly used for the production of biodiesel from biowastes. Various biowastes such as sludge waste, animal fat waste, waste cooking oil, inedible oils, and microbial oils have been used for generating biodiesel [26–29].

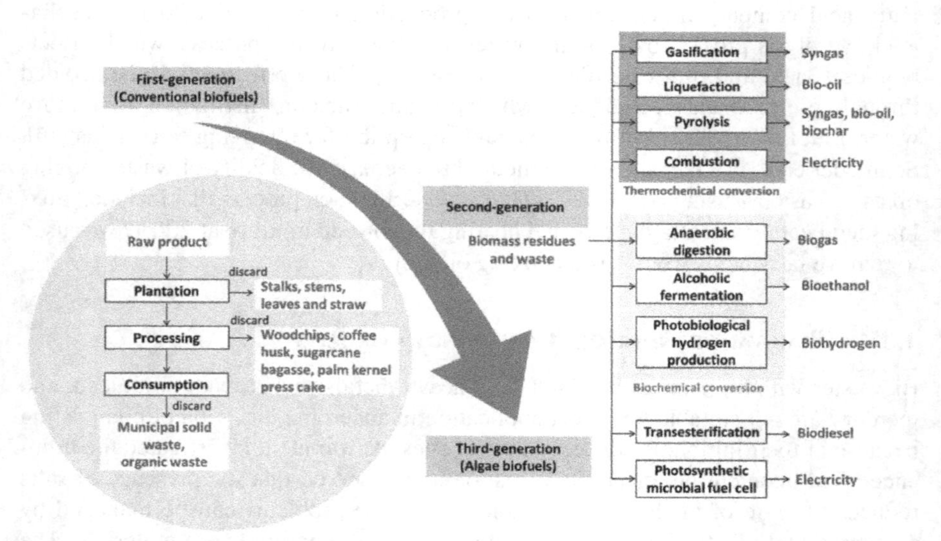

FIGURE 1.1 Schematic diagram illustrating the production of biofuel from biomass residues and wastes and their conversion pathways in producing bioenergy. (Adapted with permission from Ref. [22]. Copyright 2019, The Authors. Distributed under a Creative Commons Attribution License 4.0 (CC BY).)

Waste cooking oil can be directly transformed into biodiesel via transesterification reactions. Ho et al. developed a two-step synthetic route for biodiesel production from waste cooking oil using an esterification process [30]. In the first step, an acidic catalyst was used, followed by the application of alkaline catalysts to understand their effect on the characteristics of the produced biodiesel [30]. Similarly, a series of hydrothermal carbonization and sulfonation of xylose was employed by Tran et al. to develop a non-polluting catalyst for generating biodiesel from waste cooking oil [31]. The synthesized catalyst produced biodiesel with a yield of 85%. A solid acidic catalyst was synthesized by Gardy et al. by applying a series of functionalization methods [32]. The solid acidic catalyst efficiently produced 96% fatty acid methyl ester from waste cooking oil in 2.5 h. Moreover, in the biodiesel production process, glycerol is obtained as a by-product and can be utilized as a commercial product to earn extra profit from the biodiesel plant.

During the wastewater treatment process, sewage sludge is produced as waste that has a large lipid content and fatty acids for biodiesel generation. More lipids are present in activated sludge compared to primary sludge due to the presence of microbes in the activated sludge, which are important in the biological treatment of wastewater. The microorganisms are heterotrophic and consume organic matter to obtain energy for the preparation of lipids [33]. The microbes containing a large amount of lipids generate an increased amount of biodiesel. There are two types of biodiesel production processes: (i) one-stage method (simultaneous lipid extraction and esterification) and (ii) two-stage method (lipid extraction followed by esterification). A two-stage method was applied by Pastore et al. for producing biodiesel using dried-up sludge [34]. Kwon et al. worked on the production of biodiesel and found that sewage sludge has the potential to generate 980,000 L/ha.year of biodiesel, which is much higher than that produced by any other feedstocks/biowastes [35]. For example, algae and soybean oil generate 446 and 2,200 L/ha.year, respectively.

Animal fat waste contains glycerol and fatty acids as the main components. Through the transesterification process, extracted fat is used to generate biodiesel in the same way as other wastes/feedstocks. Different research groups have reported efficient biodiesel production using various animal fats such as beef tallow, pork lard, chicken, fish, and sheep [36,37]. A transesterification process by applying charcoal and CO_2 at atmospheric pressure and high temperature (350 °C–500 °C) was a newly designed method for biodiesel production with a transformation efficiency of 98.5% in 1 min [38]. A catalytic method to produce biodiesel using chicken fat via esterification using methanol was proposed by Shi et al. [39]. The free fatty acids in chicken fat were transformed into biodiesel using a membrane catalyst.

Biodiesel can also be produced from microbial oil [40]. Microbes can collect triacylglycerol in strained conditions. When microbial triacylglycerol is removed from microbes and subjected to transesterification, biodiesel is produced. Biodiesel can be produced using three types of processes based on microbial systems: in vivo, in vitro, and semi-in vivo/semi-in vitro. The in vivo approach comprises microbes that generate all the important basic components needed to generate biodiesel, i.e., fatty acids, alcohol, and lipase. In in vitro methods, microbial cells generate lipase and release it into the medium, while other constituents, i.e., fatty acid and alcohol, are utilized to produce biodiesel. The semi-in vivo/semi-in vitro approach involves the lipase and fatty acids or alcohol prepared by microorganisms with the insertion of other required components [40]. Biodiesel of a wide range of properties is produced when different oils are used.

1.4.2 Biogas from Biowaste

Biological materials can be broken down in the absence of oxygen by microorganisms through anaerobic digestion [41,42]. Biogas which is composed of methane and other rare gases can be produced via the anaerobic digestion process. Various microorganisms carry out stepwise anaerobic digestion that takes place by hydrolysis, acidogenesis, acetogenesis, and methanation. During hydrolysis, microbes produce hydrolases, which degrade complex polymeric biomass materials into simpler molecules. Simpler molecules are further fermented into volatile fatty acids and gases, such as H_2 and CO_2. The produced materials are then reduced into acetic acid employing acetogenic bacteria. Methanogenic bacteria ferment products formed in different steps to produce water, CO_2, and methane [5]. The anaerobic digestion process can apply to many biological wastes such as animal, agricultural, food, and municipal wastes.

Landfill disposal of waste is a threat to the environment as it generates uncontrolled biogases such as CH_4 and CO_2. CH_4 can trap 20 times more heat than CO_2 if discharged into the environment. Anaerobic digestion of wastes can be used for the controlled production of CH_4 as an energy gas. Parameters, such as temperature and pH, can significantly affect the anaerobic digestion process. Bacteria, which play an important role in the production of CH_4, are generally operative in the pH range of 6.5–7.2, and the rest of the fermentative bacteria are functional in a wide pH range of 4.0–8.5 [43]. Temperature affects the metabolism of microorganisms, and hence their composition. The solubility and degradability of volatile fatty acids increase with temperature, which thus enhances their degradation process. Anaerobic digestion processes are classified based on various parameters such as feeding type (fed-batch, continual, and batch), moisture content (wet and dry), and temperature (mesophilic and thermophilic). The theoretical production of biogas is affected by the protein, fat, and carbohydrate contents of the waste. The yield of anaerobic digestion can be improved by co-digestion with other substrates [44,45]. Co-digestion using different substrates improves the anaerobic digestion process due to the synergism effect introduced by co-substrates. The anaerobic digestion technology is a viable option in managing biowaste as well as energy challenges, at the same time generating income from the formed by-products.

1.4.3 Bioelectricity from Biowaste

Electricity can be generated using biowastes as raw materials in microbial fuel cells. Microbial fuel cells utilize microorganisms as bioreactors for transforming chemical energy stored in biowastes into electrical energy in an anaerobic controlled environment. Structurally, microbial fuel cells are made of a cathode and an anode that are linked to each other by a salt bridge or membrane. The organic content in biowastes is metabolized by microorganisms, which multiply and grow and form multiple intermediate products. During metabolism, electrons and protons are generated via redox reactions. In an anaerobic environment, electrons are carried to the redox mediator by microorganisms, from where they can be conveyed to electrodes during oxidation. As electrons reach the electrode, it flows via an external circuit. At the same time, protons reach the cathodic chamber by diffusional movement across the solution and merge with O_2 ions to form H_2O. The anode's potential diminishes and therefore causes potential differences across the electrodes, giving rise to an electrical current during the substrate's oxidation [46].

The microbes and the substrates can affect the current generation. In microbial fuel cells, electron transfer is carried out by direct electron transfer and mediated electron transfer [9,47]. Direct electron transfer involves nanowires known as transmembrane proteins, and mediated electron transfer utilizes a mediator that helps transfer electrons [48]. Bacterial species such as *Aeromonas, Escherichia, Saccharomyces, Candida, Clostridium, Klebsiella,* and *Shewanella* were investigated and proven to be capable of generating electricity in microbial fuel cells [49–53]. Different electron mediators such as natural red and potassium ferricyanide are reported to improve the microbial fuel cells' performance [54]. When an exogenous mediator is added, it enhances the microbial fuel cells' efficiency, but they are poisonous and can hamper the growth of microbes.

1.4.4 Bioalcohol from Biowaste

There is a high demand for alcohol in the automobile and chemical industries for fueling and solvent purposes. Alcohol is produced via different biological techniques. For a long time, alcohol production was mostly dependent on crops and posed a potential food scarcity likelihood. Production of alcohol from biowastes is a rewarding solution for maintaining food security. Alcohol production has been addressed by using various biowastes in the following sections. Fermentation of biological material is a standard method for producing alcohol. Waghmare et al. used dried banana skin as a raw material with *Saccharomyces cerevisiae* for the production of ethanol [55]. Waste from rice winery was used by Vu and Kim to obtain ethanol utilizing *Saccharomyces cerevisiae* KV25 [56]. Usually, the accumulation of ethanol in the course of fermentation hinders microbial growth and their multiplication, resulting in a disturbance in ethanol formation. To tackle the issue, Huang et al. applied a vacuum recovery technique to generate ethanol using food waste via maintaining a bioreactor vacuum environment [57]. The obtained ethanol in the bioreactor was heated to its boiling temperature and tapped as a condensate. Hossain et al. obtained ethanol by using pineapple waste, analyzed it, and asserted satisfactory properties in compliance with ASTM requirements for use as a transportation fuel [58]. Figure 1.2 shows a scheme for bioethanol production using biowastes.

Butanol can be also used as a fuel to power engines without a need for further refinement [60]. It is positioned after ethanol on a fermentation scale. Butanol can be produced from various biowastes after pretreatment. Physical and chemical processes are two common methods employed for butanol production. Cheng et al. pretreated rice straw and sugarcane bagasse followed by a hydrolysis process in the presence of cellulose to generate biobutanol [61]. The hydrolysate was passed through

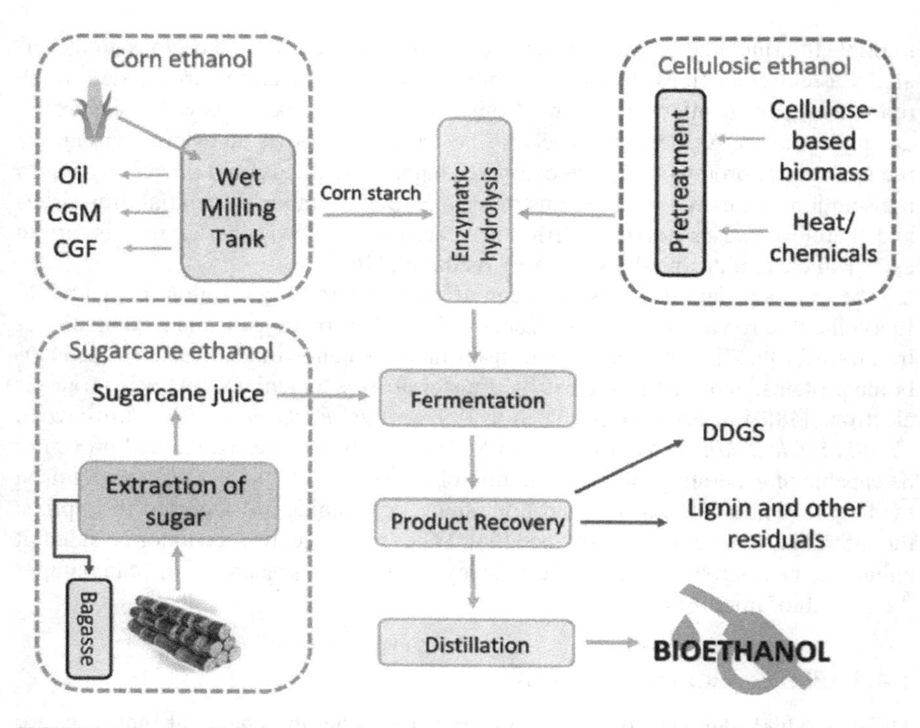

FIGURE 1.2 Bioethanol production using biowastes. (Adapted with permission from Ref. [59]. Copyright 2019, Royal Society of Chemistry.)

the fermentation process with *Clostridia* sp., and the respective yield of butanol from rice straw and bagasse was 2.93 and 1.95 g/L. Orange peels were also used for butanol production via steam explosion in hydrolysis and fermentation of the hydrolysate using *Clostridium acetobutylicum* NCIM 2877 [62]. This process resulted in a high yield of 19.5 g/L of butanol. Other alcohols such as isobutanol, 1,3-propanediol, and 2,3-butanediol can also be produced via a fermentation process.

1.4.5 Electrochemical Energy from Biowastes

Energy is one of the important parts of modern society for its proper functioning and advancement. Energy devices such as batteries and supercapacitors are used in many areas such as in electronic devices, appliances, defense, automobiles, and biomedicals. The main focus of the current research in energy is to reduce the cost by utilizing eco-friendly materials for high-performance energy devices for a sustainable future. Using biowastes for electrochemical energy devices is an interesting approach for value-added applications as well as waste management. The use of biowastes for various electrochemical devices is briefly covered in the following sections. Among various energy storage devices, supercapacitors have certain advantages such as fast charging time, safety, and high electrochemical cyclic stability. Electrochemical charge storage in supercapacitors can be via electrochemical double layer (EDL), redox, and hybrid (combination of EDL and redox) mechanisms. Redox mechanism applies to the materials that are redox-active such as metal oxides, metal sulfides, and conducting polymers, while EDL is the main charge storage mechanism for carbon-based materials such as activated carbons, carbon nanotubes, and graphite. Biowastes can be converted into high-performance carbon for use in supercapacitor devices [63–67]. The electrochemical performance of carbon-based materials depends on many factors such as surface area, pore size, and doping. Many surface characteristics and thus the electrochemical behavior of carbons derived from biowastes can be modified by physical and chemical activations [68,69]. Physical activation involves the treatment of carbon with gases such as steam, while chemical activation requires activating agents such as $CaCl_2$ and KOH. For example, Bhoyate et al. fabricated supercapacitors using waste tea leaves and observed almost 100% capacitance retention up to 5,000 cycles of charge–discharge studies [66]. Waste from the citrus industries such as orange peel was used as a source of high-performance carbon for energy storage applications [65]. The chemical treatment of orange peel with KOH created meso- and microporous carbon, which was used as electrode material for supercapacitors (Figure 1.3). The surface area of the carbon derived from orange peel improved from 0.852 to 1391 m^2/g after chemical activation, which resulted in a significant improvement in charge storage capacity from 115 to 407 F/g. Other biowastes such as coffee grounds, banana peels, and jute fibers were also used after chemical activation for the fabrication of supercapacitors [63,67,70].

Li-ion batteries are the most popular device for storing and delivering power to many electronic devices, automobiles, defense equipment, and emergency power backups due to their high energy density and stable performance. Li-ion batteries use many different types of materials as cathode and anode. Lithium-based metal oxides such as lithium manganese oxide are used as the cathode, while carbon-based materials such as graphite are applied as the anode. The structure and morphology of the electrode materials play an important role in determining the performance and life span of a battery. The charge (energy) in Li-ion batteries is stored via the intercalation

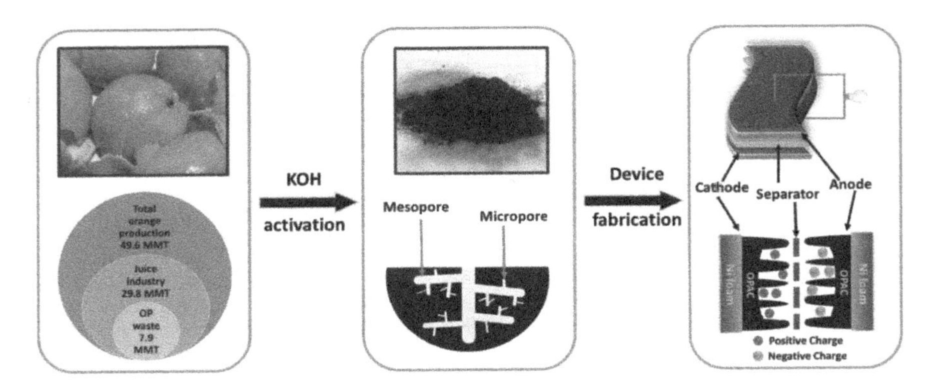

FIGURE 1.3 Activated carbon derived from the orange peel as supercapacitor material. (Adapted with permission from Ref. [65]. Copyright 2017, The Authors. Distributed under a Creative Commons Attribution License 4.0 (CC BY).)

mechanism, and thus, surface-modified carbons from the biowastes can be used as electrode materials [71]. For example, corn starch was used for the fabrication of Li-ion batteries [72]. Figure 1.4 shows the structure of a Si–C nano-/microstructure synthesized using corn starch as the anode material for Li-ion battery. The low stability of Si as an anode material, which is an issue for its commercial applications, was significantly improved after synthesizing Si–C. The composite electrode showed a high capacity of 1,800 mAh/g, outstanding cycling stability with capacity retention of 80% over 500 cycles, and fast charge–discharge capability. The high electrochemical performance might be derived from the combination of a structurally stable and high-conductive carbon from biowastes with Si.

Li–S batteries are designed to provide better performance than existing Li-ion batteries. Typical Li-ion batteries ($LiCoO_2$/graphite) show an energy density of about 400 Wh/kg, whereas Li–S batteries could provide four times higher energy density [73]. The high energy density of Li–S batteries is due to the conversion of sulfur to lithium sulfide, which involves two-electron transfers per sulfur atom ($S_8 + 16\ Li^+ + 16e^- \rightarrow 8Li_2S$) [73]; however, the low conductivity of sulfur and lithium sulfide could prevent Li–S batteries to reach their theoretical energy density. Also, the conversion of sulfur to lithium sulfide brings about 80% volume change, which can cause structural instability of the electrode [74]. A conducting material with high

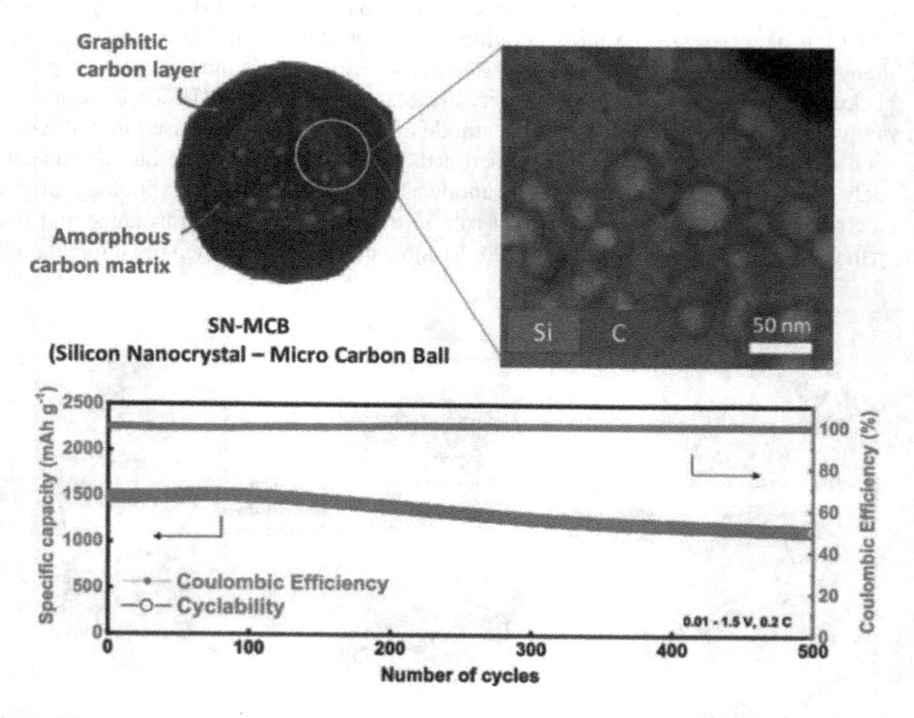

FIGURE 1.4 Silicon–carbon hybrid composite particles using corn starch biowaste material for Li-ion batteries. (Adapted with permission from Ref. [72]. Copyright 2020, American Chemical Society.)

porosity such as activated carbon can be used to host sulfur, which will not only increase the conductivity, but also provide structural stability. Babu and Ramesh used sugarcane waste as a source of carbon to host sulfur for Li–S batteries [73]. Melamine was used as a source of nitrogen to dope the carbon from sugarcane. The nitrogen-doped carbon showed a high reversible capacity of 1,169 mAh/g along with 77% capacity retention after 200 cycles.

Metal–air batteries are emerging as advanced batteries due to their lightweight, high energy density, and robust performance and have the potential to meet the current energy demand. Metal–air batteries use pure metal as an anode and ambient air as the cathode where oxygen reduction reactions (ORRs) take place. The performance of metal–air batteries is inferior to the theoretically predicted performance due to sluggish ORRs; therefore, efficient electrocatalysts are often required to improve the performance of metal–air batteries. Many of the current electrocatalysts are derived from precious noble metals; hence, there is a need for low-cost electrocatalysts. Biowastes find their applications in this area, too. For example, waste sweet potato vines were used for the synthesis of an efficient electrocatalyst for the ORR process [75]. S and N co-doped carbon from waste sweet potato vines displayed excellent ORR activities compared to the commercial Pt/C catalyst (Figure 1.5). An efficient electrocatalyst for Zn–air batteries was prepared by co-doping of carbon from garlic stems with N and S via the self-activation pyrolysis process [76]. The synthesized electrocatalyst showed improved ORR activities with the onset and half-wave potentials of 0.89 and 0.80 V vs. RHE along with high inhibition toward CO poisoning. The fabricated Zn–air battery displayed a discharge voltage of 1.19 V at a current density of 10 mA/cm^2, a high power density of 95 mW/cm^2, and better stability than the commercial Pt/C electrocatalyst.

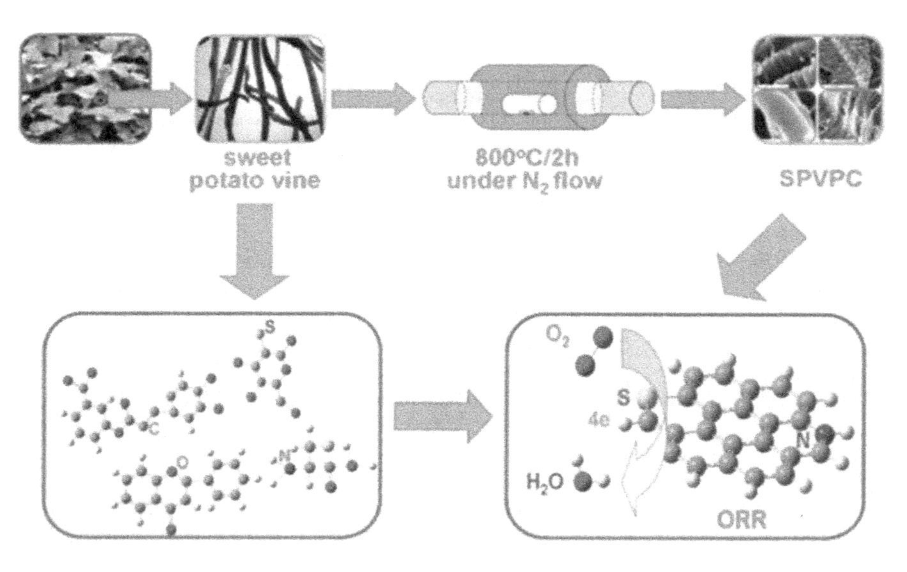

FIGURE 1.5 Use of wasted sweet potato vines for the synthesis of an electrocatalyst. (Adapted with permission from Ref. [75]. Copyright 2015, Elsevier.)

1.5 CONCLUSIONS

Even though diverse approaches and technologies have been invented and applied in converting biowastes to bioenergy, various challenges are faced by the industries in efficiently harnessing energy from wastes. There is a race mismatch between the industry and laboratory researches. Numerous laboratory-scale investigations have been conducted and reported on the potential bioenergy production from various biowastes. But the industrial-scale uptake for technological transfer and commercialization has been at a slow pace. A high collaboration among all stakeholders is inevitable to push and make BtB technology a well-established and reliable renewable energy technology for a sizeable societal consumption. Biowastes contain various components that require pretreatment to make them viable for undergoing biological processes for bioenergy production. This requirement increases costs and processing time, which classify them as low-priority materials for energy production. However, because of its sustainability, low carbon content, and environmental and socioeconomic benefits, bioenergy remains essential renewable energy, which needs further enhancement in techniques to reduce the operation cost. The bioenergy production process also leads to side products such as biofertilizers and glycerol, which are commercially valuable.

REFERENCES

1. Bhatia SK, Joo HS, Yang YH (2018) Biowaste-to-bioenergy using biological methods – A mini-review. *Energy Convers Manag* 177:640–660.
2. Simcock R, Cavanagh J, Robinson B, Gutierrez-Gines MJ (2019) Using biowastes to establish native plants and ecosystems in New Zealand. *Front Sustain Food Syst* 3:85 (1–14).
3. Wang S, Jena U, Das KC (2018) Biomethane production potential of slaughterhouse waste in the United States. *Energy Convers Manag* 173:143–157.
4. Qureshi AS, Khushk I, Naqvi SR, Simiar AA, Ali CH, Naqvi M, Danish M, Ahmed A, Majeed H, Mir Jatt AN, Rehan M, Nizami AS (2017) Fruit waste to energy through open fermentation. *Energy Procedia* 142:904–909.
5. Bhatia SK, Kim SH, Yoon JJ, Yang YH (2017) Current status and strategies for second generation biofuel production using microbial systems. *Energy Convers Manag* 148:1142–1156.
6. 2016 WEC (2016) World Energy Resources 2016. World Energy Counc 2016:6–46.
7. Pokhrel D, Viraraghavan T (2004) Treatment of pulp and paper mill wastewater – A review. *Sci Total Environ* 333:37–58.
8. Bajpai P (2015) Management of Pulp and Paper Mill Waste – Pratima Bajpai – Google Books.
9. Islam KMN (2016) Municipal solid waste to energy generation in Bangladesh: Possible scenarios to generate renewable electricity in Dhaka and Chittagong City. *J Renew Energy* 2016:1–16.
10. Gooding CH, Meeker DL (2016) Review: Comparison of 3 alternatives for large-scale processing of animal carcasses and meat by-products. *Prof Anim Sci* 32:259–270.
11. Taher H, Al-Zuhair S, AlMarzouqui A, Hashim I (2011) Extracted fat from lamb meat by supercritical CO_2 as feedstock for biodiesel production. *Biochem Eng J* 55: 23–31.

12. Canoira L, Rodríguez-Gamero M, Querol E, Alcántara R, Lapuerta M, Oliva F (2008) Biodiesel from low-grade animal fat: Production process assessment and biodiesel properties characterization. *Ind Eng Chem Res* 47:7997–8004.
13. Kumar P, Barrett DM, Delwiche MJ, Stroeve P (2009) Methods for pretreatment of lignocellulosic biomass for efficient hydrolysis and biofuel production. *Ind Eng Chem Res* 48:3713–3729.
14. Mulakhudair AR, Hanotu J, Zimmerman W (2017) Exploiting ozonolysis-microbe synergy for biomass processing: Application in lignocellulosic biomass pretreatment. *Biomass and Bioenergy* 105:147–154.
15. Çanakcı M, Özsezen AN (2010) Atık mutfak yağlarının alternatif dizel yakıtı olarak değerlendirilmesi. *Gazi Univ J Sci* 18:81–91.
16. Refaat AA, Refaat AA (2010) Archive of SID Different techniques for the production of biodiesel from waste vegetable oil. *Int J Environ Sci Tech* 7:183–213.
17. Kulkarni MG, Dalai AK (2006) Waste cooking oil – An economical source for biodiesel: A review. *Ind Eng Chem Res* 45:2901–2913.
18. Kombe GG, Temu AK, Rajabu HM, Mrema GD, Kansedo J, Lee KT (2013) Pre-treatment of high free fatty acids oils by chemical re-esterification for biodiesel production—A review. *Adv Chem Eng Sci* 03:242–247.
19. Anwar N, Wang W, Zhang J, Li Y, Chen C, Liu G, Zhang R (2016) Effect of sodium salt on anaerobic digestion of kitchen waste. *Water Sci Technol* 73:1865–1871.
20. Abdulrazak S, Hussaini K, Sani HM (2017) Evaluation of removal efficiency of heavy metals by low-cost activated carbon prepared from African palm fruit. *Appl Water Sci* 7:3151–3155.
21. Gunatilake SK (2015) Methods of removing heavy metals from industrial wastewater 1:12–18.
22. Lee SY, Sankaran R, Chew KW, Tan CH, Krishnamoorthy R, Chu D-T, Show P-L (2019) Waste to bioenergy: A review on the recent conversion technologies. *BMC Energy* 1:1–22.
23. Pollardo AA, Lee H shik, Lee D, Kim S, Kim J (2017) Effect of supercritical carbon dioxide on the enzymatic production of biodiesel from waste animal fat using immobilized Candida antarctica lipase B variant. *BMC Biotechnol* 17:1–6.
24. Bhatia SK, Kim J, Song H-S, Kim HJ, Jeon J-M, Sathiyanarayanan G, Yoon J-J, Park K, Kim Y-G, Yang Y-H (2017) Microbial biodiesel production from oil palm biomass hydrolysate using marine Rhodococcus sp. YHY01. *Bioresour Technol* 233:99–109.
25. Jain SK, Kumar S, Chaube A (2011) Technical sustainability of biodiesel and its blends with diesel in C.I. engines: A review. *Int J Chem Eng Appl* 2:101–109.
26. Bhatia SK, Yi D-H, Kim Y-H, Kim H-J, Seo H-M, Lee J-H, Kim J-H, Jeon J-M, Jang K-S, Kim Y-G, Yang Y-H (2015) Development of semi-synthetic microbial consortia of Streptomyces coelicolor for increased production of biodiesel (fatty acid methyl esters). *Fuel* 159:189–196.
27. Peng YP, Amesho KTT, Chen CE, Jhang SR, Chou FC, Lin YC (2018) Optimization of biodiesel production from waste cooking oil using waste eggshell as a base catalyst under a microwave heating system. *Catalysts* 8:1–16.
28. Gardy J, Hassanpour A, Lai X, Ahmed MH, Rehan M (2017) Biodiesel production from used cooking oil using a novel surface functionalised TiO_2 nano-catalyst. *Appl Catal B Environ* 207:297–310.
29. Vafakish B, Barari M (2017) Biodiesel Production by transesterification of tallow fat using heterogeneous catalysis. *Kem u Ind Chem Chem Eng* 66:47–52.
30. Ho K-C, Chen C-L, Hsiao P-X, Wu M-S, Huang C-C, Chang J-S (2014) Biodiesel production from waste cooking oil by two-step catalytic conversion. *Energy Procedia* 61:1302–1305.

31. Tran TTV, Kaiprommarat S, Kongparakul S, Reubroycharoen P, Guan G, Nguyen MH, Samart C (2016) Green biodiesel production from waste cooking oil using an environmentally benign acid catalyst. *Waste Manag* 52:367–374.
32. Gardy J, Osatiashtiani A, Céspedes O, Hassanpour A, Lai X, Lee AF, Wilson K, Rehan M (2018) A magnetically separable SO_4/Fe-Al-TiO_2 solid acid catalyst for biodiesel production from waste cooking oil. *Appl Catal B Environ* 234:268–278.
33. Cea M, Sangaletti-Gerhard N, Acuña P, Fuentes I, Jorquera M, Godoy K, Osses F, Navia R (2015) Screening transesterifiable lipid accumulating bacteria from sewage sludge for biodiesel production. *Biotechnol Reports* 8:116–123.
34. Pastore C, Lopez A, Lotito V, Mascolo G (2013) Biodiesel from dewatered wastewater sludge: A two-step process for a more advantageous production. *Chemosphere* 92:667–673.
35. Kwon EE, Kim S, Jeon YJ, Yi H (2012) Biodiesel production from sewage sludge: New paradigm for mining energy from municipal hazardous material. *Environ Sci Technol* 46:10222–10228.
36. Mata TM, Mendes AM, Caetano NS, Martins AA (2014) Properties and sustainability of biodiesel from animal fats and fish oil. *Chem Eng Trans* 38:175–180.
37. Alptekin E, Canakci M, Sanli H (2014) Biodiesel production from vegetable oil and waste animal fats in a pilot plant. *Waste Manag* 34:2146–2154.
38. Kwon EE, Seo J, Yi H (2012) Transforming animal fats into biodiesel using charcoal and CO_2. *Green Chem* 14:1799–1804.
39. Shi W, Li J, He B, Yan F, Cui Z, Wu K, Lin L, Qian X, Cheng Y (2013) Biodiesel production from waste chicken fat with low free fatty acids by an integrated catalytic process of composite membrane and sodium methoxide. *Bioresour Technol* 139:316–322.
40. Bhatia SK, Bhatia RK, Yang Y-H (2017) An overview of microdiesel — A sustainable future source of renewable energy. *Renew Sustain Energy Rev* 79:1078–1090.
41. Kadam R, Panwar NL (2017) Recent advancement in biogas enrichment and its applications. *Renew Sustain Energy Rev* 73:892–903.
42. Angelidaki I, Treu L, Tsapekos P, Luo G, Campanaro S, Wenzel H, Kougias PG (2018) Biogas upgrading and utilization: Current status and perspectives. *Biotechnol Adv* 36:452–466.
43. Appels L, Baeyens J, Degrève J, Dewil R (2008) Principles and potential of the anaerobic digestion of waste-activated sludge. *Prog Energy Combust Sci* 34:755–781.
44. Huang X, Yun S, Zhu J, Du T, Zhang C, Li X (2016) Mesophilic anaerobic co-digestion of aloe peel waste with dairy manure in the batch digester: Focusing on mixing ratios and digestate stability. *Bioresour Technol* 218:62–68.
45. De la Rubia MA, Villamil JA, Rodriguez JJ, Borja R, Mohedano AF (2018) Mesophilic anaerobic co-digestion of the organic fraction of municipal solid waste with the liquid fraction from hydrothermal carbonization of sewage sludge. *Waste Manag* 76: 315–322.
46. Rahimnejad M, Adhami A, Darvari S, Zirepour A, Oh S-E (2015) Microbial fuel cell as new technology for bioelectricity generation: A review. *Alexandria Eng J* 54:745–756.
47. Sayed ET, Barakat NAM, Abdelkareem MA, Fouad H, Nakagawa N (2015) Yeast extract as an effective and safe mediator for the baker's-yeast-based microbial fuel cell. *Ind Eng Chem Res* 54:3116–3122.
48. TerAvest MA, Rosenbaum MA, Kotloski NJ, Gralnick JA, Angenent LT (2014) Oxygen allows Shewanella oneidensis MR-1 to overcome mediator washout in a continuously fed bioelectrochemical system. *Biotechnol Bioeng* 111:692–699.
49. Li S-W, He H, Zeng RJ, Sheng G-P (2017) Chitin degradation and electricity generation by Aeromonas hydrophila in microbial fuel cells. *Chemosphere* 168:293–299.
50. Han TH, Cho MH, Lee J (2014) Indole oxidation enhances electricity production in an E. coli-catalyzed microbial fuel cell. *Biotechnol Bioprocess Eng* 19:126–131.

51. Rossi R, Fedrigucci A, Setti L (2015) Characterization of electron mediated microbial fuel cell by Saccharomyces Cerevisiae. *Chem Eng Trans* 43:337–342.
52. Lee Y-Y, Kim TG, Cho K (2016) Enhancement of electricity production in a mediatorless air–cathode microbial fuel cell using Klebsiella sp. IR21. *Bioprocess Biosyst Eng* 39:1005–1014.
53. Bhatia SK, Lee B-R, Sathiyanarayanan G, Song H-S, Kim J, Jeon J-M, Kim J-H, Park S-H, Yu J-H, Park K, Yang Y-H (2016) Medium engineering for enhanced production of undecylprodigiosin antibiotic in Streptomyces coelicolor using oil palm biomass hydrolysate as a carbon source. *Bioresour Technol* 217:141–149.
54. Sund CJ, McMasters S, Crittenden SR, Harrell LE, Sumner JJ (2007) Effect of electron mediators on current generation and fermentation in a microbial fuel cell. *Appl Microbiol Biotechnol* 76:561–568.
55. Waghmare AG, Arya SS (2016) Utilization of unripe banana peel waste as feedstock for ethanol production. *Bioethanol* 2:146–156.
56. Vu VH, Kim K (2009) Ethanol production from rice winery waste – Rice wine cake by simultaneous saccharification and fermentation without cooking. *J Microbiol Biotechnol* 19:1161–1168.
57. Huang H, Qureshi N, Chen MH, Liu W, Singh V (2015) Ethanol production from food waste at high solids content with vacuum recovery technology. *J Agric Food Chem* 63:2760–2766.
58. Hossain ABMS, Fazliny AR (2010) Creation of alternative energy by bio-ethanol production from pineapple waste and the usage of its properties for engine. *African J Microbiol Res* 4:813–819.
59. Gavahian M, Munekata PES, Eş I, Lorenzo JM, Mousavi Khaneghah A, Barba FJ (2019) Emerging techniques in bioethanol production: From distillation to waste valorization. *Green Chem* 21:1171–1185.
60. Li Q, Du W, Liu D (2008) Perspectives of microbial oils for biodiesel production. *Appl Microbiol Biotechnol* 80:749–756.
61. Cheng C-L, Che P-Y, Chen B-Y, Lee W-J, Lin C-Y, Chang J-S (2012) Biobutanol production from agricultural waste by an acclimated mixed bacterial microflora. *Appl Energy* 100:3–9.
62. Joshi SM, Waghmare JS, Sonawane KD, Waghmare SR (2015) Bio-ethanol and biobutanol production from orange peel waste. *Biofuels* 6:55–61.
63. Using HS, Kahol P, Gupta R (2019) Waste Coffee Management : Deriving Nitrogen-Doped Coffee-Derived Carbon. C 44 Waste Coffee Management : Deriving Nitrogen-Doped Co ff ee-Derived Carbon
64. Mensah-Darkwa K, Zequine C, Kahol PK, Gupta RK (2019) Supercapacitor energy storage device using biowastes: A sustainable approach to green energy. *Sustain* 11:414 (1–22).
65. Ranaweera CK, Kahol PK, Ghimire M, Mishra SR, Gupta RK (2017) Orange-peel-derived carbon: Designing sustainable and high-performance supercapacitor electrodes. *Journal of Carbon Research* 3:25.
66. Bhoyate S, Ranaweera CK, Zhang C, Morey T, Hyatt M, Kahol PK, Ghimire M, Mishra SR, Gupta RK (2017) Eco-friendly and high performance supercapacitors for elevated temperature applications using recycled tea leaves. *Glob Challenges* 1:1700063.
67. Zequine C, Ranaweera CK, Wang Z, Dvornic PR, Kahol PK, Singh S, Tripathi P, Srivastava ON, Singh S, Gupta BK, Gupta G, Gupta RK (2017) High-performance flexible supercapacitors obtained via recycled jute: Bio-waste to energy storage approach. *Sci Rep* 7:1–12.
68. Li T, Ma R, Lin J, Hu Y, Zhang P, Sun S, Fang L (2020) The synthesis and performance analysis of various biomass-based carbon materials for electric double-layer capacitors: A review. *Int J Energy Res* 44:2426–2454.

69. Bai Q, Li H, Zhang L, Li C, Shen Y, Uyama H (2020) Flexible solid-state supercapacitors derived from biomass konjac/polyacrylonitrile-based nitrogen-doped porous carbon. *ACS Appl Mater Interfaces* 12:55913–55925.
70. Liu B, Zhang L, Qi P, Zhu M, Wang G, Ma Y, Guo X, Chen H, Zhang B, Zhao Z, Dai B, Yu F (2016) Nitrogen-doped banana peel-derived porous carbon foam as binder-free electrode for supercapacitors. *Nanomaterials* 6:4–13.
71. Wang Y, Song Y, Xia Y (2016) Electrochemical capacitors: Mechanism, materials, systems, characterization and applications. *Chem Soc Rev* 45:5925–5950.
72. Kwon HJ, Hwang JY, Shin HJ, Jeong MG, Chung KY, Sun YK, Jung HG (2020) Nano/microstructured silicon-carbon hybrid composite particles fabricated with corn starch biowaste as anode materials for Li-ion batteries. *Nano Lett* 20:625–635.
73. Babu DB, Ramesha K (2019) Melamine assisted liquid exfoliation approach for the synthesis of nitrogen doped graphene-like carbon nano sheets from bio-waste bagasse material and its application towards high areal density Li-S batteries. *Carbon N Y* 144:582–590.
74. Manthiram A, Fu Y, Chung SH, Zu C, Su YS (2014) Rechargeable lithium-sulfur batteries. *Chem Rev* 114:11751–11787.
75. Ouyang T, Cheng K, Gao Y, Kong S, Ye K, Wang G, Cao D (2016) Molten salt synthesis of nitrogen doped porous carbon: A new preparation methodology for high-volumetric capacitance electrode materials. *J Mater Chem A* 4:9832–9843.
76. Ma Z, Wang K, Qiu Y, Liu X, Cao C, Feng Y, Hu P (2018) Nitrogen and sulfur co-doped porous carbon derived from bio-waste as a promising electrocatalyst for zinc-air battery. *Energy* 143:43–55.

Part 2

Municipal Waste for Energy

2 Operational Tools and Techniques for Municipal Solid Waste Management

Zobaidul Kabir
The University of Newcastle

Mahfuz Kabir
Bangladesh Institute of International and Strategic Studies (BIISS)

M. Ashiqur Rahman
Western Sydney University

Mofijur Rahman
University of Technology Sydney (UTS)

CONTENTS

DOI: 10.1201/9781003178354-4

2.1 INTRODUCTION

Because of the rapid urbanization, increasing population, economic growth of the nation, industrialization, and increasing quality of lifestyle, the amount of solid waste is increasing globally [1]. Globally, the generation of MSW was around 2,010 million tonnes (MT) in 2016 and the generation of MSW continues to increase. The World Bank has estimated that the production of MSW will be approximately 2,200 MT each year until 2025 globally [2]. It has been estimated that the amount of MSW worldwide may reach up to 3,400 MT by 2050 [1]. This means the production of MSW will increase about 70% within three decades only [3]. Globally, the per capita generation of MSW per day is now 0.74 kg on average. It has been estimated that the average per capita waste generation per day will increase from 0.74 to 1.41 kg in 2050. By 2050, the generation of waste per person per day is projected to increase by 19% in high-income countries. In developing countries, the generation of waste per person per day will increase by approximately 40% [4]. Figure 2.1 shows the per capita MSW generation per day of the top 11 countries.

Urban areas are the key sources of MSW, and with the rapid urbanization and growing population, the quantity of MSW production is also increasing, especially in developing countries. Currently, 55% of the world's population – approximately 4.2 billion people – live in urban areas [6]. Given the rapid urbanization and the generation of huge quantity of MSW, the cities in developing countries are facing immense challenges relating to waste management [7]. The management of MSW in a sustainable way becomes essential for municipality authorities to make the urban places livable for residents [8]. However, the developed countries are more advanced in the application of tools and technologies to manage the MSW sustainably than developing countries.

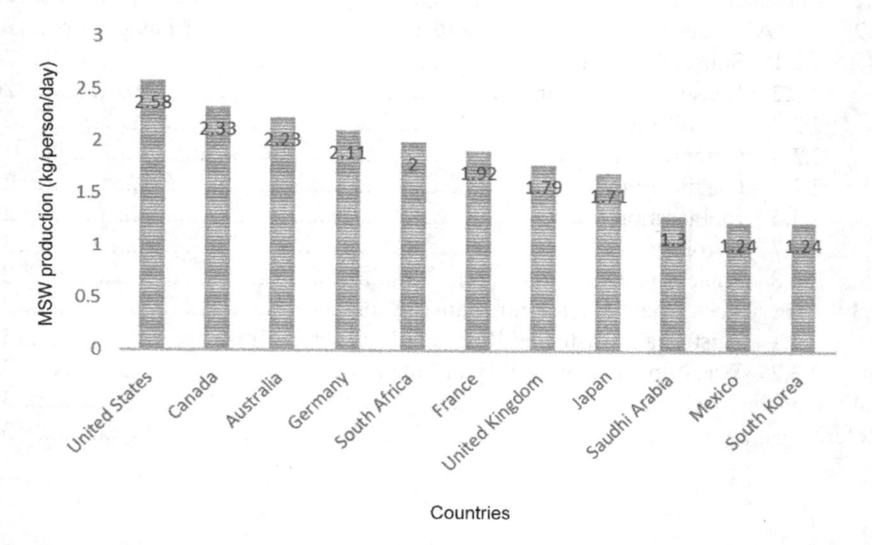

FIGURE 2.1 Per capita MSW generation (in kg) per day in top 11 countries, produced by the authors based on the data from Ref. [5].

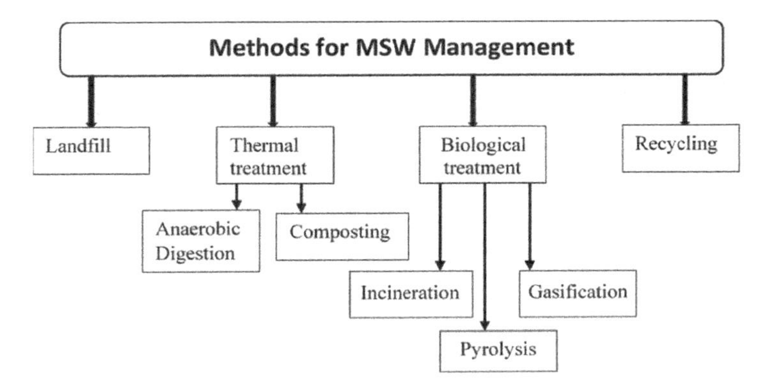

FIGURE 2.2 A schematic view of methods for MSW treatment.

The literature shows that there are advanced tools and techniques (technologies) for MSW treatment, and each of the techniques may have advantages and disadvantages. These tools or techniques may range from source reduction to recycling and waste-to-energy (WtE) technologies although open dumping of MSW is prevalent particularly in the most developing countries as mentioned earlier. The techniques such as reduction, reuse, or recycling are more or less common to minimize MSW. Over the years, some advanced techniques or technologies have evolved to manage MSW efficiently. These technologies used for the recovery of resources and the generation of energy from MSW can be characterized into two main procedures. These include (i) thermochemical process that involves gasification, incineration, and pyrolysis, and (ii) biological process that involves anaerobic digestion, composting, and landfill [9]. Among these methods, it is imperative to use the ones that are suitable to the country given its socioeconomic and geophysical context in addition to the nature of the MSW [9]. Figure 2.2 shows a schematic overview of some advanced methods for MSW management.

This chapter aims to provide an overview of operational tools and techniques used for MSW management. This chapter is divided into four sections. First, the chapter provides an overview of waste management tools and techniques following an introduction. The third section provides experiences from developed countries on MSW management practices using advanced technologies. This is followed by a conclusion.

2.2 AN OVERVIEW OF AVAILABLE TOOLS AND TECHNIQUES FOR MSW MANAGEMENT

Several tools and operational techniques are available for management interventions of solid waste. The preference in the use of operational tools and techniques for solid waste management can be arranged as a minimization hierarchy as shown in Figure 2.3, where source reduction is the most preferred option and disposal is the least. The fundamental argument for waste minimization is that it is far better to reduce waste in the first place than to cope inadequately with their aftermaths [10].

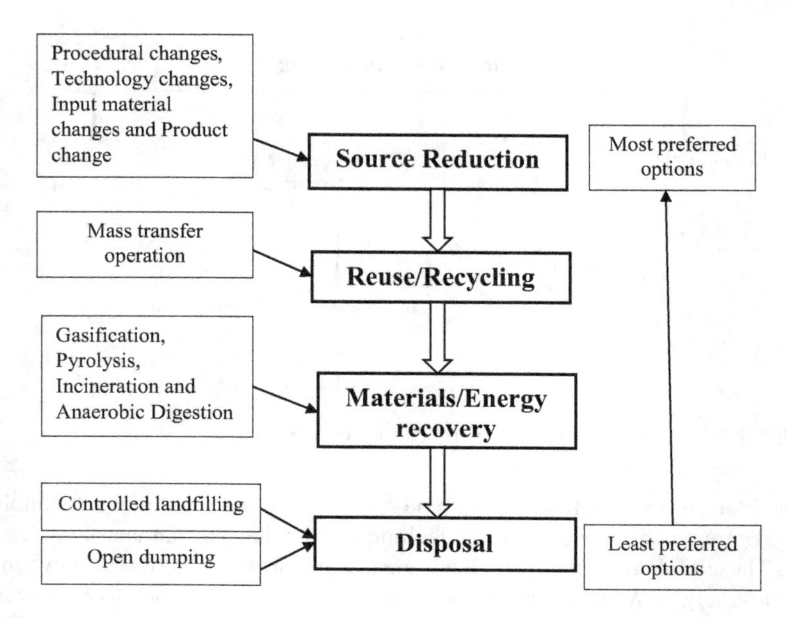

FIGURE 2.3 Minimization hierarchy of municipal solid waste.

The philosophy behind the waste minimization hierarchy is that it is crucial in developing policies for using appropriate tools and techniques and that the solutions are easier to deal with at the front end of the situation rather than at the back end.

2.2.1 Source Reduction

The reduction in waste at sources denotes considerations in the design, manufacture, purchase, or use of materials or products, including packaging to reduce their amount or toxicity before they enter the waste stream [11]. The practice of source reduction can be performed at the production or waste generation level. For example, at the household level, source reduction can be best carried out through selective buying habits, that is buying products that last long and reuse of products and materials until they are worn out or unusable [11].

2.2.2 Reuse and Recycling

Waste reuse occurs when a product or material is reused in its original form, but not necessarily for the same application [12]. For example, used papers can be reused as paper packets or as packaging material and glass bottles can be washed and used many times. In Australia, there is a trend for reusing used materials as they change hands through the much popular Revolve, the Sunday Markets, Garage Sales, Cash Converters stores, Salvation Army stores, or similar stores and online platforms such as Gumtree and Market Place.

Waste recycling involves the conversion of the product or material to a simpler form, which is then manufactured to a new or similar product [13]. Recycling may avoid the use of a valuable and huge amount of land required for landfill space. Recycling also saves energy and natural resources, offers beneficial products from materials, and even makes a monetary return [13]. For example, inorganic (non-biodegradable) wastes such as aluminum cans, glass, or plastic bottles can be recycled as raw materials for making new or similar products and papers can be recycled for making cardboard, packaging materials, or new papers.

2.2.3 Landfilling

Sanitary landfilling is an improved waste disposal technique, which refers to waste compaction, exclusion of moisture and leachate and extraction of gas from waste, and daily application of soil cover on the dumped waste to reduce air pollution and spread of diseases [14]. The requirements for the treatment of solid waste before disposal depend on its composition. For example, sanitary landfills with impermeable liners constructed of clay or polyethylene can be a good option to avoid groundwater contamination by leaching waste minerals [14]. Furthermore, high- and upper-middle-income countries are operating controlled landfills and going beyond recycling adopting advanced technologies [15]. The landfills remain one of the sources of greenhouse gases (GHGs) emissions and other toxic gases although controlled landfills may release these gases less than the traditional landfills. Methane is naturally released from the landfill sites and is a powerful greenhouse gas that may contribute to climate change significantly by absorbing heat from the atmosphere.

2.2.4 Composting

Composting is one of the key techniques of biologically treating MSW. In this process, the organic part of the waste is decomposed in a controlled atmosphere. Almost 70% of MSW is organic materials, particularly in developing countries, and overall, the percentage varies from 40 to 70 [16]. Composting of MSW has become a popular technique in both developed and developing countries where the development and use of the technique are widespread [16]. It has been identified that compost as a soil conditioner and manure may have an infinite market globally. Compost is a source of nitrogen, phosphorus, potassium, and other nutrients for plants and soil. These qualities made composting a suitable option for the treatment of a huge quantity of biodegradable solid wastes [15]. During the composting of MSW, odor, bioaerosol, dust, noise, and leachate may be generated, and these may pollute the surrounding environment of composting facilities [15] if proper measures are not taken.

2.2.5 Gasification

Gasification is a thermal breakdown of MSW in the absence of oxygen [17]. The chemical reaction process in the gasification method depends on temperature (above >650 °C), steam pressure, and concentration of O_2 in the reactor. In general, the gasification process is a combination of three stages: (i) gasifier for the production of syngas; (ii) cleaning of syngas through the removal of pollutants and harmful compounds such as tar from syngas; and (iii) the recovery of energy from syngas using a gas engine [17]. The transformation of organic materials to combustible gas or syngas is the key part of the process. In addition to syngas, liquids such as oils and tars and solids such as char (fixed carbon and ashes) are also produced from gasification [17].

The syngas produced from gasification mainly contains hydrogen, carbon monoxide, carbon dioxide, and methane. The by-products or residues of gasification are char or slag, oils, ash, and water. Gasification also generates hydrocarbons, inert gases, tar, and gas pollutants. Tar is heavier hydrocarbons that liquefy at temperatures in the range of 250 °C–300 °C [18]. The structures of syngas usually depend on the operational parameters of the gasifier. These parameters may include the features of feedstock, level of air, steam, air, CO_2, plasma, temperature, pressure, and catalyst type. Major chemical products such as ammonia, methanol, and hydrogen can be produced from syngas [19]. One advantage is that the syngas produced from gasification is used for valuable chemical products and other forms of fuel energy. The synthesis gas or syngas can be used for industries such as chemical industries, fuel industries, and power plants. Furthermore, the gasification method may generate electricity at 685 kWh/tonne. Gasification may generate more electricity than other methods, for example pyrolysis and incineration.

During the burning of MSW, gasification may release toxic gases. These may include CO_2, N_2O, CO, CH_4, NOx, and non-methane volatile organic compounds (NMVOCs). The gases, for example CO_2, CO, N_2O, CH_4, may contribute to global warming. Other gases, for example NOx and CO, affect photochemical oxidant formation (POF) and HCl, SO_2, NOx, and H_2S may cause acidification (AC). Overall, the release of toxic gases from gasification may affect public health including breathing difficulties, and disrupt the entertaining activities of residents [20].

2.2.6 Incineration

Incineration is the most common method throughout the world for energy generation from MSW. This is an exothermal process that produces and releases energy from the system to its surroundings usually in the form of heat. Also, this process consists of the thermal breakdown of MSW, where an excess of air may occur and lead to the generation of a flue gas including CO_2, O_2, N_2, and water vapor. The raw or unprocessed waste is used as feedstock for incineration. An adequate quantity of air is needed in this process to oxidize the feedstock. Also, the combustion of waste requires a high temperature of 850 °C to convert the waste materials into carbon dioxide, water, and solid residue called bottom ash [21]. Table 2.1 shows the products and by-products of MSW treatment using pyrolysis.

There were 1,179 MSW incineration facilities capable of power generation around the world by 2015, and most of them are in developed countries. However, evidence shows that some developing countries such as Ethiopia, Bangladesh, and India have recently taken initiatives to adopt incineration plants although with small-scale capacity. The key advantage of this method is that it may significantly decrease the volume of disposed waste by up to 90% and 70% of the mass recovery can be converted from waste to energy in the form of electricity [23]. Another key advantage is that incineration facilities require relatively less amount of land than other methods

TABLE 2.1
The Products and By-products of MSW Treatment Using Incineration

Products	Net Energy Generated (Exported to the Grid/District Heating)	Electricity and Heat
Electricity (KWh/t)	519	–
Heat (MJ/t)	–	1,785
Recovered Materials		
Ferrous metals (kg/t)	22.1	26.6
Bottom ash (kg/t)	219.4	180.8
Hazardous Waste/Pollutants		
Air pollution residue (kg/t)	313	20.7
Carbon dioxide (CO_2) (kg/t)	452	452
Carbon monoxide (CO) (g/t)	48.7	23.9
Sulfur dioxide (SO_2) (g/t)	95	75
Nitrogen oxide (N_2O) (kg/t)	1.7	23.1
Hydrochloride (HCl) (g/t)	32.3	32.9
NH_3 (g/t)	17.9	0.4
Particulates (PM_{10}) (kg/t)	17.9	4.9
Dioxin (furans) (ng/t)	92.3	36.2

Source: Adapted from Ref. [22] where the first author of this chapter is the first author of the relevant source.

of MSW treatment such as landfills. The GHGs produced are released into the atmosphere through smokestacks. Residues such as bottom ash and fly ash are generated from incineration plants. These ashes can be useful for road construction [21]. Among the residues, bottom ash may take up to 80%–90% and the remaining is fly ash along with other pollutants [23]. The key elements of bottom ash include Si, Al, Fe, K, Na, Ca, Mg, and Cl [23]. The incineration of waste also produces some undesirable outputs as residue, including bottom ash, and ferrous and non-ferrous products which are harmful to human health and the environment, and therefore, proper measures are necessary to manage these pollutants.

BOX 1: WASTE-TO-ENERGY TECHNOLOGY IN ETHIOPIA [24]

Recently, an incineration plant facility has been installed in Addis Ababa, the capital city of Ethiopia. The capacity of this WtE plant is 50 MW and is established on a landfill site with 5.3 ha of land to supply MSW to the incineration plant. Around US $118.5 million was spent to establish this plant with an annual capacity of 350,000 tonnes of MSW. This is a pioneer project in sub-Saharan Africa, and it is expected that Ethiopia will be able to provide electricity around 330 days in a year without interruption. This method is an effective approach for MSW management, and some similar projects are under consideration after feasibility studies. Despite some challenges such as (i) low calorific value of MSW, (ii) low power output (as low as 44%), (iii) lack of local technical experts, (iv) underdeveloped system of waste management, and (v) relatively higher cost of collecting waste that may not be affordable to all residents, the plant is an iconic technology for Ethiopia's MSW management in Addis Ababa. Steps have been taken to maintain the waste stream collaboratively by the Ethiopian Electric Power Corporation and the city's administration. This plant will create hundreds of jobs for waste collection from the city in addition to operating and maintaining the plant.

2.2.7 Pyrolysis

Pyrolysis is a thermal process that converts MSW into various materials in the absence of oxygen to generate heat and electricity. The recovery of energy from the treatment of MSW using pyrolysis technology may be up to 80% [25]. Based on the transfer of heat at various rates and other parameters such as residence time, particle size, and temperature, there are three different types of reactions in pyrolysis. These reactions include slow pyrolysis (production of charcoal), fast pyrolysis (production of bio-oil), and flash pyrolysis (production of gases such as syngas) [26]. In the pyrolysis process, a high temperature that is from 300 °C to 800 °C is required for the degradation of MSW in an inert environment [26]. In the pyrolysis process, the pre-treatment of MSW is necessary before the inlet point of the pyrolizer where glasses and inert materials such as sand, soil, concrete, and rock are removed. This process begins with the breakdown of pre-treated MSW at 300°C in heated chambers. This starts with the consumption of initially present oxygen, and thereafter, an oxygen-free atmosphere is maintained [26]. The temperature may then be increased up to 800 °C. Biochar or bio-oil is the key product generated from pyrolysis; however, the amount of production depends on residence time, final temperature, and heating rate [27]. Furthermore, gaseous products, for example syngas – a combination of methane, carbon monoxide, and hydrogen in addition to a wide variety of volatile organic compounds (VOCs), are generated. The syngas generated from pyrolysis is usually utilized for the production of power and heat [27].

Pyrolysis technology is considerably important for the treatment of MSW, and the recovery of materials and energy is high. Therefore, this method is receiving increasing consideration not only for its high performance of energy production, but also for some advantages over incineration or landfill technologies. For example, incineration and landfill technologies require finding suitable and new locations for land and transportation of MSW [28]. The pyrolysis technology does not need to face these challenges. The pyrolysis-involved process can generate energy in a cleaner way than, for example, the incineration of MSW since the energy produced from the pyrolysis process contains a lower amount of nitrogen oxides and sulfur oxides. This is because the syngas in the pyrolysis process is produced in an inert atmosphere and is washed or cleaned before its combustion [28]. Furthermore, in addition to the emission of reduced GHGs, the quality of solid residue such as char can be expected to be better produced from the pyrolysis of MSW.

The main disadvantages of the pyrolysis technology include the following: (i) It has a more complex product stream compared to other waste management technologies and (ii) due to the high concentration of CO, further treatment of gas is necessary without which it cannot be collected directly in the cabin. Pyrolysis generates toxic residues such as inert mineral ash, inorganic compounds, and unreformed carbon. Also, there is a potential to generate toxic air emissions such as acid gases, dioxins and furans, nitrogen oxides, and particulates. These pollutants may affect air and human health [29].

2.2.8 Anaerobic Digestion

In the anaerobic digestion (AD) process, the organic fraction of MSW is used as feedstock and the key function of the process is to decompose the organic part of MSW. Microorganisms decompose the organic part of MSW in the absence of oxygen. There are different stages of the AD process, including acidogenesis, acetogenesis, and methanogenesis, and microorganisms involved in these different stages require a suitable atmosphere to produce and increase the yield of end products [30]. Special reactors are required for this process that may operate at appropriate conditions including well-maintained temperature and pH level [31]. Usually, the level of pH ranges between 6.7 and 7 according to the use of microorganisms in the corresponding stage of the process and different ranges for the temperature at the same time. In most cases, mesophilic or thermophilic temperatures are followed since these situations are economically feasible [32]. The well-mixed organic feedstock is placed in the digester for 5–10 days. Given the diverse types of feedstock, there are also different types of reactors used to generate and recover energy from MSW. For example, a continuously stirred tank reactor is used for food waste, while plug flow and batch reactors are used for other types of organic waste [32].

The key products produced from the AD process include biogas, fiber, and liquid digestate. The biogas is the dominating product generated from the AD process that contains 50%–80% methane, 20%–50% carbon dioxide, and a little amount of sulfide and ammonia [31,32]. This biogas generated from the AD process can be used as an alternative to natural gas to generate combined heat and power (CHP) although heat and power production efficiency (around 5.5–7.5 kWh/m^3) from biogas is less as compared to natural gas due to lower caloric value (two-thirds of natural gas) [31,32]. The other two products (liquid digestate and fiber) generated in the AD process are used as raw materials in the fertilizer industry [31].

2.3 EXPERIENCES FROM SELECTED INNOVATIVE APPROACHES

2.3.1 Australia's Waste and Resource Recovery Infrastructure

Australia targets to recover 80% of its waste nationally by 2030. Blue Environment [33] reports that in 2018–2019, Australia generated an amount of around 61.5 MT of "core" waste, which is the waste managed by the waste and resource recovery sector.

This core waste included 12.6 MT of municipal solid waste, 21.9 MT of commercial and industrial waste, and 27 MT of construction and development waste. Further, it has been stated that in 2018–2019, the national resource recovery rate was 63% and the recycling rate was 60% [34]. There has been a positive trend of the national resource recovery rate in Australia, which increased from about 50% in 2006–2007 to 61% in 2016–2017 and then to 63% in 2018–2019 [33].

Australia's National Waste Report 2013 outlines the solid waste management system where rejected resources are collected from the source of waste generation. The wastes are collected using suitable vehicles that can effectively transport waste over a reasonable remote area for disposal or recovery. The MSW is then cleared from the vehicles usually for on-site disposal to landfill or sent to facilities (for example, recycling) to recover resources or materials. This first stop can be called a transfer station from where a huge amount of waste is loaded to railroad vehicles for transport to a remote site for processing using technologies or disposal [35]. The 2013 Report further states that Australia has 806 resource recovery facilities, where discarded materials are sorted and processed to raw materials for industries using biological, mechanical, and thermal transforming technologies. Box 2.2 provides a brief description of Australia's main types of resource recovery facilities. Besides, in 2018–2019 Australia utilized 2.1 MT of core waste for energy recovery through landfill gas collection, solid recovered fuels, and anaerobic digestion of food-derived waste.

Australia has significantly increased the investment in resource recovery infrastructure during the last decade [36]. For example, in the 2020–2021 budget the government announced an expenditure of A$190 million to establish a Recycling Modernisation Fund, which along with co-funding from the states, territories, and industries is expected to generate A$600 million of recycling investment. The federal government will deliver the fund via the states and territories, where their contributions will be matched dollar for dollar [35]. However, the Department of Agriculture, Water and the Environment [35] has recognized one of the main challenges to enhance facilities to the recovery of resources is to get the approval of plans from the government for infrastructure relating to resource recovery. Planning for the facility needs to address several commercial and environmental issues, for which a strong and workable business case needs to be demonstrated along with widespread community consultation [35].

BOX 2.2: BRIEF DESCRIPTION OF AUSTRALIA'S MAIN RESOURCE RECOVERY FACILITIES

Material Recovery Facilities (MRF): Commingled waste and the materials that can be recycled are collected from residential, commercial, and industrial areas and sorted. Then the waste including recyclable materials is organized according to the type and separated mechanically using appropriate technologies. For more processing, the accumulated materials of waste are then recycled using recovering techniques (usually recycling facilities). The technologies used for sorting materials from commingled waste may vary. The range of technologies may include small loaders equipped with a bucket to move waste to railroads as well as fully automated tools for classification activities.

Alternative Waste Treatment (AWT) Facilities: Biodegradable organic waste is separated from MSW and stored for treatment by the AWT facilities. The use of these facilities enhances the degradation of organic wastes. Two key facilities or technologies such as anaerobic digestion and aerobic digestion generate gases and fertilizers or soil conditioners, respectively. The gases produced from the anaerobic digestion system are rich in methane and therefore suitable for energy generation. It is ensured that the feedstock for both processes is free of non-organic materials. Therefore, the sorting of MSW is done carefully. Although the development of the market for AWT is at an early stage, the potential of generating energy from organic waste using AWT is highly regarded.

Recycling Facilities: Plastics, scraps, metals, timber, glass, paper, and cardboard are separated from MSW suitable for reprocessing using the recycling facilities in Australia. Usually, the recycled products are exported to other countries.

Thermal Waste Technologies: These technologies used in Australia process combustible wastes such as household waste after separating the materials for recycling. The combusting of waste using these technologies generates heat and steam for electricity production. The residues such as bottom ash can be used for other useful purposes, for example cement for construction.

2.3.2 Waste-to-Energy Facility in Singapore

As waste volume grows in Singapore, they aim to maximize resource recovery with adequate infrastructure. In 2016, Singapore generated 7.8 MT of waste, of which the recycling rate was 60%, while by the end of 2030, the estimated volume of waste generation is 12.6 MT, of which 70% is the recycling rate [37]. Singapore is thriving toward a "zero waste nation" with approaches such as waste minimization and prevention, recycling, maximizing resource and energy recovery, and minimizing landfilling demand. In Singapore, 28% and 72% of the total waste are generated from households and industries, respectively. The total waste generated per day is 21,350 tonnes. There is a systematic collection and transportation of all MSW, and among them, 37% of waste is incinerable and directly goes to incineration plants. Among the 67% of MSW, only 3% of waste goes to controlled landfills, and the rest go for recycling. Among the materials generated from incineration and recycling, ash is 1,595 tonnes per day and metals recovered are 118 tonnes per day. Importantly, the generation of electricity from MSW is 2,508 MWh per day. Therefore, Singapore is an example of an almost zero waste country [37] made by an improved and innovative MSW management system.

Singapore's waste-to-energy facility adopted innovative technologies to maximize energy recovery and minimize land use and ash residue. Its new Integrated Waste

Management Facility (IWMF) has been considered one of the major energy recovery plants in the world. This integrated plant involves a facility to recover energy from waste and an incineration facility to treat sludge [38]. The construction of the first phase of IWMF is targeted to complete by 2025. This innovative co-located facility will have 2×4 combustion lines and can digest around 2.5 million tonnes of MSW yearly. The IWMF will treat different waste streams, such as incinerable waste (5,800 tonnes/day), household recyclables (250 tonnes/day), source-segregated food waste (400 tonnes/day), and dewatered sludge (800 tonnes/day) from Tuas Water Reclamation Plant (TWRP) [38].

The key benefits from the IWMF will be in terms of optimized land use, maximized energy and resource recovery, minimized environmental impacts, and co-location with TWRP, jointly known as Tuas Nexus (water–energy–waste nexus) [38]. The estimated amount of excess electricity that this facility will be able to generate is about 200 MW, which is adequate to power 300,000 four-room Housing & Development Board apartments in Singapore [38]. This massive environment-friendly project on water–energy–waste nexus will incur a total cost of S$9.5 billion to the Singapore Government [38], of which S$1.5 billion will be spent for IWMF.

2.4 CONCLUSIONS

This chapter aims to provide an overview of the tools and techniques for municipal solid waste management. Given the increasing urbanization along with a high density of population, the amount of MSW is causing environmental pollution and affects livability and sustainable natural resource management. There are tools and operational techniques for MSW management. Each of the techniques, however, has advantages and disadvantages. The tools and techniques include minimization of MSW through reduction, reuse, and recycle. Also, advanced technologies such as landfill, pyrolysis, anaerobic digestion, incineration, and composting were discussed. Two case studies relating to the use of tools and techniques for MSW management in developed countries were analyzed. The case studies include MSW management in Singapore and Australia. While developed countries are adopting advanced technologies for the management of MSW and the generation of renewable energy, developing countries have not responded yet to effectively harness the benefits of the advanced techniques.

REFERENCES

1. Kaza, S., Yao, L., Bhad-Tata, P., and Woerden, F.V. (2018) What a Waste 2.0: A Global Snapshot of Solid Waste 2050. World Bank Report, The World Bank. Washington.
2. Campusano, F., Brown, R.C., and Martínez, J.D. (2019) Auger reactors for pyrolysis of biomass and wastes. *Renewable Sustainable Energy Review*, 102:372–409.
3. Khandelwal, H., Dhar, H., Thalla, A.K., and Kumar, S. (2019) Application of life cycle assessment in municipal solid waste management: a worldwide critical review. *Journal of Cleaner Production*, 209:630–654.
4. World Bank (2020a) Trend in Solid Waste Management, What a Waste 2: A Global Snapshot of Solid Waste Management to 2050, World Bank Report, New York. Available at https://datatopics.worldbank.org, accessed on 5/5/2021 (Authors).

5. Statista (2020) Available at https://www.statista.com/statistics/689809/per-capital-msw-generation-by-country-worldwide/, accessed on 5/5/2021.
6. World Bank (2020b) United Nations Population Division. World Urbanization Prospects: 2018 Revision, New York. Available at https://data.worldbank.org/indicator/SP.URB.TOTL.IN.ZS, accessed on 5/5/2021.
7. Sharma, M., Joshi, S., and Kuman, A. (2020) Assessing enablers of e-waste management in circular economy using DEMATEL method: An Indian perspective. *Environmental Science and Pollution Research*, 27:13325–13338.
8. Rodic, L., and Wilson, D.C. (2017) Resolving governance issues to achieve priority sustainable development goals related to solid waste management in developing countries. *Sustainability*, 9(404):1–18; doi:10.3390/su9030404.
9. Khan, I., and Kabir, Z. (2020) Waste-to-energy generation technologies and the developing economies: A multi-criteria analysis for sustainability assessment. *Renewable Energy*, 150:320–333.
10. Malav, L., Yadav, K.M., Gupta, N., Kumar, S., Sharma, G.K., and Krishnan, S. (2020) A review on municipal solid waste as a renewable source for waste-to energy project in India: Current practices, challenges, and future opportunities. *Journal of Cleaner Production*, 277:123227.
11. US EPA (1989) Recycling, in Decision Makers Guide to Solid Waste Management, Chapter-6: 59–75, United States Environmental Protection Agency. Available at https://nepis.epa.gov, accessed on 7/7/2021.
12. Blight, G.E., and Mbande C.M. (1996) Some problems of waste management in developing countries. *The Journal of Solid Waste Technology and Management*, 23(1):19–27.
13. Miller, G.T. (2000) *Living in the Environment: Principles, Connections, and Solutions* (11th edition). Brooks/Cole Publishing Company, Pacific Grove, CA.
14. DeAraujo, A.S.F., De Melo, W.J., and Singh, R.P. (2010) Municipal Solid Waste compost amendments in agricultural soil: Changes in microbial biomass. *Reviews in Environmental Science and Bio/Technology*, 9:41–49. (need to be replaced by landfill)
15. Gajalakshmia, S., and Abbasi, S.A. (2008) Solid waste management by composting: State of the art. *Critical Reviews in Environmental Science and Technology*, 38(5):311–400.
16. Wei, Y., Li, J., Shi, D., Liu, G., Zhao, Y., and Shimaokac, T. (2017) Environmental challenges impeding the composting of biodegradable municipal solid waste: A critical review. *Resources, Conservation and Recycling*, 122:51–65.
17. Seo, Y., Alam, M.T., and Yang, W. (2018) Gasification of municipal solid waste, in Yun, Y., (ed.) *Gasification for Low-Grade feedstock*, Chapter-7: 115–141, IntechOpen, London, UK. doi: 10.5772/intechopen.73685.
18. Indrawan, N., Kumar, A., Moliere, M., Sallam, K.A., Raymond, L., and Huhnke, R.L. (2020) Distributed power generation via gasification of biomass and municipal solid waste: A review. *Journal of the Energy Institute*, 93:2293–2313.
19. Hameed, Z., Aslam, M., Khan, M., Khan, Z., Maqsood, K., Atabani, E.A. et al. (2021) Gasification of municipal solid waste blends with biomass for energy production and resources recovery: Status, hybrid technologies and innovative prospects. *Renewable and Sustainable Energy Reviews*, 136:10375.
20. Zaman, A.U. (2010) Comparative study of municipal solid waste treatment technologies using life cycle assessment method. *International Journal of Environmental Science Technology*, 7(2):225–234.
21. Jeswani, H., Smith, R.W., and Azapagic, A. (2013) Energy from waste: Carbon footprint of incineration and landfill biogas in the UK. *The International Journal of Life Cycle Assessment*, 18(1):218–229.
22. Kabir, Z., and Khan, I. (2020) Environmental impact assessment of waste to energy projects in developing countries: General guidelines in the context of Bangladesh. *Sustainable Energy Technologies and Assessments*, 37:100619.

23. Huang, Y., Chen, J., Shi, S., Li, B., Mo, J., and Tang, Q. (2020) Mechanical properties of municipal solid waste incinerator (MSWI) bottom ash as alternatives of subgrade. *Advances in Civil Engineering*, 2020:1–11. Available at https://doi.org/10.1155/2020/9254516, accessed on 5/6/2021.
24. Mubeen, I., and Buekens, A. (2019) Energy from waste: Future prospects toward sustainable development, in Kumar, S., Kuman, R., and Pandey, A., (eds.) *Biotechnology and Bioengineering: Waste Treatment Processes for Energy Generation*, Chapter-14: 283–305, Elsevier, USA.
25. Chen, D., Yin, L., Wang, H., and He, P. (2015) Reprint of: Pyrolysis technologies for municipal solid waste: A review. *Waste Management*, 37:116–136.
26. Ouda, O.K.M., Raza, S.A., Nizami, A.S., Rehan, M., Al-Waked, R., and Korres, N.E. (2016) Waste to energy potential: A case study of Saudi Arabia. *Renewable Sustainable Energy Review*, 61:328–340. doi:10.1016/j.rser.2016.04.005.
27. Qazi, W.A., Abushammala, M.F.M., Azam, M.H., and Younes, M.K. (2018) Waste-to-energy technologies: A literature review. *Journal of Solid Waste Technology Management*, 44:387–409. doi:10.5276/JSWTM.2018.387.
28. Chen, D., Yin, L., Wang, H., and He, P. (2014) Pyrolysis technologies for municipal solid waste: A review. *Waste Management*, 34(12):2466–2486. doi:10.1016/j.wasman.2014.08.004.
29. Jahirul, M.L., Rasul, M.G., Chowdhury, A.A., and Ashwath, N. (2012) Biofuel production through bio-mass pyrolysis – A technological review. *Energies*, 5:4952–5001.
30. Farooq, A., Haputta, P., Silalertruksa, T., Shabbir, H., and Gheewala, S.H. (2021) A framework for the selection of suitable waste to energy technologies for a sustainable municipal solid waste management system. *Frontiers in Sustainability*. doi:10.3389/frsus.2021.681690.
31. Slorach, P.C., Jeswani, H.K., Cuéllar-Franca, R., and Azapagic, A. (2019) Environmental sustainability of Anaerobic Digestion of household food waste. *Journal of Environmental Management*, 236:798–814.
32. Mutz, D., Hengevoss, D., Christoph, H., and Gross, T. (2017) Waste-to-Energy Options in Municipal Solid Waste Management –A Guide for Decision Makers in Developing and Emerging Countries, Technical Report (57 pages), GIZ, Germany.
33. Blue Environment (2020) National Waste Report 2020, Prepared for Department of Agriculture, Water and the Environment, Government of Australia, viewed on 23 June 2021. Available at https://www.environment.gov.au/system/files/pages/5a160ae2-d3a9-480e-9344-4eac42ef9001/files/national-waste-report-2020.pdf.
34. Keys, H. (2021) Waste Management Review, National Waste Report 2020. Australian Government, viewed on 23 June 2021. Available at https://wastemanagementreview.com.au/national-waste-report-2020-released/.
35. Department of Agriculture, Water, and the Environment (2021) National Waste Report 2013Factsheet, Australian Government, viewed on 20 June 2021. Available at https://www.environment.gov.au/protection/waste/publications/national-waste-reports/2013/infrastructure.
36. Commonwealth of Australia (2020) Budget 2020–21: Budget Measures, Budget Paper No. 2, viewed on 21 June 2021. Available at https://parlinfo.aph.gov.au/parlInfo/search/display/display.w3p;query=Id%3A%22publications%2Ftabledpapers%2Fe6270c26-a7e4-42eb-a392-b55a9008c2d0%22.
37. NEA (2021) Waste Statistics and Recycling Rate, National Environmental Agency, Singapore, viewed 23 June 2021.
38. NEA (2018) Integrated Waste Management Facility: Meeting Singapore's Long-Term Waste Management Needs, Technical Report, National Environmental Agency, Singapore, viewed 23 June 2021. Available at https://www.nea.gov.sg/docs/default-source/resource/iwmf.pdf.

3 Municipal Waste for Energy Production

Mahfuz Kabir
Bangladesh Institute of International and Strategic Studies (BIISS)

Zobaidul Kabir
The University of Newcastle

CONTENTS

3.1 INTRODUCTION

Municipal solid waste (MSW) mainly stems from households, commerce, and trade. The usual options for managing the MSW are landfilling, combustion, recycling, and composting. The most traditional method of MSW management, especially in the developing world, is dumping in landfill sites or burning in open space, which aims at diminishing the volume of the waste [1]. However, disposing of the MSW in the landfill causes massive environmental hazards, such as polluting groundwater, spreading toxic vapors in the atmosphere, and emitting greenhouse gases. Piled up MSW in old landfills also pollutes land and water and gives birth to vector populations that include mosquitoes, flies, and pests, which increases vector-borne diseases. In addition, such traditional means of managing MSW overlook the potential for generating various energy resources, which could be significantly beneficial for the economy and environment, especially for developing countries [2].

The MSW mainly consists of dry wastes (e.g., plastics, cans, newspapers, metals, glass bottles, and wood materials) and wet wastes (e.g., vegetables, meat, kitchen waste, yard and garden waste, eggshells, and leftover food) [3,4]. The MSW is generally classified as (i) biodegradable waste, (ii) recyclable materials, (iii) inert waste, (iv) waste electrical and electronic equipment (WEEE), (v) composite waste, (vi) hazardous and toxic matters, and (vii) biomedical waste. Based on the characteristics, the MSW has further been classified as organic (biodegradable) and inorganic

DOI: 10.1201/9781003178354-5

(non-degradable) [5]. The MSW is mainly generated from urban, industrial, and rural areas [6]. Globally, 2.01 billion tonnes of MSW was generated in 2016, one-third of which was not managed to ensure the environmental safety standards, and landfilling was the leading method to reduce the volume of MSW. It is expected to be 3.40 billion tonnes by 2050 if the MSW continues to increase at the current rate. At present, only 13.5% of the MSW is reused, while 5.5% is used for composting. About 40% of the collected MSW was either burned or dumped in open space in 2016, which was environmentally harmful [7].

Developed countries, such as the USA and Canada, and countries of the European Union generate 34% of the global MSW even though they have only about 16% share of the world population. The USA alone generated 292.4 million tonnes of MSW in 2018. The most notable types of MSW in the USA are paper and paperboard (23.05%), food (21.59%), and yard trimmings (12.11%) [8]. China and India produce 27% of the global MSW together, while they jointly have more than 36% of the world's population (Figure 3.1). In developing countries, 55%–80% of MSW is produced from households, 10%–30% from commercial activities, and the rest from institutions, streets, etc. [2]. However, they do not manage more than 93% of their MSW in an environmentally safe and productive manner. According to the projection of Kaza et al. [9], the generation of MSW would be twofold in South Asia and threefold in sub-Saharan Africa by 2050, which would account for about 35% of the total generation of MSW of the world.

The proper management of MSW has some strong and non-trivial benefits. These include, inter alia, reducing the greenhouse gases emission, saving fossil fuels, no requirement of separate landfill area, speedy and easy disposal, and production of clean energy products. MSW is traditionally disposed of through composting, landfilling, and incineration. However, sustainable development principles suggest that these wastes can be used as a potential source of energy or in any other productive

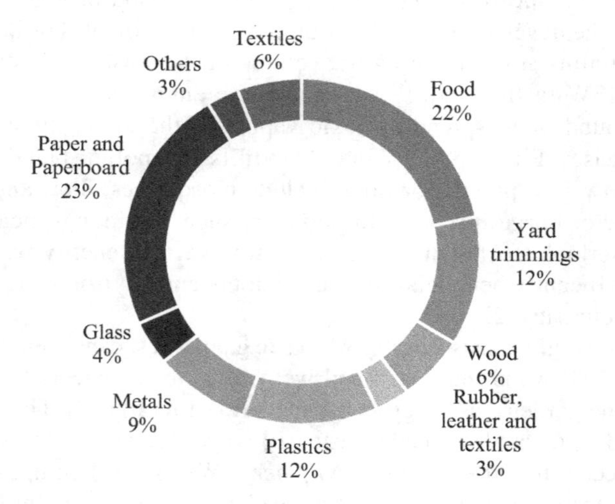

FIGURE 3.1 The composition of MSW in the USA in 2018. Based on the data of US EPA, 2020 [7].

means. The present chapter aims at discussing the technologies of MSW-to-energy. Thus, the rest of the chapter is organized as follows: Section 3.2 presents various existing MSW-to-energy techniques and emerging technologies. Section 3.3 discusses the potential techniques and experiments to improve the efficiency and economic viability of energy projects. Section 3.4 briefly presents the selected try cases of MSW-to-energy initiatives and their successes. Finally, concluding remarks have been made.

3.2 TECHNIQUES OF GENERATING ENERGY FROM MSW

There are mainly two types of techniques to convert the MSW into energy products: thermochemical and biochemical. Thermochemical technologies, such as gasification, pyrolysis, and liquefaction, can be utilized to generate energy from the MSW. The thermochemical processes convert the MSW into char, synthetic gas, and bio-oil through processing the waste with high heats in a zero- or low-oxygen environment [9]. Conversely, biochemical conversion processes include anaerobic digestion (AD) and composting, which are utilized to produce mainly biogas (i.e., biomethane) as energy products. Table 3.1 shows the MSW profile of the countries/economies by income level and geographical location.

The main technologies to recover energy from MSW are thermochemical (incineration, pyrolysis, and gasification) and biochemical (biomethanation) processes. Currently, there are more than 1,700 waste-to-energy plants across the world, most of which are located in the Asia-Pacific (62%), followed by Europe (33%) and North America (4.5%) [11]. The non-recyclable MSW is also used in producing refuse-derived fuel (RDF), which is good renewable energy and an environment-friendly alternative to coal for use in boilers.

Incineration: The main thermochemical processes are conventional and hydrothermal incineration as well as oxidation. Incineration processes combustible materials of the waste at approximately 850°C. Moving grate, fixed grate, rotary kiln, and fluidized bed are some of the noted technologies of incineration. Electricity is generated directly from the energy produced in this process [12]. Syngas or synthesis gas that includes hydrogen can be produced through gasification. It produces H_2, clean energy, with a better heating value (141.7 MJ/kg) [13]. Figure 3.2 shows energy production technologies for MSW management.

Conventional incineration burns the organic fraction of MSW (OFMSW) to generate heat with high temperatures for combined heat and power plants. Electricity can also be produced from the heat released through incineration. Incineration significantly decreases the weight of MSW by 80%–85% and volume by 95%–96%. However, traditional incineration releases many hazardous pollutants to the atmosphere, such as particulates, dioxins, furan, hydrocarbons, CO_2, N_2, water vapor, and benzene-like compounds. Hydrothermal flame incineration addresses this problem by preventing the generation of toxic by-products during the production of energy [14].

Hydrothermal incineration and oxidation: Hydrothermal flame incineration or oxidation can be utilized instead of conventional incineration. This incineration has significant benefits such as the near-complete conversion of stable recalcitrant

TABLE 3.1
MSW Profile of the Countries/Economies by Income Level and Geographical Location (Reproduced with permission from Ref. [10]. Copyright (2017) Emerald.)

		Income Group				Geographical Location						
Indicator	**Unit/Type**	**L**	**LM**	**UM**	**H**	**MENA**	**SSA**	**LAC**	**NA**	**SA**	**ECA**	**EAP**
MSW production rate, 2016	Percent	5	29	32	34	6	9	11	14	17	20	23
	Annual MT	93	586	655	683	129	174	231	289	334	392	468
	Daily per capita (kg)	0.40	0.53	0.69	1.58	0.81	0.46	0.99	2.21	0.52	1.18	0.56
Forecasted MSW production rate, 2025	Annual MT	143	827	835	781	177	269	290	342	466	440	602
	Daily per capita (kg)	0.43	0.63	0.83	1.71	0.90	0.50	1.11	2.37	0.62	1.30	0.68
Forecasted MSW production rate, 2050	Annual MT	283	1,233	1,004	879	255	516	369	396	661	490	714
	Daily per capita (kg)	0.56	0.79	0.99	1.87	1.06	0.63	1.30	2.50	0.79	1.45	0.81
Growth rate of MSW production (2016–2050)	Percent	204	110	53	29	75	197	60	37	98	25	53
Collection of MSW (%)	Total	39	51	82	96	82	44	84	100	51	90	71
	Urban	48	71	85	100	90	43	85	100	77	96	77
	Rural	26	33	45	98	74	9	30	99.7	40	55	45
Type of MSW (%), 2016	Organic	56	53	54	32	58	43	52	28	57	36	53
	Paper and cardboard	7	12.5	12	25	13	10	13	28	10	18.6	15
	Plastic	6.4	11	11	13	12	8.6	12	12	8	11.5	12
	Glass	1	3	4	5	3	3	4	4.5	4	8	2.6
	Metal	2	2	2	6	3	5	3	9.3	3	3	3
	Rubber and leather	0	0.5	1	4	2	0	0.5	9	2	0.4	0.4
	Wood	0.6	1	1	4	1	0.4	0.5	5.6	1	1.6	2
	Others	27	17	15	11	8	30	15	3.6	15	21	12

(Continued)

TABLE 3.1 (*Continued*)
MSW Profile of the Countries/Economies by Income Level and Geographical Location (Reproduced with permission from Ref. [10]. Copyright (2017) Emerald.)

		Income Group				Geographical Location						
Indicator	**Unit/Type**	**L**	**LM**	**UM**	**H**	**MENA**	**SSA**	**LAC**	**NA**	**SA**	**ECA**	**EAP**
Methods of management (%), 2016	Open dumping	93	66	30	2	52.7	69	26.8	0	75	25.6	18
	Sanitary landfilling	3	18	54	39	34	24	68.5	54.3	4	25.9	46
	Recycling	3.7	5.5	4	29	9	6.6	4.5	33.3	5	20	9
	Composting	0.3	10	2	6	4	0.4	0.2	0.4	16	10.7	3
	Incineration	0	0.5	10	22	0.3	0	0	12	0	17.8	24
	Others	0	0	0	2							

Key: Categories by income categories are high income (H), upper middle income (UM), lower middle income (LM), and low income (L). By geographical location, they are classified as East Asia and Pacific (EAP), Europe and Central Asia (ECA), South Asia (SA), North America (NA), Latin America and Caribbean (LAC), sub-Saharan Africa (SSA), and the Middle East and North Africa (MENA).

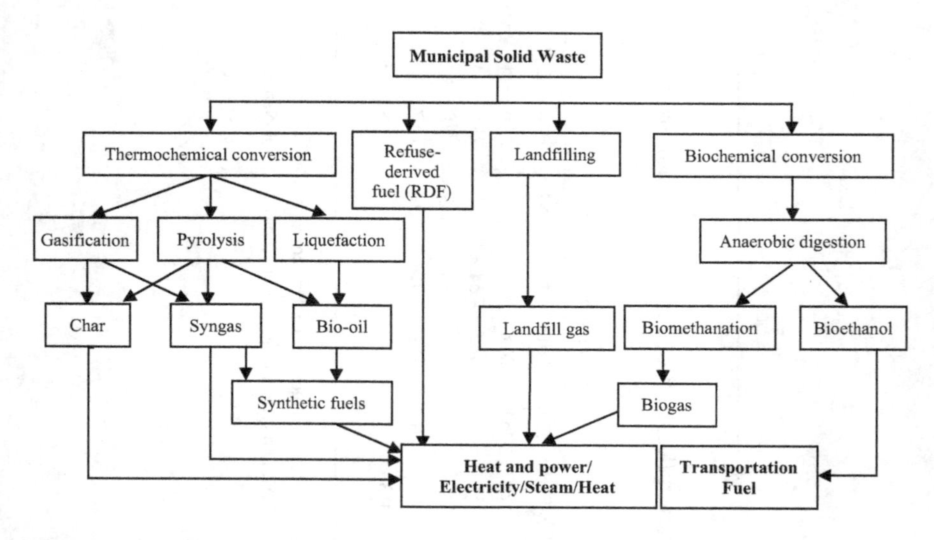

FIGURE 3.2 Energy production technologies for MSW management.

compounds, high reaction rates, high energy recovery, and no or insignificant hazardous by-products and toxic residues [15]. In pyrolysis, the MSW undergoes thermal degradation between the temperature of 300 °C and 800 °C in an inert environment. Glass, inert materials, and metals are removed before the thermal decomposition of MSW. Then, it is processed in heated chambers without oxygen at 300 °C. Then, the temperature is increased up to 800 °C in a non-reactive environment, which produces biochar or bio-oil along with some minor methane and hydrogen [16]. Slow pyrolysis produces charcoal, while fast pyrolysis generates bio-oil. Because of the high energy and resource recovery as well as high energy efficiency, pyrolysis is considered to be a viable alternative to incineration [17].

AD/biomethanation: Energy can be recovered from wet and biodegradable MSW such as food waste, garden and crop residue, and waste from pruning, by utilizing AD or biomethanation. Biomass is transformed into biogas (which consists of 55%–60% of methane) in the AD in closely monitored conditions. Biogas can be used directly as a fuel or utilized to generate electricity through a generator. Biogas is extracted by utilizing an anaerobic digester or reactor and from landfills (landfill gas or LFG). However, the generation of LFG takes a much longer time than biomethanation at thermophilic conditions. Therefore, the thermophilic process of generating biogas is recommended for commercialization [18]. Biomethanation is a cost-effective method of generating energy from MSW because it requires less human labor. It is a low-cost process that needs less maintenance, less land area, and low-cost infrastructure [19].

Bioethanol production: Bioethanol can be produced through hydrolysis, fermentation, and distillation, which also produces hydrogen as a by-product. About 329.75 m^3 bioethanol can be produced per day using 11,558 tonnes of feedstock of degradable MSW that comprises 50.89% organic fraction. A microbial fuel cell (MFC), a bio-electrochemical tool, can also be generated from MSW, which can be used for producing electricity. Some next-generation biochemical conversion of solid organic fraction and leachate of MSW to produce biohydrogen is dark and photo-fermentation, direct and indirect bio-photolysis, microbial electrolysis cells, and microbial electrohydrogenesis cells [14].

Pyrolysis: Pyrolysis is an important thermochemical technique, which operates in the absence of oxygen, to convert OFMSW into high-energy bio-oil, char, or gases. Nevertheless, the quality and quantity of energy products generated through pyrolysis depend on temperature and heating rate; vapor residence time; rate of inert gas flow; size and type of reactor; and moisture, particle size, and composition of elements in the OFMSW [14].

Liquefaction: Liquefaction is a viable MSW-to-liquid thermochemical technology, which is utilized to produce high-quality and energy-intensive bio-oil with high pressures [20].

Mutually reinforcing interactions of various components of MSW (especially plastics) can be utilized in pyrolysis to produce fuels that comprise heating values that are comparable to traditional fossil fuels. The commercialization of pyrolysis of MSW is required to obtain high-grade fuel. Advanced post-treatment techniques must be applied to scale up the process of pyrolysis [21]. There is a significant untapped potential of MSW for producing primary energy that can save a substantial amount of fossil fuel in the Balkans [22]. An examination of the power generation process of a very small power plant (VSPP) was conducted in Thailand based on MSW from an organic Rankine cycle (ORC) with an incinerator. The main heat source of the VSPP was the RDF type 3 (RDF-3) from a mixed fuel of the hazardous medical wastes and MSW. Energy and exergy efficiencies of 0.91% and 0.89%, respectively, were found in the VSPP-MSW system. The results suggest that the levelized energy cost is slightly higher than that of the feed-in tariff (FiT) for integrated solid waste management in Thailand. However, this technique is less expensive in terms of the disposal cost of hazardous medical waste in the country [23]. Optimization of methane production rate can result in better management of methane production from the landfill cells along with lessening the operational time of the landfill.

3.3 IMPROVED AND EMERGING TECHNOLOGIES OF MSW-TO-ENERGY

Different MSW-to-energy technologies are in place to recover energy and extract valuable resources from MSW. These techniques are also important in lessening environmental hazards and reducing traditional MSW management costs, such as transportation, landfilling, and dumping. Commercialization and scaling up of the projects bring a positive net return.

Thermochemical technologies are utilized to burn MSW to generate heat and gases, which can be used to produce electricity. Shahnazari et al. [24] selected incineration, gasification, plasma, and pyrolysis as thermochemical technologies to assess the viability of electricity generation from the MSW. The results demonstrated that plasma gasification is the optimal thermochemical technique [24]. Gasification is an important technique to generate energy from MSW. This technique is more efficient and environmentally friendlier than incineration. However, some hazardous pollutants and adverse by-products are also generated in the gasification technique, which include halogen, alkaline, and tar that are detrimental to environment and cause operational problems. Therefore, it is important to decrease such pollutants at the operational level through improvising the technologies and utilizing optimal solutions.

In recent times, the generation of hydrogen has been gaining strong ground as an energy source all over the world, which is mainly because of being cleaner and environment-friendly energy as it does not release pollutants while burning. It contains a high density of energy of 122 MJ/kg and burns fast with a high octane number. At present, almost all H_2 production at industrial scale is based on natural gas and coal through gasification, natural reforming, or their combination. However, recent developments in the H_2 production from biomass wastes (such as wheat, food grains, sugarcane, and algae) demonstrate its strong prospect to substitute fossil fuels. The interface of experimental parameters was examined on the optimum H_2 production as a quality energy product from the gasification of MSW. The results suggest that utilizing air as an agent in the gasification of MSW was effective in generating H_2 and total gas yield [25].

The joint production of ethanol and biogas is possible from MSW in biorefinery, which can be a viable substitute to fossil fuels and would be a value-added source of energy. Therefore, the cost-effective production of biofuels from various types of MSW, e.g., lignocellulosic waste, can be operated at a commercial scale. The biorefinery method can maximize the production of energy from the OFMSW. The OFMSW is a suitable input to produce ethanol and biomethane. Pre-treatment of the OFMSW is necessary to make bioenergy production efficient. Hydrothermal pre-treatment demonstrates promising results to improve bioenergy production because it does not require any chemicals, nor generate any waste. In addition, the suspension remaining after the segregation of ethanol and the solid residues after the enzymatic hydrolysis can efficiently be transformed into biomethane through AD [19]. The energy potential of MSW was examined from its segregation. It was found that 368–770 kW electricity can be produced per tonne of processed MSW in their studied gasification system. They suggested low-cost pre-treatment of MSW through segregation for developing a sustainable route of electricity production via gasification and combustion. If the organic materials are segregated, energy production can be doubled by the gasifier. Segregated organic matters can also be used for biodigestion and/or processing for the production of gases that have high aggregate financial returns [26].

Gasification of MSW with different biomass blends has been found as effective among thermochemical conversion techniques to maximize the production of energy, recover resources, and reduce pollution significantly. Currently, hybrid gasification systems are available that include fuel oxidation, plasma torch, or biochemical conversion that improve the efficiency of the process, increase generation of energy, ensure cost-effectiveness, enhance quality and production of syngas, and change the configuration of gaseous yields from MSW. Thus, the gasification of MSW can be a viable option to decrease toxic elements and unsafe gases in the atmosphere [27]. Thermochemical treatments, e.g., gasification, hydrogenation, or pyrolysis of plastic MSW, reduce the negative environmental effects compared to landfilling.

These techniques can be utilized to produce char, syngas, and value-added bio-oil of high calorific value. These can be utilized both as fuels for the internal feeding of the plant or as a substitute to traditional fossil fuel [28].

The AD process produces digestate, which can be utilized as a solid biofuel or input for producing activated carbon. Nevertheless, the solid by-products of the AD of a wet fraction of MSW (WFMSW) cannot be used as energy as it is still considered as waste. Magdziarz et al. [29] examined hydrothermal carbonization (HTC) as a pre-treatment technique for the digestate obtained by the AD of the WFMSW. They found that the HTC process changed the physical and chemical properties of the hydrochars when compared to those of the raw materials. The hydrochar demonstrated the best combustion parameters and physical properties at a temperature of 200 °C and 60min residence time during HTC, which was found to be optimal for energy consumption [29]. Co-pyrolysis using microwave heating efficiently resolves several shortcomings of traditional pyrolysis, such as reduced oxygen content and thickness, and augmented calorific value of liquid fuel. This new technique can be a viable means to generate an environment-friendly and sustainable third-generation biofuel from MSW [30]. OFMSW can be processed through hydrothermal carbonization (HTC), a thermochemical process, to generate energy and fatty acid. In this process, chemical extraction removed up to 61% of the hydrochar, which would be a potential means of recovering condensation of fuel molecules and fatty acids in the solid hydrochar. Their results imply that HTC can be utilized to convert the OFMSW into a dry solid fuel and extract biodiesel and biofuel precursors. This technique can be a viable alternative to traditional AD that merely produces only methane from the MSW [31].

3.4 GOOD PRACTICES AND POTENTIAL OF MSW-TO-ENERGY

MSW generation rate is increasing alarmingly in cities of Asia and other parts of the world in line with the growth of urbanization and commercial activities. In addition, adequate and traditional practices of managing waste, such as dumping and landfilling, are making MSW management a considerably challenging task, which has been causing negative externalities on public health and the environment. However, there are good practices to generate energy and other valuable resources from MSW, some of which are in place and some are emerging through experiments. Several studies were conducted to examine the energy potential of MSW.

Sohoo et al. conducted a biochemical methane potential test on the MSW generated in Karachi, Pakistan, where sample fresh synthetic waste was used. They found that about 63 MW of electricity can be produced by constructing biowaste digestion plants with power generation facilities in Karachi, which would contribute 2.1% share to the daily power supply and reduce 21% power shortage. Moreover, these plants have the potential for earning about US$2.6 million per annum by selling the electricity produced by utilizing the MSW [32]. Dalmo et al. analyzed the energy recovery potential from MSW of 33 landfills at 32 cities in São Paulo state, Brazil. Thermochemical and biochemical techniques were considered to examine the energy potential. The thermochemical processes included incineration and gasification, while landfill gas (LFG) and AD were the biochemical techniques. Moreover, hybrid combinations such as incineration–AD and gasification–AD were also incorporated in the analysis. It was found that the combination of incineration and AD had the highest potential for electricity generation (8,051,623 MWh/year) [33].

MSW management is a persistent economic and environmental challenge in Malaysia as more than 80% of MSW is currently being disposed of at the country's landfills and dumping sites. Therefore, the country needs to achieve sustainable management of MSW through recovering energy. Yong et al. [34] examined the feasibility of energy and biofertilizer production from OFMSW via AD. OFMSW is about half of the total MSW generated in the country. The results reveal that this half of the OFMSW can be used via AD to produce 3,941 MWh of electricity per day [34]. Biogas systems have been constructed based on solid waste landfills and their pipeline network at Bantan subdistrict in Thailand. This project has a total biogas capacity of 1,000 m^3/day in two villages. The cooking energy cost has decreased from US $12 per month on LPG to only US $3 per month [35].

Cudjoe et al. examined the potential of electricity generation of biogas produced from OFMSW in 31 provinces of China over the period 2004–2018 using LFG-to-energy and AD methods. Their findings demonstrate that the potential of electricity production of AD process was higher in all provinces than that of LFG-to-energy project. Even though both projects are found to be feasible in all provinces, the AD project was highly feasible, with higher net present value and lower levelized cost of energy production. LFG-to-energy and AD technology have the potential to reduce global warming by 71.5% and 92.7%, respectively. It shows that AD is more beneficial than LFG project in terms of reducing global warming [36]. Integration of LFG with biomass gasification provides benefits in energy generation from MSW as found in a landfill located in Reggio Emilia province in Italy. LFG is generated from MSW fuels with four internal combustion engines that have an overall power generation capacity of 2 MW. The electricity produced from this project is sold to the grid, and the thermal power is utilized for the heating of an industrial greenhouse section to produce basil. The filtered fraction of MSW is sold to the market for electricity production in large-scale boiler-based power plants. Syngas is merged with LFG and then used in gas engines. The result shows that the project is commercially viable and contributes to managing MSW in a sustainable manner [37]. Tursunov and Abduganiev examined two MSW-to-energy projects in Urta-Chirchik district of the Tashkent Region, which are based on mixed MSW and wood waste. Bomb calorimeter and ASTM standards were used to study energy recovery from the projects. The energy content was found to be 2,479.34 kcal/kg for mixed MSW and 2,190.02 kcal/kg for wood waste [38].

Energy production from MSW is considered as a solution to achieve sustainable waste management and enhance the country's energy security situation in Mauritius. Neehaul et al. [39] found AD as the prioritized technology in terms of social acceptance, while energy recovery from MSW utilizing incineration technology has been identified as an economically viable technique in the Mauritian economy [39]. A study was conducted to select appropriate MSW-to-energy technology in Dhaka city, given the fact that waste disposal is related to soil, water, and air pollution, and there is an acute shortage of land and its steeply rising prices in places within and nearby Dhaka have given rise to major problems to develop landfill sites. Therefore, they selected Mirpur, Dhaka, for the conversion of household wastes into energy. Pyrolysis, plasma gasification (PG), and AD were examined to understand the suitability of the energy generation methods. This result demonstrated that PG was the optimal technology to convert MSW into energy [40].

Mondal et al. analyzed the optimized performance of an MSW-fired combined cycle power plant in India to provide electricity in an urban municipality through assessing its energy, exergy, economy, and environmental aspects. They found that the optimal values of exergy efficiency and electricity production cost are nearly 39% and $ 0.085/kWh, respectively. The study also demonstrated that the plant can deliver 1,810kW electricity and save US $270 of environmental damage cost per year compared to the traditional landfilling of MSW. The results suggest that MSW-to-power projects are economically viable if they are combined cycle plants that utilize externally heated clean air turbines with a bottoming organic vapor turbine [41]. Cudjoe et al. examined the electric potential of LFG generated from MSW in the Beijing-Tianjin-Hebei region in China based on the data of MSW disposed in landfills in this region over the period 2004–2018. They evaluated the potential to generate methane from the MSW using Landfill Gas Emission Model (LandGEM), version 3.02. The results suggest that the total potential of electricity production from LFG is 12,525.2 GWh in the region with the highest potential in Beijing. However, the economic viability of the projects is found to be sensitive to the discount rates [42]. These good practices and experiments suggest that different MSW-to-energy technologies have good potential as an alternative to fossil fuel and cost-effective means of energy security across the world.

3.5 CONCLUSIONS

Because of the negative aspects of environmental hazards related to the traditional management of MSW, most countries look forward to reducing the use of landfills for managing the MSW. Energy production from MSW has contributed to the economy and environment significantly, and it is one of the most effective means of tapping its potential. Given its sizable amount, MSW can be used as an input to produce renewable secondary energy resources instead of being kept for polluting the environment, causing serious threats to human and animal health, and reducing land space. Heat, electricity, and transportation fuel can be produced from the energy extracted from the MSW. Thus, the MSW-to-energy helps obtain additional energy, resolve the problems of environmental pollution, and release land from use as a traditional landfill area. MSW is usually less reactive to burning than coal. However, its reactivity can be improved through pre-treatments to reduce non-combustible materials (e.g., oxygen and ash content). Improved thermochemical conversion of MSW is required to increase the performance of the MSW-to-energy projects. Sorting of MSW is required as an approach to improving the feasibility of co-pyrolysis and other advanced thermochemical processes by placing the desired quantity and type of MSW as feedstock in the energy generation plants. In addition, more research and development on MSW-to-energy projects is necessary to maximize the amount and quality of products as well as to ensure their commercial viability.

REFERENCES

1. Balasubramanian S, Tyagi R, Surampalli R, Zhang TC (2012) Green biotechnology for municipal and industrial wastewater treatment. In: *Green Chemistry for Environmental Remediation*, John Wiley & Sons, pp. 627–660. doi: 10.1002/9781118287705. Chapter 20.
2. Balasubramanian S, Gandhi V, Yadav B, Tyagi RD (2021) Sustainable production of bioadsorbents from municipal and industrial wastes in a circular Bioeconomy context. In: *Biomass, Biofuels, Biochemicals: Circular Bioeconomy—Current Developments and Future Outlook*, Elsevier, pp. 639–668. https://doi.org/10..1016/B978–0–12–821878–5.00019–2.
3. Kumar AP, Janardhan A, Viswanath B, Monika K, Jung JY, Narasimha G (2016) Evaluation of orange peel for biosurfactant production by Bacillus licheniformis and their ability to degrade naphthalene and crude oil. *Biotech* 6(1): 43. doi: 10.1007/s13205-015-0362-x.
4. Herbert L (2007) Centenary History of Waste and Waste Managers in London and South East England. Chartered Institution of Wastes Management: 1–52. Available at: https://www.ciwm.co.uk/Custom/BSIDocumentSelector/Pages/DocumentViewer.aspx?id=QoR7FzWBtitMKLGdXnS8mUgJfkM0vi6KMAYwUqgqau3ztZeoed%252bsdmKIqDzPOm8yAXgBZR%252fn1fYhL%252bTNdjUq9g2xwY63C2g8GcAQQyfpf3SImIrrED%252bTfsUM91bKsogr.
5. Alemayehu E (2004) Solid and Liquid Waste Management, Ethiopia Public Health. Available at: https://www.cartercenter.org/resources/pdfs/health/ephti/library/lecture_notes/health_extension_trainees/ln_hew_solid_waste_final.pdf.
6. Buenrostro O, Bocco G, Cram S (2001) Classification of sources of municipal solid wastes in developing countries. *Resour. Conserv. Recycl.* 32(1): 29–41. doi: 10.1016/S0921–3449(00)00094–X.
7. Themelis NJ, Mussche C (2014) Energy and Economic Value of Municipal Solid Waste (MSW), Including Non-recycled Plastics (NRP), Currently Landfilled in the Fifty States. Earth Engineering Center, Columbia University. Available at: http://citeseerx.ist.psu.edu/viewdoc/summary?doi=10.1.1.645.7112.
8. United States Environmental Protection Agency (US EPA) (2020) Advancing Sustainable Materials Management: 2018 Fact Sheet—Assessing Trends in Materials Generation and Management in the United States, December 2020, Washington DC: US EPA. Available at: https://www.epa.gov/sites/production/files/2021-01/documents/2018_ff_fact_sheet_dec_2020_fnl_508.pdf.
9. Kaza S, Yao L, Bhada-Tata P, Van Woerden F (2018) What a Waste 2.0: A Global Snapshot of Solid Waste Management to 2050. Urban Development. Washington, DC: World Bank. Available at: https://openknowledge.worldbank..org/handle/10986/30317.
10. Sharma KD, Jain S (2020) Municipal solid waste generation, composition, and management: The global scenario. *Soc. Res. J.* 16(6): 917–948. doi: 10.1108/SRJ-06-2019-0210.
11. Malav LC, Yadav KK, Gupta N, Kumar S, Sharma GK, Krishnan S, Rezania S, Kamyab H, Pham QB, Yadav S, Bhattacharyya S, Yadav VK, Bach QV (2020) A review on municipal solid waste as a renewable source for waste-to-energy project in India: Current practices, challenges, and future opportunities. *J. Clean Prod.* 277: 123227. https://doi.org/10.1016/j.jclepro..2020.123227.
12. Kalyani KA, Pandey KK (2014) Waste to energy status in India: A short review. *Renew. Sustain. Energy Rev.* 31: 113–120. https://doi.org/10.1016/j.rser.2013.11.020.
13. Okolie JA, Nanda S, Dalai AK, Kozinski JA (2020) Hydrothermal gasification of soybean straw and flax straw for hydrogen-rich syngas production: Experimental and thermodynamic modeling. *Energy Convers. Manag.* 208: 112545. https://doi.org/10.1016/j.enconman..2020.112545.

14. Sarangi PK, Nanda S (2020) Biohydrogen production through dark fermentation. *Chem. Eng. Technol.* 43: 601–612. https://doi.org/10.1002/ceat..201900452.
15. Reddy SN, Nanda S, Hegde UG, Hicks MC, Kozinski JA (2015) Ignition of hydrothermal flames. *RSC Adv.* 5: 36404–36422. https://doi.org/10..1039/C5RA02705E.
16. Agarwal M, Tardio J, Mohan SV (2013) Critical analysis of pyrolysis process with cellulosic based municipal waste as renewable source in energy and technical perspective. *Bioresour. Technol.* 147: 361–368. doi: 10.1016/j.biortech.2013.08.011.
17. Chen D, Yin L, Wang H, He P (2015) Reprint of: Pyrolysis technologies for municipal solid waste: A review. *Waste Manag.* 37: 116–136. https://doi.org/10.1016/j.wasman.2015.01.022.
18. Hosseinalizadeh R, Izadbakhsh H, Shakouri HG (2021) A planning model for using municipal solid waste management technologies – Considering energy, economic, and environmental impacts in Tehran-Iran. *Sust. Cit. Soc.* 65: 102566. https://doi.org/10.1016/j.scs..2020.102566.
19. Mahmoodi P, Karimi K, Taherzadeh MJ (2018) Hydrothermal processing as pretreatment for efficient production of ethanol and biogas from municipal solid waste. *Biores. Tech.* 261: 166–175. https://doi.org/10.1016/j.biortech.2018.03.115.
20. Nanda S, Berruti F (2021) A technical review of bioenergy and resource recovery from municipal solid waste. *J. Haz. Mat.* 403: 123970. https://doi.org/10.1016/j.jhazmat..2020.123970.
21. Sipra AT, Gao N, Sarwar H (2018) Municipal solid waste (MSW) pyrolysis for bio-fuel production: A review of effects of MSW components and catalysts. *Fuel. Pro. Tech.* 175: 131–147. https://doi.org/10.1016/j.fuproc.2018.02.012.
22. Adamović VM, Antanasijević DZ, Ćosović AR, Ristić MĐ, Pocajt VV (2018) An artificial neural network approach for the estimation of the primary production of energy from municipal solid waste and its application to the Balkan Countries. *Waste Manag.* 78: 955–968. https://doi.org/10.1016/j.wasman..2018.07.012.
23. Yatsunthea T, Chaiyat N (2020) A very small power plant – Municipal waste of the organic rankine cycle and incinerator from medical and municipal wastes. *Ther. Sci. Eng. Prog.* 18: 100555. https://doi.org/10.1016/j.tsep..2020.100555.
24. Shahnazari A, Rafiee M, Rohani A, Nagar BB, Ebrahiminik MA, Aghkhani MH (2020) Identification of effective factors to select energy recovery technologies from municipal solid waste using multi-criteria decision making (MCDM): A review of thermochemical technologies. *Sust. Ener. Tech. Ass.* 40: 100737. https://doi.org/10.1016/j.seta..2020.100737.
25. Chen G, Jamro IA, Samo SR, Wenga T, Baloch HA, Yan B, Ma W (2020) Hydrogen-rich syngas production from municipal solid waste gasification through the application of central composite design: An optimization study. *Int. J. Hyd. Ener.* 45(58): 33260–33273. https://doi.org/10.1016/j.ijhydene..2020.09.118.
26. Lopes EJ, Okamura LA, Maruyama SA, Yamamoto CI (2018) Evaluation of energy gain from the segregation of organic materials from municipal solid waste in gasification processes. *Ren. Ener.* 116: 623–629. https://doi.org/10.1016/j.renene.2017.10.018.
27. Hameed Z, Aslam M, Khan Z, Maqsood K, Atabani AE, Ghauri M, Khurram MS, Rehan M, Nizami AS (2021) Gasification of municipal solid waste blends with biomass for energy production and resources recovery: Current status, hybrid technologies and innovative prospects. *Renew. Sust. Ener. Rev.* 136: 110375. https://doi.org/10.1016/j.rser..2020.110375.
28. Banu JR, Sharmila VG, Ushani U, Amudha V, Kumar G (2020) Impervious and influence in the liquid fuel production from municipal plastic waste through thermochemical biomass conversion technologies—A review. *Sci. Tot. Env.* 718: 137287. https://doi.org/10.1016/j.scitotenv..2020.137287.

29. Magdziarz A, Mlonka-Mędrala A, Sieradzka M, Aragon-Briceño C, Pożarlik A, Bramer EA, Brem G, Niedzwiecki Ł, Pawlak-Kruczek H (2021) Multiphase analysis of hydrochars obtained by anaerobic digestion of municipal solid waste organic fraction. *Renew. Ener.* 175: 108–118. https://doi.org/10.1016/j.renene..2021.05.018.
30. Mahari WAW, Azwar E, Foong SY, Ahmed A, Peng W, Tabatabaei M, Aghbashlo M, Park YK, Sonne C, Lam SS (2021) Valorization of municipal wastes using co-pyrolysis for green energy production, energy security, and environmental sustainability: A review. *Chem. Eng. J.* 421: 129749. https://doi.org/10.1016/j.cej..2021.129749.
31. Ischia G, Fiori L, Gao L, Goldfarb JL (2021) Valorizing municipal solid waste via integrating hydrothermal carbonization and downstream extraction for biofuel production. *J. Clean Prod.* 289: 125781. https://doi.org/10.1016/j.jclepro..2021.125781.
32. Sohoo I, Ritzkowski M, Heerenklage J, Kuchta K (2021) Biochemical methane potential assessment of municipal solid waste generated in Asian cities: A case study of Karachi, Pakistan. *Renew. Sust. Ener. Rev.* 135: 110175. https://doi.org/10.1016/j.rser..2020.110175.
33. Dalmo FC, Simão NM, de Lima HQ, Jimenez ACM, Nebra S, Martins G, Palacios-Bereche R, Sant'Ana PHdM (2019) Energy recovery overview of municipal solid waste in São Paulo State, Brazil. *J. Clean Prod.* 212: 461–474. https://doi.org/10.1016/j.jclepro..2018.12.016.
34. Yong ZJ, Bashir MJK, Hassan MS (2021) Biogas and biofertilizer production from organic fraction municipal solid waste for sustainable circular economy and environmental protection in Malaysia. *Sci. Tot. Env.* 776: 145961. https://doi.org/10.1016/j.scitotenv..2021.145961.
35. Damrongsak D, Chaichana C (2020) Biogas initiative from municipal solid waste in Northern Thailand. *Ener. Rep.* 6: 428–433. https://doi.org/10.1016/j.egyr..2019.11.098.
36. Cudjoe D, Han MS, Nandiwardhana AP (2020) Electricity generation using biogas from organic fraction of municipal solid waste generated in Provinces of China: Techno-economic and environmental impact analysis. *Fuel Proc. Tech.* 203: 106381. https://doi.org/10.1016/j.fuproc..2020.106381.
37. Pedrazzi S, Santunione G, Minarelli A, Allesina G (2019) Energy and biochar co-production from municipal green waste gasification: A model applied to a landfill in the North of Italy. *Ener. Conv. Manag.* 187: 274–282. https://doi.org/10.1016/j.enconman.2019.03.049.
38. Tursunov O, Abduganiev N (2020) A comprehensive study on municipal solid waste characteristics for green energy recovery in Urta-Chirchik: A case study of Tashkent region. *Mater. Today: Proceedings* 25(1): 67–71. https://doi.org/10.1016/j.matpr..2019.11.108.
39. Neehaul N, Jeetah P, Deenapanray P (2020) Energy recovery from municipal solid waste in Mauritius: Opportunities and challenges. *Env. Dev.* 33: 100489. https://doi.org/10.1016/j.envdev..2019.100489.
40. Rahman SS, Azeem A, Ahammed F (2017) Selection of an appropriate waste-to-energy conversion technology for Dhaka City, Bangladesh. *Int. J. Sustain. Eng.* 10: 99–104. https://doi.org/10.1080/19397038.2016.1270368.
41. Mondal P, Samanta, Zaman SKA, Ghosh S (2021) Municipal Solid Waste Fired Combined Cycle Plant: Techno-economic Performance Optimization Using Response Surface Methodology. *Ener. Conv. Manag.* 237: 114133. https://doi.org/10.1016/j.enconman..2021.114133.
42. Cudjoe D, Han MS, Chen W (2021) Power generation from municipal solid waste landfilled in the Beijing-Tianjin-Hebei region. *Energy* 217: 119393. https://doi.org/10.1016/j.energy..2020.119393.

4 A Brief History of Energy Recovery from Municipal Solid Waste

Debra R. Reinhart
University of Central Florida

Aditi Podder
Brown and Caldwell

Stephanie C. Bolyard
Florida State University

CONTENTS

DOI: 10.1201/9781003178354-6

4.1 INTRODUCTION

Municipal solid waste (MSW) is composed primarily of organic materials (~46%–70%, by weight) [1,2] and consequently has a high energy content (6.2–23.7 MJ/kg dry basis net calorific value) [3]. However, the recovery of this energy has historically been relatively inefficient, costly, and largely limited to conversion of biodegradable fractions (~40%–60% of waste by weight) [4] into methane (CH_4) within landfills and thermal conversion within incinerators. In some parts of the world, anaerobic digestion, gasification, and pyrolysis have also been practiced to recover energy from MSW. This chapter will describe the historical development of energy recovery from MSW through biological and thermal processes. The chapter focuses on US waste management; however, many of the conclusions drawn apply globally. Section 4.9 considers the worldwide aspects of waste generation and characteristics on energy recovery from waste as a function of country wealth.

4.2 HISTORY OF MSW DISPOSAL

The need for MSW disposal dates to prehistoric times; waste disposal became problematic with civilization and urbanization as increased production rates and population density led to unsightly litter, odors, and adverse health impacts. Combustion of waste has long been practiced to control these problems and to achieve significant volume reduction; however, this often results in a transfer of pollutants and greenhouse gases to the atmosphere. Landfilling remains a common disposal approach and, in the past 100 years, has been accomplished with greater attention to the control of liquid and gas emissions, although landfills are still a potential source of odors and greenhouse gas emissions. Figure 4.1 provides details regarding the relative tonnages managed using various disposal processes over the past nearly 50 years in the USA; the amounts of waste recycled, combusted, and landfilled have largely plateaued since the early 2000s. Waste avoidance and recycling have been encouraged to minimize landfilling and combustion and reduce the use of raw materials in products. For example, recycling plastic waste globally can save the equivalent of

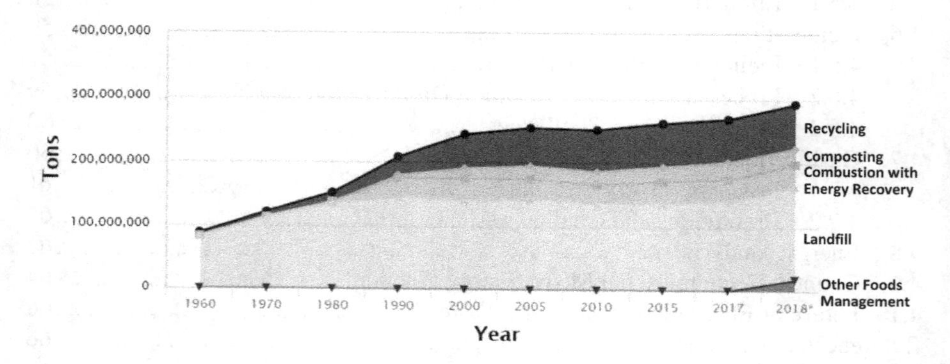

FIGURE 4.1 Municipal solid waste management in the USA between 1960 and 2018 [6].

3.5 billion barrels of oil [5] and aluminum recycling reduces energy used during the extraction of aluminum from bauxite by 95%. Converting waste plastics/polymers into value-added chemicals (or upcycling) ensures a more energy-efficient production of chemical feedstocks compared to raw fossil fuels. In addition, a small amount of waste is composted; however, aerobic composting results in net energy consumption. Because there is no direct opportunity for energy recovery during recycling or composting, neither will be discussed further in this chapter.

4.3 THERMAL AND BIOLOGICAL ENERGY CONVERSION PROCESSES

Figure 4.2 provides a high-level tracking of the fate of the energy contained in products over their lifetime, including final disposal. There are many challenges associated with energy recovery from waste, including its heterogeneity, moisture content, and seasonality. The specific energy content of waste is approximately half of that of fossil fuels such as gasoline, diesel fuel, and kerosene and is equal to lower-value fuels such as coal and biomass. As seen in Figures 4.3, much of this energy content is associated with plastics, textiles, and paper. Some types of plastic waste (e.g., polyethylene, polypropylene, and polystyrene) have calorific values close to conventional fuels. The relative abundance of these materials has changed in the USA over the past 50 years as illustrated in Figure 4.3a. Plastics have increased at a compound annual rate of 8.4% since 1950 [7], largely because of the upsurge in the amount of packaging used, while the fraction of paper, yard waste, and metal has declined, perhaps due

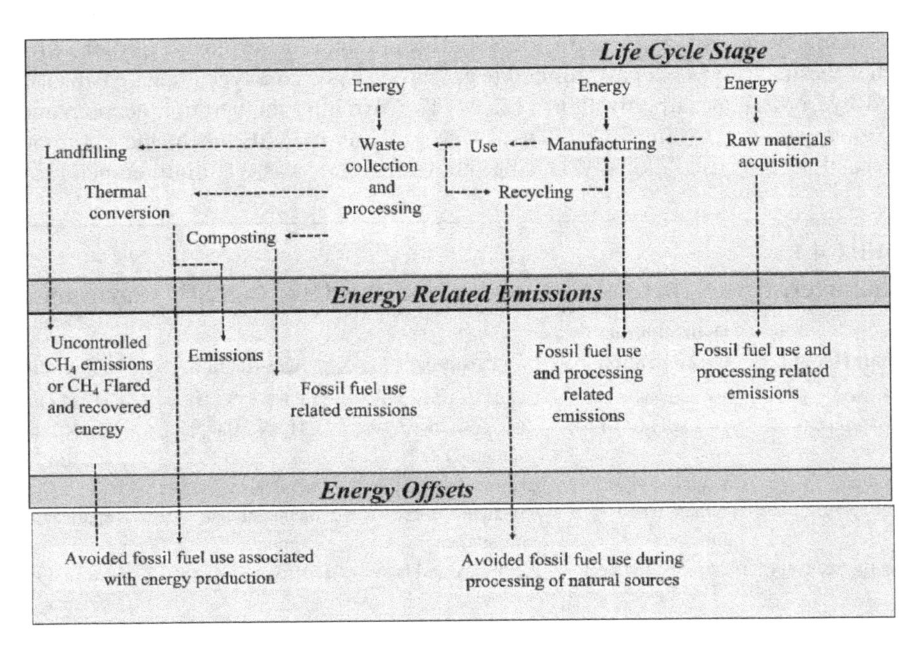

FIGURE 4.2 Energy related to waste management tracked through waste life cycle stage.

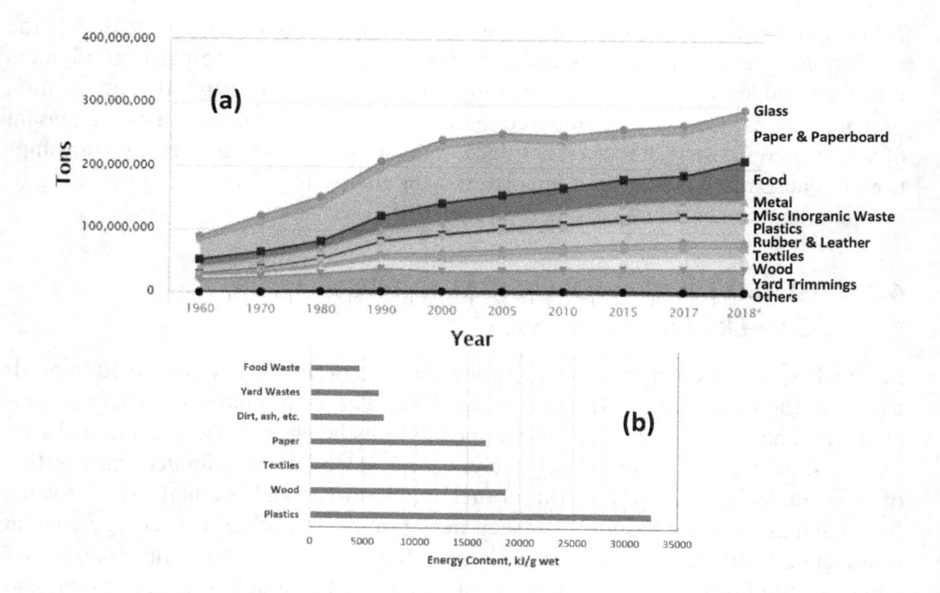

FIGURE 4.3 (a) Distribution [6] and (b) energy content of US MSW components.

to waste avoidance by consumers. The composition of waste is also closely aligned with the economy, as discussed further in this chapter.

Thermal conversion promotes the transfer of energy from waste to final products including steam, oil, and gas. Incineration occurs under a surplus of oxygen, while other thermal processes have limited (e.g., pyrolysis) or no oxygen (e.g., gasification and hydrothermal carbonization) (Table 4.1). Currently, incineration accompanied by energy recovery (often referred to as waste-to-energy (WtE)) is the most common thermal process used globally. During the oxidation of waste at high temperatures,

TABLE 4.1
Characteristics of Thermal Conversion Processes Used in MSW Management

Characteristic	Hydrothermal Carbonization	Pyrolysis	Gasification	Incineration
Primary by-products	Char and gas (primarily CO_2)	CO, H_2, hydrochar, ash, coke, pyrolysis oil, and H_2O	CO, H_2, CO_2, H_2O, CH_4, N_2, slag, and ash	CO_2, H_2O, O_2, N_2, slag, and ash
Product recovery	Chemicals, gasoline, methanol, and ammonia	Chemicals, gasoline, methanol, and ammonia	Chemicals, gasoline, methanol, and ammonia	Not applicable
Energy recovery	Electricity and heat	Electricity and heat	Electricity and heat	Electricity and heat
Temperature range (°C)	280–300	300–1,300	600–800	800–1,200

heat is released and is typically captured in the form of steam. The steam can then be used to power turbines to generate electricity or to directly heat homes or businesses. The efficiency of energy recovery is impacted by the design and operation of the thermal conversion facility and waste characteristics including heat value, moisture content, and heterogeneity.

MSW contains significant amounts of biodegradable waste including paper products, food waste, yard waste, textiles, and leather. Although plastics are organic, they typically are not biodegradable. Anaerobic conversion of organic wastes is a microbially mediated process where the biodegradable organic waste is decomposed in the absence of oxygen to produce biogas. The digested solids can be used as soil amendments after further processing. A series of metabolic reactions are involved in the process, which can be divided into four different stages: hydrolysis, acidogenesis, acetogenesis, and methanogenesis. A separate group of microorganisms is responsible for carrying out each of the steps. Ultimately, the energy content of waste materials is transferred to CH_4, which can be captured and used as a fuel source. The biological processes occur in a landfill, but can be better managed in anaerobic digestion reactors where they can also be controlled to produce other simple organics such as ethanol or methanol.

4.4 WASTE-TO-ENERGY – LANDFILLING

The history of landfills dates back to as early as 3000 BC (the likely first landfill was discovered in Crete). These early landfills were pits dug in the ground and filled with waste that was covered with minerals, rocks, and soil. It was discovered more than a century ago that open air and uncontrolled disposal of waste were not ideal due to liquid and gaseous emissions. Engineered systems were needed to protect human health and the environment from these emissions. These engineered containment systems included bottom liners to protect the groundwater and cover systems to control the infiltration of liquid after closure. In addition, gaseous emissions were controlled through gas collection systems and interim soil covers. As engineers began to focus on the operation of well-engineered landfills, it became apparent that the exclusion of liquids was retarding the biodegradation of waste. Their designs were then refined through the introduction of bioreactor landfills, which involved adding liquids intentionally to the waste. The proactive management of liquids increased the rate of waste degradation and gas production, saved landfill space, and even improved leachate quality.

4.4.1 Landfill Gas Production

Landfill gas is produced during the anaerobic decomposition of MSW. The type of waste responsible for the generation of gas is the biodegradable fraction. This fraction includes paper products, food waste, yard waste, textiles, and leather and typically accounts for ~40%–60% of the waste by weight [4]. The decomposition of these materials occurs via complex processes that are discussed in Section 4.3 and follow first-order kinetics. Landfill gas is primarily composed of CO_2 (40%–60%, by volume) and CH_4 (45%–60%, by volume), but also includes trace gases such as nitrogen, oxygen, ammonia, sulfides, hydrogen, carbon monoxide, and non-methane organic compounds.

Various factors affect landfill gas generation, including landfill operation, the methane potential of the waste, and the time since placement [8]. Gas generation can be predicted using various first-order models that use inputs that rely on site-specific conditions such as the amount of waste, type of waste, and climate conditions. One input parameter, the methane generation potential (L_o), is the practical amount of methane generation expected and is dependent on waste composition. Waste composition varies from country to country and is affected by economic development (e.g., developed vs. developing country) and lifestyle [9]. A study by Krause et al. [10] determined experimentally that the L_o of mixed MSW ranged from 35 to 167 m^3 CH_4/mg MSW. As waste degrades, the substrate that is available to be converted into landfill gas is consumed. Over time, these older wastes have a lower L_o relative to fresh waste that has not undergone anaerobic decomposition. Therefore, in a landfill, it is expected that the newly placed waste will contribute more to landfill gas generation and the volume of gas declines over time. Gas generation is affected by the amount of moisture that is present in the waste. Typically, MSW has a moisture content of approximately 20% with a typical range of 15%–40% [11]. The more the moisture that is present in the waste mass, the greater the rate of gas production. Conventional landfills are typically operated to minimize the amount of liquid that enters a cell. Landfills can be operated as bioreactors where liquids are intentionally added to increase the overall moisture content to 35%–65% [12], which is optimal for anaerobic decomposition and serves to increase the rate of gas generation.

Landfill gas is collected by a series of horizontal and vertical wells placed directly into the waste mass. These wells are connected to a system that controls gas extraction by placing a vacuum at the wellhead. The efficiency of gas collection varies over time as new waste is placed within the influence area of the gas collection wells. Gas collection efficiency values range from 50% to 95% with a suggested average of 75% [13–15]. The collection efficiency can be improved by a well-operated and well-maintained gas collection system and the use of a final geomembrane cover. Table 4.2 outlines the improvement to collection efficiency related to cover type. Gas that is not collected will either be oxidized within the soil cover, or escape as fugitive emissions. Oxidation with the soil cover can range from 14% to 55% of influent methane depending on soil type and CH_4 flux through the cover with an average of about 20% [17].

TABLE 4.2
Landfill Gas Collection Efficiency Ranges by Cover Type [16]

Cover Type	Collection Efficiency Range
Landfill or portions of a landfill that are under daily soil cover	50%–70% (mid-range default = 60%)
Landfill or portions of a landfill that contain an intermediate soil cover	54%–95% (mid-range default = 75%)
Landfills that contain a final soil and/or geomembrane cover systems	90%–99% (mid-range default = 95%)

4.4.2 Energy Recovery and Utilization

A landfill gas collection system is used to capture landfill gas to reduce greenhouse gas emissions and odors and to utilize CH_4 beneficially. Examples of beneficial use of landfill gas are electricity generation, direct use of landfill gas to replace another fuel (e.g., natural gas), and pipeline natural gas. Approximately 70% of the landfill gas energy projects in the USA are operated to produce electricity with the remaining 30% split between direct use and generation of renewable natural gas. Electricity generation from landfill gas relies on technologies such as microturbines, fuel cells, and internal combustion engines. Combined heat and power projects have been used at landfills as well, which generate both thermal energy and electricity. The use of landfill gas for electricity requires the removal of siloxane and water. Upgrading landfill gas for direct use on-site or nearby to replace another fuel or as a pipeline gas requires additional processing to remove impurities (e.g., CO_2, N_2, and O_2). The direct use of landfill gas is typically limited to facilities within 5 miles of a landfill. If there is not enough landfill gas being collected or the quality is not optimal for beneficial use, then an on-site flare is required to control the gas.

4.4.3 Limitations and Challenges

Although the utilization of landfill gas for energy is relatively well established, limitations and challenges remain that must be considered and overcome to improve the overall efficiency of recovering energy. Waste heterogeneity and delay of gas well installation after waste placement can reduce the overall collection efficiency. Highly biodegradable wastes (e.g., food wastes) degrade rapidly, and in many cases, gas generation occurs prior to the installation of a gas collection system. The prediction of gas generation using models may not align with gas generation measured in the field. There are challenges with utilizing these models; however, it is important to compare field-acquired data not only to track the recovery of landfill gas, but also to track the stabilization of landfills following closure. More accurate models and techniques need to be developed to improve the accuracy of landfill gas generation quantification and prediction. The recent decline in the price of natural gas has made landfill gas extraction and beneficial use less competitive. Therefore, a favorable purchase price by utilities or a conveniently located direct-use facility is needed to ensure the use of landfill gas. There are multiple initiatives to direct organics away from landfills. Europe has already successfully implemented bans on untreated organics going to landfills. The change in waste streams will impact the materials available to produce landfill gas. Further, landfill gas cannot be directly utilized to produce energy without some form of processing. There is a need to simplify the processing of landfill gas to upgrade the energy content and direct use of the gas.

4.5 ANAEROBIC DIGESTION

Anaerobic digestion (AD) has recently gained attention as a means to divert organic wastes from landfills because of its potential for production of renewable bioenergy from biodegradable waste products such as sewage sludge, brewery and food wastes, municipal and industrial wastewater, residues from agriculture and wood industry, and livestock manure [18]. In the AD process, biodegradation of waste is carried out in a sealed reactor, where optimal conditions are provided for the maximum activity of microorganisms. To improve the digestion or increase the biogas production yield, co-digestion is occasionally carried out, where multiple organic materials such as food waste, crop residues, manure, sludge, and biosolids are combined and can even be treated at wastewater treatment facilities in a single reactor system.

The process of generating flammable gas from the decomposition of organic matter was first discovered by Jan Baptista van Helmont in the 17th century. Early in the 18th century, CH_4 was detected in the gas produced during the AD of cattle manure by Sir Humphry Davy. AD plants were installed in India and England in 1859 and 1895, respectively, and the biogas was utilized to fuel street lamps [19,20]. In the 19th century, developments in microbiology guided researchers to identify microbial populations responsible for anaerobic degradation and to determine optimal conditions to promote maximum efficiencies of these populations. The proper understanding of the AD process enabled facility installation globally toward the latter part of the 19th century in the USA, Germany, Denmark, China, and India. Digesters are low-tech facilities (although with inconsistent waste stream they can be difficult to operate) with simple designs and therefore have proven to be effective in developing countries at a small scale in providing biogas for cooking and lighting. Administrative support from local government regarding land use and waste disposal policies was also useful in the installation and application of this process.

4.5.1 Limitations and Challenges

Some of the advantages that have increased the interest regarding the implementation of anaerobic digestion center around reduced greenhouse gas emissions compared to landfilling and the potential for CH_4 gas utilization [21]. However, despite these benefits, there are some challenges encountered in the successful implementation of this process. For example, anaerobic digestion process is often rate-limited by either the hydrolysis, or the methanogenesis step. Hydrolysis becomes a rate-limiting step when refractory wastes such as fruit peels, wood debris, and green wastes are introduced in the digestion process. These lignocellulosic waste materials have complex molecular structures, thus slowing down the degradation process. Optimum pH within the range of 6.7–8 is required for successful methanogenesis, which becomes the rate-limiting step if acid accumulates during the acetogenesis step, reducing the overall pH of the system [22]. Methanogens are slow-growing microorganisms, and therefore, a rapid change in feedstock composition (refractory waste to more readily biodegradable waste) or a change in feedstock flow rate can shock the system. Temperature also plays a key role in maintaining an appropriate bacterial population in the digester. The digestion process typically is carried out under mesophilic (ranging from 20 °C to 44 °C) or thermophilic (45 °C–80 °C) temperature regimes [23]. Therefore, the vital operational parameters must be controlled to ensure ideal reactor operating conditions, where maximum bacterial activity would be promoted. Other limitations in the implementation of the AD process include the heterogeneous composition of waste with variable degradation rates, moisture content of the feedstock, and the presence of compounds that are toxic or inhibitory to the microbial population [24].

4.6 INCINERATION

Thermal conversion processes are an alternative and preferred method to landfilling according to the waste management hierarchy. These processes are an effective way to reduce the overall volume (80%–90% reduction) and mass (70% reduction) of the waste while recovering renewable energy, fuels, and other by-products (e.g., chars, tars, and oils). The first waste incinerator in the world was constructed in the UK in the 1870s followed by a US incinerator built in 1885. Early on, there was little information on the environmental impacts of the liquid discharges and air emissions from incinerators. As regulations were put in place limiting the release of air emissions, some incinerators were closed to avoid the costly addition of air pollution control units. Currently, incinerators are well-operated and well-controlled facilities that are successful in MSW management. Over time, different types of incinerator designs have been developed based on how the waste is introduced and moved through the combustion chamber, including simple, rotary kiln, fixed or moving grate, and fluidized bed incinerators.

4.6.1 Incineration Process Basics

Incineration chemically oxidizes waste to thermally convert MSW in the presence of either a stoichiometric or excess amount of air. The need for supplemental fuel will depend on the lower heating value of the waste, which is dictated by the overall composition. The primary characteristics that affect the need for supplemental fuel are heating value, moisture content, and ash content. The typical operating temperature range is 790 °C–980 °C. The primary end products of the incineration process are N_2, CO_2, water vapor, ash, and heat. Air pollution control is needed to capture dust, acid gases, volatile organic compounds, and nitrogen oxide. Particulates are captured as fly ash that is transported in the flue gases. Acid gases are generated from acids and acid precursors that can form sulfur dioxide, hydrochloric acid, and nitrogen oxides. Lastly, dioxins are generated and mercury in waste is volatilized through the combustion of MSW and needs to be controlled.

The incineration of MSW has many advantages including the significant reduction in the volume and weight of the waste, the ability to control air emissions, the production of non-putrescible (does not decay or produce odors) ash, and, lastly, the production of energy that can financially offset capital costs.

4.6.2 Process Design and Operation Optimization over Time

Time, temperature, and turbulence are the three important operational parameters that need to be carefully monitored and controlled to ensure an acceptable level of thermal conversion efficiency is achieved. Over time, it has become apparent that for efficient combustion to occur, the waste and gas must reach a high enough temperature and also the fuel (waste) and oxygen must be properly mixed. The heat released from the oxidation process adds to the system to help ensure the proper temperature is maintained for complete combustion. If the overall temperature gets too high, excess nitrous oxide can be produced and damage can occur in the system. Turbulence is important to make sure there is enough contact in the system between the waste, combustion gases, and oxygen. Incinerators have evolved into WtE facilities that typically recover heat as steam. When MSW is combusted, energy is released that can be captured to produce energy. The generated energy can be recovered using various types of systems including heat-only, power-only, steam-only, combined steam and power, or combined heat and power systems. The overall efficiency of these systems ranges from 75% to 92% with the power-only system having the lowest efficiency of 25%–35% [25].

4.6.3 Limitations and Challenges

Although incineration has multiple advantages, there are limitations and challenges. Incineration is capital-intensive, and operating costs are greater than less complex approaches to waste management such as landfilling. In addition, to make sure an incinerator is operating efficiently, the facility needs skilled operators to have the knowledge necessary to ensure the process is running properly. Fly ash and bottom ash are end products of the combustion process and must be managed properly. Some facilities have an on-site monofill for ash disposal while others need to dispose of these materials off-site. There have been advancements in the reuse of ash, but there are still environmental concerns because of the presence of heavy metals and dioxins. Emissions from the process are inevitable and require extensive air pollution controls. Lastly, waste management personnel not only have to overcome operational challenges, but also have to deal with public perception. The "not-in-my-backyard" syndrome is certainly applicable to incinerators. There are opportunities to engage with the public to provide additional educational opportunities to better understand that these systems are well engineered to protect human health and the environment.

4.7 GASIFICATION AND PYROLYSIS

For efficient conversion of MSW to energy, emerging technologies such as gasification and pyrolysis have gained attention from the scientific community in recent years. Gasification and pyrolysis are processes where organic matter from MSW is decomposed by heating the waste in the presence of a small amount of air or no air as compared to other waste conversion technologies such as incineration. Table 4.1 details MSW thermal conversion technologies and their gaseous, solid, and liquid products. Although all processes described aim at the thermal decomposition of organic matter, their operating temperature, which is linked to oxygen levels in the process, and the products from the processes vary. Reacting the materials at high temperatures (>700 °C) in the presence of a controlled amount of oxygen without combustion enables these processes. In these processes, organic matter can be converted into a gas mixture of CO, H_2, and CO_2. Gasification and pyrolysis are extremely efficient ways of using biomass to produce energy, compared to incineration [26]. They are flexible technologies where existing gas-fueled devices (e.g., ovens, furnaces, and boilers) can be retrofitted and syngas can directly replace fossil fuels. Gasification-generated energy is cheaper and more efficient than the steam turbine process used in incineration. MSW can be reduced by as much as 75% through these processes, reducing the potential emissions the waste would have created in a landfill. These technologies are cleaner than incineration and, in general, do not pose toxicity threats. Gasification and pyrolysis processes are described in more detail in the next section.

4.7.1 Processes Overview

Gasification is a complicated process and is a thermochemical route to produce synthesis gas or syngas (primarily H_2 and CO) from renewable sources. Gasification was first industrially applied in the early 19th century [27]. "Town" gas was produced from coal which contained almost 50% H_2; the remaining fraction is a mixture of CH_4, CO, and CO_2. It was used for lighting and heating in both the USA and Europe. Currently, 117 gasification plants have been constructed around the world, of which 36% are in use [28]. The gasification process is carried out in the presence of steam or CO_2 as gasification agents, with oxygen content below the stoichiometric requirements, within the temperature range of 730 °C–1,680 °C [29]. These conditions promote the conversion of organics to syngas. Inorganic compounds become bottom ash or slag depending on the gasification condition. The gasification process does not produce energy from MSW through direct combustion; rather, a mixture of waste, steam, and oxygen is used as an input to a gasifier and the applied heat and pressure break the chemical bonds of the waste. The obtained syngas can be used directly in internal combustion engines or the production of compounds substituting for natural gas, fertilizers, and transportation fuels [30]. Typically, before combusting the syngas, pollutants are removed to limit emission issues.

Gasification studies are carried out mostly on platform molecules, biomass, and sorted organic waste fractions with considerable caloric value [31]. Sometimes, syngas of specific composition is produced based on their downstream use; yield is increased by fine-tuning the process. For example, CH_4 and tar reforming improves the quality of syngas and reduces the emission of organics with wastewater that originates from cooled raw gas. Reforming CH_4 before the gas is cooled further allows the removal of the formed CO_2 and other inorganic gaseous impurities. Catalytic waste processing is often opted for this type of technology [31]. For biomass conversion, there are two catalytic groups. The first group of catalysts, which are also known as primary catalysts, are usually added directly to the biomass. This process is carried out either by wet impregnation of the biomass, or by dry mixing the catalyst with it [32]. The addition of catalysts also reduces the content of tar. The second group of catalysts is added downstream of the gasifier in a second reactor. This allows the use of more optimal operating conditions than those used for the first group of catalysts where they are directly added to the biomass. These catalysts are active in reforming CH_4 and hydrocarbons [32].

Although gasification is a mature technology, MSW gasification is still under development. Two consecutive processes are typically required to achieve successful MSW gasification toward valuable syngas. First, pyrolysis occurs at temperatures below 630 °C, and at that point, char is produced while volatile components are released [32]. Furthermore, under high pressure, these compounds can undergo gasification with steam or combustion with air or pure oxygen at 780 °C–1,700 °C. A mixture of CO and H_2 is obtained with only trace or no nitrogen when gasification with pure oxygen takes place. When gasification is performed with steam, that is described as steam reforming. Like MSW incineration, fixed bed and fluidized bed reactors are used for MSW gasification. However, to deal with MSW with variable composition, larger gasifiers are preferred to achieve better operational performance. Although the gasification process requires several technical advancements and developments, due to diversification in power generation, more stringent environmental regulations, and higher landfilling fees, this process is favored over many conversion techniques by the scientific communities.

Plasma gasification is also an efficient process in the conversion of MSW to energy or other valuable products that result in greater electricity production and lower emissions when compared to the conventional gasification process. However, this process is associated with high implementation costs, mainly due to the high operating temperature (1,230 °C–1,730 °C). A plasma arc (>5,300 °C) is typically used to produce plasmas of oxygen, steam, or air. If the process is implemented with heterogeneous catalysis, for example, a combination of non-thermal plasma and a selective catalyst, the effective use of plasma can be guaranteed. Also, this combination would be associated with lower temperature requirements, higher synergy potential, reduction in activation energy through the use of catalyst, and enhancements in the conversion of reactants, which would allow high yield to valuable target products [33].

The pyrolysis process was used in the preindustrial era in the production of charcoal from wood that was essential in the extraction of iron from iron ore. The pyrolysis process evolved and was industrially applied in the latter part of the 19th century. The development of modern petrochemical industries was based on the production of kerosene through the pyrolysis process [34]. Like gasification, pyrolysis involves the conversion of waste into energy by heating (above 430 °C) under pressure at controlled conditions, however in the absolute absence of air. The process produces char, pyrolysis oil, and syngas, all of which can be used as fuels. The by-product char can be used to reform volatile products such as syngas and oil into more valuable materials, which would enhance pyrolysis efficiency [34].

4.7.2 Limitations and Challenges

MSW composition is heterogeneous, and therefore, various undesired compounds such as halogens, heavy metals, sulfur dioxide, tars, and alkaline compounds are likely to be present in the product stream from the gasification and pyrolysis processes. These products are often considered as harmful as their exposure is associated with health, environmental, and technical or installation complications. Steam gasification can be utilized with advantages associated with its short residence time and relatively low tar formation in a fluidized bed. However, fluctuating syngas $CO:H_2$ ratios and typically low yield in the overall biomass conversion to syngas limit the use of this technique. Further, to have better control over the process, consistent MSW composition is required with MSW pre-processing steps, whereas more commonly used thermal conversion technologies can burn raw, unprocessed, and unsorted MSW, making pyrolysis and gasification less acceptable.

Both technologies offer an opportunity for implementation, but are still relatively costly, require highly specialized skills to manage and operate, and need secure waste receipt rates (feedstock) to ensure their viability. Another barrier relates to enabling legislation regarding land zoning for the siting of such facilities. However, the technologies are still relatively new and a limited number of plants are in operation around the world, although this number is anticipated to grow substantially in the future.

4.8 ENERGY ANALYSIS

Waste generation, composition, and management practices reported in Figures 4.1 and 4.3 were used to estimate the energy recovery potential for the USA from landfills since 1960 following the procedures outlined in Amini et al. [36]. These calculations required many assumptions; therefore, the results indicate trends rather than exact values. Assumptions were made regarding the efficiency of landfill gas collection over time, waste composition, waste quantities, the energy content of wastes, and the efficiency of internal combustion engines (used to convert gas to electricity). Figure 4.4 illustrates the tonnages of various MSW components reaching landfills (4.4a) and the annual energy production based on the composition of the waste landfilled at the time (4.4b). Energy potential increased dramatically from 1960 to 1990 as waste production rates in the USA soared and landfill gas collection efficiency improved.

Despite the decline in the paper waste landfilled, the landfill energy recovery potential remained high as the fraction and tonnage of food waste increased. In 2018, energy from landfill gas could power 1 million households in the USA, although this value is less than 0.05% of the recent total US energy consumption. Amini et al. [35] published the power density for various energy sources, defined as the rate of flow of energy produced per unit horizontal land area requited [36]. They reported that the average power density of landfill gas to energy was higher than that of renewable energy resources such as biomass, hydro, wind, and ocean and had comparable density to photovoltaics, geothermal, and tidal.

Figure 4.4 also provides an estimate of US incineration tonnage and the potential energy recovery based on Figures 4.1 and 4.3. Again, these calculations required numerous assumptions (e.g., energy content of waste, composition and quantity of waste, conversion efficiency to electricity); however, they illustrate the trends over time. The increase in the number of thermal processing facilities with energy recovery that occurred in the late 1980s was particularly impactful. The dramatic rise in the amount of plastics reaching incineration facilities also resulted in a greater energy potential, which can be seen in Figure 4.4c where incinerated waste-specific energy content is reported. Few new incineration facilities have been constructed in the USA over the past several decades; therefore, the total tonnages and energy yield have stagnated. It is estimated that WTE facilities provide less than 0.08% of the annual energy demand in the USA and power the equivalent of nearly 3 million households.

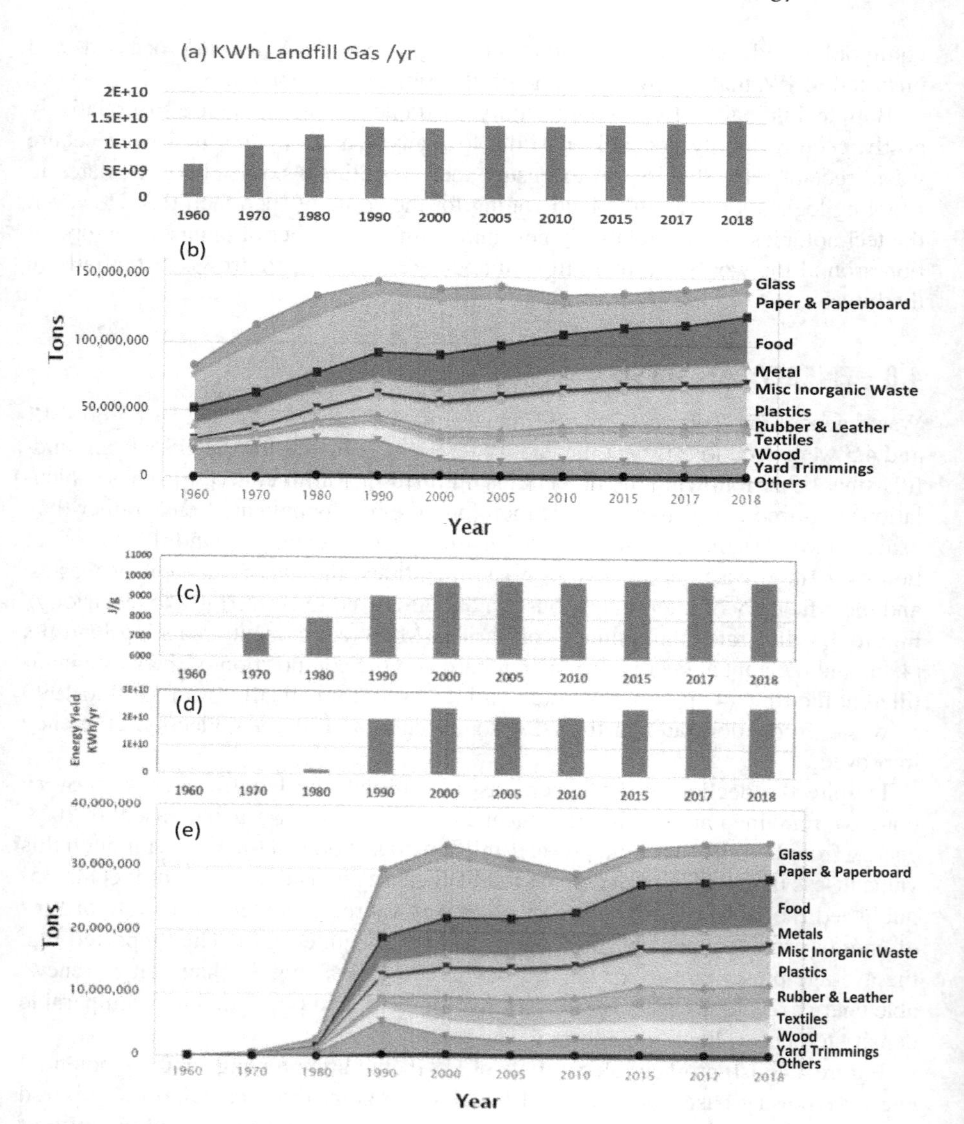

FIGURE 4.4 (a) The amount of energy provided by US landfill gas collection over the same time, (b) tonnages of US waste components landfilled over time, (c) the specific energy of waste, (d) the energy yield from combusted US waste, and (e) the US tonnage for US MSW components combusted from 1960 through 2018.

4.9 COUNTRY ECONOMIES AND MSW ENERGY POTENTIAL

Waste disposal practices vary among countries as shown in Figure 4.5a, where lower economies tend to favor open dumps and burning over practices with higher capital cost and energy recovery potential [37]. The amount of MSW generated is positively correlated with per capita waste generation [38] presumably as the consumption of

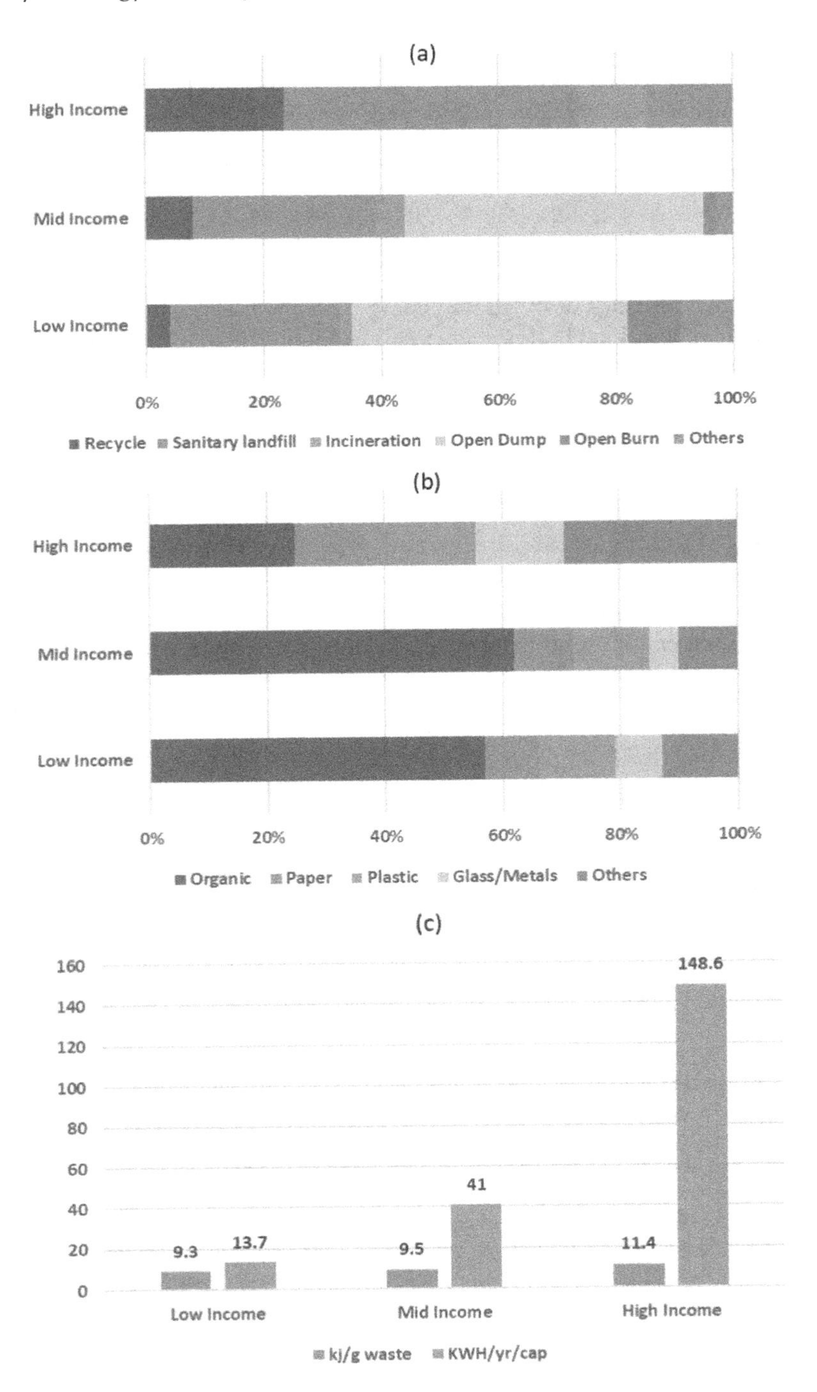

FIGURE 4.5 The effect of country income on (a) waste management options, (b) waste composition, and (c) waste energy potential per waste weight and energy production per year per capita.

disposable and packaged goods increases with wealth. This consumption pattern is also reflected in the composition of the waste as seen in Figure 4.5b [37] and its energy content as seen in Figure 4.5c (derived from Ref. [39]). These factors combine to reveal that MSW generated in wealthier countries has significantly greater energy recovery potential than in poorer countries (Figure 4.5c).

4.10 FUTURE OF ENERGY RECOVERY FROM WASTE

There is a long history of extracting energy from waste in landfills and thermal conversion, as described in this chapter, albeit inefficiently and only justified as an offshoot of their primary purpose, waste management. More efficient thermal conversion is likely from gasification and pyrolysis, discussed herein, or technologies such as hydrothermal carbonation and plasma arc combustion, which are under development. The potential for resource and energy recovery from waste has been recognized by the waste industry and is a focus of the public and private investment. However, separation of waste into specific categories of materials (e.g., biodegradable, combustible, plastic, fiber, food, yard waste, and metals) will always be necessary for process efficiency and contamination continues to be problematic for many waste streams. Further, as products and materials evolve, designers must begin with the end in mind, an example of this is the rapid development of electric vehicles and the batteries they depend on. Nevertheless, it is imagined that, in future, landfilling will be avoided and a global circular economy will truly be achieved.

REFERENCES

1. M. Albanna, Anaerobic digestion of the organic fraction of municipal solid waste, in: *Manag. Microb. Resour. Environ.*, 2013, Springer, Dordrecht. doi:10.1007/978-94-007-5931-2_12.
2. D. Baxter, T. Al Seadi, AD of the organic fraction of MSW: System overview for source and central separated waste (2013). https://www.ieabioenergy.com/wp-content/uploads/2013/09/AD-of-the-organic-fraction-of-MSW-Baxter.pdf.
3. I. Ozbay, E. Durmusoglu, Energy content of municipal solid waste bales, *Waste Manag. Res.* 31 (2013) 674–683. doi:10.1177/0734242X13485866.
4. M.A. Barlaz, R.K. Ham, D.M. Schaefer, Methane production from municipal refuse: A review of enhancement techniques and microbial dynamics, *Crit. Rev. Environ. Control.* 19 (1990) 557–584. doi:10.1080/10643389009388384.
5. R. Geyer, J.R. Jambeck, K.L. Law, Production, use, and fate of all plastics ever made, *Sci. Adv.* 3 (2017) 25–29. doi:10.1126/sciadv.1700782.
6. US EPA, National overview: Facts and figures on materials, Wastes and Recycling, United States Environ. Prot. Agency (2021). https://www.epa.gov/facts-and-figures-about-materials-waste-and-recycling/national-overview-facts-and-figures-materials.
7. S.L. Wong, N. Ngadi, T.A.T. Abdullah, I.M. Inuwa, Current state and future prospects of plastic waste as source of fuel: A review, *Renew. Sustain. Energy Rev.* 50 (2015) 1167–1180. doi:10.1016/j.rser.2015.04.063.
8. M.A. Barlaz, A.P. Rooker, P. Kjeldsen, M.A. Gabr, R.C. Borden, Critical evaluation of factors required to terminate the postclosure monitoring period at solid waste landfills, *Environ. Sci. Technol.* 36 (2002) 3457–3464. doi:10.1021/es011245u.

9. A. Lagerkvist, H. Ecke, T.H. Christensen, Waste characterization: Approaches and methods, in: T.H. Christensen (Ed.), *Solid Waste Technol. & Manag.*, 2010, Wiley, Chichester, UK. doi: 10.1002/9780470666883, Ch 5.
10. A. Krause, T.A.M. Pugh, A.D. Bayer, M. Lindeskog, A. Arneth, Impacts of land-use history on the recovery of ecosystems after agricultural abandonment, *Earth Syst. Dyn.* 7 (2016) 745–766. doi:10.5194/esd-7-745-2016.
11. Y. Hui, W. Li'ao, S. Fenwei, H. Gang, Urban solid waste management in Chongqing: Challenges and opportunities, *Waste Manag.* 26 (2006) 1052–1062. doi:10.1016/j.wasman.2005.09.005.
12. T. Tolaymat, H. Kim, P. Jain, J. Powell, Moisture addition requirements for bioreactor landfills, *J. Hazard. Toxic Radioact. Waste.* 17 (2013) 360–364.
13. M.A. Barlaz, J.P. Chanton, R.B. Green, Controls on landfill gas collection efficiency: Instantaneous and lifetime performance, *J. Air Waste Manag. Assoc.* 59 (2009) 1399–1404. doi:10.3155/1047-3289.59.12.1399.
14. US EPA, Background information document for updating AP42 Section 2.4 for estimating emissions from municipal solid waste landfills, Natl. Risk Manag. Res. Lab. Air Pollut. Prev. Control Div. U.S. EPA. September (2008) 249.
15. H.R. Amini, D.R. Reinhart, K.R. Mackie, Determination of first-order landfill gas modeling parameters and uncertainties, *Waste Manag.* 32 (2012) 305–316. doi:10.1016/j.wasman.2011.09.021.
16. SWICS, Current MSW industry position and state-of-the-practice on LFG collection efficiency, methane oxidation, and carbon sequestration in landfills (2009). https://www.scsengineers.com/scs-white-papers/current-msw-industry-position-and-state-of-the-practice-on-lfg-collection-efficiency-methane-oxidation-and-carbon-sequestration-in-landfills/.
17. J. Chanton, K. Liptay, Seasonal variation in methane oxidation in a landfill cover soil as determined by an in situ stable isotope technique, *Global Biogeochem. Cycles* 14 (2000) 51–60. doi:10.1029/1999GB900087.
18. C. Sawatdeenarunat, K.C. Surendra, D. Takara, H. Oechsner, S.K. Khanal, Anaerobic digestion of lignocellulosic biomass: Challenges and opportunities, *Bioresour. Technol.* 178 (2015) 178–186. doi:10.1016/j.biortech.2014.09.103.
19. W. McCabe, J. Eckenfelder (Eds.) *Biological Treatment of Sewage and Industrial Wastes*, 1957, Reinbold Publishing, New York.
20. P.J. Meynell, *Methane: Planning a Digester*, Sochen Books, Prison Stable Court, Dorset, Clarington, 1976.
21. M. Čater, L. Fanedl, Š. Malovrh, R. Marinšek Logar, Biogas production from brewery spent grain enhanced by bioaugmentation with hydrolytic anaerobic bacteria, *Bioresour. Technol.* 186 (2015) 261–269. doi:10.1016/j.biortech.2015.03.029.
22. J. Ma, C. Frear, Z.W. Wang, L. Yu, Q. Zhao, X. Li, S. Chen, A simple methodology for rate-limiting step determination for anaerobic digestion of complex substrates and effect of microbial community ratio, *Bioresour. Technol.* 134 (2013) 391–395. doi:10.1016/j.biortech.2013.02.014.
23. D.P. Van, T. Fujiwara, B.L. Tho, P.P.S. Toan, G.H. Minh, A review of anaerobic digestion systems for biodegradable waste: Configurations, operating parameters, and current trends, *Environ. Eng. Res.* 25 (2020) 1–17. doi:10.4491/eer.2018.334.
24. A. Anukam, A. Mohammadi, M. Naqvi, K. Granström, A review of the chemistry of anaerobic digestion: Methods of accelerating and optimizing process efficiency, *Processes* 7 (2019) 1–19. doi:10.3390/PR7080504.
25. T. Hulgaard, J. Vehlow, Incineration: Process and technology, *Solid Waste Technol. Manag.* 1 (2010) 363–392. doi:10.1002/9780470666883, Ch 26.
26. Waste to energy: Incineration, gasification and pyrolysis, (n.d.). http://sustainable.org.za/userfiles/incineration(1).pdf.

27. R.W. Breault, Gasification processes old and new: A basic review of the major technologies, *Energies* 3 (2010) 216–240. doi:10.3390/en3020216.
28. USEPA, Domestic and global usage of gasification technology (2016), https://archive.epa.gov/epawaste/hazard/wastemin/web/html/gasdom.html.
29. V. Marcantonio, A.M. Ferrario, A. Di Carlo, L. Del Zotto, D. Monarca, E. Bocci, Biomass steam gasification: A comparison of syngas composition between a 1-d matlab kinetic model and a 0-d aspen plus quasi-equilibrium model, *Computation* 8 (2020) 1–15. doi:10.3390/computation8040086.
30. U.S.E.I. Administration, Biomass—Renewable energy from plants and animals (2020). https://www.eia.gov/energyexplained/biomass/.
31. I.S. Pieta, W.S. Epling, A. Kazmierczuk, P. Lisowski, R. Nowakowski, E.M. Serwicka, Waste into fuel—Catalyst and process development for MSW valorisation, *Catalysts* 8 (2018) 1–16. doi:10.3390/catal8030113.
32. M. Sharifzadeh, N. Shah, Synthesis, integration, and intensification of solid oxide fuel cell systems : Process systems engineering perspective, in: M. Sharifzadeh (Ed.), *Des. Oper. Solid Oxide Fuel Cells Syst. Eng. Vis. Ind. Appl.*, Woodhead Publishing Series in Energy, 2020, Elsevier, Amsterdam: pp. 185–215.
33. A. Sanlisoy, M.O. Carpinlioglu, A review on plasma gasification for solid waste disposal, *Int. J. Hydrogen Energy* 42 (2017) 1361–1365. doi:10.1016/j.ijhydene.2016.06.008.
34. P. Basu, Pyrolysis, in: *Biomass Gasification, Pyrolysis Torrefaction* (Third Ed.), 2018, Elsevier, Amsterdam: pp. 155–187. https://doi.org/10.1016/B978–0–12–812992–0.00005–4.
35. H.R. Amini, D.R. Reinhart, Regional prediction of long-term landfill gas to energy potential, *Waste Manag.* 31 (2011) 2020–2026. doi:10.1016/j.wasman.2011.05.010.
36. V. Smil, *Energy Transitions: History, Requirements, Prospects*, 2010, Praeger/ABC CLIO, Santa Barbara.
37. G. Cerminara, R. Cossu, Waste input to landfills, in: R. Cossu, R. Stegmann (Eds.), *Solid Waste Landfilling – Concepts, Process. Technol.*, 2019, Elsevier, Amsterdam.
38. D. Hoornweg, P. Bhada-Tata, *What a Waste : A Global Review of Solid Waste Management*, 2012, World Bank, Washington, DC. https://openknowledge.worldbank.org/handle/10986/17388, License: CC BY 3.0 IGO.
39. P.A. Vesilind, W.A. Worrell, D.R. Reinhart, *Solid Waste Engineering*, 2002, Brooks/Cole, Pacific Grove, CA .

5 Materials and Energy from Waste Plastics

A Catalytic Approach

Shadab Shahsavari
Department of Chemical Engineering, Varamin- Pishva Branch, Islamic Azad University, Varamin, Iran.

Gita Bagheri
Department of Chemical Engineering, Shahryar Branch, Islamic Azad University, Shahryar, Iran.

Zahra Shokri and Shahin Shahsavari
Process Engineering Department, Faculty of Chemical Engineering, Tarbiat Modares University, Tehran, Iran.

CONTENTS

DOI: 10.1201/9781003178354-7

5.1 PYROLYSIS–CATALYSIS OF WASTE PLASTICS

Since the mercantile, industrial, and domestic useful life of plastics can be less than 1 year to more than 40 years, most plastics enter the waste stream. Mechanical recycling includes fragmenting, washing, disinfecting, drying, and pelletizing of the plastics to produce recycled substance. Thermochemical recycling is mainly done through catalysis–pyrolysis and gasification to produce more valuable products such as gasoline, liquid fuel, and aromatic chemicals [1,2]. Many studies have been conducted on the production of fuels and syngas from catalysis–pyrolysis of wastage plastics; however, there has been lesser research into the production of hydrogen gas from wastage plastics through catalysis–pyrolysis mechanism conjoined with catalytic steam amending. Plastics found in municipal solid wastage mainly include high-density polyethylene, polypropylene, low-density polyethylene, polystyrene, polyethylene terephthalate, and polyvinyl chloride.

5.1.1 Hydrogen Gas Production from Wastage Plastics

Hydrogen gas is a valuable commodity that is widely used in the production of ammonia for the refining of fertilizers and oil, for the production of methanol and cyclohexane, and as a raw material for the pharmaceutical and plastics industries [3]. In recent years, hydrogen gas has been derived from fossil fuels and its main source of production is natural gas. This process includes steam amending of methane in the presence of catalysts at a pressure of 0.3–2.5 MPa and temperature of 700 °C–1,000 °C. Typically, the first step includes the pyrolysis of plastic waste at ~500 °C, which generates a mixed series of vapor and hydrocarbon gases from the thermal demotion of wastage plastics. The derived pyrolysis gases are then transferred to a catalytic reactor (with temperature 800 °C), and hydrogen is produced in the presence of a catalyst.

Sharuddin et al. [4] investigated the catalysis–pyrolysis of various types of usual wastage of plastics. They reported the effect of temperature, catalyst type, reactor type, pressure, residence time, flow rate, and type of carrier gas on the compound of the percentage of product oil. Miandad et al. [5] conducted a study on the catalytic pyrolysis process of plastic wastage for the production of oil fuels and gaseous products and investigated the impact of various catalysts and operational conditions on the efficiency and properties of the production. Al-Salem et al. [6] made an investigation into the two-stage mechanism of catalysis–pyrolysis of waste plastics and found that the interaction of the pyrolysis and catalyst process modifies the contact between the pyrolysis products and the catalyst in the following catalytic step and minimizes heat and mass transfer problems.

5.1.1.1 Reactor Design for Hydrogen-Rich Gas Production from Wastage of Plastics

Various types of reactors have been considered for the catalysis–pyrolysis mechanism for hydrogen production. In several studies, a two-step fixed bed reactor process has been used for investigating the effect of operational parameters and different types of catalysts [3,7]. The first step includes the pyrolysis of plastic materials at a constant heating rate to a terminal temperature of pyrolysis (typically 500 °C–700 °C). Namioka et al. [8] used a two-stage reactor system, with pyrolysis being the first step and the passage of the steam through an amended unit with a catalyst bed being the second step. In the first stage, volatile materials were produced and then pass through to the second step and the Ru/γ-Al_2O_3 catalyst was used to reform catalytic steams at 580–680 °C. They indicated that the Ru/γ-Al_2O_3 catalyst is more efficient at producing hydrogen compared to Ni-based catalysts because the amending temperature is ~200 °C lower than the temperature applied for nickel catalysts [8,9]. Ovsyannikova et al. [10] prepared a reactor similar to that of Namioka, where polyethylene is continuously fed into the first stage of the reactor to produce gas, and the product gases of pyrolysis were passed through a fixed catalyst bed of NiO/γ-Al_2O_3 catalyst, and then the product gases were passed through a cyclone, condensing process, filter and dryer to clean the product gases. The optimum temperature of catalysis–pyrolysis mechanism for hydrogen production was determined to be between 700 °C and 900 °C.

Czernik and French [11] used a fluidized bed pyrolysis process at 650 °C, which was continuously fed with plastic material, and the derivative pyrolysis gases were passed through a cyclone, and then they were fed into a second fluidized bed reactor using a Ni-based catalyst (C11-NK) at 850 °C, and 34 g of H_2 per 100 g of polypropylene was produced. The product gas composition consisted of ~70 vol% H_2, ~16 vol% CO_2, ~11 vol% CO, and lower concentrations of hydrocarbons. Makibar et al. [12] employed fluidized bed steam gasification of plastic materials. They used catalytic steam amending in a fluidized bed reactor combined with a catalyst absorption-enhanced reactor, and they showed that fully continuous reactors were highly efficient at producing hydrogen. Barbarias et al. [13–15] used a two-stage continuous reactor consisting of a vibrating feeder to transfer plastic material to a conical bed reactor at 500 °C, resulting in rises in heat transfer and the contact surface between solid plastic and gas. The gaseous products from the process were passed through a condenser and coalescence filter for purification [16].

5.1.1.2 The Effect of Operational Parameters on the Level of Hydrogen Production from Plastic Wastages

Operational parameters such as catalyst type, catalyst temperature, plastic type, steam input, and catalyst-to-plastic ratio are all considered in the calculations to maximize hydrogen production and minimize catalyst deactivation by catalyst sedimentation or deposits of coke.

Barbarias et al. [14] demonstrated the influence of plastic type on the amount of hydrogen production, and they reported 37.3 g H_2 per 100 g of polyethylene, 34.8 g H_2 per 100 g of polypropylene, 29.1 g H_2 per 100 g of polystyrene, and 18.2 g H_2 per 100 g of polyethylene terephthalate. In other studies, the effect of plastic type on the hydrogen yield was investigated and the yield was reported to be 26.6 and 26.0 g H_2 per 100 g polypropylene and high-density polyethylene, respectively, while for polystyrene, the yield of H_2 was only 18.5 g H_2 per 100 g. Namioka et al. [8] reported a slight difference in the amount of hydrogen produced from polystyrene compared to polypropylene at a catalyst temperature of 630 °C. Hydrogen yields were 26.2 and 33.3 g H_2 per 100 g polystyrene at 580°C temperature, respectively, and a similar range was reported for polypropylene.

However, Kumagai et al. [17,18] used a two-stage reactor system with Ni/Mg/Al catalysts to produce hydrogen from polyamide and polyurethane and to remove HCN using calcium oxide adsorption. Dou used a two-way fluidized bed reactor together with an adsorption reactor with CaO to produce a hydrogen-rich gas with a minimum HCl content [19].

5.1.1.3 The Effect of Catalyst Type on the Level of Hydrogen Production from Waste Plastics

Nickel catalysts have widely been used in pyrolysis–catalysis for hydrogen production. Many studies have been performed using transition metal-based catalysts or the addition of metal promoters to the nickel to increase the hydrogen yield [10]. Ruthenium catalysts show a higher catalytic activity compared to nickel-based catalysts, and ruthenium has been used in the mechanism of hydrogen production from waste plastics [8,9]. However, the addition of Cu to a Ni–Al catalyst reduces the amount of hydrogen produced. The addition of cerium to nickel-based catalysts reduces the formation of carbonaceous coke on the catalyst [20].

Yao et al. [21] investigated the effect of β-type and Y-type zeolite (ZSM-5) using a two-stage mechanism in a fixed bed reactor. They showed that the ZSM-5 catalyst resulted in the highest hydrogen yield, but the narrower pores of the Y-type zeolite produced less hydrogen and led to higher catalyst deactivation.

5.1.1.4 The Effect of Catalyst Temperature on Hydrogen Production from Waste Plastics

The effect of catalyst temperature on the amount of hydrogen produced has been reported by several researchers. Increasing the catalyst temperature from 600 °C to 900 °C leads to a significant increase in the amount of hydrogen produced from polypropylene using the pyrolysis–catalysis process with Ni-CeO_2-Al_2O_3 as a catalyst [22]. The increase in the yield from 5.7 g H_2 per 100 g at 600 °C to 22.3 g H_2 per 100 g at 900 °C and low catalyst temperatures, performed an incomplete correction of pyrolysis gases. Using a different catalyst (Ni-CeO_2-ZSM-5) in the temperature range of 600 °C–900 °C, Wu et al. investigated the effect of catalyst temperature on pyrolysis steam reforming [23].

Barbarias et al. [13] investigated the pyrolysis stage of high-density polyethylene at 500 °C in a spouted bed reactor, and the catalysis stage at 600 °C, 650 °C, and 700 °C in a fluidized bed reactor. They reported that the hydrogen yield (98%) increased with increasing temperature up to 700 °C due to kinetic reforming (37.3 g H_2 per 100 g of plastic), and they also showed an increase in CO production with increasing catalyst temperature. He et al. [24] demonstrated an improvement in gas yield with increasing temperature and showed a yield of 86.5 wt% at 900 °C, thus increasing H_2 and CO_2 and decreasing CO, CH_4, and C_2 hydrocarbons. They concluded that an increase in temperature enhances the total gas yield. Park et al. [9] examined the pyrolysis phase temperature from 380 °C to 550 °C and the catalysis phase temperature from 580 °C to 680 °C. They showed that higher temperatures not only increased the rate of carbon conversion, but also increased carbon coke deposits on the catalyst and reduced the activity of the catalyst. Using a Ni/ZSM-5 catalyst, Yao et al. examined the yield of hydrogen from polyethylene at 650 °C–850 °C and showed that the hydrogen yield ranged from 8.8 g per 100 g of plastic at 650 °C to 13.3 g at 850 °C.

5.1.2 Carbon Nanotubes Production from Waste Plastics

Many studies have been performed to produce carbon nanotubes (CNTs) using plastic wastages as a source of hydrocarbons through chemical vapor deposition route, and the effects of reactor design factors, the type of catalyst employed for carbon nanotubes growth, and the impact of operational parameters on the efficiency of carbon nanotubes have been reviewed. CNTs have a great deal of interest due to their characteristics and potential use in a wide range of applications. Carbon nanotubes are cylindrical hollow tubes of carbon with nanoscale diameters and long lengths. Carbon nanotubes are chemically and physically stable and have great electrical conductivity and high tensile strength.

Carbon nanotubes are generated by the process of chemical vapor deposition (CVD), where carbon-rich gases interact with catalyst materials to produce CNTs at high temperatures (600 °C–1,200 °C). Kumar [25] reported the mechanism of carbon nanotubes via chemical vapor deposition. The hydrocarbon decomposes on the nanoparticles to release hydrogen and carbon which solves into the metal then distributes through the particle. When there is an incapable interaction between the catalyst support and metal, carbon nanotubes precipitate out and the metal is removed from the surface. The carbon nanotubes grow and remove the metal on the tip of the carbon nanotube, and the mechanism is known as the tip growth process. If there is a strong interaction between metal particles and the support, the precipitate of carbon cannot remove the particle, and instead, precipitation is accomplished on the upper metal surface. Various carbon sources for the production of CNTs have been reported, including methane; aromatic hydrocarbons such as toluene, benzene, xylene [26–28], coal [28], and animal fat [29]; and biomass such as coconut oil [30]. Typical catalysts include transition metals, such as nickel, cobalt, and iron, and organometallic catalysts, such as ferrocene. The production of carbon nanotubes from wastage plastics will be complex compared to using pure hydrocarbons. The use of wastage plastics for the production of carbon nanotubes has been researched by various groups. Bazargan and McKay [31] investigated the production of carbon nanotubes from plastics through different catalytic and thermal processes.

5.1.2.1 The Effect of Operational Parameters on the Production of Carbon Nanotubes from Waste Plastics

Much research has been done on the production of carbon nanotubes made of plastics using a mechanism that performs both pyrolysis and catalyst in a reactor, but it is possible to better control the operating parameters in both gas production and chemical deposition when each of the processes to be performed in a separate reactor is provided. Further research is focused on ongoing processes for the production of carbon nanotubes.

Kong and Zhang [32] applied an autoclave reactor to carry out the pyrolysis of polyethylene with ferrocene to support the dispersion of the iron. The reactor was heated up to 700 °C. The production of carbon consisted of more than 80% carbon nanotubes with diameters between 20 and 60 nm and also 5% helical carbon nanotubes. Another research using polypropylene–nickel as a catalyst found that the carbon nanotubes constructed had larger diameters in the range of 160 nm with multiple walls. In the absence of nickel, carbon spheres were organized instead of carbon nanotubes. Kukovitskii et al. [33] used polyethylene to produce carbon nanotubes in a two-step process with a nickel catalyst in a quartz reactor at 420 °C–450 °C. Higher temperatures (700 °C–800 °C) of nickel catalysts improved carbon nanotubes and reduced metal impurities. Yen et al. used a combined system to form carbon nanotubes [34]. They used a mixture of polycarbosilane and polyethylene waste as an input to the reactor and heated it to 800 °C–950 °C, followed by a liquid fluid stream.

Acomb et al. used a separate two-stage fixed bed reactor [35], which makes it possible to better control process parameters such as temperature, gas flow rate, catalytic temperature, and the amount of inlet gas. To investigate the effect of catalyst temperature on the efficiency of carbon nanotubes, Fe–Al_2O_3 catalyst was used in the temperature range of 700 °C–900 °C and it was shown that the efficiency and uniformity of produced structure at 700 °C were higher than the amorphous carbon produced at 900 °C.

5.2 NANOCATALYSTS IN WATER TREATMENT

Nanosized zero-valent iron (nZVI) particles, titanium dioxide (TiO_2), nanostructured iron oxide (Fe_3O_4) adsorbent, magnetoferritin nanoparticles, silver (Ag) nanoparticles, and nitrogen-doped titanium dioxide (nitrogen-doped TiO_2) are considered as the popular nanocatalysts used in water treatment.

5.2.1 Zero-Valent Iron Nanoparticles as Catalysts

In recent years, the use of metals as an influential factor in the remediation of polluted water has widely been investigated. Among these metals, more attention has been paid to the use of zero-valent iron because zero-valent iron nanoparticles are abundant, inexpensive, non-toxic, highly reactive, and highly effective in groundwater remediation. Zero-valent iron nanoparticles are an inexpensive, non-toxic electron donor to nitrate, and currently, it is one of the popular metal reducers. Numerous studies have shown that zero-valent iron nanoparticles are highly effective in decontaminating aqueous solutions. Zero-valent iron nanoparticles have a greater surface area and are more reactive than iron (II) particles. However, the strong tendency of these particles to antagonize and oxidize has made their use difficult. The aggregation feature of nanoparticles creates a mass of iron particles; therefore, the affinity of these materials is greatly reduced, resulting in a decrease in the efficiency of these particles as well [36].

Nanocatalysts such as semiconductor materials, zero-valent metals, and bimetallic nanoparticles are widely used for reducing environmental pollutants such as pesticides and azo dyes, and this is due to their high specific surface area and shape-related properties. For in situ treatment, zero-valent iron nanoparticle powder is mixed with water and then injected into the intended site. In on-site treatment, zero-valent iron nanoparticles are embedded in fields with activated carbon, zeolite, carbon nanotubes, and so on. Zero-valent iron nanoparticles simultaneously act as adsorbents and reducing agents [37]. Zero-valent iron nanoparticles are used to process a wide range of common environmental pollutants such as chlorinated methane, chlorinated benzene, organic dyes, trihalomethanes, arsenic, and nitrate, and heavy metals such as mercury, nickel, and silver. Furthermore, zero-valent iron nanoparticles can reduce contaminants and 2-dichlorobiphenyl [38].

5.2.2 Titanium Dioxide as Catalysts

Metal oxide nanomaterials such as titanium dioxide and cerium oxide (CeO_2) are used as catalysts to accelerate the reactions, leading to the reduction of organic pollutants in ozonation processes. Titanium dioxide acts as a photocatalytic reducing agent as well as an adsorbent and can purify water in situ and on-site. The photocatalytic properties of titanium dioxide have attracted remarkable attention than its catalytic properties because titanium dioxide has the potential for producing free radicals in the presence of ultraviolet light, water, and oxygen. These radicals decompose various contaminants in the water into less toxic carbon compounds. Titanium dioxide photocatalysts are capable of removing pollutants that include polychlorinated biphenyl, benzenes, and chlorinated alkanes.

The use of titanium dioxide nanoparticles increases the specific surface area and photocatalytic activity compared to micrometer particles. Titanium dioxide is used either in nanopowder form in suspensions, or as granular filters in water treatment processes [39]. Each of these systems has advantages and disadvantages, and they differ in the speed and efficiency of water treatment. For example, the yield of suspended titanium dioxide nanopowders is higher than that of other systems. A good solution to reducing the loss of these nanoparticles is to use nanocrystalline microspheres. These nanocrystals are suspended by air bubbling in water and settle on the water side [40].

Doping of titanium dioxide nanoparticles with metals such as silicium increases the decomposition of organic compounds. This is because of the increased production of hydroxyl radicals. For example, doping of titanium dioxide nanoparticles with silicium improves their function by increasing the specific surface area and crystallinity of the nanoparticles. Besides, doping of titanium dioxide nanoparticles with nitrogen and iron (III) particles increases their efficiency in reducing azo and phenol dyes, respectively. Replacement of bulk titanium dioxide with titanium dioxide nanotubes reduces toluene as much as possible. Moreover, these nanotubes are more efficient than titanium dioxide nanoparticles in reducing organic compounds. Almost all organic contaminants can be removed using nanoscale titanium dioxide. It is very hydrophilic and can absorb environmental pollutants and heavy metals such as arsenic. Factors influencing the efficiency of titanium dioxide nanoparticles in the water treatment process include the quality and purity of nanoparticles, UV intensity, water pH, the presence of oxygen, and the concentration of existing contaminants [41].

5.2.3 Nanostructured Iron Oxide as Catalysts

In recent years, nanostructured iron oxide AD33 has been used to remove arsenic from various environments. It has a combination of catalytic and absorption properties derived from iron oxide and also has the ability to convert arsenic to less toxic substances. It is possible to remove more than 99% of arsenic in water using this nanostructured oxide, and AD33 can simultaneously remove these non-toxic compounds from water. Not only AD33 can reduce arsenic, but it can also be used to reduce contaminants such as lead, zinc, chromium, and copper. The nanostructured iron oxide AD33 used to reduce pollution is not environmentally hazardous and can be buried in the ground after use. The main limitation of these nanocatalysts is the need to replace them frequently.

5.2.4 Magnetic Nanoparticles as Catalysts

Magnetic nanoparticles are used as adsorbents and nanocatalysts in the water treatment process. These nanoparticles are used to generate energy for reverse osmosis. These systems use magnetic nanoparticles to generate the osmotic pressure required to conduct water through a filtration membrane. Similarly, magnetic nanoparticles with forward osmosis capability are used in desalination. The main advantage of using these nanoparticles as catalysts is their long life span and reusability. Paramagnetic nanoparticles can be used to remove contaminants from water, and magnetic adsorbents can be removed from the system after the contaminants are absorbed. Magnetite (Fe_3O_4) is the strongest magnetic species. The reactive surface of iron oxides allows the absorption of more impurities in water. These nanoparticles release a variety of superoxide radicals, reactive oxygen, hydrogen peroxide, hydroxyl radicals, and singlet oxygen, which can decompose DNA in bacterial cells as well as a variety of proteins [42]. Therefore, magnetic nanoparticles can exert their antibacterial effects through both physical adsorption and chemical disinfection. And the magnetic properties of magnetic nanoparticles cause them to be easily removed in the vicinity of an external magnetic field.

Silver (Ag) is a metal that exhibits relatively weak oxidation behaviors with low reactivity, but its antimicrobial properties are extremely high due to the increased specific surface area and can exhibit antibacterial activity as a result of interactions with the surface of bacterial cells [43].

5.2.5 Other Nanomaterials as Catalysts

Other nanomaterials used as catalysts include oxidized nanostructured films (ZnO), silver nanoparticles and amidoxime fibers, palladium–copper/alumina (gamma) bimetallic nanocomposites, palladium nanoparticles, bimetallic nanoparticles (palladium/silver), manganese (MnO_2) nanoparticles, and monoenzymes [44]. Oxidized nanostructured films are used to reduce organic pollutants (4-chlorocatechol); silver nanoparticles and amidoxime fibers are used to reduce organic dyes; palladium–copper/alumina (gamma) bimetallic nanocomposite are used for nitrate reduction; manganese oxide nanoparticles are used to accelerate the mineralization of organic dyes; and monoenzyme nanoparticles are used to clean a wide range of contaminants [45].

5.3 BIOCATALYSTS FOR CONVERTING KERATIN WASTE

The use of biocatalysts to convert keratin waste into valuable products is a very suitable and environmentally friendly method for converting waste into valuable derivatives, including chemicals and biofuels. Keratin is the main constituent of the vertebrate epidermal appendages and is also a waste product generated from feathers, nails, hair, and scales [46,47]. After cellulose and chitin, keratin is the third most abundant polymer in nature and is highly resistant to chemical and biological hydrolysis. Today, tens of millions of tons of keratin waste is produced annually in the world, with a crude protein content of more than 85%, which is a good supply of protein. But similar to cellulose recycling, the recycling of keratin is fraught with problems. Keratinous materials are divided into α-keratin and β-keratin and contain a large number of disulfide bonds, hydrogen bonds, and hydrophobic interactions that are difficult to decompose.

Keratinase is a proteolytic enzyme that degrades keratin and is found in fungi and bacteria, which can convert keratin into soluble proteins, peptides, and amino acids.

Keratinases are the only group of proteases that can bind to keratin materials and break down amino acids at specific sites, reducing the high degree of disulfide crosslinks and then hydrolyzing them. Keratinase has a great potential and ability in the hydrolysis of keratin; therefore, it is used as a suitable biocatalyst for the conversion of keratin materials [48]. However, the actual and optimal use of keratinase is not yet known and various studies have been conducted to increase keratinase performance. The chemical and physical properties of keratin make it resistant to the hydrolysis of proteases such as pepsin, trypsin, and papain [49].

After detachment from living cell tissue during hydrolysis, almost all keratinases are not able to function, which can be attributed to low keratinase activity. In addition, it is a time-consuming process. For example, Yang et al. recombinantly expressed keratinase KerK in Bacillus amyloliquefaciens K11, which after a 60-hour fermentation process, resulted in an extracellular activity that reached 1,500 U/mL [50]. Furthermore, Su et al. employed a recombinant *B. subtilis* keratinase mutant M7 to obtain 3,040 U/mL of extracellular keratinase by continuous fermentation for 32 h in a bioreactor of 15 L [50]. Zheng Peng was able to express the keratinase gene from *Bacillus licheniformis* BBE11-1 so well that it was possible to improve the recombinant keratinase KerZ1 activity by up to 45.14 KU/mL by placing a promoter and screening ribosome binding sites. In addition, by optimizing the real-time control of temperature, dissolved oxygen, feed, and pH, it allowed the extracellular activity of KerZ1 to reach 426.60 KU/mL in fermentation in a bioreactor of 15 L, accounting for a 3,552-fold increase compared to wild-type keratinase [50].

5.4 CATALYSTS FOR BIOFUELS PRODUCTION FROM WASTE BIOMASS

Biofuels are used as an alternative to fossil fuels and as renewable and non-toxic fuels. The main purpose of using biofuels is to reduce the emission of toxins that cause severe problems for the environment and humans. Biofuels are produced from edible oils and non-edible vegetable oils such as waste cooking oil, jatropha oil, and microalgae and animal fats using esterification or sterilization of triglycerides.

Pyrolysis is a thermochemical process through which biofuels are produced from biomass under high pressure and temperature, and it is one of the oldest methods of biomass production in which the compounds are placed in a closed reactor at a high temperature between 350 and 650 with low oxygen content and liquid (bio-oil), solid (biochar), and gas (syngas) are classified. Based on residence time, temperature, particle size , fast and flash are classified. To increase productivity, catalysts play a major role in increasing the quality and speed of the reaction. They use various catalysts based on acid, alkali, and enzymes to produce biofuels. The selection of catalysts is an important step in improving the performance of biofuels [51].

Potassium hydroxide (KOH), sodium hydroxide (NaOH), and sulfuric acid (H_2SO_4) are catalysts used on an industrial scale. But the problem with using these catalysts is that a lot of energy is needed to separate the catalysts from the product and it is not possible to reuse them. Acid and base catalysts are used for low-grade raw materials that cause soap formation and reactor corrosion and hydrolyze triglycerides to form new diglycerides that promote stable soaping and reduce biofuel performance. Heterogeneous catalysts are mostly used to transesterify oils to produce biofuels. The activity of the catalyst is highly dependent on the presence of metals such as Ca, K, Mg, and Na, which provide the basic properties of the catalyst and increase the number of active sites. Due to the enzyme-based catalyst, lipase is used to convert lipids to biofuels [52].

Chemical pathways mainly include biochemicals, including anaerobic digestion, transesterification, fermentation, hydrothermal carbonation, pyrolysis, hydrothermal gasification, and hydrothermal liquefaction. In recent years, biochar-based catalysts have been used to convert triacylglycerol to biofuels. Biochar is a major carbon fuel produced that is obtained through pyrolysis, HTC, and hydrothermal liquefaction [53].

Reasons for using biochar as a green catalyst include its large surface area, reusability, biodegradability, low cost, stability, environmentally friendly structure, and high mechanical and thermal stability, which show a better activity during simultaneous reactions. The presence of minerals and functional groups at the biochar level improves its performance, and its physicochemical properties can be modified by various methods.

REFERENCES

1. Williams, P.T., Hydrogen and carbon nanotubes from pyrolysis-catalysis of waste plastics: A review. *Waste and Biomass Valorization*, 2020: pp. 1–28.
2. Borrelle, S.B., et al., Opinion: Why we need an international agreement on marine plastic pollution. *Proceedings of the National Academy of Sciences*, 2017 114(38): pp. 9994–9997.
3. Birol, F., The Future of Hydrogen: Seizing Today's Opportunities. IEA Report prepared for the G, 2019 20.
4. Sharuddin, S.D.A., et al., A review on pyrolysis of plastic wastes. *Energy Conversion and Management*, 2016, 115: pp. 308–326.
5. Miandad, R., et al., Catalytic pyrolysis of plastic waste: A review. *Process Safety and Environmental Protection*, 2016, 102: pp. 822–838.
6. Al-Salem, S., et al., A review on thermal and catalytic pyrolysis of plastic solid waste (PSW). *Journal of Environmental Management*, 2017, 197: pp. 177–198.

7. Spath, P.L. and M.K. Mann, *Life Cycle Assessment of Hydrogen Production Via Natural Gas Steam Reforming*, 2000, National Renewable Energy Lab, Golden, CO.
8. Namioka, T., et al., Hydrogen-rich gas production from waste plastics by pyrolysis and low-temperature steam reforming over a ruthenium catalyst. *Applied Energy*, 2011, 88(6): pp. 2019–2026.
9. Park, Y., et al., Optimum operating conditions for a two-stage gasification process fueled by polypropylene by means of continuous reactor over ruthenium catalyst. *Fuel Processing Technology*, 2010, 91(8): pp. 951–957.
10. Ovsyannikova, I., et al., Study of structural and mechanical properties of granulated alumina supports using X-ray microprobes. *Applied Catalysis*, 1989, 55(1): pp. 75–80.
11. Czernik, S. and R.J. French, Production of hydrogen from plastics by pyrolysis and catalytic steam reform. *Energy & Fuels*, 2006, 20(2): pp. 754–758.
12. Makibar, J., et al., Performance of a conical spouted bed pilot plant for bio-oil production by poplar flash pyrolysis. *Fuel Processing Technology*, 2015, 137: pp. 283–289.
13. Barbarias, I., et al., A sequential process for hydrogen production based on continuous HDPE fast pyrolysis and in-line steam reforming. *Chemical Engineering Journal*, 2016, 296: pp. 191–198.
14. Barbarias, I., et al., Valorisation of different waste plastics by pyrolysis and in-line catalytic steam reforming for hydrogen production. *Energy Conversion and Management*, 2018, 156: pp. 575–584.
15. Barbarias, I., et al., Pyrolysis and in-line catalytic steam reforming of polystyrene through a two-step reaction system. *Journal of Analytical and Applied Pyrolysis*, 2016, 122: pp. 502–510.
16. Sharma, S.S. and V.S. Batra, Production of hydrogen and carbon nanotubes via catalytic thermo-chemical conversion of plastic waste. *Journal of Chemical Technology & Biotechnology*, 2020, 95(1): pp. 11–19.
17. Kumagai, S., et al., Removal of toxic HCN and recovery of H_2-rich syngas via catalytic reforming of product gas from gasification of polyimide over Ni/Mg/Al catalysts. *Journal of Analytical and Applied Pyrolysis*, 2017, 123: pp. 330–339.
18. Kikuchi, E., et al., Steam reforming of hydrocarbons on noble metal catalysts (part 1): The catalytic activity in methane-steam reaction. *Bulletin of the Japan Petroleum Institute*, 1974, 16(2): pp. 95–98.
19. Dou, B., et al., Fluidized-bed gasification combined continuous sorption-enhanced steam reforming system to continuous hydrogen production from waste plastic. *International Journal of Hydrogen Energy*, 2016, 41(6): pp. 3803–3810.
20. Miyazawa, T., et al., Catalytic performance of supported Ni catalysts in partial oxidation and steam reforming of tar derived from the pyrolysis of wood biomass. *Catalysis Today*, 2006, 115(1–4): pp. 254–262.
21. Yao, D., et al., Investigation of nickel-impregnated zeolite catalysts for hydrogen/syngas production from the catalytic reforming of waste polyethylene. *Applied Catalysis B: Environmental*, 2018, 227: pp. 477–487.
22. Wu, C. and P.T. Williams, Effects of gasification temperature and catalyst ratio on hydrogen production from catalytic steam pyrolysis-gasification of polypropylene. *Energy & Fuels*, 2008, 22(6): pp. 4125–4132.
23. Wu, C. and P.T. Williams, Hydrogen production from the pyrolysis – Gasification of polypropylene: Influence of steam flow rate, carrier gas flow rate and gasification temperature. *Energy & Fuels*, 2009, 23(10): pp. 5055–5061.
24. He, M., et al., Syngas production from catalytic gasification of waste polyethylene: Influence of temperature on gas yield and composition. *International Journal of Hydrogen Energy*, 2009, 34(3): pp. 1342–1348.
25. Kumar, M., *Carbon nanotube synthesis and growth mechanism. Carbon nanotubes-synthesis, characterization, applications*, IntechOpen, 2011: pp. 147–170.

26. Kong, J., A.M. Cassell, and H. Dai, Chemical vapor deposition of methane for single-walled carbon nanotubes. *Chemical Physics Letters*, 1998, 292(4–6): pp. 567–574.
27. Wei, Y., et al., Effect of catalyst film thickness on carbon nanotube growth by selective area chemical vapor deposition. *Applied Physics Letters*, 2001, 78(10): pp. 1394–1396.
28. Das, N., et al., The effect of feedstock and process conditions on the synthesis of high purity CNTs from aromatic hydrocarbons. *Carbon*, 2006, 44(11): pp. 2236–2245.
29. Suriani, A., et al., Vertically aligned carbon nanotubes synthesized from waste chicken fat. *Materials Letters*, 2013, 101: pp. 61–64.
30. Paul, S. and S. Samdarshi, A green precursor for carbon nanotube synthesis. *New Carbon Materials*, 2011, 26(2): pp. 85–88.
31. Bazargan, A. and G. McKay, A review – Synthesis of carbon nanotubes from plastic wastes. *Chemical Engineering Journal*, 2012, 195: pp. 377–391.
32. Kong, Q. and J. Zhang, Synthesis of straight and helical carbon nanotubes from catalytic pyrolysis of polyethylene. *Polymer Degradation and Stability*, 2007, 92(11): pp. 2005–2010.
33. Kukovitskii, E., et al., Carbon nanotubes of polyethylene. *Chemical Physics Letters*, 1997, 266(3–4): pp. 323–328.
34. Yen, Y.-w., M.-D. Huang, and F.-J. Lin, Synthesize carbon nanotubes by a novel method using chemical vapor deposition-fluidized bed reactor from solid-stated polymers. *Diamond and Related Materials*, 2008, 17(4–5): pp. 567–570.
35. Acomb, J.C., C. Wu, and P.T. Williams, Effect of growth temperature and feedstock: Catalyst ratio on the production of carbon nanotubes and hydrogen from the pyrolysis of waste plastics. *Journal of Analytical and Applied Pyrolysis*, 2015, 113: pp. 231–238.
36. Pan, G., et al., Immobilization of non-point phosphorus using stabilized magnetite nanoparticles with enhanced transportability and reactivity in soils. *Environmental Pollution*, 2010, 158(1): pp. 35–40.
37. Zhang, H., et al., Synthesis of nanoscale zero-valent iron supported on exfoliated graphite for removal of nitrate. *Transactions of Nonferrous Metals Society of China*, 2006, 16: pp. s345–s349.
38. Shokoohi, R., et al., Optimizing laccase-mediated amoxicillin removal by the use of Box–Behnken design in an aqueous solution. *Desalination and Water Treatment*, 2018, 119: pp. 53–63.
39. Nakasaki, K., et al., Characterizing the microbial community involved in anaerobic digestion of lipid-rich wastewater to produce methane gas. *Anaerobe*, 2020, 61: p. 102082.
40. Parker, N. and A.A. Keller, Variation in regional risk of engineered nanoparticles: Nano TiO_2 as a case study. *Environmental Science: Nano*, 2019, 6(2): pp. 444–455.
41. Zhou, X.-h., et al., Aggregation behavior of engineered nanoparticles and their impact on activated sludge in wastewater treatment. *Chemosphere*, 2015, 119: pp. 568–576.
42. Tran, N., et al., Bactericidal effect of iron oxide nanoparticles on Staphylococcus aureus. *International Journal of Nanomedicine*, 2010, 5: p. 277.
43. Davoudi, M., et al., Antibacterial effects of hydrogen peroxide and silver composition on selected pathogenic enterobacteriaceae. *International Journal of Environmental Health Engineering*, 2012, 1(1): p. 23.
44. Odularu, A.T., Metal nanoparticles: Thermal decomposition, biomedicinal applications to cancer treatment, and future perspectives. *Bioinorganic Chemistry and Applications*, 2018, 2018(3): pp. 27–39.
45. Akinsiku, A.A., et al., Modeling and synthesis of Ag and Ag/Ni allied bimetallic nanoparticles by green method: Optical and biological properties. *International Journal of Biomaterials*, 2018, 2018: p. 10–26.
46. Bray, D.J., et al., Complete structure of an epithelial keratin dimer: Implications for intermediate filament assembly. *PLoS One*, 2015, 10(7): p. e0132706.

47. Reddy, M.R., et al., Effective feather degradation and keratinase production by Bacillus pumilus GRK for its application as bio-detergent additive. *Bioresource Technology*, 2017, 243: pp. 254–263.
48. Daroit, D.J., A.P.F. Corrêa, and A. Brandelli, Production of keratinolytic proteases through bioconversion of feather meal by the Amazonian bacterium Bacillus sp. P45. *International Biodeterioration & Biodegradation*, 2011, 65(1): pp. 45–51.
49. Peng, Z., et al., Effective biodegradation of chicken feather waste by co-cultivation of keratinase producing strains. *Microbial Cell Factories*, 2019, 18(1): pp. 1–11.
50. Yang, L., et al., Construction of a rapid feather-degrading bacterium by overexpression of a highly efficient alkaline keratinase in its parent strain Bacillus amyloliquefaciens K11. *Journal of Agricultural and Food Chemistry*, 2016, 64(1): pp. 78–84.
51. de Oliveira, A.L.L., et al., Waste of Nile Tilapia (Oreochromis niloticus) to biodiesel production by enzymatic catalysis—Optimization using factorial experimental design. *Industrial & Engineering Chemistry Research*, 2021, 60(9): pp. 3554–3560.
52. Nguyen, X.P., et al., Biomass-derived 2, 5-dimethylfuran as a promising alternative fuel: An application review on the compression and spark ignition engine. *Fuel Processing Technology*, 2021: p. 106687.
53. Chi, N.T.L., et al., A review on biochar production techniques and biochar based catalyst for biofuel production from algae. *Fuel*, 2021, 287: p. 119411.

6 Elucidating Sustainable Waste Management Approaches along with Waste-to-Energy Pathways

A Critical Review

Asmita Mishra and Hammad Siddiqi
Indian Institute of Technology (IIT) Kharagpur

B.C. Meikap
Indian Institute of Technology (IIT) Kharagpur
University of Kwazulu-Natal (UKZN)

CONTENTS

DOI: 10.1201/9781003178354-8

6.1 INTRODUCTION

Billions of tonnes of waste are produced worldwide each day. The larger the city's size (population and density), the higher the waste generated. Improper waste disposal by a single person can affect the entire nation. Therefore, countries have contributed to cleaning their communities and the environment by every individual and institution. The current municipal solid waste management practice is geared toward the maximum possible collection and consequent disposal, with or partially without treatment or processing. As a result, unlike waste management techniques in nature, an open cycle is adopted for waste management. Therefore, there is an urgent need to change the prototype of waste management from the open cycle to the closed cycle to accomplish the following sustainable development aims: (i) minimizing the rapid natural resources depletion; (ii) declining the environmental impact associated with different waste management elements; and (iii) improving safety and public health [1]. The most extensively recognized and practiced concept for solid waste management is integrated solid waste management. The objectives of integrated solid waste management include regulations on environment and health, social acceptability, and economic reliability. In essence, the development and implementation of a waste management plan involve a local activity that selects the perfect blend of alternatives and technologies to alter municipal solid waste management needs and fulfill legislative mandates [2]. Environmental, social, economic, and institutional dimensions should be considered in the decision making on waste handling. The inclusive attribute lies in the balance between the above four dimensions and can be at different levels.

The waste hierarchy is a crucial component in managing waste based on the principles of the environment. It has widely been used in developed countries. There is much criticism of this hierarchy as an open system. However, the closed system approach aims to profoundly change the pattern of material flows in a community, resulting in zero waste. The closed system offers a comprehensive solution with an end-to-a-pipe approach, which reassures waste diversification employing resource recovery and recycling, and an orientation toward waste disposal. As conventional primary energy sources such as coal, oil, and gas are on their way to extinction, organic wastes can be seen as one of the environmentally friendly options for renewable energy [3]. The standard techniques used for waste-to-energy conversion include physical, thermochemical, biochemical, and bio-electrochemical. Better control, high energy regeneration, and diverse ecological compatibility provide precedence over other technologies to thermochemical methods. The common thermochemical conversions include gasification, hydrogenation, combustion, liquefaction, and pyrolysis.

In the context of this chapter, a comprehensive attempt is made to outline the different sustainable waste management approaches. The first section summarizes the various types of wastes with their generation patterns and the adverse effects associated with their improper disposal. Further, the subsequent section discusses the optimum approach to the minimization and effective utilization of waste. Finally, the last section highlights the different waste-to-energy pathways with conventional practices and the latest technologies for the proper energy reclamation from the trash.

6.2 WASTES AND THEIR TYPES

The global annual municipal solid waste generation is around 2.01 billion tonnes. And by 2050, global waste would amount to 3.4 billion tonnes. However, at least 33% of the waste produced could be reduced with sustainable management [4]. The standard categories of wastes generated are represented in Figure 6.1.

6.2.1 AGRICULTURAL WASTE

Agricultural waste, commonly referred to as agri-waste, includes animal waste, food waste, crop waste, toxic agricultural waste, and hazardous waste. Their economic value, however, is less than the cost of collecting, transporting, and processing. The waste composition is dependent on the type and system of farming. This waste may be in liquid, slurry, or solid form. However, the pesticides, herbicides, and insecticides introduced during crop cultivation will eventually contribute to toxic and hazardous waste in the end. Agricultural waste estimates are exceptional, but they contribute a considerable percentage toward the total waste, particularly in the developed world. The agricultural sector has naturally increased waste, agricultural crop residues, and agricultural by-products with the expansion of production [5].

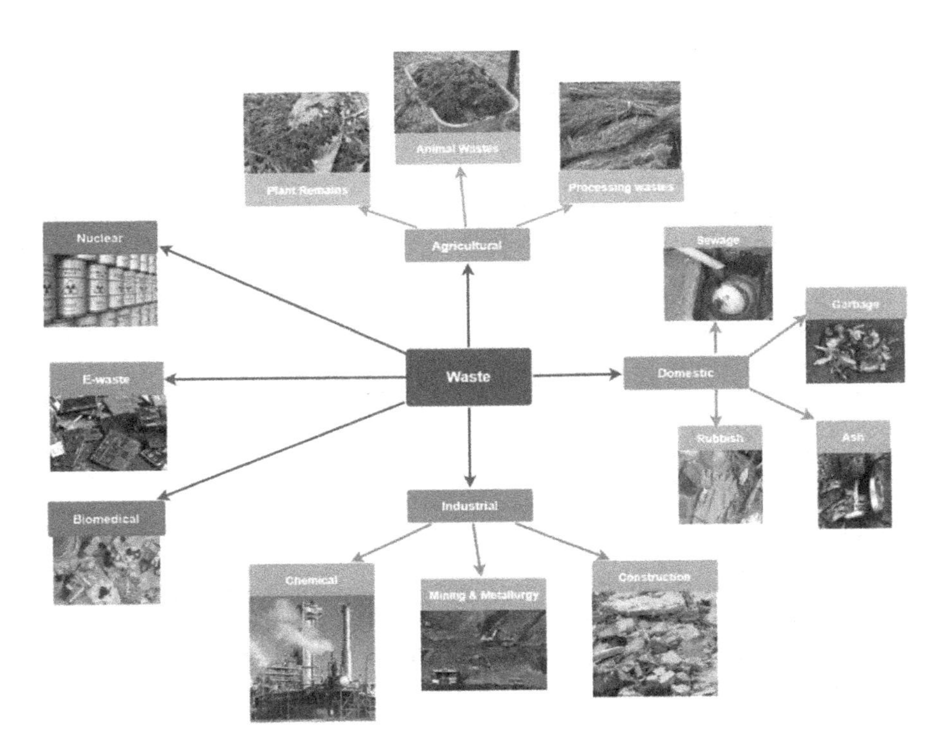

FIGURE 6.1 Standard categories of waste generation.

6.2.2 Domestic Waste

Household/domestic waste comprises kitchen waste, papers, plastic, rubbers, textile, sanitary waste, etc. These wastes are generally a mixture of the other types of waste with lesser volume and toxicity as the trash is diverse in nature, so sorting of litter is essential. The collection, disposal, and maintenance of waste generated by local governments are generally within their jurisdiction. On the contrary, local governments typically have no authorities and resources available to provide satisfactory and economically viable services. A fair distribution of responsibilities, administration, and income between the national government and local authorities will depend on efficient and effective solid waste management [6]. However, the poor management of household waste affects human health due to direct exposure. Due to improper sanitary waste management, common diseases can lead to water-borne diseases such as jaundice, typhoid, and cholera.

6.2.3 Industrial Waste

With the evolution of industrial growth, there is an addition in many new industries such as construction, mining and metallurgical, chemical, and automobile. Emerging capitalist activities generate considerable waste, which is even hazardous, toxic, corrosive, and reactive. The waste resulting from industrial activities can be of any form, say solid, liquid, or gaseous, and can pose health and environmental risks if improperly disposed of in the environment.

The standard form of industrial liquid wastes includes dirty water, rinse water, waste detergents, and organic liquids. The incredible variety of contaminants present in the industrial liquid waste running into oceans, rivers, or lakes has many ecological effects. Businesses and factories must implement wastewater treatment plants to prevent wastewater flow into significant water areas because it can harm the environment and its people if it flows untreated. Because of the incredible variety of contaminants, there are many ecological effects. A scrap in one industry could be a feedstock for another. We can significantly decrease the waste disposal costs by recycling as much industrial waste. With increasing environmental concerns, industries and enterprises worldwide must comply with the guidance and standards for industrial disposals and waste management, such as those published by Environmental Protection Agency (EPA) and the Occupational Safety and Health Administration (OSHA).

6.2.4 Biomedical Waste

According to World Health Organization (WHO), medical waste is coined as "the waste produced during treatment, immunization, and diagnosis of human beings and animals." Improper handling and disposal of medical waste present infection risks to healthcare personnel and the general public due to the spread of microorganisms into the environment from healthcare facilities [7]. A large amount of medical waste is being disposed of each year. The rapid increase in the medical waste demands for safer waste disposal. The WHO has classified 20% of the medical waste as hazardous for its infectious, toxic, and radioactive nature [8]. However, due to pathogenic concerns, infectious medical waste is unsuitable for disposal as municipal solid waste.

6.2.4.1 The Risks Associated with Biomedical Waste

The highest risk is associated with the scavenger healthcare workers and then the general public. Exposure to biomedical waste causes microbial infections. Further, blood-borne pathogen infections can be caused by improper handling of sharps. Furthermore, cytotoxic and radioactive biomedical waste leads to dizziness, headache, vomiting, and tissue destruction. It can also trigger genotoxicity, death, and ecological disturbances. Besides, it may even lead to malignancy, fatal malformations, and cardiac and respiratory disorders [9].

6.2.5 E-Waste

The reliance on electric and electronic equipment (EEE) exponentially increases with industrialization, techno-economic advancement, and a luxurious lifestyle. However, with technological advancement, there is a reduction in the effective life span of EEE products. The point at which the EEE products are discarded and/or considered outdated is defined as their end of life. And thus they are disposed of as electronic waste (e-waste). The global e-waste monitor reported an increase in the e-waste generation from 44.75 MT (million metric tonnes) in 2016 to 53.6 MT in 2019 [10]. In 2019, the average worldwide per capita e-waste generated was 7.3 kg with 2.9, 5.6, 13.1, and 16.2 kg in Africa, Asia, America, and Europe, respectively [11].

The e-wastes are a mixture of 30% organics, 30% ceramic, and 40% inorganics. The presence of hazardous materials increases the complexity of the treatment of e-waste. The common hazardous materials present in e-waste include (i) brominated flame retardants and (ii) heavy metals (lead, mercury, cadmium, nickel, and hexavalent chromium) [12].

Thus, the e-waste management is highly challenging. Besides, the processing of e-waste and recovery of resources is critical. Only 17.4% of e-waste is collected formally. And the remaining 82.6% of e-waste is illegally traded or dumped and landfilled [11]. A substantial amount of e-waste is discarded in municipal landfills or stored in a warehouse for preliminary recycling [13,14]. Even though e-waste can help obtain livelihood and business opportunities, its management endures a shortage of infrastructure and scientific strategies [15]. The informed recycling methods such as incineration and acid treatment generate dioxins, heavy metals, and furans [16]. Despite several attempts, sustainable management's ideal roadmap includes collection, segregation, storage, transportation, treatment method, and regulation yet to be established [17].

6.2.6 Nuclear Waste

Nuclear waste is a distinguished type of waste that can be either solid, liquid, or gas containing radioactive materials and emitting harmful ionizing radiation together with attended heat. These wastes are broadly classified into three categories: (i) high-level nuclear waste, (ii) transuranic waste, and (iii) low-level nuclear waste. High-level nuclear waste is highly radioactive with a very long life; for example, Pu (239) possesses a half-life of 24,000 years. The transuranic waste is generated either from the processing of spent nuclear fuel or during nuclear weapon fabrication, for example Pu (94), Am (95), Cm (96), and Np (93). The low-level liquid waste possesses a lower radiation level and shorter half-life [18].

The USA generated around 47,000 tonnes of high-level nuclear waste in 2003. It was estimated that by 2035 this amount would increase to 105,000 tonnes [18,19]. Globally, around 270,000 tonnes of high-level waste was temporally stored in the reactor site in 2004. According to the World Nuclear Association (2007), nearly 12,000 tonnes of high-level nuclear waste is produced, out of which 9,000 tonnes is disposed of and only 3,000 tonnes is reprocessed [18]. The IAEA estimates that 370,000 tonnes of heavy metal from used fuel has been discharged since the first nuclear power plant's operation. The IAEA further estimates that approximately 22,000 m^3 of solid high-level nuclear wastes are disposed of currently; geological disposal is the best solution for nuclear waste [20].

6.3 SUSTAINABLE WASTE MANAGEMENT APPROACHES

The waste hierarchy is a criterion for analyzing operations that focus on socio-environmental aspects ranging from the most favorable to the least favorable actions and resources and energy consumption. The hierarchy sets priorities for the program on a sustainable basis. Waste management cannot be elucidated sustainably only through technical solutions of end of pipe. Instead, an integrated methodology is needed. The mandate of preference for action to diminish and manage waste is given through the hierarchy of waste management and is generally shown as a pyramid in a diagrammatic way. The objective of the waste pyramid is to derive maximum practical advantages and to produce minimal waste. There are many advantages associated with the effective utilization of waste hierarchy. It contributes toward conserving energy and resources and simultaneously minimizing pollutants and GHG emissions with job generation and green technology development as another benefit (Figure 6.2).

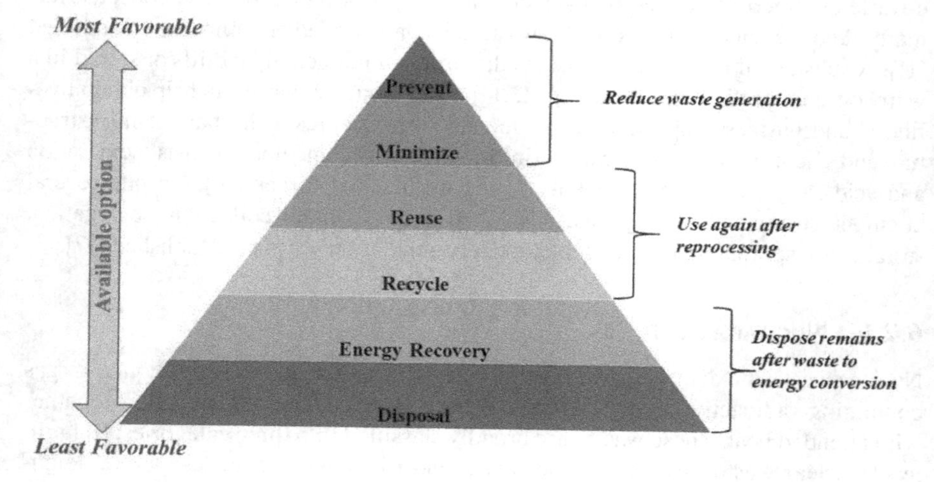

FIGURE 6.2 Waste management hierarchy.

6.4 WASTE-TO-ENERGY TECHNOLOGY

The most sustainable and essential aspect of waste management is the co-generation of energy and high-value products and their disposal. This approach produces renewable energy and simultaneously has zero or negative carbon cycle with lower climatic and environmental impacts. There are different conventional methods available that have extensively been used in the recent past. Furthermore, some latest and developing technologies are available for optimum and efficient energy reclamation due to proactive research in this field. A detailed overview of both these types of technologies is discussed in the subsequent section.

6.4.1 Conventional Methods

The conventional methods broadly classify a wide range of waste-to-energy pathways categorized as physical, thermochemical, and biochemical based on the conversion mode [21]. These methods can treat a diverse range of wastes with the reclamation of energy and added materials. The feedstock with different waste-to-energy pathways and the products obtained from each conversion method are shown in Figure 6.1. Primarily, heat, electricity, fuel oil, fuel gas, biogas, etc., are the recovered products from these technologies [22]. The most basic method includes the physical technique in which grinding and briquette will give the refuse-derived fuel (RDF), which directly provides heat energy on combustion. This RDF can act as a supplementary fuel in the boilers to generate steam and electricity. However, due to inefficient energy reclamation, more developed technologies are preferred, giving green energy and a positive environmental impact. The advanced technologies include applying the thermal effect with a chemical catalyst and various microorganisms to generate more refined and developed renewable energy sources (Figure 6.3).

The thermochemical conversion includes incineration, pyrolysis, gasification, and hydrothermal liquefaction, where temperature plays a crucial role in converting waste into a wide range of products [23,24]. Better control of the process parameters, high energy recovery, and diverse ecological compatibility give thermochemical methods predominance over other technologies. However, significant capital and operational costs with substantially high energy input are the severe drawbacks of these methods, limiting their usage. The process description of different technologies and the advantages and disadvantages associated with each of them are shown in Table 6.1. Pyrolysis is the most beneficial and highly implemented technique due to its flexible and more accessible operating conditions [25].

Further, the overall diversity of the products obtained from this process makes it favorable [26,27]. The liquid product produced, known as fuel oil, has a high heating value (~40 MJ/kg) and can be further processed to get petroleum alternatives [28]. Subsequently, the uncondensed fuel gas can also be utilized in or outside the processes to reclaim energy and generate electricity, improving the overall efficiency. Finally, the char has widespread adsorbent-based applications due to its unique surface morphology and hydrophobic chemistry. Another widely implemented technique is using microorganisms to decompose the waste and generate gaseous products and added chemicals. Anaerobic digestion and hydrolysis fermentation are the two most common biochemical methods that give biogas and bioethanol as major converted

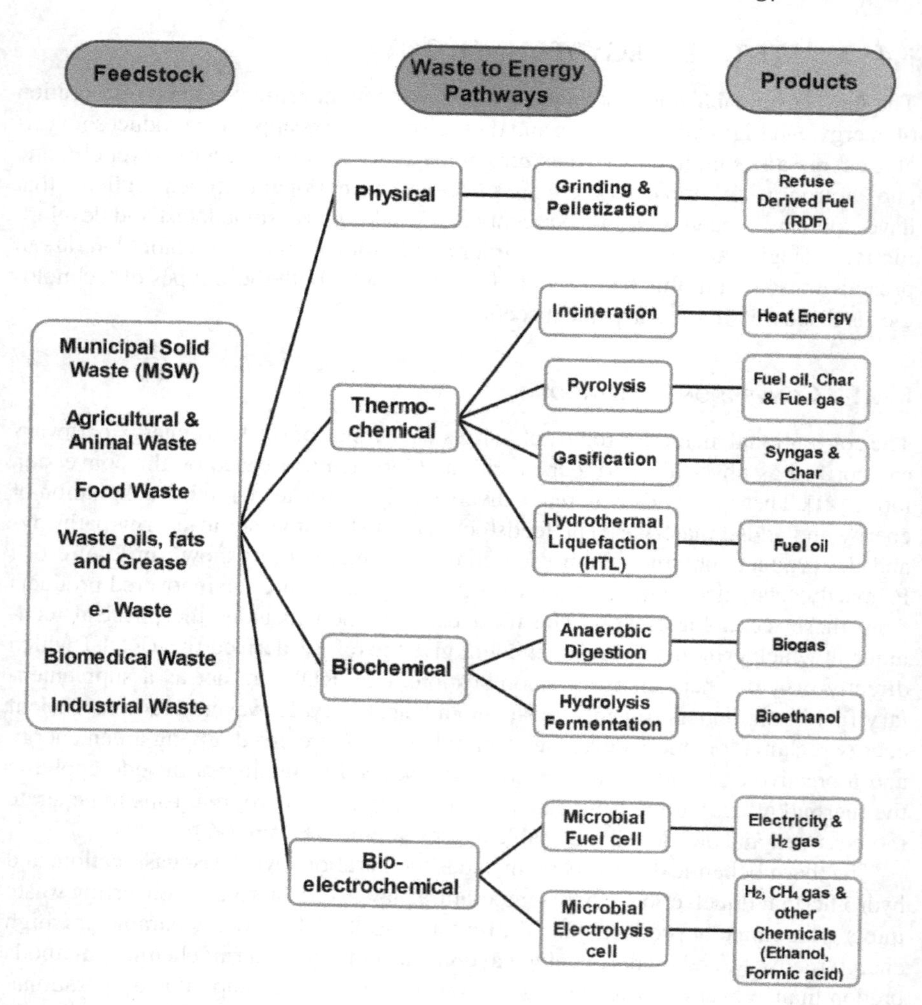

FIGURE 6.3 Different waste-to-energy pathways available along with their feedstock and the desired product.

products. These methods are not very capital-intensive and are suitable for a wide variety of substrates [29]. However, contamination risks associated with culturing and inhibition of microorganisms may be detrimental its use.

6.4.2 Future Trends and Developing Technology

The drawbacks of the conventional methods eventually lead to profound research to develop more eco-friendly technology with lower carbon footprints and zero waste. The eco-friendly technology also includes bio-thermochemical methodology where microbial fuel cells and electrolysis cells are promising technologies in the waste-to-energy nexus [22]. Microbial fuel cells utilize electrodynamically active

TABLE 6.1
Process Description along with the Advantages and Disadvantages Associated with the Various Waste-to-Energy Technologies [25,28,31]

S. No.	Waste-to-Energy Technology	Process Description	Advantages	Disadvantages
			Physical Method	
1.	Grinding and pelletization	This method includes direct combustion of waste to generate thermal energy after the mechanical treatment of drying, grinding, and briquetting.	• Very low operational, maintenance, and capital costs. • Can act as a supplementary fuel with conventional fossil fuels such as coal in boilers, kilns, etc. • Better homogeneity in size gives proper combustion.	• Energy inefficient as direct combustion leads to wastage of energy. • Creates air pollution with emission of harmful pollutants such as SO_x, NO_x, and CO_x if burnt in open space.
			Thermochemical Method	
2.	Incineration	Thermal degradation of organic elements in waste at a controlled temperature (>800°C) and in the presence of oxygen inside the furnace to generate heat and energy.	• Mass and volume reductions of waste up to 70% and 90%, respectively, are achievable. • Energy efficiency is high for heat generation and co-generation of steam and electricity.	• Generation of gaseous pollutants along with intensely dangerous heavy metals, PAH, etc.
3.	Pyrolysis	It is an endothermic destructive distillation process in an oxygen-depleted environment to give a diverse range of solid, liquid, and gaseous products, i.e., char, fuel oil, and fuel gas.	• Offers a broader range of products. • Lower temperature requirement (400°C–700°C). • Accessible operational parameters such as lower pressure requirement (0.1–0.2 MPa). • High calorific value of products. • Low environmental emissions.	• High oxygen content and other undesirable properties in fuel oil require further upgradation. • High capital and operational costs with substantially intensive setup.

(Continued)

TABLE 6.1 (*Continued*)
Process Description along with the Advantages and Disadvantages Associated with the Various Waste-to-Energy Technologies [25,28,31]

S. No.	Waste-to-Energy Technology	Process Description	Advantages	Disadvantages
4.	Gasification	It is a thermal degradation in the controlled oxygen atmosphere to produce mainly the gaseous product "syngas" and residual char.	• The high heating value of syngas with a combination of hydrocarbon gases (H_2, CO, CO_2, CH_4, and low molecular weight hydrocarbons). • Better emission control than combustion as higher operating parameters facilitate easier pollutant removal.	• Higher operating temperature (800°C–1,200°C). • Cost-intensive technology. • High operational and maintenance costs.
5.	Hydrothermal liquefaction	It is a moderate-temperature (~300°C) and high-pressure (~20 MPa) thermal disintegration process that mainly produces liquid fuel.	• Fuel oil produced has high energy density and less moisture and oxygen content. • No energy required for drying of feedstock as wet biomass can be directly utilized.	• Higher pressure requirement makes it a capital-intensive process with very high operational and maintenance costs. • Subcritical conditions cause rapid corrosion. • Lesser product diversity with further upgradation required to convert to premium liquid transport fuels.
			Biochemical Method	
6.	Anaerobic digestion	The biological decomposition of organic waste in an oxygen-depleted condition produces biogas containing CH_4 and CO_2.	• Contains a higher CH_4 concentration as compared to CO_2, which increases its energy content. • Suitable for high water content and diverse feedstock with a significant change in substrates. • Nutrient recovery is possible, which decreases waste capacities.	• Only suitable for feedstock containing high organic matter. • Sometimes leakage of gases such as CH_4, CO_2, NO_x, and SO_x can lead to severe atmospheric pollution. • High concentration and low degradation of lignin in the sludge create disposal problems.

(*Continued*)

TABLE 6.1 ***(Continued)***
Process Description along with the Advantages and Disadvantages Associated with the Various Waste-to-Energy Technologies [25,28,31]

S. No.	Waste-to-Energy Technology	Process Description	Advantages	Disadvantages
7.	Hydrolysis fermentation	It is the biochemical fermentation using enzymes to convert cellulose-rich waste into bioethanol. It includes sucrose hydrolysis followed by enzymatic reaction and distillation to produce bioethanol.	• Lesser CO_2 emissions as compared to other technologies. • Low capital cost. • Good bioethanol yield.	• Applicable only for cellulose- and starch-rich waste. • Associated risk of contamination and inhibitory effects.
			Bio-Electrochemical Method	
8.	Microbial fuel cell	It is a bio-electrochemical setup that utilizes electrochemically active microorganisms to convert waste directly into electricity.	• Suitable for a wide range of organic substrates. • Very low or negligible emissions from the process. • Multiple usages as with wastewater treatment, seawater desalination, H_2 production, etc.	• The electric current generated is potentially low. • Highly temperature-dependent process as its efficiency decreases with lowering of temperature.
9.	Microbial electrolysis cell	It has the reverse process of the microbial fuel cell where electric current is applied along with electrochemically energetic microorganisms to produce H_2 or CH_4 and value-added chemicals.	• Higher H_2 production and efficiency as compared to other processes with less energy requirement. • Suitable for a wide range of organic wastes.	• Cost-intensive system. • Substrate-dependent yields. • High resistance to the current flow.

microorganisms as a catalyst to generate direct electricity and H_2 gas from a wide range of organic substrates. Fuel cells are based on a sustained release of an electron at anode due to substrate oxidation by microbe and subsequent consumption at the cathode. The cathode compartment contains a proton exchange membrane where oxygen combines with electron and proton diffuses through the membrane [30]. The higher the microbe's metabolic energy rate, the more the potential that is generated in the cell. The series connection of these cells in large numbers will facilitate a higher electric current.

Conversely, microbial electrolysis cells work on the principle of converting waste into H_2 and added chemicals such as CH_4, HCOOH, and C_2H_5OH by applying electric energy. Here, the electron does not combine with oxygen as the cathode is not exposed to air. This method offers higher H_2 recovery than other processes with a lower electric potential requirement for microbial electrolysis [31]. However, the risks associated with microbe contamination and the inhibition of its activity at lower temperatures with less electric current generation are some of the limitations and future research prospects for these methods.

Another sustainable approach and the most sought-to-be-developing method is biological production of H_2 from waste. Various metabolic activities of bacteria can achieve this sustainable approach to synthesize bio-hydrogen, which has nearly three times high energy yield than the existing fossil fuels. Further, zero GHG emissions make it an ideal renewable energy source. The biological synthesis route offers several advantages over conventional chemical methods, such as environmental sustainability, flexibility over substrate selection, and low energy requirement [32]. The biological process is mainly divided into two types: photobiological techniques (light dependent) and dark fermentation (light independent), based on the conditions maintained during anaerobic fermentation [33]. Various phototrophic microorganisms are used in the photobiological processes. These phototrophic microorganisms ferment the organic carbon substrate into bio-hydrogen, CO_2, and other organic acids. However, the absence of light facilitates the decomposition of the organic substrate into bio-hydrogen in the dark fermentation processes. It involves a complex process where large molecules undergo a series of chemical transformations to generate bio-hydrogen. The H_2 yield is primarily dependent on the substrate type; for example, the carbohydrate-rich group gives higher bio-hydrogen production [22]. However, researchers' continuous effort is to improve the process's overall efficiency and achieve higher bio-hydrogen yields.

6.5 CONCLUSIONS

An essential aspect of sustainable waste management is the awareness among individuals, businesses, plants, and factories to adhere to government waste disposal regulations, irrespective of the type of waste. Standard waste treatment practices can ensure that pollutants released into the environment are kept to the lowest level. That's why every person and organization should respect society's interests and play its part in the fight against global pollution. The development of feasible, sustainable waste management solutions represents one of the significant challenges for society. These solutions should be capable of achieving social, economic, and environmental benefits. Optimum avoidance of waste generation is still far from being attained. However, 3R approach should be implemented with source reduction and maximum attainable reuse and recycle. Organic wastes can be used as a feedstock for the production of energy via various waste-to-energy pathways. It requires adequate segregation of waste that can be transformed into a resource. Practical and convenient solutions targeted according to the area of interest must be designed. Besides appropriate waste management programs, environmentally friendly and profitable waste treatment solutions can be obtained. Although the holistic waste-to-energy concept appears to be very promising, a highly efficient positive energy system supported by a thorough life cycle assessment is required for practically scaling up the technology. Therefore, the interconnection of thermochemical processes in the co-production of energy and valuable by-products presents a vital strategy for maximizing organic waste utilization and raising the entire techno-economic chain's potential income.

REFERENCES

1. Zia H, Devadas V (2008) Urban solid waste management in Kanpur: Opportunities and perspectives. *Habitat Int* 32:58–73.
2. Singh WR, Kalamdhad AS (2016) Role of urban local bodies and opportunities in municipal solid waste management. In: Sarma AK, Singh VP, Kartha SA, Bhattacharjya RK (eds) *Urban Hydrology, Watershed Management and Socio-Economic Aspects*. Springer International Publishing, Cham, pp. 341–351.
3. Mishra A, Kumari U, Turlapati VY, Siddiqi H, Meikap BC (2020) Extensive thermogravimetric and thermo-kinetic study of waste motor oil based on iso-conversional methods. *Energy Convers Manag* 221:113–194.
4. Kaza S, Yao L, Bhada-Tata P, Van Woerden F (2018) *What a Waste 2.0: A Global Snapshot of Solid Waste Management to 2050*. Washington, DC: World Bank.
5. Obi FO, Ugwuishiwu BO, Nwakaire JN (2016) *Agri Wastes* 35:957–964.
6. Yoada RM, Chirawurah D, Adongo PB (2014) Domestic waste disposal practice and perceptions of private sector waste management in urban Accra. *BMC Public Health* 14:1–10.
7. Mohee R (2005) Medical wastes characterisation in healthcare institutions in Mauritius. *Waste Manag* 25:575–581.
8. Birchard K (2002) Out of sight, out of mind...the medical waste problem. *Lancet* 359:56.
9. Capoor MR, Bhowmik KT (2017) Current perspectives on biomedical waste management: Rules, conventions and treatment technologies. *Indian J Med Microbiol* 35:157–164.
10. Balde CP, Forti V, Gray V, Kuehr R, Stegmann P (2017) The Global E-waste Monitor 2017: Quantities, Flows and Resources.
11. Forti V, Baldé CP, Kuehr R, Bel G (2020) The Global E-waste Monitor 2020.
12. Kaya Y, Martin ND (2016) Managers in the global economy: A multilevel analysis. *Sociol Q* 57:232–255.
13. Herat S, Agamuthu P (2012) E-waste: A problem or an opportunity? Review of issues, challenges and solutions in Asian countries. *Waste Manag Res* 30:1113–1129.
14. Li J, Lopez N. BN, Liu L, Zhao N, Yu K, Zheng L (2013) Regional or global WEEE recycling. Where to go? *Waste Manag* 33:923–934.
15. Lakshmi S, Raj A, Jarin T (2017) A review study of e-waste management in India. *Asian J Appl Sci Technol* 1:33–36.

16. Dai Q, Xu X, Eskenazi B, Asante KA, Chen A, Fobil J, Bergman Å, Brennan L, Sly PD, Nnorom IC, Pascale A, Wang Q, Zeng EY, Zeng Z, Landrigan PJ, Bruné Drisse MN, Huo X (2020) Severe dioxin-like compound (DLC) contamination in e-waste recycling areas: An under-recognized threat to local health. *Environ Int* 139:105731.
17. Adanu SK, Gbedemah SF, Attah MK (2020) Challenges of adopting sustainable technologies in e-waste management at Agbogbloshie, Ghana. *Heliyon* 6:e04548.
18. Hasan SE (2007) Chapter 4 International practice in high-level nuclear waste management. *Dev Environ Sci* 5:57–77.
19. Stuckless JS, Levich RA (2016) The road to Yucca Mountain—evolution of nuclear waste disposal in the United States. *Environ Eng Geosci* 22 (1):1–25. https://world-nuclear.org/information-library/nuclear-fuel-cycle/nuclear-wastes/radioactive-waste-management.aspx
20. (2020) Radioactive waste management. *World Nuclear Association*. https://world-nuclear.org/information-library/nuclear-fuel-cycle/nuclear-wastes/radioactive-waste-management.aspx
21. Sharma A, Pareek V, Zhang D (2015) Biomass pyrolysis – A review of modelling, process parameters and catalytic studies. *Renew Sustain Energy Rev* 50:1081–1096.
22. Beyene HD, Werkneh AA, Ambaye TG (2018) Current updates on waste to energy (WtE) technologies: A review. *Renew Energy Focus* 24:1–11.
23. Foster W, Azimov U, Gauthier-Maradei P, Molano LC, Combrinck M, Munoz J, Esteves JJ, Patino L (2021) Waste-to-energy conversion technologies in the UK: Processes and barriers – A review. *Renew Sustain Energy Rev* 135:110226.
24. Siddiqi H, Bal M, Kumari U, Meikap BC (2020) In-depth physiochemical characterization and detailed thermo-kinetic study of biomass wastes to analyze its energy potential. *Renew Energy* 148:756–771.
25. Mishra A, Siddiqi H, Kumari U, Behera ID, Mukherjee S, Meikap BC (2021) Pyrolysis of waste lubricating oil/waste motor oil to generate high-grade fuel oil : A comprehensive review. *Renew Sustain Energy Rev* 150:111446.
26. Siddiqi H, Kumari U, Biswas S, Mishra A, Meikap BC (2020) A synergistic study of reaction kinetics and heat transfer with multi-component modelling approach for the pyrolysis of biomass waste. *Energy* 204:117933.
27. Siddiqi H, Biswas S, Kumari U, Bindu VNVH, Mukherjee S, Meikap BC (2021) A comprehensive insight into devolatilization thermo-kinetics for an agricultural residue : Towards a cleaner and sustainable energy. *J Clean Prod* 310:127365.
28. AlQattan N, Acheampong M, Jaward FM, Ertem FC, Vijayakumar N, Bello T (2018) Reviewing the potential of Waste-to-Energy (WTE) technologies for Sustainable Development Goal (SDG) numbers seven and eleven. *Renew Energy Focus* 27:97–110.
29. Sikarwar VS, Pohořelý M, Meers E, Skoblia S, Moško J, Jeremiáš M (2021) Potential of coupling anaerobic digestion with thermochemical technologies for waste valorization. *Fuel* 294:120533.
30. Logroño W, Ramírez G, Recalde C, Echeverría M, Cunachi A (2015) Bioelectricity generation from vegetables and fruits wastes by using single chamber microbial fuel cells with high andean soils. *Energy Procedia* 75:2009–2014.
31. Kadier A, Simayi Y, Abdeshahian P, Azman NF, Chandrasekhar K, Kalil MS (2016) A comprehensive review of microbial electrolysis cells (MEC) reactor designs and configurations for sustainable hydrogen gas production. *Alexandria Eng J* 55:427–443.
32. Pathak VV., Ahmad S, Pandey A, Tyagi VV., Buddhi D, Kothari R (2016) Deployment of fermentative biohydrogen production for sustainable economy in Indian scenario: Practical and policy barriers with recent progresses. *Curr Sustain Energy Reports* 3:101–107.
33. Carolin Christopher F, Kumar PS, Vo DVN, Joshiba GJ (2021) A review on critical assessment of advanced bioreactor options for sustainable hydrogen production. *Int J Hydrogen Energy* 46:7113–7136.

7 Biomass Downdraft Gasifier

State of the Art of Reactor Design

Nathada Ngamsidhiphongsa and Phuet Prasertcharoensuk
Chulalongkorn University

Yaneeporn Patcharavorachot
King Mongkut's Institute of Technology Ladkrabang

Amornchai Arpornwichanop
Chulalongkorn University

CONTENTS

DOI: 10.1201/9781003178354-9

7.1 INTRODUCTION

Biomass-derived synthesis gas (syngas) has attracted much attention in recent years because of its versatility and greenhouse gas neutrality. It plays an important role in replacing fossil fuels in terms of utilizing the existing downstream technologies with some modifications, such as replacing the diesel and gasoline fuels used in internal combustion engines (ICEs) and gas turbines (GTs). Syngas has grown in popularity not only in the energy production and chemical fields such as those related to hydrogen, diesel, gasoline, methanol, and ammonia [1], but also for polygeneration, including the generation of heat and electricity, for example through combined heat and power (CHP) technology, and other by-products such as biofuels or char [2]. This could literally be the way to utilize biomass more efficiently and produce zero waste in a circular economy system. Syngas can be produced through gasification technology to provide more sustainable feedstocks. This technology has regional, environmental, climatic, and social benefits because the use of biomass and waste requires labor; therefore, it increases local employment and living standards.

The gasification process converts the solid biomass into producer gas at high temperatures. The rapid development of gasifiers during the last two decades has globally distributed and accelerated the scale of operation, leading to a greater energy capacity. In contrast to a scattering of successful commercial small-scale gasification plants, product price competition, the nature of biomass, and the complexity of the process dominate this achievement [2]. In 2020, about 30 countries were operating 686 gasifiers for syngas production, but only 14% were biomass and waste gasification plants with a small capacity [3]. The standard gasifier design for these uses the downdraft gasifier with a cold gas filter. The scale of these operations has increased from 180 kW_e to 1 MW_e; the actual plants appear to apply only high-grade biomass, such as hardwood [4]. Accordingly, several designs for downdraft gasifiers have been developed for use with various raw materials and in an attempt to reduce the tar content in the gasifier in situ.

The wide biomass variety and topology play a challenging role in gasifier design and operation [5]. The moisture content of fresh biomass should be restricted below 20% to obtain uniform temperature across the oxidation zone [1,6]. The tar in the producer gas has significantly limited its use because it causes operational difficulties for the downstream process. The limitation for the tar in ICEs is 100 mg/Nm^3 and is even lower for GTs [7]. It is common to remove the tar using a cold gas filter, while other strategies use hot gas treatment [4]. These have a significant impact from an economic standpoint because they are expensive technologies. Therefore, there has been great interest in a primary tar reduction approach, which involves design modification and adjustment of the operating conditions. This chapter addresses the design of a biomass gasifier, particularly a downdraft gasifier, including the basic calculations and specific sizing reported in the literature. Moreover, commercial gasifiers have recently been developed. The aim of this chapter is to collect and analyze the data on the existing designs of downdraft gasifiers that have been progressed.

7.2 DOWNDRAFT BIOMASS GASIFICATION PROCESS

A downdraft gasifier uses concurrent flows of fuel and air streams. The producer gas is forced downward, which leads to the name of this device. A downdraft gasifier can be operated by either pressurized or suction operations, with suction gasifiers being more practical because they clean up the gas train [6]. The gasifying agent used for gasification can be either air, oxygen, steam, or a mixture. Therefore, the choice of the gasifying agent used strongly determines the final producer gas composition and its heating value. The producer gas from gasification with steam contains more hydrogen per unit of carbon, given a higher gas heating value. Oxygen as a gasifying agent demonstrates the highest heating value of the producer gas besides the higher capital and operating costs. Adding air along with steam can be a more popular choice [1]. During the operation of a gasifier, the reactions can be roughly separated into four zones: drying, pyrolysis, oxidation, and reduction, according to the diagram in Figure 7.1.

In the drying zone, the biomass releases moisture at above 100 °C, as a consequence of heat generated from the oxidation zone. Then, pyrolysis occurs rapidly as the temperature continues to increase above 700°C. The dry biomass breaks down into volatile gases and condensable vapor, tar, and solid residues of almost pure carbon materials, known as char. At this stage, air is injected through air blowers and immediately combusted with the volatile gas and a slight amount of char in the oxidation zone. Heat is liberated by the exothermic combustion reactions and contributes to all the other zones, which mostly involve endothermic reactions. The gasifier is a self-adjusting system that maintains the burned level of the char. The undermost zone is a reduction zone; hot combustion gases flow through the hot char bed and are

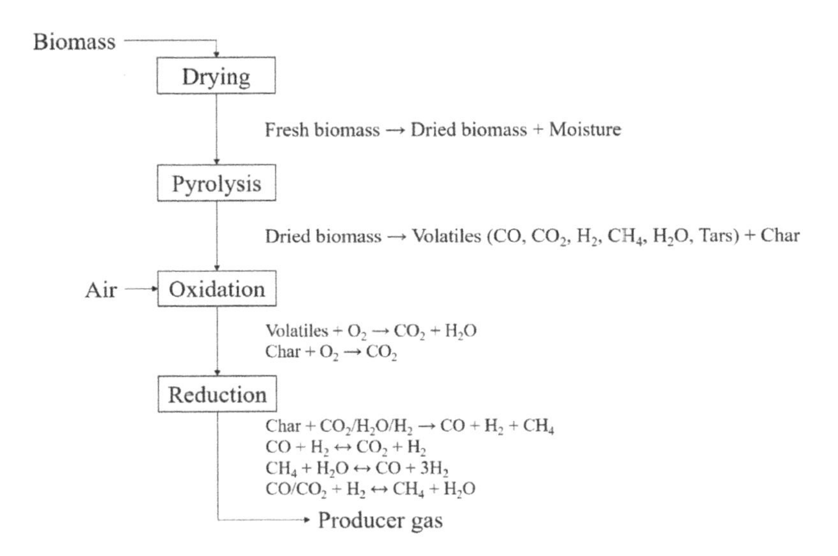

FIGURE 7.1 Schematic diagram of the gasification process.

partially reduced to fuel gas storing energy in the form of chemical bonds, which mainly contain CO and H_2. The producer gas is rapidly cooled as the endothermal reactions occur. The char is disintegrated into a smaller powder. The tar escaping from the oxidation zone may crack at temperatures above 850 °C [1,6]. Biomass char is considered a by-product in polygeneration of this system [1,4,8].

7.3 PRELIMINARY CALCULATION FOR DESIGNING DOWNDRAFT GASIFIERS

The volumetric flow rate of the producer gas (V_g) in Nm^3/s is a vital parameter in designing any gasifier. It relates to the required thermal output of the gasifier (Q) in MW_{th} and the desired heating value of the producer gas (LHV_g) in MJ/Nm^3, which are first identified. The output of the downdraft gasifiers is generally in the range of 10 kW_{th}–1 MW_{th} [1]; for example, the output of the gasifier in [9,10] was 20 kW_{th} and in [11–13] was 70–80 kW_{th}. The lower heating value of the gas can be primarily assumed to be 4–7 MJ/Nm^3 for air-blown gasifiers. The gas production rate can be calculated as follows [1]:

$$V_g = \frac{Q}{\mathrm{LHV}_g} \tag{7.1}$$

As can be seen in Figure 7.1, the biomass consumption rate ($\dot{m}_f$) in kg/s can also be related to the thermal output of the gasifier through an energy balance equation which is expressed as follows:

$$\dot{m}_f = \frac{Q}{\mathrm{LHV}_f \eta_{\mathrm{gef}}} \tag{7.2}$$

where LHV_f is the lower heating value of the biomass on a wet basis in MJ/kg and η_{gef} is the gasifier efficiency (%), typically about 68% [4]. The LHV_f can be calculated by subtracting the vaporization energy of water from the higher heating value on a dry basis (HHV_d), according to Equation 7.3 [14], while the value of HHV_d can be calculated as shown in Equation 7.4 [15].

$$\mathrm{LHV}_f\left(\mathrm{MJ/kg}\right) = \mathrm{HHV}_d\left(1-M\right) - 2.444M - 21.839H_d\left(1-M\right) \tag{7.3}$$

$$\mathrm{HHV}_d\left(\mathrm{MJ/kg}\right) = 0.3491C_d + 1.1783H_d + 0.1005S_d - 0.1034O_d - 0.00151N_d - 0.0211A_d \tag{7.4}$$

where M is the moisture and H_d is the hydrogen mass fraction. The subscript d represents the dry basis. Similarly, C_d, H_d, O_d, N_d, S_d, and A_d are the mass fractions of carbon, hydrogen, oxygen, nitrogen, sulfur, and ash on a dry basis, respectively.

Afterward, the air flow rate ($\dot{m}_a$) in kg/s required for the gasification process can be determined from the previous fuel consumption rate or feeding rate and a dimensionless index named the equivalence ratio (ER) through Equation 7.5. The ER is defined as the ratio between the actual and the stoichiometric air-to-fuel flow rates;

for downdraft gasifiers, the best yield for ER equals to 0.25 [6]. The stoichiometric air-to-fuel requirement (m_{stoich}) is the amount of air required per unit mass of fuel in $\text{kg}_{\text{air}}/\text{kg}_{\text{fuel}}$ for a complete combustion reaction, expressed as in Equation 7.6 [1].

$$\dot{m}_a = m_{\text{stoich}}\text{ER} \times \dot{m}_f \tag{7.5}$$

$$m_{\text{stoich}} = \left(\frac{32}{12}C_d + 8H_d + S_d - O_d\right)\frac{1}{0.23} \tag{7.6}$$

7.4 DESIGN OF DOWNDRAFT GASIFIER

Among the different gasifiers, downdraft gasifiers claim to produce the lowest amount of tar in the producer gas, with values of approximately 0.015–3 g/Nm^3 [1]. Therefore, it has the most potential to be used as a stand-alone instrument, without post-gas treatment. In addition, its simple construction and shorter ignition time to warm up the gasifier, which lowers the capital cost compared to other designs, make this technology worth the investment. The standard downdraft gasifier consists of a biomass feeding system; firebox, which is a fuel magazine where the drying and pyrolysis occur; hearth zone; and ash removal system. The various designs are discussed below.

7.4.1 Imbert-Type Downdraft Gasifier

This gasifier was designed by an inventor named Imbert. It is characterized by a constricted hearth or throat at the lower half of the gasifier, as shown in Figure 7.2. The char from the pyrolysis zone travels downward, passing through the hot zone where the air injection nozzles are located. The char layer can absorb heat; therefore, volatile gases are forced to pass through the highest temperature region. The throat exhibits enhancement in fluid mixing and tar cracking reactions [6]. The various throat designs include the classic, V-shape, and flat-plate throats, as shown in Figure 7.2. The two latter hearths have superior char and ash accumulation, which acts as a hot filtration medium for the producer gas, enhances the tar cracking, and increases the efficiency.

The method for determining the size of an Imbert-type downdraft gasifier is next presented. The hourly load of the gasifier on a volume basis can be expressed as the producer gas produced from the gasifier per cross-sectional area of the throat (Nm^3/h cm^2) as apparent in Equation 7.7. The calculated gas production rate obtained from the previous section is set as an input, while the typical value of the hearth load based on the successful downdraft gasifiers is in the range of 0.1–0.9 [6]. For the Imbert-type gasifier, a minimum value of 0.35 is recommended [6,16–18]. The hearth cross-sectional area achieved by this equation is translated to a throat diameter, a starter for sizing the downdraft gasifier.

$$\text{Hearth load} = \frac{V_g}{A_{\text{th}}} \tag{7.7}$$

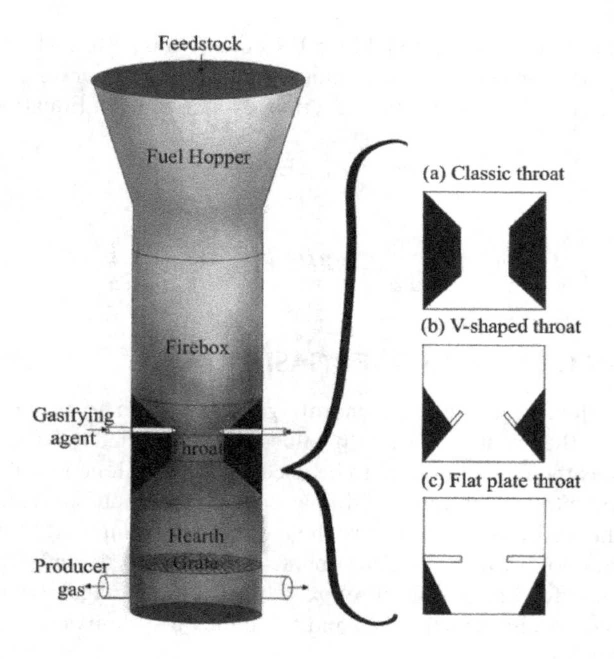

FIGURE 7.2 Schematic diagrams of three constriction hearths for Imbert downdraft gasifier.

Therefore, the throat diameter can be calculated with Equation 7.7; the throat diameters of the existing Imbert-type gasifier are in the range of 0.15–0.30 m [6]. The firebox diameter is assumed to be three and a half times the throat diameter [18]. It is generally restricted to 1.5 m to ensure that air jet can adequately penetrate the bed [1]. The height of the firebox can be expressed as follows [18]:

$$H_{\text{firebox}} = \frac{\dot{m}_f h}{\rho_f \frac{\pi}{4} D_{\text{firebox}}^2} \tag{7.8}$$

where h is the operation time (h) and ρ_f is the bulk density of the fuel. High bulk density biomass lowers the firebox volume and reduces the time to refill the feedstock. In spite of low bulk density biomass, it may create bridging and channeling, burn char rapidly in the reduction zone, and result in a low heating value of the producer gas [19].

In the case of the classic throat, its diameter is generally reduced to the throat diameter by a cone structure with an inclination of 60° to the horizontal. It helps to limit the bridging effect resulting from an angle that is higher than the angle of repose of the wood chips, which is 45° at ambient temperature [16–18,20]. For smaller angles, a higher cumulative conversion efficiency may be achieved, even though a longer reduction zone is required. The specifications of the V-shape throat are described thoroughly in [1,6]. The flat-plate throat is rarely seen [9,10].

Afterward, the total air nozzle area can be estimated from the throat area as of 4%–7% [6] or 5%–9% [18]. The air nozzles are injected from the periphery of the downdraft gasifier. The diameter at which the air nozzle is located is approximately two- to two-and-a-half-fold greater than the diameter of the throat [18]. The air nozzles are distributed equally at the level approximately one-third of the gasifier from the bottom; they should be placed 0.1–0.15 m above the throat [6]. It is recommended that an odd number of air nozzles should be used to prevent the injected air from directly impinging on an opposite nozzle, leaving a dead zone in between. Eventually, the diameter of the reduction zone can be assumed equal to the diameter of the firebox [18], while the height should be more than 0.1 m [6].

This gasifier design has a limitation on how much it can be scaled up. A larger gasifier diameter builds a higher pressure drop to ensure satisfactory air penetrating toward the center; this fluid stream is operated by a gas blower, and its capacity is limited [6]. The fuel particle size is also involved, as the larger diameter of the gasifier requires a larger size. The maximum size for the downdraft gasifier is recommended as one-eighth of the throat diameter [19]. As a result of poor temperature distribution and flow mixing; consequently, more tar will remain in the product stream.

Table 7.1 lists the sizes of the existing downdraft gasifiers. These can be grouped into three categories. The first group is the Imbert-type with a classic throat, as shown in Figure 7.2 [11,13,18,21–24]. It is the most widely used, and therefore, specific

TABLE 7.1
Existing Imbert-Type Downdraft Gasifier Designs

Design	Throat Type	Feedstock	Gasifying Agent (No. of Nozzles)	Desired Power Output	Diameter (m)	Height (m)	Throat Diameter (m)	Air Nozzle Diameter (m)	References
					Dry Gas Composition (%)				
					CO	H_2	CO_2	CH_4	
Imbert	Classic	Wood chips	Air (8)		0.60	2.50	0.20	0.01	[20]
					24.04	14.05	14.66	2.02	
Imbert	Classic	Wood chips and hazelnut shell	Air (3)		0.300	1.095	0.100	0.01	[24]
					22.00	12.50	12.00	3.5	
Imbert	Classic	Hazelnut shell	Air	5 kW_e	0.305	0.81	0.135	-	[19]
					19.89	11.86	11.25	2.47	
Imbert	Classic	Wood chips	Air (12)	80 kW_{th}	0.92	2.11	0.10	0.06	[11]
					19.10	15.50	11.40	1.10	
Imbert	Classic	Wood sawdust and sunflower pellets	Air (4)	85 kW_e	0.55	1.60	0.30	-	[25]
					21.20	17.20	12.60	2.50	
Imbert	Flat-plate	Wood chips	Air (5)	15 kW_e	0.60	1.02	0.09	0.01	[10]
					17.40	13.20	12.40	0.80	

design details are available. The second group uses a cone-shaped firebox, along with the classic throat type, and inclination air nozzles to penetrate the air near the throat as much as possible [19,25]. The last one uses a flat-plate throat [9,10]. According to the benefit of this throat, it seems to be the most promising one, but its precise details need to be studied further.

7.4.2 Stratified Downdraft Gasifier

This type of gasifier is a throatless gasifier, which is modified to overcome the complexity of both the Imbert gasifier designs and the topology of the biomass. It contains a constant-diameter cylinder, where the reactions take place sequentially, as described in Section 7.2. The biomass and air enter from the top uniformly and continuously, preventing bridging or channeling [6]. Some types of biomass, such as rice husk, which is fine and of low bulk density, work better with this gasifier type than with the Imbert-type gasifier. Nevertheless, the pelletization of the fuel is recommended to improve the efficiency of the downdraft instead of feeding the raw biomass directly [26]. The char and ash accumulate at the bottom and absorb heat, providing for the char gasification and tar cracking reactions [1].

The brief details for the existing design of the stratified gasifier comprise of its cross-sectional area and height. The gasifier cross-sectional area is calculated from the specific gasification rate (SGR), which is defined as follows:

$$A = \frac{\dot{m}_f}{\text{SGR}} \tag{7.9}$$

The SGR was assumed to be 110 kg/h m^2 [27]. The definition of SGR is, in fact, a mass basis of hearth load as mentioned in Equation 7.7. As if the volume basis of hearth load is being used, the hearth load of 0.09 Nm^3/h cm^2 is recommended [16]. Then, the height of the gasifier is calculated by adding 10% to the height provided for the grate:

$$H = 1.1\frac{\dot{m}_f h}{\rho_f A} \tag{7.10}$$

Theoretically, this gasifier design can be scaled to a larger size because the biomass and air are mixed uniformly, which seems to make it a promising design. However, some biomass including a stringy type, such as sawdust or rice hulls, may create bridges. In addition, the ash and char powders at the bottom readily escape with the gas product and plug the train, resulting in a lower gasifier efficiency [6]. Nevertheless, extensive designs based on the stratified downdraft gasifiers subdue these problems by adding more air supply stages, as described in Section 7.4.3.3. Figure 7.3 shows a stratified downdraft gasifier with three air supply stages [28]. Gautam et al. [29] reported that the tar produced by a gasifier having two air stages was in the range of 340–680 mg/Nm^3, which was higher than the amounts produced by a throat gasifier and by the same gasifier using wood chips. Table 7.2 demonstrates the existing stratified gasifiers and the producer gas composition.

TABLE 7.2
Existing Stratified Downdraft Gasifier Designs

Design	Air Supply Stage	Feedstock	Gasifying Agent (No. of Nozzles)	Desired Power Output	Diameter (m)	Height (m)	Air Nozzle above Grate (m)	Air Nozzle Diameter (m)	References
					Dry Gas Composition (%)				
					CO	H_2	CO_2	CH_4	
Stratified	One	Agricultural residues	Air (4)	236 kW_{th}	1.00	3.50	0.40	0.014	[27]
Stratified	Two	Wood chips	Air	190 kW_e	1.154	6.205	-	-	[8]
					16.60	16.10	13.80	2.30	
Stratified	Three	Coconut shell	Air	11.44 kW_e	0.20	1.76	1.36/0.59/0.23	-	[28]
					21.30	13.50	11.80	1.50	
Stratified	Three	Corn stalk	Air (1/9/9)		0.42	1.05	0.43/0.70/0.36	-	[30]
					18.68	12.89	16.03	2.10	

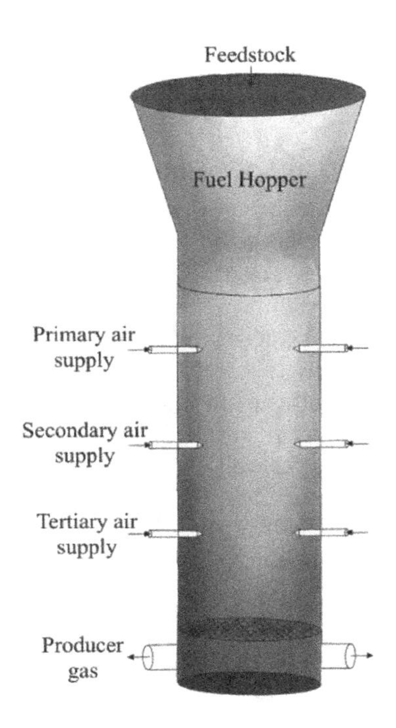

FIGURE 7.3 Schematic diagram of a stratified gasifier.

7.4.3 Modified Downdraft Gasifier Designs

Tar reduction and gasifier efficiency improvement are key factors in developing the gasifier design. One of the primary tar reduction approaches, an ideal method to minimize tar formation instead of removing them from the process which may cause disposal waste, is to modify the gasifier design. The collection of the modification is described as (i) internal recycling of the pyrolysis gas, (ii) dividing the gasifier into two stages, (iii) supplying more air stages, (iv) adjusting the throat diameter, and (v) extending the reduction zone length.

7.4.3.1 Internal Recycling of Pyrolysis Gas

Susanto and Beenackers [31] developed a gasifier with internal tar recycling, as shown in Figure 7.4a. The pyrolysis gas was recycled to combust with air in the central line injection nozzle from the top of the throatless downdraft gasifier. This design was developed to scale up a wood chips downdraft gasifier. The ratio of the recycle gas to the gasifying agent was controlled at 0.6, which gives the fuel gas with tar concentration below 100 mg/Nm^3. However, a further increase in the recycle ratio above 0.9 causes a lower gasifier efficiency. The optimum recycling ratio is chosen in the range of 0.6–0.9 for reasonable tar content in the fuel gas and the satisfactory efficiency of the gasifier.

Recently, a low-tar biomass (LTB) gasifier has been developed, which is a throatless downdraft gasifier with an inverted magazine for internal recirculation to combust the pyrolysis gas with air coming from the top center line of the gasifier, as shown in Figure 7.4b. The pyrolysis gas both flows upward into the magazine and recirculates colliding the canopy through the top of the magazine, resulting in a better mixing capacity, and no cool zone occurred, compared to the conventional air injection technique. It can reduce the tar formation to around 10–30 mg/Nm^3 [32].

7.4.3.2 Separating Gasifier into Two Stages

A two-stage downdraft gasifier consists of two chambers connected at 90°. The first part involves the drying and pyrolysis on a screw conveyer with external heat. The volatile gas and char are then delivered to a second vertical reactor with a constriction hearth and preheated air injection. A proven two-stage gasifier from Denmark is shown in Figure 7.4c. It is claimed that it can be scaled up to 400 kW, with lower tar content in the producer gas of approximately 15 mg/Nm^3 [13]. This technology is, up to date, built up to 1.5 MW$_{th}$ and continues to further increase the scale. It was evaluated to advance in polygeneration producing heat, power, and biofuel based on large-scale combined biomass gasification, solid oxide fuel cells, and biofuel production system [33].

7.4.3.3 Supplying More Air Stages

A gasifier having two air supply stages is also called a two-stage gasifier. The first air supply stage helps increase the temperature of the drying and pyrolysis zones. The second air injection supports the oxidation reactions. The ratio of the two air-stage flows can be weighed and optimized [34]. For multiple gasifying agents, a mixture of air and oxygen/steam in variety of fractions has also been investigated [35]. The use of three air supply stages for a throatless downdraft gasifier is presented in Figure 7.3. The primary air stage is located near a drying zone; the secondary and tertiary air stages are at the oxidation and reduction zones, respectively. The flow rate of each air stage was investigated. Bhattacharya et al. [28] reported that the lowest tar content of 28 mg/Nm^3 can be achieved. The primary air stage exhibited the influence on the reduction of tar in the producer gas; as if its amount is too small, the phenomena of three stages will be

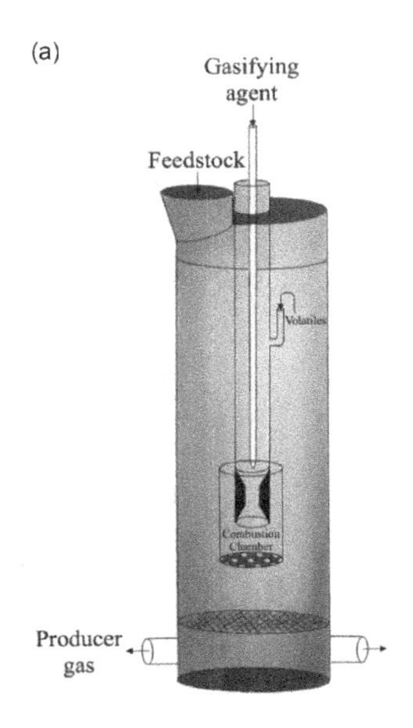

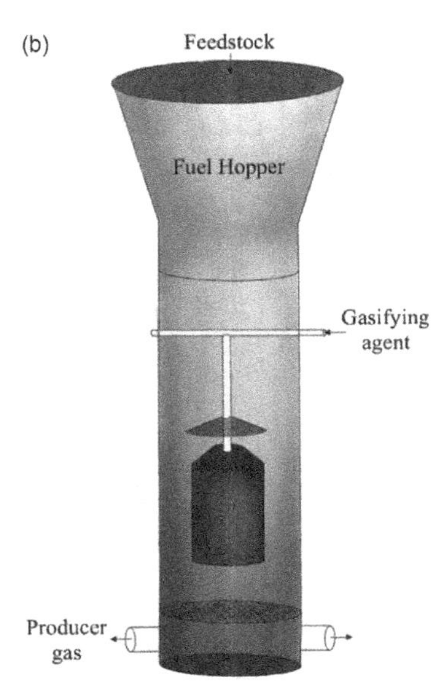

FIGURE 7.4 Schematic diagram of gasifier designs (a) with internal tar recycling, (b) LTB, (c) two-stage downdraft, (d) Imbert downdraft with gas recirculation, (e) Tarpo, (f) MIGH, and (g) GP750.

(Continued)

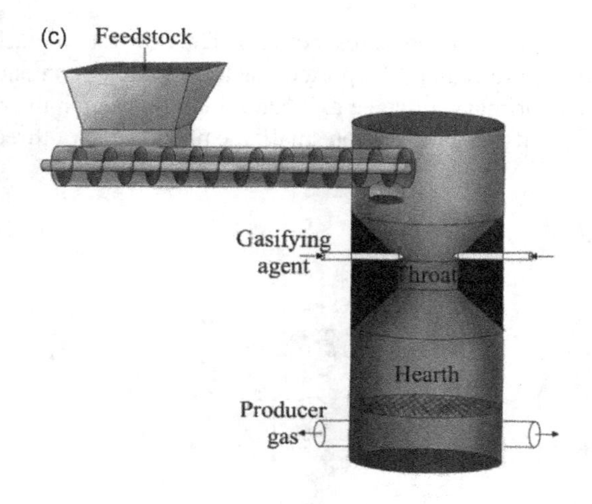

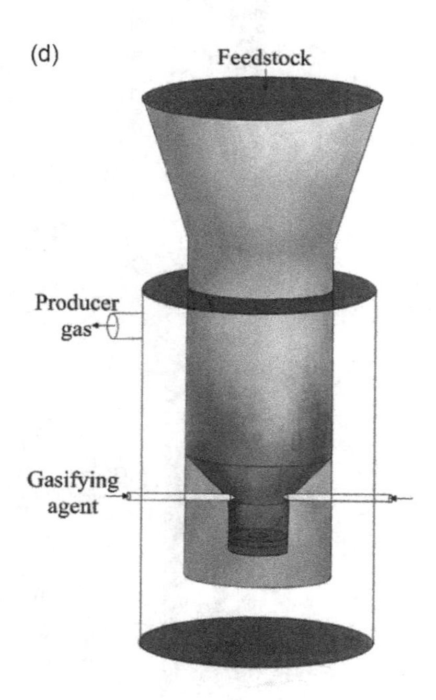

FIGURE 7.4 (*CONTINUED*) Schematic diagram of gasifier designs (a) with internal tar recycling, (b) LTB, (c) two-stage downdraft, (d) Imbert downdraft with gas recirculation, (e) Tarpo, (f) MIGH, and (g) GP750.

(*Continued*)

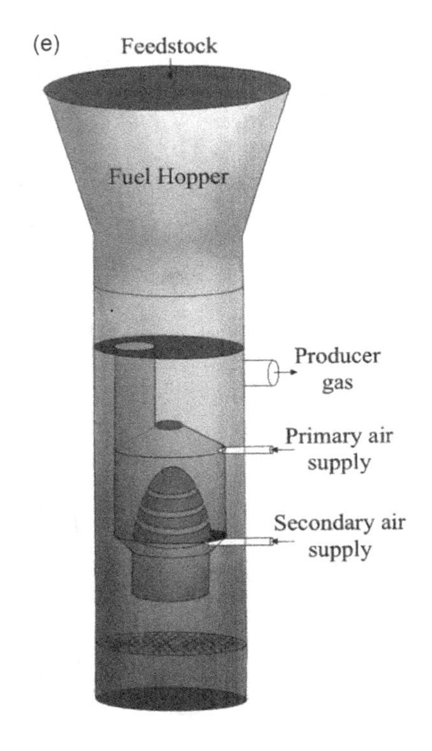

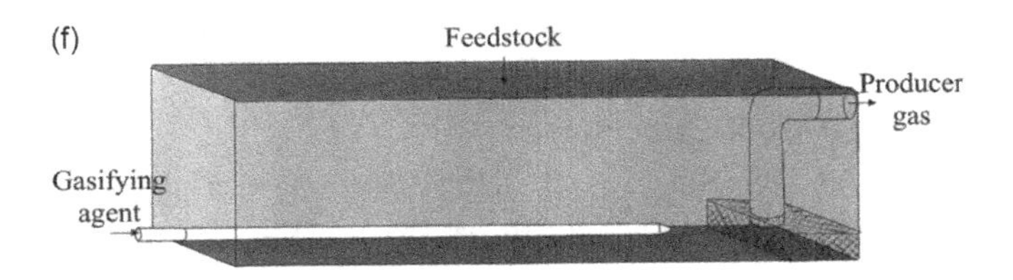

FIGURE 7.4 (*CONTINUED*) Schematic diagram of gasifier designs (a) with internal tar recycling, (b) LTB, (c) two-stage downdraft, (d) Imbert downdraft with gas recirculation, (e) Tarpo, (f) MIGH, and (g) GP750.

(*Continued*)

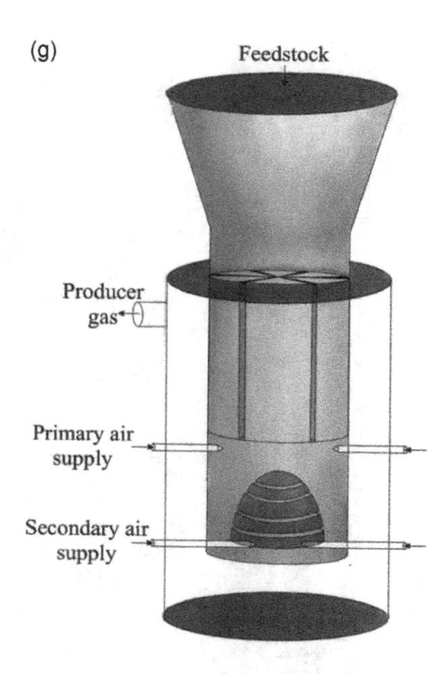

FIGURE 7.4 (*CONTINUED*) Schematic diagram of gasifier designs (a) with internal tar recycling, (b) LTB, (c) two-stage downdraft, (d) Imbert downdraft with gas recirculation, (e) Tarpo, (f) MIGH, and (g) GP750.

invisible, reducing to a two-stage gasifier. It appears that biomass feeding rate is a key parameter affecting the quality of producer gas. Despite tar reduction with increasing equivalence ratio, the gas yield increases with the increase in the feeding rate, and it affects the gas quality due to shorter gas residence time in the gasifier [30].

7.4.3.4 Adjusting Throat Diameter

Montuori et al. [22], Bunchan et al. [21], Prasertcharoensuk et al. [36], and Ngamsidhiphongsa et al. [37] studied the throat size to reduce the tar formation in the producer gas. The ratio of the throat to gasifier diameters ($D_{th/rx}$) was used as a normalized parameter. An optimum diameter ratio of 0.67 was obtained from the experiment of classic Imbert downdraft gasifier [21], while a value of 0.4 was achieved through a computational fluid dynamics modeling [36]. For the flat-plate downdraft gasifier, Ngamsidhiphongsa et al. [37] optimized the $D_{th/rx}$ value. It is in a very low range of about 0.15 because the opening of the flat-plate throat is very small compared to other throats. It can be concluded that the throat diameter is another major parameter affecting fuel gas quality and gasifier efficiency. A larger throat diameter can reduce a pressure drop in bed with greater bed stabilization; however, it reduces the cumulative efficiency of the gasification [22]. As a result, tar concentration increases [37].

7.4.3.5 Extending Reduction Zone Length

The extension of the reduction zone increases the cracking of the tar and gives a longer residence time to accelerate the Boudouard and water–gas reactions occurring in the gasification process. In the case of recirculating gas upward in the outer ring, the producer gas temperature may continue to reduce as a result of the increase in gasifier efficiency. As shown in Figure 7.4d, Machin et al. [10] claimed that the tar content in the producer gas was reduced to 10 mg/Nm^3. This can also be found in Ref. [38].

7.5 STATUS OF DOWNDRAFT GASIFIER DESIGNS

The advancement of small-scale gasification plants toward more sustainable and higher gasification efficiencies is still challenging, even though gasification technology is recognized as a well-known technology. This is a result of the evolution of the raw materials used and attempts to minimize the secondary tar reduction approaches such as a post-gas cleaning equipment. Novel designs of downdraft gasifiers have been developed, as discussed in the previous sections. Some exclusive designs can be found, as exemplified by the systems produced by Tarpo [39] and Wildfire Energy [40].

Moreover, the end-use technology extends not only to CHP and ICEs, but also to polygenerative systems or biorefineries [4]. This seems to be a present challenge in many types of research [8,33]. One very interesting by-product from the gasification process is the char, which has a well-developed porosity in the range of 40%–50% [1] with high carbon content and surface area, about 600 m^2/g. The disposal of char or its utilization directly as a soil amendment is not recommended because of its high metal and polycyclic aromatic hydrocarbon (PAH) contents [4]. However, it can be used as a valuable fuel, gas adsorption material, tar cracking catalyst, filtering medium, or polymer initiating material, hence reducing the disposal and discharging costs. This ensures that the downdraft gasifier is a powerful technology for sustainability, circular economy, and commercialization that is worth the investment.

7.5.1 Multi-stage Downdraft Gasifier by Tarpo

A multi-stage vertical downdraft gasifier was presented by Tarpo [39]. As shown in Figure 7.4e, dried biomass is constricted immediately to more than twofold the gasifier diameter. Then, the primary air injection is introduced at one side, forming an impasse of two branches where oxidation reactions occur. The other side transfers the pyrolysis product to the reduction zone mixed with the secondary injected gas and the gas produced from the oxidation zone, which slips through an inverted conical-shaped grate. The producer gas recirculates through the outer ring of the gasifier located at the middle part of the gasifier before exiting.

7.5.2 Moving Injection Horizontal Gasification (MIHG) Technology by Wildfire Energy

This small-scale novel gasifier was designed with a capacity of up to 400 kW_e, encouraging a circular economy. They claim that pretreatment and feeding systems are unnecessary for these raw materials. The MIHG [40], as shown in Figure 7.4f, consists of a batch horizontal rectangular reactor filled with waste feedstock, with a long air injection nozzle placed at the bottom. The air nozzle slowly departed in a horizontal direction from innermost as the bed of char occurs and fuel gas is produced.

7.5.3 GP750 Gasifier Design

This novel gasifier design (GP750) [41] was developed in 2020 (Figure 7.4g). This gasifier design is a combination of two air supply stages and twin-fired downdraft gasifiers with producer gas recirculation. It was designed for the polygeneration of low-tar high-quality producer gas and char. The gasifier was developed for a small-scale capacity (0.7 MW) with a tar content of 5–50 mg/Nm^3. It can be used directly in ICEs for heat and electricity generation and can produce biochar as a by-product.

7.6 CONCLUSIONS

The downdraft biomass gasifier shows high performance in producing the lowest tar in the producer gas compared to other types of gasifiers. The most up-to-date designs have progressed toward in situ tar reduction for ICEs and GTs. Modification of gasifier designs has shown less tar content in the producer gas, including designs for recycling the pyrolysis gas, separating the gasifier into two stages, supplying more air stages, adjusting the throat diameter, and extending the reduction zone. These proved to be successful prototypes for high-quality producer gas, which can be directly used in downstream engines. The last section showed the latest small-scale commercial plants using downdraft gasifiers. Innovative designs can be found for producing stable and high-quality producer gases while keeping the tar concentration adequately low and obtaining a satisfactory gasifier efficiency value. A novel design for a downdraft gasifier could also be established based on local raw materials such as biomass and waste, using the provided data as a basis. This would implement biomass gasification as a more commercial prospect.

ACKNOWLEDGMENTS

The support of the Thailand Science Research and Innovation Fund Chulalongkorn University (CU_FRB65_bcg (28)_142_21_08) and the Research Chair Grant (National Science and Technology Development Agency) is gratefully acknowledged. N. Ngamsidhiphongsa gratefully acknowledges the Royal Golden Jubilee Ph.D. Programme (Thailand Science Research and Innovation and National Research Council of Thailand).

REFERENCES

1. Basu, P. 2018. *Biomass Gasification, Pyrolysis and Torrefaction*. Academic Press, London, United Kingdom.
2. Thomson, R., Kwong, P., Ahmad, E., and Nigam, K.D.P. 2020. Clean syngas from small commercial biomass gasifiers; a review of gasifier development, recent advances and performance evaluation. *Int J Hydrog Energy* 45: 21087–21111.
3. Jafri, Y., Waldheim, L., and Lundgren, J. 2020. Emerging gasification technologies for waste & biomass. *IEA BioEnergy*: 1–80. http://www.ieatask33.org/app/webroot/files/file/publications/Emerging%20technologies/Emerging%20Gasification%20Technologies_final.pdf (accessed May 20, 2021).
4. Patuzzi, F., Basso, D., Vakalis, S., Antolini, D., Piazzi, S., Benedetti, V., Cordioli, E., and Baratieri, M. 2021. State-of-the-art of small-scale biomass gasification systems: An extensive and unique monitoring review. *Energy* 223: 120039.

5. Baratieri, M., and Prando, D. 2015. Biomass for polygeneration and district heating. In *Handbook of Clean Energy Systems*. John Wiley & Sons, Ltd, Chichester, United Kingdom.
6. Reed, T.B., and Das, A. 1988. *Handbook of Biomass Downdraft Gasifier Engine Systems*. The Biomass Energy Foundation Press, Golden, Colorado, USA.
7. Anis, S., and Zainal, Z.A. 2011. Tar reduction in biomass producer gas via mechanical, catalytic and thermal methods: A review. *Renew Sustain Energy Rev* 15: 2355–2377.
8. Ma, Z., Zhang, Y., Zhang, Q., Qu, Y., Zhou, J., and Qin, H. 2012. Design and experimental investigation of a 190 kWe biomass fixed bed gasification and polygeneration pilot plant using a double air stage downdraft approach. *Energy* 46: 140–147.
9. Erlich, C., and Fransson, T.H. 2011. Downdraft gasification of pellets made of wood, palm-oil residues respective bagasse: Experimental study. *Appl Energy* 88: 899–908.
10. Machin, E.B., Pedroso, D.T., Proenza, N., Silveira, J.L., Conti, L., Braga, L.B., and Machin, A.B. 2015. Tar reduction in downdraft biomass gasifier using a primary method. *Renew Energy* 78: 478–483.
11. Jayah, T.H., Aye, L., Fuller, R.J., and Stewart, D.F. 2003. Computer simulation of a downdraft wood gasifier for tea drying. *Biomass Bioenergy* 25: 459–469.
12. Sharma, A.K. 2009. Experimental study on 75 kWth downdraft (biomass) gasifier system. *Renew Energy* 34: 1726–1733.
13. Henriksen, U., Ahrenfeldt, J., Jensen, T.K., Gøbel, B., Bentzen, J.D., Hindsgaul, C., and Sørensen, L.H. 2006. The design, construction and operation of a 75 kW two-stage gasifier. *Energy* 31: 1542–1553.
14. van Loo, S., and Koppejan, J. 2008. *The Handbook of Biomass Combustion and Co-firing*. Twenty University Press, London, United Kingdom.
15. Channiwala, S.A., and Parikh, P.P. 2002. A unified correlation for estimating HHV of solid, liquid and gaseous fuels. *Fuel* 81: 1051–1063.
16. Kishore, V.V.N. 2010. *Renewable Energy Engineering and Technology Principles and Practice*. The Energy and Resources Institute (TERI), London, United Kingdom.
17. Siva Kumar, S., Pitchandi, K., and Natarajan, E. 2008. Modeling and simulation of down draft wood gasifier. *J Appl Sci* 8(2): 271–279.
18. Salem, A.M., and Paul, M.C. 2018. An integrated kinetic model for downdraft gasifier based on a novel approach that optimizes the reduction zone of gasifier. *Biomass Bioenergy* 109: 172–181.
19. Dogru, M., Howarth, C.R., Akay, G., Keskinler, B., and Malik, A.A. 2002. Gasification of hazelnut shells in a downdraft gasifier. *Energy* 27: 415–427.
20. Zainal, Z.A., Rifau, A., Quadir, G.A., and Seetharamu, K.N. 2002. Experimental investigation of a downdraft biomass gasifier. *Biomass Bioenergy* 23: 283–289.
21. Bunchan, S., Poowadin, T., and Trairatanasirichai, K. 2017. A study of throat size effect on downdraft biomass gasifier efficiency. *Energy Procedia* 138: 745–750.
22. Montuori, L., Vargas-Salgado, C., and Alcázar-Ortega, M. 2015. Impact of the throat sizing on the operating parameters in an experimental fixed bed gasifier: Analysis, evaluation and testing. *Renew Energy* 83: 615–625.
23. Garcia-Bacaicoa, P., Bilbao, R., Arauzo, J., and Salvador, M.L. 1994. Scale-up of downdraft moving bed gasifiers (25–300 kg/h) – Design, experimental aspects and results. *Bioresour Technol* 48: 229–235.
24. Olgun, H., Ozdogan, S., and Yinesor, G. 2011. Results with a bench scale downdraft biomass gasifier for agricultural and forestry residues. *Biomass Bioenergy* 35: 572–580.
25. Simone, M., Barontini, F., Nicolella, C., and Tognotti, L. 2012. Gasification of pelletized biomass in a pilot scale downdraft gasifier. *Bioresour Technol* 116: 403–412.
26. Yoon, S.J., Son, Y., Kim, Y., and Lee, J. 2012. Gasification and power generation characteristics of rice husk and rice husk pellet using a downdraft fixed-bed gasifier. *Renew Energy* 42: 163–167.

27. Rathore, N.S., Panwar, N.L., and Vijay Chiplunkar, Y. 2009. Design and techno economic evaluation of biomass gasifier for industrial thermal applications. *Afr J Environ Sci Technol* 3(1): 6–12.
28. Bhattacharya, S.C., Hla, S.S., and Pham, H.L. 2001. A study on a multi-stage hybrid gasifier-engine system. *Biomass Bioenergy* 21: 445–460.
29. Gautam, G., Adhikari, S., Thangalazhy-Gopakumar, S., Brodbeck, C., Bhavnani, S., and Taylor, S. 2011. Tar analysis in syngas derived from pelletized biomass in a commercial stratified downdraft gasifier. *BioResources* 6(4): 4652–4661.
30. Guo, F., Dong, Y., Dong, L., and Guo, C. 2014. Effect of design and operating parameters on the gasification process of biomass in a downdraft fixed bed: An experimental study. *Int J Hydrog Energy* 39: 5625–5633.
31. Susanto, H., and Beenackers, A.A.C.M. 1996. A moving-bed gasifier with internal recycle of pyrolysis gas. *Fuel* 75: 1339–1347.
32. Rahman, M.D.M., Henriksen, U.B., Ahrenfeldt, J., and Arnavat, M.P. 2020. Design, construction and operation of a low-tar biomass (LTB) gasifier for power applications. *Energy* 204: 117944.
33. Gadsbøll, R.Ø., Clausen, L.R., Thomsen, T.P., Ahrenfeldt, J., and Henriksen, U.B. 2019. Flexible TwoStage biomass gasifier designs for polygeneration operation. *Fuel* 166: 939–950.
34. Martínez, J.D., Lora, E.E.S., Andrade, R.V., and Jaén, R.L. 2011. Experimental study on biomass gasification in a double air stage downdraft reactor. *Biomass Bioenergy* 35: 3465–3480.
35. de Sales, C.A.V.B., Maya, D.M.Y., Lora, E.E.S, Jaén, R.L., Reyes, A.M.M., González, A.M., Andrade, R.V., and Martínez, J.D. 2017. Experimental study on biomass (eucalyptus spp.) gasification in a two-stage downdraft reactor by using mixtures of air, saturated steam and oxygen as gasifying agents. *Energy Convers Manage* 145: 314–323.
36. Prasertcharoensuk, P., Hernandez, D.A., Bull, S.J., and Phan, A.N. 2018. Optimisation of a throat downdraft gasifier for hydrogen production. *Biomass Bioenergy* 116: 216–226.
37. Ngamsidhiphongsa, N., Ponpesh, P., Shotipruk, A., and Arpornwichanop, A. 2020. Analysis of the Imbert downdraft gasifier using a species-transport CFD model including tar-cracking reactions. *Energy Convers Manage* 213: 112808.
38. Janajreh, I., and Shrah, M.A. 2013. Numerical and experimental investigation of downdraft gasification of wood chips. *Energy Convers Manage* 65: 783–792.
39. Brynda, J., Skoblia, S., Beňo, Z., Pohořelý, M., and Moško, J. 2016. Application of staged biomass gasification for combined heat and power production. In *Proceedings of the Environmental Technology and Innovations*: 143–148. CRC Press, Leiden, The Netherlands.
40. Wildfire Energy. 2001. MIHG gasification technology. http://www.wildfireenergy.com.au/mihg-technology (accessed May 20, 2021).
41. Brynda, J., Skoblia, S., Pohořelý, M., Beňo, Z., Soukup, K., Jeremiáš, M., Moško, J., Zach, B., Trakal, L., Šyc, M., and Svoboda, K. 2020. Wood chips gasification in a fixed-bed multi-stage gasifier for decentralized high-efficiency CHP and biochar production: Long-term commercial operation. *Fuel* 281: 118637.

8 Food-Based Waste for Energy

Shadab Shahsavari
Islamic Azad University

Zahra Shokri
Tehran University of Medical Sciences

Gita Bagheri
Islamic Azad University

CONTENTS

DOI: 10.1201/9781003178354-10

8.1 INTRODUCTION

Access to alternative energy sources has recently been investigated due to serious global problems. The reduction in the fossil fuel supply due to global population growth and subsequent increase in energy demand, worldwide climate change due to the augmentation of CO_2 amount in the atmosphere, and the increasing amount of wastes due to rapid industrialization, urbanization, and population growth have strongly menaced the world stability. The conversion of food waste into renewable energy may reduce greenhouse gas emissions and prevent climate change and global warming. Different types of food wastes involving industrial (dairy wastes, sugar refinery wastes, confectionary wastes, vegetable and fruit residues, and wastes from tanneries and slaughterhouses), agricultural (animal and plant wastes), and residential wastes (garden and kitchen wastes) have the potential for conversion into renewable energy sources. However, food waste-to-energy processes have faced various challenges and drawbacks due to high moisture contents, low carbon-to-nitrogen ratio, low calorific value, high heterogeneous composition, low volatile solids, and deficient waste management, which should be further investigated for efficient conversion treatment and yield enhancement [1]. In this chapter, various methods used for the conversion of food wastes into beneficial energy resources are reviewed, and in the following sections, the useful end products from food waste are categorized.

8.2 CURRENT CONVERSION TECHNOLOGIES FOR WASTE TO ENERGY

8.2.1 Biological Technology

8.2.1.1 Composting

In composting, thermophilic microorganisms are applied to naturally decomposing the organic wastes under an aerobic biochemical treatment into biocompost products that can be utilized as fertilizers. However, composting does not generate methane. The microbial activity in the composting process is affected by various parameters such as temperature, moisture content, pH, oxygen concentration, particle size, and carbon-to-nitrogen ratio. Among the organic wastes, food waste has special characteristics such as high moisture and organic content and low C/N ratio, which make it appropriate for composting. On the other hand, the composting process has various disadvantages such as greenhouse gases and odor emission, which threaten the environment and restrict the extensive use of this method in some countries [2].

8.2.1.2 Anaerobic Digestion

Anaerobic digestion (AD) is one of the oldest biological treatments in which organic wastes are decomposed by microorganisms under anaerobic conditions and turned into biogas mainly composed of methane (CH_4), carbon dioxide (CO_2), and a small amount of H_2S, H_2, and N_2. Food and agricultural wastes are an ideal substance for anaerobic digestion because they consist of destructible lipids, carbohydrates, and proteins that can be stabilized and transformed to biogas, which can provide electrical and thermal energy. AD process consists of a series of biological metabolism steps, including hydrolysis, acidogenesis, acetogenesis, and methanogenesis. In the hydrolysis step, the insoluble complex matters (lipids, carbohydrates, and proteins) are disjointed into simple molecules (long-chain fatty acids, glucose, and amino acids) by fermentative bacteria and enzymes. Hence, hydrolysis is a relatively slow step and it can determine the rate of the AD process of food waste. In acidogenesis step, dissolved composition and monomers such as fatty acids, sugars, and amino acids obtained from hydrolysis are decomposed into smaller molecular weight molecules such as volatile fatty acids (VFAs) (i.e., propionic acid, acetic acid, and butyric acid), alcohols, and various kinds of gases (CO_2, H_2, and NH_3). Subsequently, in acetogenesis the organic materials resulting from the acidogenesis step are converted into H_2, acetate, CO_2, etc., which will be used by methanogens. The final stage in the AD is methanogenesis, in which the methanogens use acetic acid, carbon dioxide, and hydrogen to generate biogas including 60%–70% of methane and nitrogen-rich fertilizer. AD of food waste encounters many economical and technical challenges, such as high capital costs, VFA accumulation, severe pH drop, low buffer volume, foaming, and process instability, which prevents its widespread use. To overcome these problems and to improve the efficiency of the AD systems, the commonly employed methods are co-digestion, various process designs (two-/multi-stage process), and the addition of micronutrients and antifoaming agents [3]. Pramanik et al. [4] reported that co-digestion, various pre-treatment techniques, and two-stage AD of food waste could help to improve the stability of the process and the yield of methane production.

8.2.1.3 Fermentation

Hydrogen and alcohol production could be carried out by the fermentation of food waste rich in carbohydrates. Fermentation of food waste consists of three steps. First, in hydrolysis (saccharification), the raw material is decomposed into glucose, then in anaerobic fermentation, glucose is turned into carbon dioxide and ethanol, and finally, the ethanol is separated and purified by distillation [5]. Ethanol fermentation by yeast is a biochemical process in cells, and the oxidation–reduction potential should be in the optimum range [6]. Since the food waste is a complex combination of cellulosic, starchy, and lipid compounds that are so difficult to be fermented by microorganisms, some pre-treatments are usually done to transform complex sugars to simple mono- or oligosaccharides, which are subsequently fermented to ethanol. The pre-treatment could be heat treatment, acid hydrolysis, and enzymatic hydrolysis.

8.2.2 Thermal and Thermochemical Technology

8.2.2.1 Incineration

In incineration, burning waste feedstock leads to heat production in heat exchangers or turbines. During incineration, the solid wastes reduce in their volume by up to 80%. Incineration is considered a widespread method of waste disposal in municipal solid waste treatments. The transportation costs are reduced by incineration because the facilities and equipment of incineration can be installed near the accumulation area. Another advantage of incineration is that waste reduction can be performed immediately, while other methods of waste reduction take more time because of the biological reaction time. However, it produces a considerable amount of incombustible ash and also toxic emissions, which contain heavy metals and dioxins and cause air pollution and many health problems. Therefore, in some countries incineration is not entirely accepted [7].

8.2.2.2 Pyrolysis

Pyrolysis process is an effective method of energy generation from food waste since it needs carbon-based wastes. The food waste comprises about 80% of volatile components that can easily be converted by pyrolysis at temperatures between 700 °C and 1,000 °C in approximately 8–12 min. The pyrolysis consists of thermal conversion of materials without an oxidizing agent at the temperature of about 400 °C–900 °C, which results in an irreversible degradation of polymer into small molecules and finally leads to the production of gaseous products (syngas), liquid (bio-oils or tar), and solid (char) according to the process conditions [1]. Pyrolysis is categorized as follows based on the organic waste conversion rate: (i) slow pyrolysis, (ii) fast pyrolysis, (iii) flash pyrolysis, and (iv) catalytic pyrolysis [8]. Slow pyrolysis leads to more solid production (char) and a little amount of pyrolysis oil and gas, and the heating rate is about 6 °C/min. The heating rate in fast pyrolysis is about 300 °C/min, which provides pyrolysis oil of high efficiency. In flash pyrolysis, the reaction takes place only in several seconds and the heating rate is higher than that of the other types of pyrolysis [9].

8.2.2.3 Gasification

In gasification, the biomass containing hydrogen and carbon (such as food waste) is oxidized and converted into gaseous products by a gasifying agent (oxygen or air) at high temperatures of about 800 °C–1,100 °C [10]. The mechanism of gasification includes four stages: (i) dehydration, (ii) pyrolysis, (iii) oxidation, and (iv) gasification. The final product named syngas mostly consists of hydrogen (H_2), carbon dioxide (CO_2), carbon monoxide (CO), water (H_2O), and also a small amount of methane (CH_4). To enrich the syngas with hydrogen, catalytic processes have been researched, and Wu et al. [11] reported the significant effect of Ni-based catalysts on hydrogen production in biomass gasification. The design of the gasifier can be categorized into three principal types: fixed bed gasifier, fluidized bed gasifier, and entrained suspension bed gasifiers. Opatokun et al. [12] investigated the pyrolysis of dry raw food waste and also biological anaerobic digested food waste. They concluded that both substrates showed significant potential for fast gasification due to volatile components. Energy content for both substrates was mainly diffused into bio-oil and biochar, while syngas provided lower energy.

8.2.2.4 Plasma Treatment

Plasma technology is a relatively novel concept in waste management and has recently been applied for the conversion of municipal solid wastes into energy. Plasma is produced by enforcing gaseous molecules to the generation of charged particles via a severe energy field. Plasma-based systems are considered an expensive method due to electricity usage in plasma generation. Therefore, its application is reasonable only when more valuable end products can be produced. Plasma-based systems are divided into plasma gasification and plasma pyrolysis [13]. In plasma pyrolysis, the thermal decomposition of feedstock is performed without oxygen, while in plasma gasification, a limited quantity of steam and oxygen is present. In both methods, the heat source is plasma and the syngas containing mainly carbon monoxide and hydrogen is the end product. Since the plasma-based process restrains the formation of char, soot, tar, and toxic gases, the syngas generated from plasma gasification or plasma pyrolysis is so cleaner than that generated from common gasification or pyrolysis.

8.2.2.5 Hydrothermal Carbonization

Hydrothermal carbonization (HTC) is another thermal technique applied to convert food wastes and related materials to an energy-rich and valuable resource. HTC is a beneficial method for converting high moisture content waste streams (such as food waste) to hydrochar. Since HTC avoids the consumption of severe energy in the drying process, it is considered the highly efficient treatment of waste to energy [1]. In comparison with other processes, HTC is a suitable process for carbon decomposition to discount climate changes. In anaerobic digestion or fermentation process, some compounds in biomass are converted to CO_2 and result in emissions, while in HTC, the original carbon of food waste stays in the hydrochar product [14]. Compared to biological processes, HTC has several benefits such as more waste capacity reduction, less treatment footprints, shorter processing time, elimination of pathogens, and no process odors. HTC of food waste has been investigated with various substrates such as rice bran [15], leftover food [16], sweet corn [17], grape, sweet potato, pomelo peels [18], sugar beet pulp [19], and lots of other food wastes. These researches were carried out in various operational conditions at the temperature of 200 °C–350 °C and time resistance of 0.2–120 h. The results illustrated that HTC of food wastes is an effective treatment for the production of hydrochar with high carbon amount (45%–93% of initial carbon) and high energy content (15–30 kJ/g dry solids).

8.2.3 Transesterification (Esterification)

In transesterification, used oils and fats are converted into biodiesel by recycling polyesters to the individual monomers [20]. In transesterification, triglycerides of fatty acids and alcohols accompany a catalyst to produce monomethyl esters called biodiesel and glycerol as by-products [21]. The amount of free fatty acid (FFA) of feedstock, the type and quantity of alcohol and catalyst, temperature, and reaction time can significantly influence biodiesel production [22]. The type of catalyst in transesterification treatment not only depends on the composition of FFA of the feedstock, but also determines the composition of the final biodiesel product [23]. Generally, two categories of catalysts, i.e., chemical or enzymatic, are applied in the transesterification reaction. However, using a chemical catalyst resulted in high yield. The process faces many disadvantages such as great energy consumption, difficulty in separation of water and inorganic salt from the product and recovery of glycerol downstream, complex alkaline wastewater treatment, and extra cost [24].

8.2.4 Bioelectrochemical Systems

In bioelectrochemical systems (BESs), microorganisms are applied as biochemical catalysts for converting complex substrates to energy products. The BESs act like a battery system since a biofilm is created on electrodes, which then oxidizes substrates and finally generates electricity or energy-rich gases [25]. The basic principle in bioelectrochemical systems is redox potential. Electrochemical reactions are carried out by a particular kind of bacteria called exoelectrogens, which have the ability to transform electrons beyond the microbial cell [26].

The usage of BES in food waste treatment mainly focuses on two methods: microbial fuel cells (MFCs) and microbial electrolysis cells (MECs). Typically, MFCs consist of two electrodes named anode and cathode, which are connected externally to a resistor and separated via a membrane. In MFCs, the generation of electric current and substrate removal occur simultaneously [27]. Lots of food waste feedstock have been treated by MFCs, such as brewery and winery wastewater; cafeteria and canteen wastes; dairy industry wastes and cheese whey; fruits, vegetables, and food wastes; animal processing and meat industry wastes; sugar refinery and distillery wastewaters; and wastes from seafood industry and edible oil industry. Among them, brewery and sugar refinery wastewaters demonstrate higher power generation from MFCs, and the others may have better results if accompanied with a pre-treatment process such as fermentation [28]. MECs are another type of BESs in which the external power source catalyzes the substrate to by-products involving H_2, CH_4, and H_2O_2 [29].

8.3 USEFUL PRODUCTS FROM FOOD WASTE

8.3.1 Gaseous-State Products

8.3.1.1 Biogas (Biomethane)

Biogas is the final product of anaerobic digestion processes, in which the organic feedstock is converted in the absence of oxygen. It generally contains 65% methane, 35% carbon dioxide, a low amount of hydrogen sulfide, and water vapor. Organic biodegradable materials such as food waste, if digested anaerobically, will not only generate a noticeable amount of biogas and manure, but also decrease the load at the landfill and prevent environmental pollution. During the recent decades, some worth researches concerning the conversion of food waste to biogas have been performed. Bouallagui et al. [30] investigated both vegetable and fruit wastes together for biogas generation similar to studies performed on rotten vegetables, fruit skins, potatoes, onion, and household solid waste. The biogas cannot be converted into the liquid phase at ambient temperature. Carbon dioxide is eliminated from the biogas, while the biogas is applied as a fuel so that the energy of the biogas is augmented. In fact, after the elimination of carbon dioxide, the enriched biogas can be easily stored and compressed into cylinders by means of easy transportation. In other words, the enriched biogas can be used instead of CNG [31]. Both natural gas and biogas are generated under anaerobic conditions, and methane is the main component. While the production of natural gas from dead biomass took million years, the biogas production from food wastes takes about 20 days. On the other hand, using biogas has some disadvantages. The atmosphere is highly polluted when burning biogas in comparison with burning natural gas. Also since the biogas contains siloxanes, silica deposits that cause damage to the engines are produced. Methane emissions are another drawback of the biogas in terms of greenhouse gas [32].

8.3.1.2 Synthetic Gas (Syngas)

Syngas is a specific gas mixture generated from the pyrolysis or gasification (thermochemical conversion of different fuels such as coals, wastes, or biomass). The actual combination of the syngas depends on the raw material, temperature and pressure condition, gasifying medium, and the retention time of feed in the defined gasification zones. Syngas can be applied for electricity generation or for eliciting hydrogen that is intended as a future fuel. It can also be used for producing some chemicals such as ammonia and methanol in the presence of a catalyst (hydrogen) in high-pressure and high-temperature conditions. Fischer–Tropsch synthesis can transform syngas into synthetic oil, such as diesel/kerosene, wax, naphtha, and gasoline. The production of syngas from biomass and organic waste has recently been considered. Syngas is generated either via direct gasification or through a two-stage process that consists of converting biomass or waste into bio-oil by pyrolysis process and subsequently gasifying it into the form of syngas. Ko et al. [33] investigated the thermal destruction of food waste at 200 °C. The carbonized solid consists of 59% carbon which is produced with an efficiency of 26%–43%, enter into a fluidized bed-type gasifier and the affected parameter for optimum conditions examined. The temperature range and the amount of oxygen and steam are the most effective parameters in syngas production. The temperature of the reaction affected the composition and the rate of syngas production. Nanda et al. [34] studied different food product residues and vegetable and fruit wastes such as sugarcane bagasse, lemon peel, orange peel, aloe vera rind, banana peel, pineapple peel, and coconut shell to generate syngas with a high amount of hydrogen via supercritical water gasification under high pressure of 23–25 MPa, high temperature of 400 °C–600 °C, biomass-to-water ratio of 1:5 and 1:10, and reaction time of 15–45 min. The results show that coconut shell generated a higher hydrogen efficiency (4.8 mmol/g) and syngas yield (15 mmol/g). Orange peels also had a higher hydrogen efficiency (0.91 mmol/g) and syngas yield (5.5 mmol/g). These significant results indicate that supercritical water gasification was advantageous in producing bioenergy from fruit waste and food residues.

8.3.1.3 Biohydrogen

Hydrogen, which is a sustainable source of energy with minimal hydrocarbons, has been considered a promising alternative for fossil fuels recently. It is an environmentally friendly and clean fuel that generates water instead of greenhouse gases during combustion. The heating value of hydrogen is 61,000 Btu/lb, while the heating value of methane is 23,879 Btu/lb, approximately one-third of hydrogen.

Hydrogen can be applied in fuel cells and converted directly into electricity. Conventional reforming methods of hydrogen production are steam reforming of hydrocarbons such as methane, thermal reforming and oxidation of fuels, desulfurization, plasma reforming, pyrolysis, ammonia reforming, and aqueous reforming. Also, hydrogen can be generated from renewable sources such as biomass and food waste. Among these techniques, hydrocarbons steam reforming, electrolysis of water, and autothermal processes are famous processes for hydrogen production, but they are not cost-effective due to high energy demand. On the other hand, nitrogen-deficient and carbohydrate-rich solid wastes such as cellulose- and starch-containing food wastes and some wastewaters of the food industry such as cheese whey, baker's yeast, and olive mill can be applied for hydrogen production via appropriate bioprocess technologies. Hydrogen production through biological methods is categorized into four groups: (i) direct biophotolysis, (ii) indirect biophotolysis, (iii) photofermentation, and (iv) dark fermentation. In all the mentioned methods, hydrogen-producing enzymes, hydrogenase, and also nitrogenase control the hydrogen production process. Dark fermentation is a commonly applied technology due to its commercial values. It proposes a significant potential in both practical uses and merging with emerging hydrogen in fuel cells [35,36]. During dark fermentation, the organic carbohydrate-rich matters are converted to biohydrogen by anaerobic bacteria which, as the name implies, are activated in the dark. Enterobacter, Clostridium, and Bacillus are the famous anaerobic bacteria commonly applied in dark fermentation for biohydrogen production. Liu et al. [37] employed a two-stage fermentation of food waste to generate hydrogen and methane, in mesophilic conditions. The highest yield of H_2 (106.4 mL H_2/g VS) and CH_4 (353.5 mL CH_4/g VS) was obtained under the food waste proportion of 85%.

8.3.2 Liquid-State Products

8.3.2.1 Biodiesel

Biodiesel or fatty acid methyl esters are produced via the transesterification reaction. Biodiesel is considered an alternative fuel that can replace petroleum diesel fuels. Since biodiesel production is an extensive method, the cost-effective feedstock should be attributed, which leads to an economically feasible process. It seems the application of lipids gained from the food waste industry, such as vegetable oils, animal fats, or waste cooking oil, as a low-cost resource for the production of biodiesel is logical and economical. The lipids obtained from food waste are transesterified with alcohol (ethanol, methanol, butanol, etc.) in the presence of lipase as catalysts. During the transesterification process, the mono-, di-, or triglycerides of animal fats or vegetable oils are transformed to fatty acid alkyl esters called biodiesel as the main product and also glycerin as the by-product. Biodiesel is produced from food waste in two steps, initially extracting lipids of food waste and then transesterification [38]. Food waste usually contains lipids in the range of 5%–30%, which can be isolated via evaporation, solvent extraction, enzymatic hydrolysis, or supercritical extraction and applied as a potential resource to biodiesel production. The authors of Ref. [39] investigated the use of lipids from food waste as a feedstock for biodiesel production. The results showed 100% biodiesel production yield when using KOH as a catalyst and a ratio of 1:10 (lipid to methanol) in 2 h and 90% biodiesel production yield with a ratio of 1:5 (lipid to methanol) at 40 °C using lipase in 24 h.

8.3.2.2 Bioethanol

Bioethanol or bioethyl alcohol is a reproducible liquid fuel that can be generated from different biomass feedstocks. However, the major resource of bioethanol production is sugarcane and corn, and food waste is considered the cheaper feedstock compared to other resources currently being applied. Bioethanol is the most used biofuel alcohol that is obtained by treating liquid and semisolid food waste containing carbohydrates. Similar to ethanol, all bioalcohols are produced by enzymes and microorganisms via the fermentation of starches, sugars, or cellulose. Methanol is currently extracted from natural gas. It is also extracted from food waste or biomass through the fermentation process. The syngas produced from the gasification process can be transformed into ethanol or other chemicals such as methane, methanol, heavy waxes, and acetic acid via the Fischer–Tropsch process, which is expensive since it needs a separate reactor for each stage. The other way is using microorganisms that can transform syngas into bioethanol. In this procedure, at first, syngas is passed through several filters for removing tar and some solid particles and more fermented by microorganisms as catalysts. Various products are obtained, such as organic acids (lactic, acetic, propionic, formic, and butyric acids) and alcohols (ethanol, methanol, butanol, and propanol). The parameters affecting the process can be optimized to maximize the ethanol production. This method is more advantageous than the fermentation process in terms of efficiency and less toxic component production. In an enzymatic reaction, the amount of CO or H_2 determines the yield of ethanol production [40].

8.3.2.3 Pyrolysis Oil (Bio-Oil)

Bio-oil is a complex combination of oxygenated hydrocarbons, char particles, and water. Bio-oil can be generated as a stable emulsion, detachable aqueous liquid, or oil phase, based on the raw material combination and the conditions of the pyrolysis process. The main products of pyrolysis are syngas, biochar, and bio-oil. Slow pyrolysis is usually used for charcoal production, whereas fast pyrolysis is applied for bio-oil production. The distribution of products in slow pyrolysis is 35% biochar, 35% syngas, and 30% bio-oil, while in fast pyrolysis, 12% biochar, 13% syngas, and 75% bio-oil are produced. During fast pyrolysis, food wastes or biomass decompose very fast; therefore, mostly aerosols and vapor composition are formed instead of gas and charcoal and, when it cooled and condensed, a viscous dark brown liquid called bio-oil is formed. Kelkar et al. [41] investigated the fast pyrolysis process of spent coffee grounds by using a conveyor screw reactor for bio-oil production. A residence time of 23–42 s and a reactor temperature of 429°C–550°C were applied. The highest yield of bio-oil of 61.7% was obtained at a residence time of 23 s and temperature of 500°C. The results demonstrated that the yield of bio-oil will increase with increasing screw speed and decreasing residence time.

8.3.3 Solid-State Products

8.3.3.1 Biochar (Hydrochar)

Biochar has a high amount of carbon and is produced in the pyrolysis process. Biochar provides various preferences over syngas and bio-oil [37]. It can be considered a soil conditioner similar to fertilizer. It can improve agricultural fertility, particularly in the case of contemptible soils. It helps reduce climate change by increasing sustainable soil carbon reserves and carbon sequestration of soil due to a reduction in the CO_2 concentration in the atmosphere. Biochar also helps to recover soil water storage capacity. The production of biochar is much more efficient than composting in terms of carbon lock. The carbon of compost is decomposed by microbial reaction in 10–20 years, while the sequestered carbon of biochar remains sustainably in the soil. The reduction in greenhouse gas emissions in comparison with nitrous oxide and methane is another advantage of using bio-oil. On the other hand, bio-oil requires further upgrading before usage due to its complex composition and high acidity. A high yield of biochar production is obtained using a low heating rate and low operational temperature. In fast pyrolysis, the cellulose and hemicellulose resources are converted to syngas and bio-oil (the main products) with a low amount of biochar production. On the contrary, in slow pyrolysis, high-lignin resources are converted to biochar with high efficiency. Hence, the selection of resources and required products determine whether slow pyrolysis is applied or not [42]. Hydrochar is also produced via hydrothermal carbonization at autogenous pressures. This process is proposed especially when wet biomass such as food waste needs to be treated. The final product is a homogeneous watery substance that is porous and brittle and called hydrochar. McGaughy and Reza [43] investigated the hydrothermal carbonization process of homogenized food waste using a batch reactor at temperatures of 260 °C, 230 °C, and 200 °C for a reaction time of 30 min. The final product was hydrochar with a heating value of 33.1 MJ/kg, a sulfur content of <0.5%, and an ash content of 1%–2%. According to the results, the yield of hydrochar production differs from 68% to 75% depending on the process temperature, and the lower yield was obtained at 260 °C and the higher yield was obtained at 200 °C.

8.3.3.2 Compost

Compost is generated from organic residues and wastes during the composting process. This biodegradation treatment starts initially with microorganisms such as fungi, bacteria, or molds under the temperature of ambient temperature to 70 °C and oxygen level of less than 5% and takes several weeks to months to complete. The optimum conditions for compost production are a carbon-to-nitrogen ratio of 20–40, natural pH, a moisture content of 50%–65%, sufficient oxygen supply, enough void space of airflow, and a large particle size. The carbon-to-nitrogen ratio of food waste is lower than 2; thus, using it alone is not appropriate for compost production. Moreover, the moisture content of food waste is about 70%–90%, which is too high to produce compost. As a result, other organic matters should be retained in balance for obtaining the optimum moisture content and C/N ratio [44]. Consequently, the food waste is usually co-composted with the other types of wastes such as green wastes and paper wastes to correct the C/N ratio.

8.4 CONCLUSIONS

In summary, food wastes are considered a promising feedstock for producing useful energy resources. Various conversion techniques for converting food waste into some beneficial products are reviewed. Each method has its advantages and also limitations.

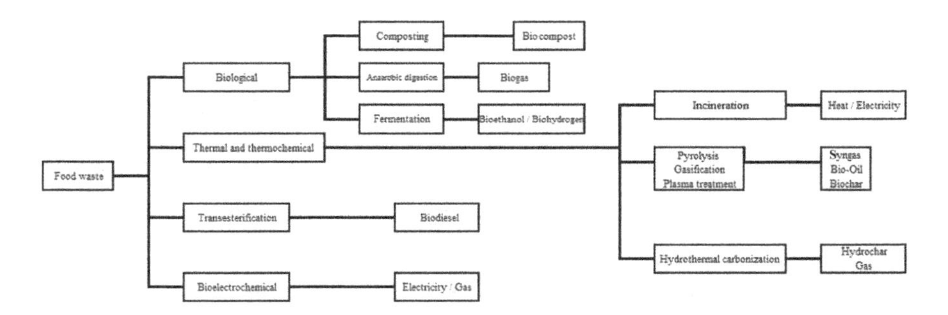

FIGURE 8.1 Summary of food waste-to-energy technologies and the useful products.

Figure 8.1 illustrates the summary of food waste-to-energy technologies and the useful products. Although composting of food wastes needs a low initial cost, it leads to greenhouse gases and odor emission and, also, it is not energy efficient due to the lack of heat or electricity generation. Biological treatments (fermentation and anaerobic digestion) are simple and of low cost compared with other processes; however, long reaction time, the necessity of some pre-treatments, and process-related odors are some drawbacks of this technology. The incineration process converts food wastes directly to heat, while in other thermal processes such as gasification, pyrolysis, and plasma technologies, initially the food wastes are converted into an intermediate energy carrier which is burnt to generate heat. This procedure leads to the efficient and clean generation of electricity which has more market acceptability. But these mentioned thermal processes generate less calorific values because of the high moisture content of food waste; thus, they are energetically unfavorable. Gasification and pyrolysis are desirable in terms of reduction in operating costs and also CO_2 emissions in comparison with incineration. In plasma technology, electricity is applied as the heat resource; thus, temperatures can be very high. Plasma treatment has various advantages including rapid process control and high system flexibility, but on the contrary, is considered an expensive method. The hydrothermal carbonation process takes less than 60 min, which is much shorter than anaerobic digestion treatment. Nevertheless, the product energy content in hydrothermal carbonation should be offset by a higher amount of input heat. Also, the utilization of efficient catalysts is necessary for applying low temperature and pressure. Generally, food waste is considered a promising resource for converting into useful products. It seems more researches are necessary for optimizing the cost and conditions of the process, maximizing yields, and minimizing reaction time. These technologies should improve to generate valuable energy resources and to solve the food waste disposal issues simultaneously.

REFERENCES

1. Pham, T.P.T., et al., Food waste-to-energy conversion technologies: Current status and future directions. *Waste Management*, 2015. **38**: pp. 399–408.
2. Melikoglu, M., C.S.K. Lin, and C. Webb, Analysing global food waste problem: Pinpointing the facts and estimating the energy content. *Central European Journal of Engineering*, 2013. **3**(2): pp. 157–164.

3. Ariunbaatar, J., et al., Pretreatment methods to enhance anaerobic digestion of organic solid waste. *Applied Energy*, 2014. **123**: pp. 143–156.
4. Pramanik, S.K., et al., The anaerobic digestion process of biogas production from food waste: Prospects and constraints. *Bioresource Technology Reports*, 2019. **8**: p. 100310.
5. Saeed, M.A., et al., Concise review on ethanol production from food waste: Development and sustainability. *Environmental Science and Pollution Research*, 2018. **25**(29): pp. 28851–28863.
6. Ma, H., et al., Stillage reflux in food waste ethanol fermentation and its by-product accumulation. *Bioresource Technology*, 2016. **209**: pp. 254–258.
7. Dahl, R., A second life for scraps: making biogas from food waste. 2015, NLM-Export.
8. Guran, S., F.A. Agblevor, and M. Brennan-Tonetta, Biofuels, bio-power, and bio-products from sustainable biomass: Coupling energy crops and organic waste with clean energy technologies. *Emerging Areas in Bioengineering*, 2018. **1**: pp. 127–161.
9. Kalyani, K.A. and K.K. Pandey, Waste to energy status in India: A short review. *Renewable and Sustainable Energy Reviews*, 2014. **31**: pp. 113–120.
10. Ruiz, J.A., et al., Biomass gasification for electricity generation: Review of current technology barriers. *Renewable and Sustainable Energy Reviews*, 2013. **18**: pp. 174–183.
11. Wu, C., et al., Hydrogen production from biomass gasification with Ni/MCM-41 catalysts: Influence of Ni content. *Applied Catalysis B: Environmental*, 2011. **108**: pp. 6–13.
12. Opatokun, S.A., V. Strezov, and T. Kan, Product based evaluation of pyrolysis of food waste and its digestate. *Energy*, 2015. **92**: pp. 349–354.
13. Ruj, B. and S. Ghosh, Technological aspects for thermal plasma treatment of municipal solid waste—A review. *Fuel Processing Technology*, 2014. **126**: pp. 298–308.
14. Titirici, M.-M., A. Thomas, and M. Antonietti, Back in the black: Hydrothermal carbonization of plant material as an efficient chemical process to treat the CO_2 problem? *New Journal of Chemistry*, 2007. **31**(6): pp. 787–789.
15. Sugano, M., et al., Extraction of valuable compounds from hydrothermally reacted rice bran and wheat bran. *Waste and Biomass Valorization*, 2012. **3**(4): pp. 381–393.
16. Wiedner, K., et al., Chemical evaluation of chars produced by thermochemical conversion (gasification, pyrolysis and hydrothermal carbonization) of agro-industrial biomass on a commercial scale. *Biomass and Bioenergy*, 2013. **59**: pp. 264–278.
17. Lu, X. and N.D. Berge, Influence of feedstock chemical composition on product formation and characteristics derived from the hydrothermal carbonization of mixed feedstocks. *Bioresource Technology*, 2014. **166**: pp. 120–131.
18. Wu, S.-C., et al., A hydrothermal synthesis of eggshell and fruit waste extract to produce nanosized hydroxyapatite. *Ceramics International*, 2013. **39**(7): pp. 8183–8188.
19. Cao, X., et al., Effects of biomass types and carbonization conditions on the chemical characteristics of hydrochars. *Journal of Agricultural and Food Chemistry*, 2013. **61**(39): pp. 9401–9411.
20. Thamsiriroj, T. and J. Murphy, How much of the target for biofuels can be met by biodiesel generated from residues in Ireland? *Fuel*, 2010. **89**(11): pp. 3579–3589.
21. Knothe, G. and L.F. Razon, Biodiesel fuels. *Progress in Energy and Combustion Science*, 2017. **58**: pp. 36–59.
22. Verma, P. and M. Sharma, Review of process parameters for biodiesel production from different feedstocks. *Renewable and Sustainable Energy Reviews*, 2016. **62**: pp. 1063–1071.
23. Lam, M.K., K.T. Lee, and A.R. Mohamed, Homogeneous, heterogeneous and enzymatic catalysis for transesterification of high free fatty acid oil (waste cooking oil) to biodiesel: A review. *Biotechnology Advances*, 2010. **28**(4): pp. 500–518.
24. Du, W., et al., Study on acyl migration in immobilized lipozyme TL-catalyzed transesterification of soybean oil for biodiesel production. *Journal of Molecular Catalysis B: Enzymatic*, 2005. **37**(1–6): pp. 68–71.

25. Wang, J., et al., A bibliometric review of research trends on bioelectrochemical systems. *Current Science*, 2015: pp. 2204–2211.
26. Sun, D., et al., Geobacter anodireducens sp. nov., an exoelectrogenic microbe in bioelectrochemical systems. *International Journal of Systematic and Evolutionary Microbiology*, 2014. **64**(10): pp. 3485–3491.
27. Logan, B., et al., 2006. Microbial fuel cells: Methodology and technology. *Environmental Science & Technology*. **40**(17): pp. 5181–5192.
28. Penteado, E.D., et al., Influence of sludge age on the performance of MFC treating winery wastewater. *Chemosphere*, 2016. **151**: pp. 163–170.
29. Samsudeen, N., T. Radhakrishnan, and M. Matheswaran, Effect of isolated bacterial strains from distillery wastewater on power generation in microbial fuel cell. *Process Biochemistry*, 2016. **51**(11): pp. 1876–1884.
30. Bouallagui, H., et al., Mesophilic biogas production from fruit and vegetable waste in a tubular digester. *Bioresource Technology*, 2003. **86**(1): pp. 85–89.
31. Leung, D.Y. and J. Wang, An overview on biogas generation from anaerobic digestion of food waste. *International Journal of Green Energy*, 2016. **13**(2): pp. 119–131.
32. Jeevahan, J., et al., Waste into energy conversion technologies and conversion of food wastes into the potential products: A review. *International Journal of Ambient Energy*, 2018: pp. 1–19.
33. Ko, M.K., et al., Gasification of food waste with steam in fluidized bed. *Korean Journal of Chemical Engineering*, 2001. **18**(6): pp. 961–964.
34. Nanda, S., et al., Gasification of fruit wastes and agro-food residues in supercritical water. *Energy Conversion and Management*, 2016. **110**: pp. 296–306.
35. Levin, D.B., L. Pitt, and M. Love, Biohydrogen production: Prospects and limitations to practical application. *International Journal of Hydrogen Energy*, 2004. **29**(2): pp. 173–185.
36. Kothari, R., et al., Fermentative hydrogen production–An alternative clean energy source. *Renewable and Sustainable Energy Reviews*, 2012. **16**(4): pp. 2337–2346.
37. Liu, X., et al., Hydrogen and methane production by co-digestion of waste activated sludge and food waste in the two-stage fermentation process: Substrate conversion and energy yield. *Bioresource Technology*, 2013. **146**: pp. 317–323.
38. Kiran, E.U., et al., Bioconversion of food waste to energy: A review. *Fuel*, 2014. **134**: pp. 389–399.
39. Karmee, S.K., et al., Conversion of lipid from food waste to biodiesel. *Waste Management*, 2015. **41**: pp. 169–173.
40. Devarapalli, M. and H.K. Atiyeh, A review of conversion processes for bioethanol production with a focus on syngas fermentation. *Biofuel Research Journal*, 2015. **2**(3): pp. 268–280.
41. Kelkar, S., et al., Pyrolysis of spent coffee grounds using a screw-conveyor reactor. *Fuel Processing Technology*, 2015. **137**: pp. 170–178.
42. Spokas, K.A., et al., Biochar: A synthesis of its agronomic impact beyond carbon sequestration. *Journal of Environmental Quality*, 2012. **41**(4): pp. 973–989.
43. McGaughy, K. and M.T. Reza, Hydrothermal carbonization of food waste: Simplified process simulation model based on experimental results. *Biomass Conversion and Biorefinery*, 2018. **8**(2): pp. 283–292.
44. de Lange, W. and A. Nahman, Costs of food waste in South Africa: Incorporating inedible food waste. *Waste Management*, 2015. **40**: pp. 167–172.

Part 3

Waste for Biochemicals and Bioenergy

9 Biowastes for Ethanol Production

Jeffin James Abraham, Christian Randell A. Arro, Ali A. El-Samak, Alaa H. Hawari, and Deepalekshmi Ponnamma
Qatar University

CONTENTS

9.1 INTRODUCTION

Reaffirming common knowledge, the 20th and 21st centuries are an acclaimed time point in our civilization filled with many technological advancements and remarkable discoveries supported by an abundance of research and development reflecting

DOI: 10.1201/9781003178354-12

upon the growing infrastructures and population of our species. Due to technological advancement, we also face dramatic demand and consumption of resources followed by the production of a detrimental amount of pollution. Taking a systematic approach, we first briefly address a possible major contributor to the issue – population growth. As of year 2021, the world population stands to exceed 7.5 billion people with a distribution of approximately 18% from China, 17% from India, and 4% from the USA followed by sub-4% proportions from other countries [1]. The significance of these easily accessible statistics lies in the topic of basic needs for a human being [2]. It is common knowledge that resources are needed to sustain and fuel a growing population, nonetheless with a magnitude of over 7.5 billion people. Regardless of the development stage of a country, be it a developing or developed nation, it stands to say that resource consumption and production of waste and environmental damage are accelerated by population growth [3].

Key statistics have indicated that the global primary energy consumption stands to more than 170,000 TWh in 2019 with a majority of the proportions from non-renewable resources such as oil, gas, and coal, as seen in Figure 9.1 [4]. Interestingly, sources have shown that the year-to-year outlook for the rate of change of global energy consumption has shown signs of decline, which can be attributed to better technology and growing energy efficiency, as well as the unfortunate COVID-19 pandemic which enforced lockdown measurements – ultimately putting industry and basic day-to-day lifestyle at a standstill, leading to an estimated drop in global energy demand by 5% [5]. However, the energy produced and consumed from non-renewable resources still consequently leads to environmentally damaging outcomes such as the emission of CO_2, a global contributor to air pollution and global climate change, and topological damages toward the environment as a result of oil spills, mining, and deforestation. As such, the interest toward renewable and sustainable resources has grown considerably in current times.

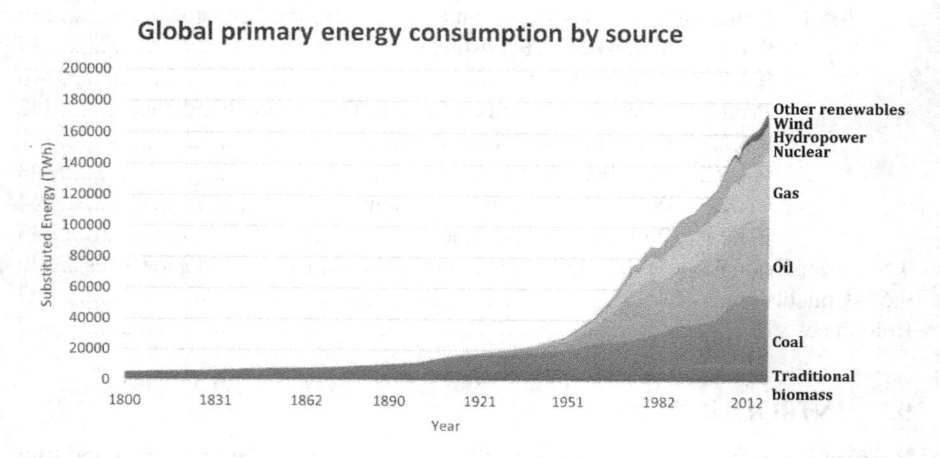

FIGURE 9.1 Key statistics acquired from BP Statistical Review of World Energy on the topic of global primary energy consumption separated by source from 1800 to 2019. (Adapted with permission from Reference [4], Copyright (2017) MIT Press.)

9.1.1 What Are Biofuels and Biomass?

Renewable energy is defined as the energy that can be produced from natural resources such that they can naturally be replenished over a period of time – including, but not limited to, solar, wind, geothermal, and more importantly biomass and biofuels, the key highlight of this chapter. Biofuels, most notably biodiesel and bioethanol, are fuels that have been derived from biological sources such as sugar, starch, organic waste, and cellulosic materials [6]. These precursor sources are referred to as biomass, which are lignocellulosic materials and are therefore heavily composed of the following components: cellulose, lignin, and even proteins. The attractive significance for the use of such precursor materials is mainly from their environmentally friendly aspect and the inexpensive availability [7]. Further elaborating on the environmental benefits of biomass precursors, they represent a stable zero CO_2 footprint as a result of CO_2 recycling from their production to their combustion and are also biodegradable. Finally, biomass precursors are inexpensive and readily available and they can be obtained from the output and wastes of other industrial and domestic processes such as the wood chippings (residue) from logging manufacturers, municipal solid waste (MSW), and agricultural and food waste (AFW) [8]. Biofuels can be categorized into two: Primary biofuels that refer to solid fuels produced from wood, peat, and pellets, which are combusted to generate energy, and secondary liquid biofuels mainly used for ground transportation. Secondary liquid biofuels can be further classified into "generations" depending on the type of feedstock used [9]:

- First generation: Uses biomass that is edible such as food crops of sugars, starch, grains, and vegetable oils.
- Second generation: Uses biomass from a variety of non-edible sources such as feedstocks with high lignocellulosic content from logging industries and other feedstock from MSW.
- Third generation: Utilizes algae-based biomass for fuel production, specifically biodiesel, bioethanol, and biomethane.

9.1.2 What Are Biowastes?

Commonly known as biodegradable waste, it is a culmination of organic or decomposable waste under aerobic conditions. A definition has been cemented by organizations such as the EU's Waste Framework Directive from 2008 stating that biowaste refers to "biodegradable garden waste, food and kitchen waste from households, restaurants, caterers and retail premises and comparable waste from food processing plants", which overlaps with their definition of MSW as follows: "Covers household waste and waste in similar nature and composition to household waste" [10]. According to the World Bank Group (2018), their review of solid waste reveals that there is a global production of 2.01 billion tons of solid waste, with only 44% belonging to food and green waste, 17% to paper and cardboard, and 2% to wood [11]. It is expected that the increase in population would further increase waste levels. With only 19% of global solid waste undergoing recycling, the utility for waste-to-energy processes and technology is ever more needed for better waste and environment management.

9.1.3 Why Bioethanol?

From waste to energy, bioethanol acts as a successful fuel candidate used in many countries, leading to a total global production of approximately 56 billion liters in 2015 [12]. Bioethanol is defined as a sustainable alcohol produced from biomass and is a popular alternative fuel source for motor vehicles. Ethanol, on its own, provides very interesting properties that make it an attractive substitute to conventional non-renewable fuels such as gasoline, especially when produced sustainably from biomass. In comparison with gasoline, an equal amount of ethanol only delivers a smaller proportion of 66% of the energy. However, ethanol provides a greater octane number of 106–110 as opposed to gasoline (91–96). This property broadens the application of bioethanol as an additive in gasoline for improved performance in motor vehicles as it allows for shorter burning time and higher compression ratios within internal combustion engines. Some important attractive characteristics of ethanol also include high combustion efficiency of 15%, negligible amount of sulfur, and less toxic by-products, namely acetic acid and acetaldehyde [12]. When combined with its biodegradable nature and production from sustainable means, bioethanol acts as a significant participant among other biofuels. Still, bioethanol faces certain difficulties such as the later discussed expense of bioethanol refinery from biowastes treatment and the food vs. fuel controversy in which feedstock and land used for bioethanol feedstock refinery could be better used for agricultural means, as such the use of edible feedstock such as sugar and grains invokes ethical issues. Still, the production of bioethanol remains within a growing state with increased annual production and demand.

9.1.4 Global Production of Biofuels and Bioethanol

According to the International Energy Agency (IEA), it is reported that in the year 2019 approximately 160 billion liters of biofuel were produced, followed by 144 billion liters in 2020 with regard to ground transportation biofuels, as seen in Figure 9.2. With the recent development of the COVID-19 pandemic, the production of biofuels did not reach the projected 2020 values; however, the prospects for biofuels remain promising with expectations of reaching up to 5.4% in contribution toward the 2025 ground transportation energy as opposed to the previous amount of 3% in 2013 [13]. Bioethanol, one of the main members within the biofuel sector, has the greatest production output among its fellow biofuel members with around 115 billion liters produced in 2019, with 59.5 billion liters from the USA and 36 billion liters from Brazil. And with increasing popularity in the Asian markets, increased production from India, Thailand, and most importantly China is said to boost the global production to 119 billion liters toward the years 2023–2025.

9.2 THE SOURCES OF BIOETHANOL

Bioethanol is considered as the predominant source of biofuel for global transportation; hence, it can be used purely in certain car designs, in addition to being blended with gasoline to form a "gasohol", which is a blend that requires no engine

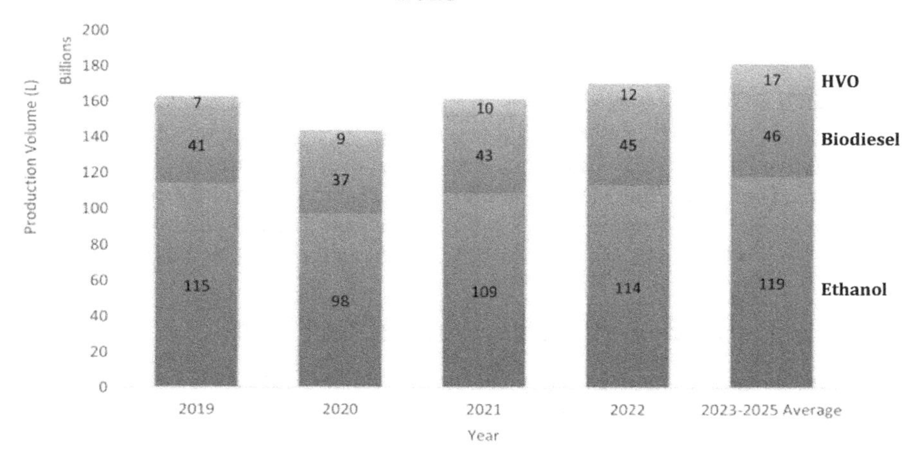

FIGURE 9.2 Global biofuel production in 2019 and forecast for 2025, IEA, Paris [13]. (Distributed under a Creative Commons Attribution License 4.0 (CC BY).)

modification to operate. The shift in interest regarding the utilization of bioethanol as fuel is motivated not only by its clean and recycled origin, but also by its clean source of energy during its application as previously mentioned in this chapter.

One of the significant benefits of bioethanol consists in the large variety of raw materials from which it can be produced. A carbon- and hydrogen-rich elemental composition with limited amounts of oxygen, nitrogen, and other organic materials is one of the main criteria for materials selection. First-generation sugar-based raw materials such as sugarcane, fruits, and sugar refinery wastes are advantageous ethanol feedstocks due to their high sugar yield and low conversion costs [12]. Additionally, sugarcane is a C4 plant known for its ability to convert solar energy into biomass; hence, it is the main ethanol feedstock in Brazil, producing up to ~80% of its bioethanol fuel. Similar C4 plants such as sugar beet and sweet sorghum have also been used in tropical countries around the world due to their high carbon assimilation (50 g/m^2/day) and their ability to accumulate high level of extractable sugars within their anatomy. The main drawback of the utilization of sugar-based raw materials in bioethanol production includes its seasonal availability and sophisticated pretreatment process. Starchy crops are another first-generation source of bioethanol production, including cereals, roots, legumes, corn, and potatoes. They are used in bioethanol production largely due to their ease of availability, easy conversion and storage, and high ethanol yield. However, the usage of starchy corps in the production of bioethanol instead of human consumption attracted large criticism due to the risk of global food shortages in certain parts of the world. Lignocellulosic biomass is possibly the most promising source of bioethanol from the second-generation classification. This is because bioethanol produced from lignocellulosic sources achieves 16 times more energy as compared to its precursor biomass pre-production. Lignocellulosic biomass can be divided into multiple categories including grass and aquatic plants, forest waste materials (hardwood and sawdust), agricultural waste (cereal straws and bagasse), and the organic component of MSW. Generally, lignocellulosic biomass is considered promising due to its abundance worldwide, as it can be utilized to form bioethanol without necessitating additional land or straining current food crop production. Hence, the possible amount of plant biomass to be used for biofuel production equates to roughly ~20×10^9 tons [12]. Lignocellulose consists of cellulose microfibers present as lignin, hemicellulose, and pectin, which present the main polymers needed to produce bioethanol. The lignocellulosic biomass provides the biomass-based ethanol industry with a continuous and concrete supply of raw materials, thereby maintaining a low-cost ethanol production in comparison with the aforementioned biomass sources.

9.3 MECHANISM OF BIOETHANOL PRODUCTION

The conversion of the preferred biomass to efficient biofuel requires a delicate and systematic chemical process that is commonly found in both first- and second-generation biomass. This mechanism is highlighted in the upcoming part of the chapter, which focuses on the hydrolysis, detoxification, and fermentation of the biomass into bioethanol.

9.3.1 Hydrolysis Process

The common component found in the first- and second-generation bioethanol sources is sugar molecule chains; these chains are hydrolyzed to form monomeric sugars, which are further fermented to produce ethanol.

This process is based on Equation 9.1:

$$(C_6H_{10}O_5)_n + nH_2O \rightarrow nC_6H_{12}O_6 \tag{9.1}$$

The hydrolysis process is categorized based on the medium in which it is carried out; hence, it has two types: The first type is acid hydrolysis, and the second type is enzymatic hydrolysis. The first-generation biomass undergoes enzymatic hydrolysis, which is considered easier due to the presence of pure cellulose, which can be easily converted to glucose. On the contrary, the second-generation biomass includes a more complicated acid-based hydrolysis process due to the presence hemicellulose and lignin, which require further treatment to release glucose.

9.3.1.1 First-Generation Hydrolysis

Certain molecules such as amylopectin and amylose are found in starch, which undergoes hydrolysis in the form of biochemical conversion to produce soluble sugars. Even though acid hydrolysis is an acceptable method of generating fermentable sugars from starch, it remains undesirable in the first-generation hydrolysis due to the necessity for acid recovery after the hydrolysis process. Additionally, the mechanism for the formation of hexose (glucose) through the acidic hydrolysis of starch produces by-products that are toxic to microbial cells, hence limiting the yeast growth and affecting the rate of ethanol fermentation.

For the aforementioned reasons, the first-generation hydrolysis is mainly carried out through enzymatic hydrolysis, which is eco-friendly, simple, and efficient in the conversion of starch into sugars. Conventional enzymatic hydrolysis is carried out in two basic steps, which are liquefaction and saccharification. The liquefaction process is carried out in the temperature range of 85 °C–105 °C in the presence of thermostable amylase to convert the starch into a gel-like substance through forming a cleavage in the glycosidic linkages in the starch molecules; this will positively influence the starch polymers conversion into sugar in the following steps. On the other hand, the saccharification is carried out at a lower temperature range of 50 °C–60 °C in the presence of glucoamylase, which is the enzyme responsible for converting dextrin into glucose [12].

9.3.1.2 Second-Generation Hydrolysis

The enzymatic hydrolysis of natural lignocellulosic biomass is an extremely slow process, due to the complex hindering structure of the raw materials, mainly cellulose and hemicellulose components, impeding the fermentation process. Therefore, a more rigorous approach is required, such as acid hydrolysis. This process is carried out under dilute or concentrated acid conditions depending on the pretreatment process of the biomass. Sulfuric acid is the most common acid used in this process. The dilute acid route is split into two stages: The first stage is the depolymerization of hemicellulose at low temperature, followed by the second stage which converts the depolymerized hemicellulose and cellulose into glucose at high temperature. Alternatively, concentrated acid hydrolysis is capable of hydrolyzing both hemicellulose and cellulose in a single step, in addition to being more effective in the complete and rapid conversion of cellulose into glucose. The acid hydrolysis process is the preferred hydrolysis process for the second-generation biomass; however, it faces a drawback of the corrosion of equipment due to the utilization of highly concentrated acid.

9.3.2 Detoxification Process

The detoxification process is required for the completion of the conversion of biomass to ethanol. The detoxification process is carried out because the pretreatment process forms aliphatic acids, acid derivatives, and phenolic compounds, and these by-products act as inhibitors for the fermenting microorganisms in addition to degrading the enzymes needed to hydrolyze the biomass feedstock. Therefore, to counteract this problem, the detoxification process is applied physically, chemically, or biologically to remove the inhibitors. A few examples include the utilization of extraction, active coal, and overliming as detoxification methods to remove the aforementioned inhibitors [14].

9.3.3 Fermentation Process

The fermentation process is a metabolic process, which involves the conversion of the hydrolyzed products (soluble sugars) of the first- and second-generation biomass in the absence of oxygen into alcohol and carbon dioxide via microorganisms such as yeasts and bacteria. The reaction below highlights the fermentation of glucose into ethanol:

$$C_6H_{12}O_6 + H_2O + \text{yeast} \rightarrow 2C_2H_5OH + 2CO_2 + H_2O + \text{heat} \quad (9.2)$$

From the above reaction, it is estimated that 100 kg of glucose is capable of yielding 48.8 kg of CO_2 and 51.1 kg of ethanol; hence during the fermentation process, 95% of the soluble sugars are converted into ethanol and CO_2 through traditional yeast. Although yeast can readily ferment glucose, it is not capable of effectively fermenting sugars produced from second-generation biomass, such as pentose and xylose, on its own. Therefore, the utilization of naturally occurring microorganisms such as *Zymomonas mobilis* in combination with yeast forms a co-culture system that is capable of fermenting pentose and xylose. The metabolic pathway used by yeast during the conversion of sugars to ethanol is known as glycolysis. Glycolysis is the process carried out in the cytoplasm of yeast cells, where one mole of glucose is catabolized to generate two moles of pyruvate; the lack of oxygen forces the pyruvate to undergo reduction reactions, resulting in the production of one mole of ethanol and another mole of CO_2 per one mole of pyruvate, thus finalizing the fermentation process (Figure 9.3) [12].

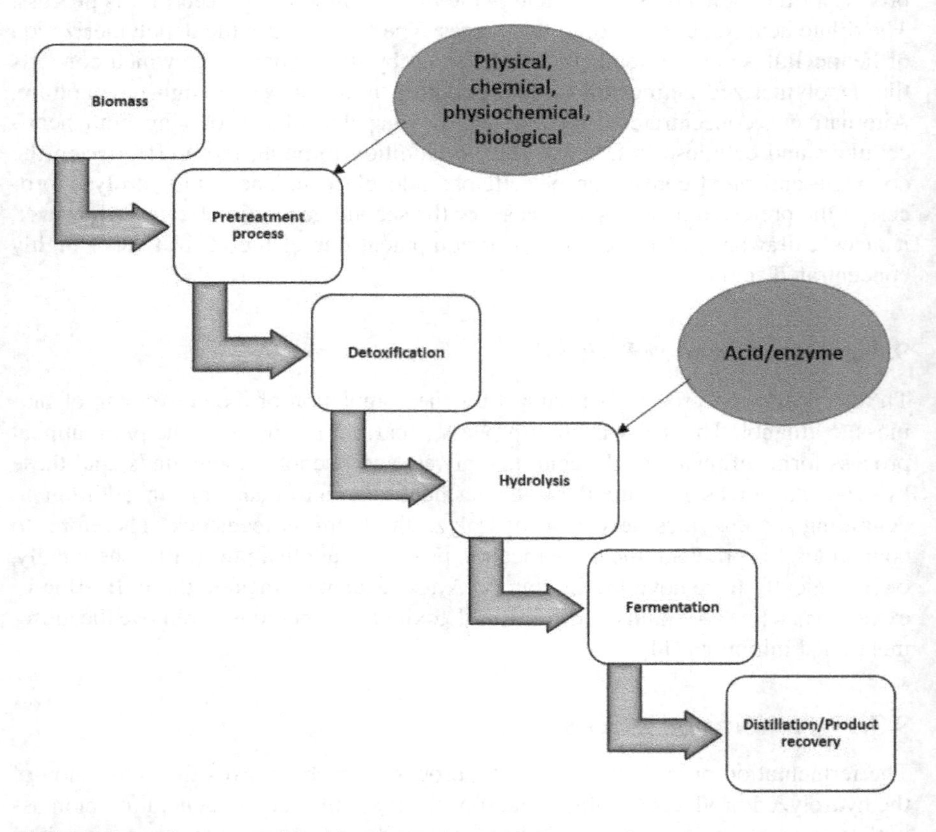

FIGURE 9.3 Summary of the biomass-to-bioethanol conversion process.

9.4 BIOETHANOL PRODUCTION SYSTEMS

The production system designs of bioethanol slightly vary according to the type of feedstock used. As explained in the previous sections, bioethanol is divided into three generations based on the type of feedstock used. The production of bioethanol occurs through the fermentation process of the available feedstocks, which mainly consist of fermentable sugars, carbohydrates, and lipids [15]. Following the fermentation process, these biomasses from the feedstock will be modified for specific energy uses such as fuels, heat generation, biogas, and electricity production. The production techniques are important to optimize the amount of biofuel produced for future applications. The production techniques between each generation of biofuel vary due to the different types of feedstock used. Other factors such as production costs and energy efficiency factor of using first-generation feedstocks have led to more studies on alternative feedstocks, resulting in second-generation feedstocks. However, during the production process of bioethanol from second-generation feedstocks, an additional pretreatment process is required to extract the fermentable sugars, which drastically increases production costs [16]. To understand the future trends and pathways required for more cost-efficient bioethanol production from biowastes, all the various production pathways of each generation feedstock need to be analyzed. Each generation of bioethanol will therefore be discussed.

9.4.1 Production Systems Based on First-Generation Feedstocks

9.4.1.1 Sugar-Based Feedstocks

The first-generation feedstock can be classified into sugar- and starch-based feedstocks. The production system line only varies slightly with the type of feedstock used. The production of bioethanol from sugar-based feedstock only requires an extract and fermentation step to produce ethanol. Comparatively, the starch-based feedstock requires an additional hydrolysis step that can convert all the starch into fermentable glucose with the help of amylolytic enzymes [12]. The cost analysis for the production of ethanol from both types of feedstock suggests that sugar-based feedstock follows cheaper production. However, due to the factors such as the requirement of controlled climatic conditions and controlled soil type, the sugar-based feedstock crops are not grown around the world [17]. The juice from sugar-based feedstock (stalks of sugar cane, etc.) is first extracted using a roller. The next step involves the purification of the extracted juice using lime, which increases the pH of the extracted mixture juice. The sources of lime mainly include calcium hydroxide and calcium saccharate. The concentration of lime added will be 2 wt.% of the total solid amount present in the extracted juice [12]. Increasing the pH of the extracted juice with lime can result in the neutralization of the organic acids present, decrease the number of colorants, and form calcium phosphate precipitates, which can be filtered out in the next step [18]. The residue material present in the extracted juice is then filtered out, and the residue removed is often identified as filter cake [12]. The next process involves fermentation of the juice for the formation of ethanol. Depending on the microorganism participating in the fermentation process, the sugar juice solution needs to be evaporated to concentrate the sugar level between 14% and 18% for the microorganism to accommodate the sugar level [19]. The syrup remaining after the concentration of the sugar is the main ingredient used for bioethanol production. The most used fermentation microorganism is yeast *Saccharomyces cerevisiae*. The ideal temperature range is between 33 °C and 35 °C, and the ideal cell density is between 8% and 17% (v/v) for the fermentation process using yeast *Saccharomyces cerevisiae* [19]. When the concentration amount of ethanol reaches about 10% (v/v) ethanol, the fermentation is stopped. The yeast typically used in the fermentation systems are recycled from the fermentation wastes by concentrating and separating them and washing them using sulfuric acid to limit the contamination from bacteria. With the recycling of the yeasts, higher cell densities are achieved, which can decrease fermentation times to 6–10 h in temperatures of 33 °C–35 °C and have high cell ethanol yields and concentrations [20]. Following the fermentation process, the remaining liquid is distilled and rectified for the production of an azeotropic solution of 95% (v/v) ethanol [19]. With the help of molecular sieves or employing azeotropic distillation with either benzene or cyclohexane, the azeotropic ethanol can be converted into absolute ethanol (high-grade or anhydrous ethanol) [21].

9.4.1.2 Starch-Based Feedstock

The starch-based feedstock as explained previously requires an additional hydrolysis step for ethanol production. However, it is necessary to note that even though the production cost of ethanol from sugar-based feedstock is cheaper than from starch-based feedstock, the overall amount of ethanol produced from starch-based feedstock is 60% compared to the 40% from sugar-based feedstock, covering all the first-generation feedstock [22]. For the necessary glucose syrup required in ethanol production, the starch present in the feedstock requires to be broken down into individual glucose units that are fermented into ethanol [19]. A hydrolytic reaction is required to break down the carbohydrates into glucose molecules with the help of a catalyst called glucoamylase enzyme, which then produces D-glucose (a glucose isomer). Once the conversion into glucose has been achieved, a similar process of fermentation, and distillation is employed to produce anhydrous ethanol [23]. However, in terms of processing the corn grains used in starch-based feedstock, two different methods are utilized depending on the required products/co-products from processing the feedstock. For producing only ethanol, the ideal process is the dry milling process, which is usually done on a smaller scale. Comparatively, the wet milling process usually transpires in corn refineries, generating other valuable products such as dextrose and high-fructose corn syrup [24]. Both the dry milling process and wet milling process reportedly have an ethanol yield of 0.395 and 0.372 L/kg, respectively, from corn grains [25].

9.4.1.2.1 Dry Milling Process

The dry milling process follows five distinct steps for ethanol production, namely milling, liquefaction, hydrolysis, fermentation, and finally distillation. Initially, the corn grains will be finely ground into a powder with the help of hammer mills, which can expedite water and enzymes in the hydrolysis step [19]. For liquefaction into starch, the fine powder grains are first heated using water at 85 °C, then mixed with alpha-amylase enzymes, and further boiled between 110 °C and 150 °C for an hour.

Additional alpha-amylase enzymes are added while cooling the mixture solution to 85 °C, and this solution is idle at this temperature for 1 h. The next step is the hydrolysis of the mixture. For this step, the mixture requires to be cooled to room temperature and then requires the addition of glucoamylase enzymes, which convert the starch into dextrose [19,20]. The grain refineries which host the dry milling plants usually prefer to reduce cost in investing in specific hydrolysis (saccharification) vessels and hence use a process known as simultaneous saccharification and fermentation (SSF). With this process, the glucoamylase enzymes used for hydrolysis are added to fermenters at the plants. With the fermentation step, the yeast ferments the sugar mixture to produce ethanol and CO_2. During the fermentation process, the mixture requires continuous stirring for uniform yeast distribution and decreased heat evolution [26]. In the dry milling plants, this step is either done in a single flow, where the process is finished in 48 h, or can occur in tandem with the continuous addition of the sugar mixture, while the already fermented broth is extracted [19]. The fermented broth is then processed for final distillation to separate the ethanol from the remaining mixture.

9.4.1.2.2 Wet Milling Process

The wet milling process usually starts with corn with shells. Initially, the shelled corns are cleaned using mechanical cleaners, which will remove all unwanted materials. The next process for the cleaned shell corns is steeping. Steeping involves soaking the shelled corns in dilute sulfuric acids in temperature of 52 °C for a time between 24 and 48 h to dissociate the protein from starch particles in the corn and eliminate other soluble materials [19]. De-germination is the next important process that separates the corn kernel into the starch and gluten parts, and the germ, which then undergoes cleaning with water and drying and other processes for corn oil extraction [27]. The remaining mixture will still contain sparse amounts of fibrous material, which will be separated from the starch and gluten mixture through washing, grinding, and screening [19]. In order to separate the gluten from the starch, multiple runs of centrifugation are required [27]. Following the availability of purified starch mix, they undergo a similar process of liquefaction, hydrolysis, fermentation, and distillation. Here, the starch mix requires pH control in the range between 5.8 and 6.2 using lime. Then alpha-amylase enzymes are added for liquefaction, where starch is broken down into soluble dextrins [19]. Additionally, small concentrations, 20–100 ppm, of calcium are added to ensure enzyme stability. As in the case of dry milling process, liquefaction, high temperature, and fast cooking under the enzyme liquefaction convert the starch mix into a 30%–40% solid starch mix. The wet milling system is a continuous process and utilizes previous resources for reuse in the conversion process. The hydrolysis process requires the use of steep water, which contains nutrients required for fermentation and pH control during the hydrolysis process. During the hydrolysis process, when the pH is controlled at 4.5 and the temperature is 65 °C, the glucoamylase enzymes break down the dextrins into glucose, which is required for ethanol production [19]. After the hydrolysis, yeast will be added to the glucose syrup mixture to start fermentation process. The conversion time for the production of ethanol and CO_2 from glucose syrup varies on the hydrolysis time and usually takes from 20 to 60 h. The glucose mixture during fermentation process has only very minute amounts of insoluble solids, which promote the fermentation times and can further help in reusing yeast for further fermentation processes [27]. With the wet milling process, the final product has an ethanol yield of about 8%–10% with respect to the volume.

9.4.2 Production Systems Based on Second-Generation Feedstock

The second-generation feedstock mainly covers agricultural and other types of wastes for use in the production of ethanol. They are mainly identified as lignocellulosic biomass, where the feedstock is mainly composed of cellulose, lignin, and hemicellulose. With second-generation feedstock, the most challenging issue comes with breaking down the lignocellulosic biomass during the pretreatment process. During pretreatment, the biomass is broken down into soluble components and separated accordingly [28]. With this process, further chemical or biological treatments are possible with the remaining solid biomass [29]. With the lignocellulosic biomass, the structure is a matrix of cellulose and lignin, which is held together by the hemicellulose [28]. To ensure smooth enzymatic hydrolysis, the pretreatment process comes in handy by breaking down the matrix and decreasing cellulose crystallinity, converting them into amorphous cellulose [30]. After the pretreatment process, the biomass undergoes similar steps of hydrolysis, fermentation, and distillation as in the case of first-generation feedstock. The main process where the second-generation feedstock mainly differs is in the pretreatment process. The pretreatment aims to prevent the loss of fermentable sugars, decrease the amount of inhibitory products, and overall make the process cost-efficient by reducing the energy requirements throughout the entire ethanol production process [28]. The pretreatment process can be physical, chemical, biological, or a mixture of any two methods. With this process, there is currently no best option, and depending on the cost, yield, etc., different methods are employed. The majority of the cost to produce ethanol from the second-generation feedstock comes from the pretreatment process, and hence, it is necessary to optimize the methods for cheaper ethanol production.

9.4.2.1 Physical Pretreatment

The physical pretreatment processes involve the use of techniques that modify the physical characteristics of the available feedstock. One of the most common physical pretreatment processes is the reduction of size using mechanical techniques such as milling, grinding, or chipping. This is mainly done for solid wastes. This process is effective for reducing cellulose crystallinity and further increases production efficiency during further processing [28,31]. The mechanical techniques usually covered include wet, dry, compression, and vibratory ball milling. Initially breaking down the large particle into smaller sizes will help during further processes. However, breaking down the feedstock into extremely fine particles can result in poor efficiency due to the formation of clumps when mixed with liquid during other pretreatment processes and hydrolysis, resulting in unwanted enzymatic reactions [28,32]. In terms of energy consumption during this mechanical processing, the main factors that need to be addressed are particle size, moisture level, and the type of feedstock used, which hence vary between feedstocks used

[31,33]. The next commonly used physical pretreatment technique is pyrolysis, where high temperatures are utilized to process the feedstock. With this process, energy consumption is low because it is an endothermic process. The feedstock is processed under temperatures higher than 300°C, which breaks down the cellulose into H_2, CO, and char [28]. Using lower temperatures will only result in an undecomposed product, which impedes the additional processes for ethanol production [30]. For microbial growth required in ethanol production, the resulting char, composed of carbon, undergoes leaching using either water or acid-producing glucose-rich water leachate [34]. The use of microwave is another pretreatment technique used for bioethanol production. The ease of use and fast heating ability are some common points that make the use of microwave viable. Two main methods, thermal and non-thermal, are utilized while using microwave in aqueous conditions. Using the thermal method, the lignocellulosic material is broken down utilizing the microwave radiation, which heats up the material using the vibration from the aqueous medium and polar bonds from the material [35]. As a by-product of this treatment, acetic acid is released from the lignocellulosic material, which makes the media acidic, beneficial for immediate hydrolysis process [28]. In the non-thermal method, the lignocellulosic material undergoes continuous changing magnetic field, which causes vibrations to the polar bonds in the materials, resulting in accelerated physical and chemical changes to the materials [36]. With the radiation released, physical characteristics such as specific surface area of the lignocellulosic material increases, the cellulose crystallinity decreases, and partial hydrolysis of the hemicellulose occurs [28]. This method is mainly identified as electron beam irradiation pretreatment technique.

9.4.2.2 Chemical Pretreatment

Chemical pretreatment processes use various chemical-based methods to convert the feedstock into hydrolysis-compatible mixtures. One of the most common chemical pretreatments is acid pretreatment with a very high rate of yield of sugar from the lignocellulosic mixture. Various concentrations of acids, in the range of 0.2%–2.5% w/w, – sulfuric, nitric, phosphoric, or hydrochloric acids – are utilized in the temperature range 130°C–210°C for this process [28,37]. This process can be done using dilute or concentrated acids mainly depending on the feedstock, for better cellulose hydrolysis [38]. The hemicellulose present in the lignocellulosic material is broken down by the acid for easier hydrolysis [37]. The acid pretreatment process, however, has its negative point in the fact of formation of inhibitors that inhibit microorganism growth during the fermentation process and hence requires detoxification before the fermentation process [28]. Like the acid pretreatment, alkaline solutions can also be utilized for the pretreatment process. The alkaline solution is mainly sodium, potassium, calcium, or ammonium hydroxides, and the alkaline pretreatment process, compared to the other processes, only requires low temperatures and pressures [28]. The lignin matrix from the lignocellulosic materials is digested using an alkaline solution, which then frees up the cellulose and hemicellulose for hydrolysis [39]. During this process, the cellulose swells, which results in decreased cellulose crystallinity. Hence, the components of the lignocellulosic materials can be separated into lignin dissolved in alkali solution, hemicellulose, and cellulose useful for ethanol and other by-product formation [28]. Another important chemical treatment process that can be utilized is the wet oxidation process. In this process, the lignocellulosic material is mixed with water and oxygen/air at temperatures greater than 120 °C [40]. Large amounts of water are required for this process as almost 1 L of water is required for 6 g of lignocellulosic material [28]. The hemicellulose from the lignocellulosic material is converted from solid to liquid phase, and sugar oligomers are the major product from the hemicellulose during hydrolysis [37]. The final process covered in chemical pretreatment processes is using organic solvents. When using organic solvents, the lignin from the lignocellulosic material is separated through use of organic solvents and water mixture and, through distillation, they can be completely separated [28]. With this pretreatment process, different organic solvents can be used, such as a mixture of 90% formic acids with 50% alcohol–water mixture and 50% pressurized CO_2, methanol, ethanol, acetic acid, or a combination of these solvents [37,41].

9.4.2.3 Physiochemical Pretreatment

The physiochemical pretreatments use branches of both physical and chemical pretreatments. The steam explosion technique is one such example of a physiochemical pretreatment process. High-pressure, 20–50 bar, and high-temperature, 160 °C–290 °C, steam is required to attack the lignocellulosic material for a few minutes, followed by rapid decompression and cooling to atmospheric pressure and room temperature [30,42]. With steam at high temperatures, the lignocellulosic material separates into fibers that are suitable for cellulase attack. The resulting material is easy to process into materials such as xylitol and ethanol [38,42]. The high yield of xylose sugar from this process is one of the main economic advantages of this process. Another example of physiochemical pretreatment is the use of the liquid hot water method, a hydrothermal technique. In this method, the hemicellulose in the lignocellulosic material is hydrolyzed using compressed hot liquid water having a pressure over saturation point [42]. The process is executed at a temperature range between 170 °C and 230 °C and pressure greater than 5 MPa for the duration of 20 min [28]. The main product of this process is hemicellulose sugars. The high xylose sugar yields is one of the prospective processes in ethanol production. However, detoxification of the residue after this process is essential due to the formation of acid that inhibits microbial growth during fermentation. The next method in physiochemical pretreatment is the ammonia fiber explosion method. The alkaline ammonia is utilized with steam explosion, breaking down the lignocellulosic material at high temperature and pressure followed by immediate pressure release. The requirements for the successful execution of the ammonia fiber explosion method include a temperature range between 60 °C and 100 °C and a time period between 5 and 30 min with respect to the lignocellulosic material saturation. Other factors that can affect this process include the amount of ammonia required, pressure, and moisture. Some of the advantages of using this method include fast process time, less particle size requirements, and prevention of inhibitor formation [28,31,43]. Disadvantages of this technique include poor conversion ability for materials having high lignin content, dissolving only limited amounts of hemicellulose, and high ammonia cost. For this technique to be viable, an efficient ammonia recovery system needs to be set up. Overall, the technique has a high cellulose and hemicellulose yield of 90%, which is beneficial for ethanol production [44]. Replacement of the ammonia with CO_2 has also been investigated as a potential pretreatment technique because it is cost-efficient and shows an improved yield of cellulose compared to the normal steam explosion technique [45].

9.4.2.4 Biological Pretreatment

Biological pretreatment involves the use of microorganisms such as brown, white, and soft rot fungi, to convert the lignocellulosic material into cellulose components by completely degrading the lignin and hemicellulose from the material [31,33,46]. Among all the microorganisms used, the white rot fungi were effective and had the best yield of ethanol. The white and soft rot deals with both cellulose and lignin, whereas the brown rot deals with only cellulose [28,46]. To avoid cellulose degradation by the fungi used and to only degrade the lignin, various bioengineered fungi are being developed [28]. Usually, the biological pretreatment processes are conducted at 25°C temperature and conducted over a long period for effective conversion. These processes are cost-efficient, consume less energy, and are relatively safe to execute. However, biological pretreatment has its demerits, mainly hampering the hydrolysis rate due to the microorganism impeding the process. This further leads to lower yield in ethanol and other by-products and hence is not an effective strategy compared to the other pretreatment processes.

The hydrolysis, fermentation, and distillation steps done after pretreatment are similar to those of the first-generation feedstock. Enzymatic hydrolysis is used for the smooth conversion of the biomass material that has undergone pretreatment into sugar monomers for fermentation. In the case of fermentation, simultaneous saccharification and fermentation techniques are preferred to bring down the production costs. Depending on the type of yeast or bacteria used, the yield of ethanol varies for various the feedstocks. The yeast *S. cerevisiae* and the bacteria *Z. mobilis* are the highest ethanol-yielding fermentation microorganisms [28,33]. Table 9.1 below summarizes the entire pros and cons of different pretreatment processes.

TABLE 9.1
Pretreatment Process Types with Their Pros and Cons

Pretreatment Process Type	Pros	Cons
Physical	Controlled particle size, low cellulose crystallinity, used with other methods.	High energy requirements, increased cost of production.
Chemical	Require less energy, high yields, improved reaction rate, low cellulose crystallinity.	Require additional detoxification before hydrolysis, high chemical cost.
Physiochemical	Hemicellulose removal/partial hydrolysis, increased reaction surface area, low cellulose crystallinity.	High energy requirements, not eco-friendly, high chemical cost.
Biological	Cost-efficient, low environmental impact.	Slow delignification rate, long production time, poor hydrolysis rate.

9.5 BRIEF EVALUATION ON THE MARKET OF BIOETHANOL PRODUCTION FROM BIOWASTES

As previously mentioned, the global production across the past years has shown tremendous support with countries such as the USA maintaining a high level of bioethanol production as the largest contributor, Brazilian national production challenging and scaling for more production, and a strongly growing interest from Asian nations such as China, India, and Thailand as this is reflective of the 2.2 billion liter growth of 2019–2020 global biofuel production [13].

This can not only be attributed to the growing demand for better environmental, pollution, and energy management, but is also driven by marketing incentives and supporting policies. Brazil exhibits such an attitude in which the introduction of the 2020 policy "RenovaBio", a marketing incentive based on the coarse market theory, where the socio-economic costs of fossil fuel utility are used to produce biofuels [13,47]. It is arguably a good strategy to promote and accommodate the nation's transition into cleaner energy – as the production costs of bioethanol are no easy undertaking. "RenovaBio" and its constituent scientific tools "RenovaCalc" allow for the assessment and comparison of greenhouse gas (GHG) emissions between fossil fuels and biofuels from an acclaimed life cycle [47]. The RenovaBio system revolves around the market trading and sale value of decarbonization credits/certification (CBIOs), more internationally known as emission reduction certificates, which represents "One metric ton of carbon save through the utilization of biofuels against fossil fuels" according to their Brazilian legislative frameworks. The IEA reports in their annual report that, by September 2020, the trade value of CBIOs reached up to USD 4.6 per CBIO, as well as report the awarding of CBIOs to 200 producers earlier in June 2020 [13,47]. Such positive feedback supports the growth and effect of the legislative action, which would benefit the country's global commitments and prestige toward international organizations such as the United Nations Framework Convention on Climate Change (UNFCCC). Such frameworks are not only limited to countries such as Brazil; on an international scale, the Sustainable Development Scenario (SDS) acts as a framework rubric designed to provide nation's transition toward cleaner energy and is prescribed by the United Nations, with the main goals being net zero emission energy to combat the growing global temperature as a consequence of GHG emissions.

With regard to bioethanol production from biowaste, such evaluations and analysis cannot be easily attained because of many contributing components from feedstock, policies, and technology. The development of new infrastructures, process costs, and transportation of biowastes still require a substantial amount of capital to initiate, albeit subsided by the low cost of biowaste feedstock procurement. The economic analysis of biowaste conversion to bioethanol was studied at many levels using different modeling approaches and scenarios. A typical approach for economic analysis would be from an operation costs standpoint, but with the abundance of different techniques, a generalized decision as to whether bioethanol production from biowaste is feasible would not be representative of the entirety of bioethanol production. An example of economic analysis can be on the usage of the thermochemical conversion processes and their commercialization. The study claimed that a feedstock composed of variously mixed domestic biowaste should be used, in which the study further comments the usage of large-scale plant with a feed of 2,000 dry-ton/day biowastes to produce 240,000 L of bioethanol per day is much more economical as compared to a small-scale plant 1,000 dry-ton/day biowastes – emphasizing the importance of plant construction [48]. In another scenario, the usage of a feedstock composed of domestic biowaste and coal for bioethanol production via hybrid processing within a dual fluidized bed gasifier was also economically analyzed [49]. Within a similar 2,000 dry-ton/day scale feed, the cost of fuel from the composition did increase the cost; however, it was shown that there was an increase in the final output of bioethanol, thus increasing the economic efficiency of this approach [49].

Different modeling methods and schemes were also used to study costs and factors outside of operations and industry plant designs. In a study to explore the weaknesses and strengths of bioethanol from biowastes in Greece, using fuzzy cognitive maps (FCMs), a generalized outlook was formed to depict major contributors and stakeholders within a biowaste-to-bioethanol industry. The analysis mapping revealed that a multifactorial system is essential in the operations of such an industry. Aside from the technical specifications and optimized operations of the biowaste processing, the FCM provides a critical perspective on the importance of effective policies to improve the bioethanol production scheme. Aside from the tangible costs required for the industry, it is emphasized that there is a significant possibility of the indirect impact of political actions toward the production of bioethanol such that political actions similarly affect social factors equally. Further elaborating, an effectively designed policy that provides a positive educational forecast, as well as beneficial initiatives and incentives, may have the ability to enhance community participation in the industry, feedstock management, and improved quality and yield of the bioethanol products [50]. This approach and understanding could be a direct explanation to the development of many proactive legislations in support of bioethanol production.

9.6 CONCLUSIONS

This chapter deals with the investigation of various conversion strategies for biomass into ethanol. The mechanism of the process uses first- and second-generation biomass, which are based on the hydrolysis, detoxification, and fermentation processes. Depending on the type of feedstock used, production techniques for first- and second-generation biofuels vary. Additional pretreatment methods required to extract bioethanol from second-generation feedstock normally enhance the production cost and are focused in the discussion. Other than the production cost, energy efficiency is also discussed throughout the chapter. The current trends of bioethanol production and the future pathways are also explained in detail for designing powerful production pathways for bioethanol production.

REFERENCES

1. S.E. Vollset, E. Goren, C.-W. Yuan, J. Cao, A.E. Smith, T. Hsiao, C. Bisignano, G.S. Azhar, E. Castro, J. Chalek, A.J. Dolgert, T. Frank, K. Fukutaki, S.I. Hay, R. Lozano, A.H. Mokdad, V. Nandakumar, M. Pierce, M. Pletcher, T. Robalik, K.M. Steuben, H.Y. Wunrow, B.S. Zlavog, C.J.L. Murray, Fertility, mortality, migration, and population

scenarios for 195 countries and territories from 2017 to 2100: A forecasting analysis for the Global Burden of Disease Study, *Lancet.* 396 (2020) 1285–1306. https://doi.org/10.1016/S0140–6736(20)30677–2.
2. E. Chiappero-Martinetti, Basic needs, in: *Encycl. Qual. Life Well-Being Res.*, Springer, Dordrecht, Netherlands, 2014: pp. 329–335. https://doi.org/10.1007/978-94-007-0753-5_150.
3. *Population Summit of the World's Scientific Academies*, National Academies Press, Washington, D.C., 1993. https://doi.org/10.17226/9148.
4. V. Smil, *Energy and Civilization*, The MIT Press, 2017. https://doi.org/10.7551/mitpress/10752.001.0001.
5. IEA, World Energy Outlook 2020 (2020). https://www.iea.org/reports/world-energy-outlook-2020.
6. H. Lund, Introduction, in: *Renew. Energy Syst.*, Elsevier, 2014: pp. 1–14. https://doi.org/10.1016/B978–0–12–410423–5.00001–8.
7. B.D. Solomon, Biofuels and sustainability, *Ann. N. Y. Acad. Sci.* 1185 (2010) 119–134.
8. M. Fatih Demirbas, M. Balat, H. Balat, Biowastes-to-biofuels, *Energy Convers. Manag.* 52 (2011) 1815–1828.
9. R.A. Lee, J.-M. Lavoie, From first- to third-generation biofuels: Challenges of producing a commodity from a biomass of increasing complexity, *Anim. Front.* 3 (2013) 6–11.
10. European Commission, Guidance for the compilation and reporting of data on municipal waste according to Commission Implementing Decisions 2019/1004/EC and 2019/1885/EC, and the Joint Questionnaire of Eurostat and OECD (2020). https://ec.europa.eu/eurostat/documents/342366/351811/Guidance+on+municipal+waste+data+collection/.
11. S. Kaza, L.C. Yao, P. Bhada-Tata, F. Van Woerden, *What a Waste 2.0: A Global Snapshot of Solid Waste Management to 2050*, World Bank, Washington, DC, 2018. https://doi.org/10.1596/978–1–4648–1329–0.
12. H. Zabed, J.N. Sahu, A. Suely, A.N. Boyce, G. Faruq, Bioethanol production from renewable sources: Current perspectives and technological progress, *Renew. Sustain. Energy Rev.* 71 (2017) 475–501.
13. IEA, Renewables 2020 (2020). https://www.iea.org/reports/renewables-2020.
14. A.K. Chandel, S.S. da Silva, O.V. Singh, Detoxification of Lignocellulose hydrolysates: Biochemical and metabolic engineering toward white biotechnology, *BioEnergy Res.* 6 (2013) 388–401. https://doi.org/10.1007/s12155-012-9241–z.
15. N.S. Mat Aron, K.S. Khoo, K.W. Chew, P.L. Show, W. Chen, T.H.P. Nguyen, Sustainability of the four generations of biofuels – A review, *Int. J. Energy Res.* 44 (2020) 9266–9282.
16. H.A. Alalwan, A.H. Alminshid, H.A.S. Aljaafari, Promising evolution of biofuel generations. Subject review, *Renew. Energy Focus.* 28 (2019) 127–139.
17. C.A. Barcelos, R.N. Maeda, G.J.V. Betancur, N. Pereira Jr., Ethanol production from sorghum grains [Sorghum bicolor (L.) Moench]: Evaluation of the enzymatic hydrolysis and the hydrolysate fermentability, *Brazilian J. Chem. Eng.* 28 (2011) 597–604.
18. S.C. Rabelo, R.M. Filho, A.C. Costa, Lime pretreatment of sugarcane bagasse for bioethanol production, *Appl. Biochem. Biotechnol.* 153 (2009) 139–150.
19. M. Vohra, J. Manwar, R. Manmode, S. Padgilwar, S. Patil, Bioethanol production: Feedstock and current technologies, *J. Environ. Chem. Eng.* 2 (2014) 573–584.
20. C.E. Wyman, Ethanol fuel, in: *Encycl. Energy*, Elsevier, 2004: pp. 541–555. https://doi.org/10.1016/B0-12-176480-X/00518–0.
21. D. Chiaramonti, Bioethanol: Role and production technologies, in: *Improv. Crop Plants Ind. End Uses*, Springer, Dordrecht, Netherlands, n.d.: pp. 209–251. https://doi.org/10.1007/978-1-4020-5486-0_8.

22. D.B. Johnston, A.J. McAloon, Protease increases fermentation rate and ethanol yield in dry-grind ethanol production, *Bioresour. Technol.* 154 (2014) 18–25.
23. S. Kumar, N. Singh, R. Prasad, Anhydrous ethanol: A renewable source of energy, *Renew. Sustain. Energy Rev.* 14 (2010) 1830–1844.
24. N.S. Mosier, K.E. Ileleji, How fuel ethanol is made from corn, in: *Bioenergy*, Elsevier, New York, USA, 2020: pp. 539–544.
25. H. Shapouri, J.A. Duffield, M.Q. Wang, The Energy Balance of Corn Ethanol: An Update (2002). https://doi.org/10.22004/ag.econ.34075.
26. V. Singh, K.D. Rausch, P. Yang, H. Shapouri, R.L. Belyea, M.E. Tumbleson, Modified Dry Grind Ethanol Process (2001). Thesis – University of Illinois. http://abe-research.illinois.edu/pubs/k_rausch/Singh_etal_Modified_Dry_grind.pdf.
27. R.J. Bothast, M.A. Schlicher, Biotechnological processes for conversion of corn into ethanol, *Appl. Microbiol. Biotechnol.* 67 (2005) 19–25.
28. N. Sarkar, S.K. Ghosh, S. Bannerjee, K. Aikat, Bioethanol production from agricultural wastes: An overview, *Renew. Energy.* 37 (2012) 19–27.
29. A. Demirbaş, Bioethanol from cellulosic materials: A renewable motor fuel from biomass, *Energy Sources.* 27 (2005) 327–337.
30. Ó.J. Sánchez, C.A. Cardona, Trends in biotechnological production of fuel ethanol from different feedstocks, *Bioresour. Technol.* 99 (2008) 5270–5295.
31. Y. Sun, J. Cheng, Hydrolysis of lignocellulosic materials for ethanol production: A review, *Bioresour. Technol.* 83 (2002) 1–11.
32. A.B. Bjerre, A.B. Olesen, T. Fernqvist, A. Plöger, A.S. Schmidt, Pretreatment of wheat straw using combined wet oxidation and alkaline hydrolysis resulting in convertible cellulose and hemicellulose, *Biotechnol. Bioeng.* 49 (2000) 568–577.
33. F. Talebnia, D. Karakashev, I. Angelidaki, Production of bioethanol from wheat straw: An overview on pretreatment, hydrolysis and fermentation, *Bioresour. Technol.* 101 (2010) 4744–4753.
34. P. Das, A. Ganesh, P. Wangikar, Influence of pretreatment for deashing of sugarcane bagasse on pyrolysis products, *Biomass and Bioenergy.* 27 (2004) 445–457.
35. Z. Hu, Z. Wen, Enhancing enzymatic digestibility of switchgrass by microwave-assisted alkali pretreatment, *Biochem. Eng. J.* 38 (2008) 369–378.
36. V. Sridar, Microwave radiation as a catalyst for chemical reactions, *Curr. Sci.* 74 (1998) 446–450.
37. C.A. Cardona, J.A. Quintero, I.C. Paz, Production of bioethanol from sugarcane bagasse: Status and perspectives, *Bioresour. Technol.* 101 (2010) 4754–4766.
38. M. Balat, H. Balat, C. Öz, Progress in bioethanol processing, *Prog. Energy Combust. Sci.* 34 (2008) 551–573.
39. A. Pandey, C.R. Soccol, P. Nigam, V.T. Soccol, Biotechnological potential of agro-industrial residues. I: Sugarcane bagasse, *Bioresour. Technol.* 74 (2000) 69–80.
40. C. Martín, H.B. Klinke, A.B. Thomsen, Wet oxidation as a pretreatment method for enhancing the enzymatic convertibility of sugarcane bagasse, *Enzyme Microb. Technol.* 40 (2007) 426–432.
41. X. Zhao, K. Cheng, D. Liu, Organosolv pretreatment of lignocellulosic biomass for enzymatic hydrolysis, *Appl. Microbiol. Biotechnol.* 82 (2009) 815–827.
42. M.A. Neves, T. Kimura, N. Shimizu, M. Nakajima, State of the art and future trends of bioethanol production, *Dyn. Biochem. Process Biotechnol. Mol. Biol.* 1 (2007) 1–14.
43. N. Mosier, Features of promising technologies for pretreatment of lignocellulosic biomass, *Bioresour. Technol.* 96 (2005) 673–686.
44. P. Alvira, E. Tomás-Pejó, M. Ballesteros, M.J. Negro, Pretreatment technologies for an efficient bioethanol production process based on enzymatic hydrolysis: A review, *Bioresour. Technol.* 101 (2010) 4851–4861.

45. C.N. Hamelinck, G. van Hooijdonk, A.P. Faaij, Ethanol from lignocellulosic biomass: Techno-economic performance in short-, middle- and long-term, *Biomass Bioenergy* 28 (2005) 384–410.
46. S. Prasad, A. Singh, H.C. Joshi, Ethanol as an alternative fuel from agricultural, industrial and urban residues, *Resour. Conserv. Recycl.* 50 (2007) 1–39.
47. S. Barros, Implementation of RenovaBio - Brazil's National Biofuels Policy, Sao Paulo (2021). https://usdabrazil.org.br/wp-content/uploads/2021/05/Implementation-of-RenovaBio-Brazils-National-Biofuels-Policy_Sao-Paulo-ATO_Brazil_02-25-2021.pdf
48. I.S. Gwak, J.H. Hwang, J.M. Sohn, S.H. Lee, Economic evaluation of domestic biowaste to ethanol via a fluidized bed gasifier, *J. Ind. Eng. Chem.* 47 (2017) 391–398.
49. Y.R. Gwak, Y. Bin Kim, I.S. Gwak, S.H. Lee, Economic evaluation of synthetic ethanol production by using domestic biowastes and coal mixture, *Fuel.* 213 (2018) 115–122.
50. A. Konti, D. Damigos, Exploring strengths and weaknesses of bioethanol production from bio-waste in Greece using Fuzzy Cognitive Maps, *Energy Policy.* 112 (2018) 4–11.

10 Waste Feedstocks for Biodiesel Production

Umer Rashid, Rose Fadzilah Abdullah, and Balkis Hazmi
Universiti Putra Malaysia

Wan Nur Aini Wan Mokhtar
Universiti Kebangsaan Malaysia

CONTENTS

10.1 INTRODUCTION

Biodiesel is one of the best alternatives for conventional fossil fuels, and its use is growing all over the world. In conjunction with global economic evaluation and the expansion of the automobile industry sectors, researchers consistently provide the

DOI: 10.1201/9781003178354-13

greatest discoveries and inventions to produce alternatives to fossil fuel. Biodiesel has gained popularity around the world because it is renewable and energy efficient as it is primarily made from crop waste and recycled resources. Besides, biodiesel is also non-toxic, biodegradable, and safe for use in sensitive environments [1]. The biodiesel produced can also be used in most diesel engines with no or only minor modifications. Furthermore, the usage of biodiesel could reduce global warming gas emissions as it emits lesser hydrocarbons (HCs), carbon monoxide (CO), carbon dioxide (CO_2), nitrous oxide (NO_x), sulfur oxide (SO_x), and particulate matter (PM) [2].

After a century of the invention by Sir Rudolf Diesel in the 1890s, the first biodiesel plant was established in Austria in 1985. In the beginning, biodiesel was produced from sugar, starch, and vegetable oil, which became known as the first generation of biodiesel. As with many other arguments about food production competition and food security concerns, the second generation has evolved to overcome the issues. The second generation of biodiesel is produced from waste vegetable oils, forest residues, and agricultural and industrial wastes. The utilization of waste materials as a feedstock has received a lot of intention because it cuts manufacturing costs by more than 80% and benefits waste management. The third generation is well known as oilgae, and the oil is derived from the algae. The fourth generation will eventually focus on carbon capture and storage, which will necessitate plant biology and metabolic engineering knowledge. The cultivation of algae by the third and fourth biodiesel generations could maintain the carbon dioxide in the atmosphere as the oil crops grow and absorb the emitted CO_2 when the fuel is combusted, which is described as "carbon neutrality" [3].

Biodiesel production in Malaysia began in 2006 when the Malaysian government announced the National Biofuel Policy, which focuses on reducing the reliance on fossil fuels and utilizing environmentally friendly and sustainable energy sources. In addition, the policy aims to stabilize the palm oil industry, with Malaysia being the second largest palm oil producer in the world after Indonesia. After the Malaysian Parliament approved the policy in 2007, the Malaysian Palm Oil Board (MPOB) under the Ministry of Plantation Industries and Commodities (MPIC) is responsible for research and development, production, and commercialization of biodiesel in Malaysia. Malaysia has allocated US$79 million for blending facilities and infrastructure and started marketing 5% blending in 2007 before increasing to 7% in 2014. Unfortunately, the domestic implementation of the 5% and 7% blends was not achieved until 2016. Malaysia pursued its efforts by announcing the 11th Malaysia Plan 2016–2020, a 5-year strategy at the end of 2015. Malaysia's goal under this approach was to raise the blending rate to 20% by 2020. Malaysia has 18 operational biodiesel plants, producing 1.25 billion liters of biodiesel using palm-based oil in 2020, and 9 petroleum blending facilities that supply 4,000 petrol stations in Malaysia. The highest amount of biodiesel was produced in 2019, which reduced in 2020 due to the COVID-19 pandemic, which slowed the economic activity. To date, Malaysia has commercialized 7% blending biofuel known as Euro 5 on April 1, 2021, replacing Euro 2M with the same blending rate [4,5].

10.2 WASTE OILS

Biodiesel derived from inedible feedstocks is an alternative for lowering production costs and overcoming the reliance on traditional feedstocks for biodiesel production. In Malaysia, waste materials such as waste cooking oil (WCO); palm fatty acid distillate (PFAD); palm oil mill effluent (POME); and fat, oil, and grease (FOG) can be beneficial to substitute palm oil-based biodiesel.

10.2.1 WCO

WCO can be defined as a by-product of fresh frying oil, as shown in Figure 10.1. WCO can be obtained from households, restaurants, food stalls, food processing factories, and selected recycling centers. It has been reported that over 0.5 million tonnes of WCO are collected in Malaysia each year, with the remainder of uncollected WCO being drained into the drains, sewage systems, and landfills. Aside from producing biodiesel, WCO has been used as a base material for soap manufacturing and has traditionally been used as a household lubricant.

FIGURE 10.1 WCO after frying activities.

FIGURE 10.2 Fat, oil, and grease from the local food stall.

10.2.2 FOG

Fat, oil, and grease (FOG) are the remaining WCO in the drains that have been mixed with water, food waste, detergents, and animal fats. FOG have a rich FFA content of more than 20% and a high moisture content. As shown in Figure 10.2, FOG are a low-density compound that always floats and accumulates on the surface of drains or sewage lines and is insoluble in water due to the high-fat composition that has a greasy and viscous liquid texture, a semisolid or solid yellow or brown color, and an unpleasant odor. Direct discharge without pretreatment may cause sewage blockage. Thus, utilizing FOG for biodiesel production can reduce the excessive amount of FOG while also saving the environment [6]. Before being subjected to biodiesel synthesis, sample pretreatment such as heating and filtration is required to eliminate water content and suspended compounds.

10.2.3 PFAD

A non-edible semisolid cake by-product known as PFAD is produced every 4%–5% of crude palm oil refining, with an estimated annual production of over eight hundred thousand (800,000) tonnes, as shown on Figure 10.3. More than 85% of PFAD comprises free fatty acids (FFAs), with the remaining made up of diglycerides, triglycerides, squalene, vitamin E, sterol, and other minerals. At room temperature, the PFAD is a greasy semisolid with a pale yellow color, which turns into a brownish liquid-like substance when heated to above 60°C. The chemical refining process generates PFAD during alkali neutralization, whereas the physical refining process deodorizes after the deacidification process.

FIGURE 10.3 PFAD at room temperature.

10.2.4 POME

POME is a colloidal suspension of wastewater produced during the sterilization and clarification process of fresh fruit bunch (FFB) and mesocarp fibers [7]. More than 60 million tonnes of POME is generated per year, with 95%–96% of water and 4%–5% of total solid, including 2%–4% suspended solids and 0.6%–0.7% oil. POME contains a high amount of degradable organic matter, which when improperly discharged without pretreatment harms aquatic life due to eutrophication and the resulting alga bloom. Meanwhile, a trace amount of POME can be further utilized for biodiesel production.

10.3 PHYSICAL AND CHEMICAL PROPERTIES OF WASTE OIL

The various feedstock sources available in Malaysia exhibit different physical and chemical properties, which are mainly attributed to their fatty acid composition, as presented in Table 10.1. The fatty acid composition is important because the degree of saturation presented significantly influenced the physicochemical properties of the biodiesel [8].

10.3.1 Moisture Content

The removal of moisture content from the oil is very important before the oil can be ready for biodiesel production. High moisture content in the oil will initiate hydrolysis, polymerization, and oxidation that produce volatile and non-volatile polymeric and monomeric compounds including triacylglycerols (TAG), diacylglycerols (DAG), and monoacylglycerols (MAG) [13]. Other than that, this situation also

TABLE 10.1
Physical and Chemical Properties of WCO, FOG, PFAD, and POME

Property	WCO [9]	FOG [10]	PFAD [11]	POME [12]
Acid number (mg KOH/g)	2.8	24–180	184.6	153.7
Saponification value (mg KOH/g)	198.3	193.6	212.08	211.7
Molecular weight (g/mol)	850.4	289.9	265.6	265.03
Moisture content (wt%)	0.3	>60.0	0.8	95.5
Lauric acid (C12:0)	0.39	-	0.18	-
Myristic acid (C14:0)	0.88	6.10	0.18	1.10
Palmitic acid (C16:0)	36.63	35.70	58.81	44.70
Steric acid (C18:0)	4.18	8.70	3.23	3.70
Oleic acid (C18:1)	42.39	34.6	30.28	40.00
Linoleic acid (C18:2n6c)	9.85	6.00	6.41	10.00

negatively affects transesterification since FFAs and high moisture content will cause soap formation, thus reducing the effectiveness of the catalyst resulting in a low biodiesel yield. The moisture content can be determined by placing 5 g of WCO in an aluminum pan and letting it dry at 100 °C. After 30 min, the sample was cooled in a desiccator and weighed. The procedures were repeated until a constant weight was achieved. The percentage was then calculated by using Equation 10.1.

$$\text{Moisture content}(\%) = \frac{M_A - M_B}{M_A} \times 100 \tag{10.1}$$

where M_A is the initial weight of WCO minus the weight of aluminum pan and M_B is the weight of dried WCO minus the weight of the aluminum pan.

10.3.2 Acid Number

The acid number measures the amount of potassium hydroxide required to neutralize the FFAs in the oil sample. Referring to the AOCS Cd 3d-63 method, 2.0 g of oil was dissolved in a mixture of isopropanol and toluene. The mixture was then titrated against 0.1 N potassium hydroxide solution with the phenolphthalein indicator. The acid number was calculated by using Equation 10.2.

$$\text{Acid number} = \frac{(A - B) \times N \times 56.11}{w} \times 100\,\text{mg KOH per g sample} \tag{10.2}$$

where A is the volume of titrant used for the sample, B is the volume of titrant used for blank, N is the normality of KOH, and w is the weight of the sample.

10.3.3 Saponification Value (SV)

The determination of SV of oil is important to measure the oil content that will form soap when reacting with an alkaline solution. Generally, the saponification value is measured through titration with an acid solution by referring to AOCS Cd 3-25.

This method requires the oil to be refluxed with potassium ethanolic solution. Thus, the SV can be calculated by using Equation 10.3. In extension, the molecular weight of the oil sample can also be determined by using Equation 10.4.

$$SV = \frac{(V_b - V_s) \times N\,\text{HCl} \times 56.1}{\text{weight of oil}} \quad (10.3)$$

where V_b is the volume of HCl used for blank, V_s is the volume used for the sample, and N is the normality of the HCl.

$$MW = \frac{56.1 \times 1{,}000}{SV} \quad (10.4)$$

10.4 PRODUCTION OF BIODIESEL FROM WASTE OIL

Esterification and transesterification are chemical reactions that produce fatty acid esters from fatty acids or triglycerides with alcohol in the presence of an appropriate catalyst, respectively [14,15]. Both reversible reactions produce fatty acid methyl esters (FAMEs) or biodiesel with water or glycerol as a by-product, as shown in Figure 10.4. The WCO has a high acidity value in comparison with the cooking oil; thus, acid pretreatment is required before continuing with transesterification with a basic catalyst to avoid saponification [16]. However, due to the advancement in

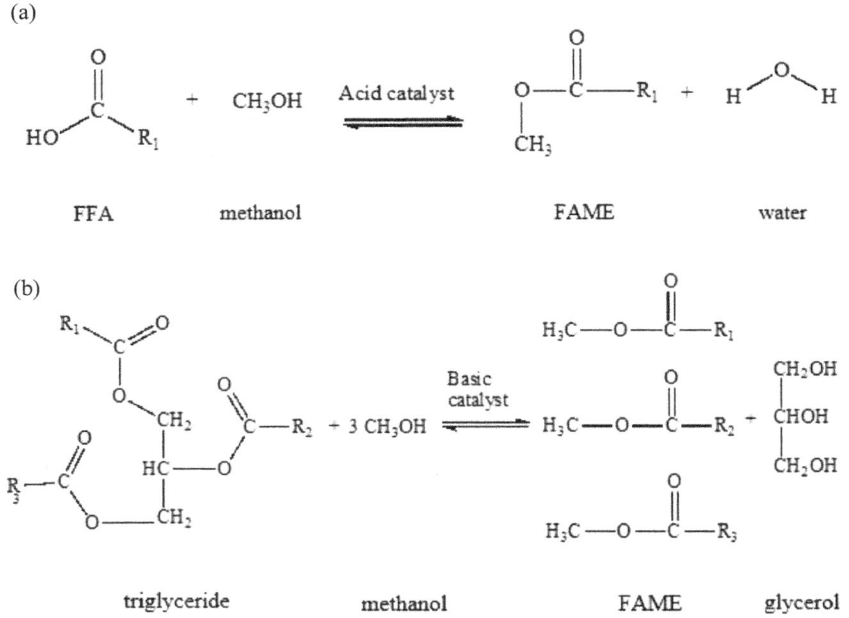

FIGURE 10.4 Schematic representation of (a) esterification and (b) transesterification reactions.

catalytic technology, a heterogeneous catalyst can be functionalized with acidic and basic functionality; thus, biodiesel feedstock with a slightly high FFA value can be converted into biodiesel in a single process step pretreatment [17,18]. Simultaneous esterification and transesterification of WCO and chicken fat using sulfonated carbon glycerol produced >90% of biodiesel yield [19].

10.5 BIODIESEL PROPERTIES

Interestingly, Malaysia had come out with Malaysia Standard of Palm Methyl Ester (MS 2008), the first national palm methyl ester standard published in 2008. This MS was adopted from both ASTM and EN/ISO standards. Along with the excellent development, the MS was revised in 2014 with few major modifications on the requirements on carbon residue, oxidation stability, monoglyceride content, and phosphorus content. At the same time, Malaysia also published an MS that specifies the requirements for Euro 5 diesel in MS 123-1:2014. The minimum/maximum limits of ASTM D6751, EN 14214, MS 2008:2014, and MS 123-1:2014 are presented in Table 10.2, and the properties are discussed.

10.5.1 Density and Kinematic Viscosity

Density is defined as the mass per unit volume at a given temperature, while kinematic viscosity measures the fluidity of the produced biodiesel. The density was determined through ASTM D4052 by using a pycnometer and was measured at 20°C. The density of biodiesel is very important as it is related to cetane number and heating value. On the other hand, the kinematic viscosity was measured by flowing a known volume of the biodiesel under gravity to pass through a calibrated glass capillary of a viscometer at 40 °C. Generally, the viscosity will increase with decreasing temperature. Both high density and high kinematic viscosity will cause many problems to the diesel engine, such as poor fuel atomization that causes engine deposits; thus, biodiesel with high density and viscosity is prohibited in cold regions.

10.5.2 Flash Point

Flash point is defined as the lowest temperature at which the vapor will ignite when an ignition source is applied and continue burning even after the source is removed. The determination of the flash point of biodiesel is important to determine its volatility and fire resistance, especially for storage and handling where special requirements can be set up, especially relating to temperature to avoid any danger during transportation. Flash point should be in the range of 93 °C–130 °C with reference to ASTM standards. The flash point was measured by referring to ASTM D56–79 method using a closed cup method. A test flame was introduced periodically at every 5 °C over the top of the open cup that contains 10 mL of the produced biodiesel. The flash point temperature was recorded as the lowest temperature at which the vapor of the oil sample ignited. The minimum limit of the ASTM requirements is 93 °C; however, biodiesel usually has a high flash point of 150 °C compared to the conventional fossil fuel, which is in the range of 55 °C–66 °C.

TABLE 10.2
List of ASTM D6751, EN 14214, MS 2008:2014, and MS 123-1:2014 Requirements for Biodiesel

Property	ASTM D6751	EN 14214	MS 2008:2014	MS 123-1:2014
Density at 15 °C (kg/m^3)	–	Max. 900	860–900	0.810–0.870
Kinematic viscosity at 40°C (mm^2/s)	1.9–6.0	Max. 5.00	3.50–5.00	1.5–5.8
Flash point (°C)	Min. 93	Min. 120	Min. 120	Min. 60
Cloud point (°C)	–	–	–	Max. 19.0
Sulfur content (mg/kg)	Max. 15.0	Max. 10.0	Max. 10.0	-
Cetane number	Min. 47.0	Min. 51.0	Min. 51.0	Min. 49.0
Sulfated ash content (% m/m)	Max. 0.02	Max. 0.02	Max. 0.02	Max. 0.01
Sediment by extraction (mass %)	–	–	–	Max. 0.01
Carbon residue on 10% bottom (mass %)	Max. 0.05	–	–	Max. 0.20
Water content (mg/kg)	Max. 200	Max. 500	Max. 500	Max. 0.05 (by distillation, vol%)
Total contamination (mg/kg)	–	Max. 24	Max. 24	–
Copper strip corrosion (3 h at 50 °C)	Max. No. 3	Class 1	Class 1	Max. 1 (3 h at 100°C)
Color (ASTM)	–	–	–	Max. 2.5
Oxidation stability, 110 °C (h)	3 h min	8 h min	–	–
Acid number (mg KOH/g)	Max. 0.50	Max. 0.50	Max. 0.50	Max. 0.25
Iodine value (g iodine/100 g)	–	Max. 120	Max. 110	–
Linolenic acid methyl ester (% m/m)	–	Max. 12.0	Max. 12.0	–
Polyunsaturated (>4 double bonds) methyl esters (% m/m)	–	–	Max. 1	–
Methanol content (% m/m)	Max. 0.20	Max. 0.20	Max. 0.20	–
Monoglyceride content (% m/m)	Max. 0.40	Max. 0.70	Max. 0.70	–
Diglyceride content (% m/m)	–	Max. 0.20	Max. 0.20	–
Triglyceride content (% m/m)	–	Max. 0.20	Max. 0.20	–
Free glycerol (% m/m)	Max. 0.02	Max. 0.02	Max. 0.02	–
Total glycerol (% m/m)	Max. 0.24	Max. 0.25	Max. 0.25	–
Group I metals (Na + K) (mg/kg)	Max. 5.0	Max. 5.0	Max. 5.0	–
Group II metals (Ca + Mg) (mg/kg)	Max. 5.0	Max. 5.0	Max. 5.0	–
Phosphorus content (mg/kg)	Max. 0.001 wt%	Max. 4.0	Max. 4.0	–
CFPP (°C)	–	–	Max. 15	–
Electrical conductivity (pS/m)	–	–	–	Min. 50
Lubricity (μm)	–	–	–	Max. 460
Ester content (% m/m)	–	Min. 96.5	Min. 96.5	–

10.5.3 Cloud Point and Pour Point

Cloud point and pour point are important parameters in low-temperature regions. The cloud point is determined by the lowest temperature at which the biodiesel starts to form particles of wax crystal and become cloudy. Pour point is described as the lowest temperature at which the oil loses its flowing characteristics before it crystallizes and becomes milky and cloudy. The pour point was determined by placing the oil in a freezer system with a thermometer to read the temperature. The oil was inspected every 5 min by tilting horizontally to check its flowing properties. The temperature was recorded as the oil is not sagged for at least 5 s, which is known as the solidification point. However, the pour points were reported to be 3 °C higher than the solidification point to represent the final temperature at which the oil loses its flowing properties.

10.5.4 Cetane Number

Cetane number indicates the autoignition quality of a fuel or can be explained as the ability of a fuel to start ignition and burn. Thus, it also relates to the ignition delay time that measures the interval between the start of injection and the start of combustion. The name cetane number was derived from its original chemical compound n-hexadecane ($C_{16}H_{34}$). From its molecular structure, cetane exists as a straight saturated hydrocarbon. High cetane numbers of more than 35 indicate lower ignition delay, which is a favorable fuel quality. However, a cetane number higher than 60 would cause excessive heating at the injector as the fuel may quickly ignite as soon as it reaches the injector nozzle. Therefore, biodiesel derived from palm, coconut, and tallow oil will have a high cetane number.

10.6 ENGINE PERFORMANCE AND EMISSIONS

The transportation sector, which uses commercial diesel, contributes considerably to atmospheric pollution by diesel engine exhaust gases, for instance HCs, CO, NO_x, SO_x, PM, and unreacted hydrocarbons (UHCs). In compression ignition (CI) engines, pure biodiesel or biodiesel blends can be used directly without any modification. This section includes in-depth information on the compression ignition engine performance and emissions running on biodiesel and its blends.

10.6.1 Engine Performance

Biodiesel performances are mainly dependent on the design of the combustion chamber of an unmodified diesel engine, and the test results are commonly different from one engine to another. Basically, there are various parameters to measure the engine performances: mechanical efficiency, brake power (BP), exhaust gas temperature, brake-specific fuel consumption (BSFC), engine torque, and brake thermal efficiency (BTE). Among these variables, BTE, BP, and BSFC are greatly impacted by the use of biodiesel blends as a fuel. By definition, BSFC is the amount of fuel needed to generate one unit of power output, and it is a metric for assessing an engine's performance.

The effect of pure biodiesel or B100 on BSFC is better than that of B0 fuel, owing to the higher density, viscosity, and smaller calorific value properties of B100 [20]. Another vital factor, BP is preferred to be used in petroleum diesel, as biodiesel has a small heating value. However, BP can be improvised by raising the fuel injection pressure [21]. Brake thermal efficiency is typically linked to the ratio between the combustion chamber (CC) and cylinder volume in an internal combustion engine or is known as compression ratio (CR); it might be higher for biodiesel blends than for conventional diesel. It occurs when the amount of fuel chemical energy indicated by the fuel's lower heating value is divided by the amount of real brake work performed in each phase [22].

Most of the reports showed that BSFC was clearly increased with the increase in biodiesel percentage in the fuel blends. An increment in BSFC was performed after consumed with B50 and B100 fuels using the John Deere 6076TF030 engine and decreased in BTE. Meanwhile, a total of 4% decline in BTE and an increase of 0.07 kg/kWh for minimal BSFC were measured in a Kirloskar mono-cylinder engine with B100 biodiesel [23]. A further study on the performance of B100 biodiesel reported that BSCF was indirectly proportional to BP when using a four-cylinder turbocharged DI engine [24]. Yildizhan et al. [25] discovered that higher CRs (14 CR and 16 CR) caused a higher BTE and lower BFSC, but the efficiency was slightly lower for the respective CR in contrast to standard diesel.

It is reported that BTE was increased by 5.4% and BP and BSFC decreased by 0.05 kW and 0.055 kg/kWh, respectively. The use of B20 fuel in a single-cylinder water-cooled engine with a rated power of 3.7 kW resulted in a good performance toward BP and BSFC, but a decrement in BTE [26]. The increased viscosity could explain why second-generation biodiesel has a lower BTE than standard diesel fuel. Overall, it can be concluded that the heating value declines as the percentage of biodiesel in the mixed fuel increases, and this value is practically lower than that of conventional diesel.

10.6.2 Exhaust Emissions

As previously stated, pure biodiesel or B100 has the highest cetane number and density compared to other diesel–biodiesel mixed fuels, although this is not a necessary criterion for running a diesel engine. The disadvantage of B100 biodiesel is that it has a lower heating value than blended diesel; double quantity of biodiesel is used in the engine to produce the equivalent heat energy of conventional diesel. Meanwhile, pure diesel or B0 is diametrically opposed to B100, whose high carbon content creates CO and CO_2 emissions [27]. Therefore, viscous biodiesel blends such as B10, B15, or B20 are suggested to control the release of pollutant gas emissions. According to Attia and Hassaneen [28], blends containing 10% biodiesel produced the best fuel economy, while blends containing 30%–50% biodiesel produced the best emissions. Table 10.1 summarizes the significant findings of the emission characteristics of biodiesel blends used in a diesel engine. In this chapter, elaborations focus more on the HC, CO, CO_2, and NO_x emissions as below.

Less ignition temperature and incomplete combustion are the primary reasons for HC emissions. Based on the physicochemical properties of biodiesel, it is clear that the oxygen content in biodiesel is higher than that of diesel fuel, resulting in more efficient fuel combustion and lower CO and HC emissions. In general, a 5% biodiesel blend is capable of reducing greenhouse gas emissions by 3% every gross tonne-kilometer. According to Lertsathapornsuk et al. [29], B100 had the greatest reduction in HC emissions, but when a mix of higher B40 was used at high CR, contrastingly, the HC emission increased [22]. Surprisingly, HC emissions for B20, B60, and B80 blends were lower than that for the commercial diesel at high CR, indicating a divergent trend. The amount of unburned HCs can be as high as 0.062 g/kWh for B100 and 0.081 g/kWh for B0 at low BP. On the other hand, the results were improved with B10 and B20 fuels, with unburnt HC levels of 66 and 64 ppm, respectively. Even most studies agreed that the decrease in the HC emissions relied on biodiesel blend percent. An et al. [30] found that a different trend appeared when using the biodiesel blend on a Euro 4 diesel engine.

Unlike HC emissions, the volume of CO emissions from engines is proportional to the physical and chemical features of fuel, such as peak temperature, engine cylinder, full combustion time, engine speed, oxygen availability, and fuel-to-air ratio [31]. In general, a higher-viscosity biodiesel mixture increases CO emissions as unmodified engines have lower atomization. CO emissions are even lower than for diesel at lower loads; however, they rise as the load increases. The drop is because biodiesel has more oxygen and less carbon in its molecule than diesel, allowing it to burn entirely [32]. Overall, as a result of high oxygen content and increased biodiesel viscosity, a complete consumption was achieved, with the result that CO emissions were reduced. The use of B50 and B100 biodiesel blends in a John Deere 6076TF030 six-cylinder engine showed a decrease in CO emissions by 18% and 14%, respectively [29]. Besides various biodiesel blend ratios, CO emissions decrease with an increase in the biodiesel percentage in the blend, but have an equal percentage of emission with diesel for B40 blend [22]. With no engine modification, 0.32–0.41 vol% reduction in CO emissions were observed at full load [21]. But, some other researchers found that the CO emissions of biodiesel blends were not differentiated, primarily because the emissions were so low and cannot be detected [33].

Fundamentally, complete combustion is defined as the production of more CO_2 and H_2O. CO_2 is supported by the fact that its life cycle is easily regulated by growing energy crops worldwide as the least harmful greenhouse gas. According to the idea, the amount of CO_2 produced by increased biodiesel fractions in diesel at lower CR, by hypothesis. The CO_2 trend is growing with regard to the diesel engine with higher loads due to higher fuel consumption and excess oxygen in the biodiesel composition [32]. Researchers agree to report that CO_2 emission is in an ascending trend for both biodiesel and pure diesel with engine load increment [34].

NO_x emissions from diesel engines have been shown to raise tropospheric ozone levels while decreasing methane levels. Controlling NO_x during the combustion phase is the most difficult task. NO_x emissions continue to rise regardless of the fuel utilized. The premium properties of biodiesel, such as sufficient oxygen content, higher cetane number, and injection and combustion advances, significantly impact NO_x emissions from biodiesel [35]. As the peak cylinder warmth is in direct variation

TABLE 10.3
Emission Variables of CI Engine Run on Diesel and Second-Generation Biodiesel Blends

Engine Specification	Composition of Fuel (Diesel: Biodiesel)	Emission Parameters				References
		HCs	CO	CO_2	NO_x	
Single-cylinder 4S, CI, AC, CR: 17.5-to-1, RS: 1,500 rpm, RP: 5.775 kW	70%:30%	↓	↓	↑	↑	[32]
Single-cylinder 4S, AC, CI, 661 cc, CR: 17.5-to-1, RS: 1,500 rpm, RP: 4.4 kW	0%:100%	↓	↓	↑	↓	[20]
Single-cylinder 4S, CI, WC, RS: 3,000 rpm	80%:20%	↓	↓	↑	↑	[37]
Single-cylinder 4S, CI, WC, CR: 18-to-1	80%:20%	↓	↓		↑	[22]
Single-cylinder 4S, CI, WC, CR: 5:1–22:1, RS: 1,500 rpm, RP: 3.7 kW	60%:40%	↑	↑	↓	↑	[38]
Single-cylinder 4S, CI, WC, CR: 17.5-to-1, RS: 1,500 rpm, RP: 5.2 kW	20%:80%	↓			↓	[39]

to the adiabatic flame temperature for NO_x emission rate, the percentage of NO_x once after high adiabatic flame and peak cylinder temperatures. According to the literature, biodiesel combustion contributes NO_x to the atmosphere, which has no significant difference compared to pure diesel performance. Overall, NO_x emissions for diesel and different biodiesel blends were reliant on the variations of engine speeds. NO_x emissions increase by 5.3%–13% compared to standard diesel in some specific conditions (Table 10.3) [20,36].

10.7 CONCLUSIONS

Biodiesel has a promising potential to be a green energy fuel to replenish depleting fossil fuels. In the Malaysian economy prospect, the demand for biodiesel from palm oil sources may boost the domestic production of palm oil, thereby increasing the commodity's market price. However, planting biodiesel feedstock based on edible palm oil competes with food production for land use. Therefore, the local waste materials such as WCO, FOG, PFAD, and POME may be beneficial alternatives to palm oil for producing biodiesel. The transesterification process is the most desired and appropriate approach among the different production procedures due to its better advantages. Also, the fatty acid profile of the generated feedstock determines the biodiesel's fuel characteristics. A reduction in the BP, BSFC, BTE values, in general, was observed when the engine characteristics were evaluated on the local second-generation biodiesel blends. Other than that, the biodiesel blends reduce HC, CO, and CO_2 emissions, while NO_x emission is higher compared to standard diesel. To that end, today's biodiesel production and engine analysis could focus on a variety of statistical and computational methodologies that can be useful in a variety of ways for better biodiesel products.

REFERENCES

1. Galadima, A.; Muraza, O. Waste Materials for Production of Biodiesel Catalysts: Technological Status and Prospects. *J. Clean. Prod.* Elsevier Ltd, August 1, **2020**, 121358.
2. Mahbub, N.; Gemechu, E.; Zhang, H.; Kumar, A. The Life Cycle Greenhouse Gas Emission Benefits from Alternative Uses of Biofuel Coproducts. *Sustain. Energy Technol. Assessments*, **2019**, *34*, 173–186.
3. Rao, A. A. An Overview on Biofuels and Their Advantages and Disadvantages. *Asian J. Chem.*, **2017**, *29* (8), 1757–1760.
4. Zulqarnain; Yusoff, M. H. M.; Ayoub, M.; Jusoh, N.; Abdullah, A. Z. The Challenges of a Biodiesel Implementation Program in Malaysia. *Processes*. MDPI AG October 1, **2020**, 1–18.
5. Khalid, N.; Nur Ahmad Hamidi, H.; Thinagar, S.; Fakhzan Marwan, N. Crude Palm Oil Price Forecasting in Malaysia: An Econometric Approach (Peramalan Harga Minyak Sawit Mentah Di Malaysia: Satu Pendekatan Ekonometrik). *J. Ekon. Malaysia*, **2018**, *52* (3), 247–259.
6. Montefrio, M. J.; Xinwen, T.; Obbard, J. P. Recovery and Pre-Treatment of Fats, Oil and Grease from Grease Interceptors for Biodiesel Production. *Appl. Energy*, **2010**, *87* (10), 3155–3161.
7. Rezania, S.; Oryani, B.; Cho, J.; Sabbagh, F.; Rupani, P. F.; Talaiekhozani, A.; Rahimi, N.; Ghahroud, M. L. Technical Aspects of Biofuel Production From. *Processes*, **2020**, 1–19.
8. Kumar, D.; Singh, B. Effect of Winterization and Plant Phenolic-Additives on the Cold-Flow Properties and Oxidative Stability of Karanja Biodiesel. *Fuel*, **2020**, *262* (November 2019), 116631.
9. Rachmadona, N.; Amoah, J.; Quayson, E.; Hama, S.; Yoshida, A.; Kondo, A.; Ogino, C. Lipase-Catalyzed Ethanolysis for Biodiesel Production of Untreated Palm Oil Mill Effluent. *Sustain. Energy Fuels*, **2020**, *4* (3), 1105–1111.
10. Abomohra, A. E. F.; Elsayed, M.; Esakkimuthu, S.; El-Sheekh, M.; Hanelt, D. Potential of Fat, Oil and Grease (FOG) for Biodiesel Production: A Critical Review on the Recent Progress and Future Perspectives. *Prog. Energy Combust. Sci.* Elsevier Ltd, November 1, **2020**, 100868.
11. Saravanan Arumugamurthy, S.; Sivanandi, P.; Pandian, S.; Choksi, H.; Subramanian, D. Conversion of a Low Value Industrial Waste into Biodiesel Using a Catalyst Derived from Brewery Waste: An Activation and Deactivation Kinetic Study. *Waste Manag.*, **2019**, *100*, 318–326.
12. Jume, B. H.; Gabris, M. A.; Rashidi Nodeh, H.; Rezania, S.; Cho, J. Biodiesel Production from Waste Cooking Oil Using a Novel Heterogeneous Catalyst Based on Graphene Oxide Doped Metal Oxide Nanoparticles. *Renew. Energy*, **2020**, *162*, 2182–2189.
13. Abduh, M. Y.; Syaripudin; Putri, L. W.; Manurung, R. Effect of Storage Time on Moisture Content of Reutealis Trisperma Seed and Its Effect on Acid Value of the Isolated Oil and Produced Biodiesel. *Energy Rep.*, **2019**, *5*, 1375–1380.
14. Tang, X.; Niu, S. Preparation of Carbon-Based Solid Acid with Large Surface Area to Catalyze Esterification for Biodiesel Production. *J. Ind. Eng. Chem.*, **2018**, *69*, 187–195.
15. Panadare, D. C.; Rathod, V. K. Applications of Waste Cooking Oil Other Than Biodiesel: A Review. *Iran. J. Chem. Eng.*, **2015**, *12* (3), 55–76.
16. Soltani, S.; Rashid, U.; Yunus, R.; Taufiq-Yap, Y. H. Biodiesel Production in the Presence of Sulfonated Mesoporous $ZnAl_2O_4$ catalyst via Esterification of Palm Fatty Acid Distillate (PFAD). *Fuel*, **2016**, *178*, 253–262.
17. Hasanudin, H.; Rachmat, A. Production of Biodiesel from Esterification of Oil Recovered from Palm Oil Mill Effluent (POME) Sludge Using Tungstated-Zirconia Composite Catalyst. *Indones. J. Fundam. Appl. Chem.*, **2016**, *1* (2), 42–46.

18. Al-Saadi, A.; Mathan, B.; He, Y. Biodiesel Production via Simultaneous Transesterification and Esterification Reactions over SrO–ZnO/Al_2O_3 as a Bifunctional Catalyst Using High Acidic Waste Cooking Oil. *Chem. Eng. Res. Des.*, **2020**, *162* (2018), 238–248.
19. Shatesh Kumar; Shamsuddin, M. R.; Farabi, M. A.; Saiman, M. I.; Zainal, Z.; Taufiq-Yap, Y. H. Production of Methyl Esters from Waste Cooking Oil and Chicken Fat Oil via Simultaneous Esterification and Transesterification Using Acid Catalyst. *Energy Convers. Manag.*, **2020**, *226* (August), 113366.
20. Nanthagopal, K.; Raj, R. T. K.; Ashok, B.; Elango, T.; Saravanan, S. V. Influence of Exhaust Gas Recirculation on Combustion and Emission Characteristics of Diesel Engine Fuelled with 100% Waste Cooking Oil Methyl Ester. *Waste Biomass Valoriz.*, **2019**, *10* (7), 2001–2014.
21. Buyukkaya, E. Effects of Biodiesel on a Di Diesel Engine Performance, Emission and Combustion Characteristics. *Fuel*, **2010**, *89* (10), 3099–3105.
22. Karthickeyan, V.; Balamurugan, P.; Senthil, R. Environmental Effects of Thermal Barrier Coating with Waste Cooking Palm Oil Methyl Ester Blends in a Diesel Engine. *Biofuels*, **2019**, *10* (2), 207–220.
23. Hirkude, J. B.; Padalkar, A. S. Performance and Emission Analysis of a Compression Ignition. Engine Operated on Waste Fried Oil Methyl Esters. *Appl. Energy*, **2012**, *90* (1), 68–72.
24. An, H.; Yang, W. M.; Maghbouli, A.; Li, J.; Chou, S. K.; Chua, K. J. Performance, Combustion and Emission Characteristics of Biodiesel Derived from Waste Cooking Oils. *Appl. Energy*, **2013**, *112*, 493–499.
25. Yildizhan, Ş.; Uludamar, E.; Çalık, A.; Dede, G.; Özcanlı, M. Fuel Properties, Performance and Emission Characterization of Waste Cooking Oil (WCO) in a Variable Compression Ratio (VCR) Diesel Engine. *Eur. Mech. Sci.*, **2017**, *1* (2), 56–62.
26. Pradhan, P.; Chakraborty, S.; Chakraborty, R. Optimization of Infrared Radiated Fast and Energy-Efficient Biodiesel Production from Waste Mustard Oil Catalyzed by Amberlyst 15: Engine Performance and Emission Quality Assessments. *Fuel*, **2016**, *173*, 60–68.
27. Hoang, A. T. Combustion Behavior, Performance and Emission Characteristics of Diesel Engine Fuelled with Biodiesel Containing Cerium Oxide Nanoparticles: A Review. *Fuel Process. Technol.*, **2021**, *218* (April), 106840.
28. Attia, A. M. A.; Hassaneen, A. E. Influence of Diesel Fuel Blended with Biodiesel Produced from Waste Cooking Oil on Diesel Engine Performance. *Fuel*, **2016**, *167*, 316–328.
29. Lertsathapornsuk, V.; Ruangying, P.; Pairintra, R.; Krisnangkura, K. Continuous Transethylation of Vegetable Oils by Microwave Irradiation. *First Thail. Conf. Energy*, **2003**, *1*, 11–14.
30. An, H.; Yang, W. M.; Chou, S. K.; Chua, K. J. Combustion and Emissions Characteristics of Diesel Engine Fueled by Biodiesel at Partial Load Conditions. *Appl. Energy*, **2012**, *99*, 363–371.
31. Arpa, O.; Yumrutaş, R.; Argunhan, Z. Experimental Investigation of the Effects of Diesel-like Fuel Obtained from Waste Lubrication Oil on Engine Performance and Exhaust Emission. *Fuel Process. Technol.*, **2010**, *91* (10), 1241–1249.
32. Abed, K. A.; El Morsi, A. K.; Sayed, M. M.; Shaib, A. A. E.; Gad, M. S. Effect of Waste Cooking-Oil Biodiesel on Performance and Exhaust Emissions of a Diesel Engine. *Egypt. J. Pet.*, **2018**, *27* (4), 985–989.
33. Lapuerta, M.; Armas, O.; Rodríguez-Fernández, J. Effect of Biodiesel Fuels on Diesel Engine Emissions. *Prog. Energy Combust. Sci.*, **2008**, *34* (2), 198–223.
34. Emaish, H.; Abualnaja, K. M.; Kandil, E. E.; Abdelsalam, N. R. Evaluation of the Performance and Gas Emissions of a Tractor Diesel Engine Using Blended Fuel Diesel and Biodiesel to Determine the Best Loading Stages. *Sci. Rep.*, **2021**, *11* (1), 1–12.

35. Xue, J. Combustion Characteristics, Engine Performances and Emissions of Waste Edible Oil Biodiesel in Diesel Engine. *Renew. Sustain. Energy Rev.*, **2013**, *23*, 350–365.
36. Gad, M. S.; Ismail, M. A. Effect of Waste Cooking Oil Biodiesel Blending with Gasoline and Kerosene on Diesel Engine Performance, Emissions and Combustion Characteristics. *Process Saf. Environ. Prot.*, **2021**, *149*, 1–10.
37. Yesilyurt, M. K.; Cesur, C.; Aslan, V.; Yilbasi, Z. The Production of Biodiesel from Safflower (*Carthamus Tinctorius* L.) Oil as a Potential Feedstock and Its Usage in Compression Ignition Engine: A Comprehensive Review. *Renew. Sustain. Energy Rev.*, **2020**, *119* (November 2019), 109574.
38. Muralidharan, K.; Vasudevan, D. Performance, Emission and Combustion Characteristics of a Variable Compression Ratio Engine Using Methyl Esters of Waste Cooking Oil and Diesel Blends. *Appl. Energy*, **2011**, *88* (11), 3959–3968.
39. Kumar, S.; Dinesha, P.; Rosen, M. A. Effect of Injection Pressure on the Combustion, Performance and Emission Characteristics of a Biodiesel Engine with Cerium Oxide Nanoparticle Additive. *Energy*, **2019**, *185*, 1163–1173.

11 Biowaste-Based Microbial Fuel Cells for Bioelectricity Generation

Bhim Sen Thapa
Kangwon National University
Indian Institute of Technology Madras

T. S. Chandra
Indian Institute of Technology Madras

CONTENTS

11.1 INTRODUCTION

The two major global concerns that require urgent attention are waste management and alternate source of energy. With the drastic rise in demand for sustainable fuel and alternate energy sources, there is an urgent need to address this problem. Over the last few years, some excellent progress has been made. One such technology that has emerged in recent times to tackle the energy and fuel requirements is microbial fuel cells (MFCs). MFCs are biocatalytic electrochemical cells that can recover the chemical energy present in organic compounds and convert it into electrical energy,

DOI: 10.1201/9781003178354-14

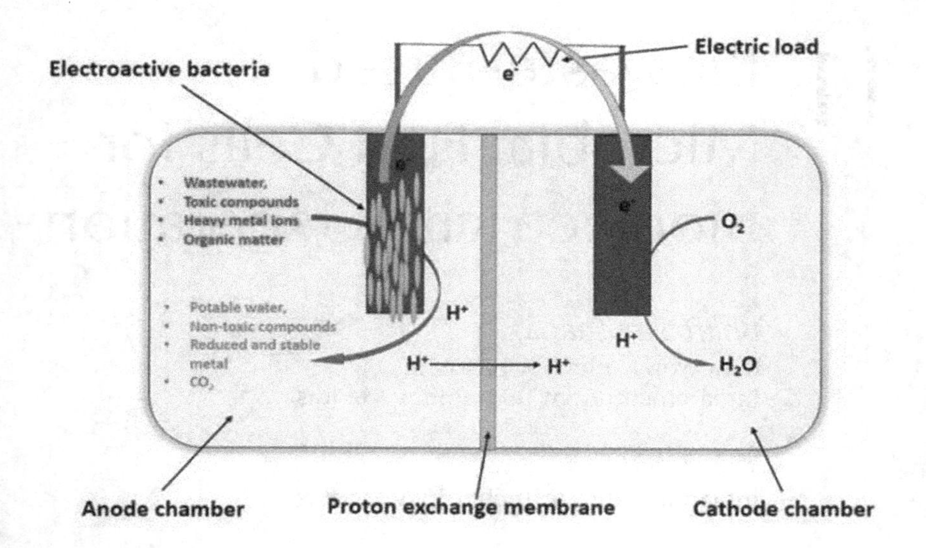

FIGURE 11.1 Pictorial representation of microbial fuel cell showing the oxidation and reduction reactions taking place in the anode and cathode.

catalyzed by microorganisms. Bacteria are the predominant group of electroactive group of microorganisms observed in MFCs; however, there are few reports on yeasts as well, but the numbers are very small. A wide range of organic and inorganic wastes are employed in MFCs and successfully treated to generate electricity. For example, bioelectricity generation from oilseed cake in a double-chamber MFC was reported [1]. Apart from bioelectricity generation, MFCs are known to produce hydrogen, methane, volatile fatty acids, and other valuable compounds. However, the potential generated solely by the MFC is too low to meet the expected quantity of these products; hence, an additional supplement, i.e., an external power source, is required to enhance the performance. In this case, the system is regarded as a microbial electrolysis cell (MEC) rather than MFC. MFC is commonly employed for treating wastewater, toxic and heavy metal pollutants, and other organic matter degradation for bioelectricity generation (Figure 11.1).

A wide variety of organic compounds have been employed in MFC, ranging from simple chemical compounds such as acetate to complex polymers. The output from the system depends on the efficiency of the system to metabolize the compounds, which largely depends on the type of inoculum and microbial community. The commonly used biowastes are wastewater that is rich in organic matter.

11.2 PRINCIPLE OF MFC

The two major components in a microbial electrochemical system (MES) are the anode and cathode compartments, which may or may not be separated by an ion-selective membrane depending on the configuration of the reactor. In the anode, where the physiological condition is anaerobic, upon oxidation of organic compounds catalyzed by bacteria, the free electrons produced are collected by the electrode. The protons are diffused from anode to cathode through a cation exchange membrane, and the electrons are transferred via an electric circuit to the cathode, thereby completing the half-cell reduction [2]. The end products formed in the cathode chamber depend on the reactant participating, which could be oxygen, potassium ferricyanide, or carbon dioxide as terminal electron acceptor.

Anodic oxidation reaction:

$$C_nH_{2n}O_n \rightarrow CO_2 + nH^+ + ne^- \tag{11.1}$$

Cathode reduction reaction:

$$O_2 + nH^+ + ne^- \rightarrow H_2O \tag{11.2}$$

The amount of organic matter present in wastewater is very high; hence, the efficiency of MFC in wastewater treatment is assessed by the extent of organic matter degradation, chemical oxygen demand (COD). It is measured as follows:

$$\text{Percentage of COD reduction} = \frac{\text{Initial COD} - \text{Final COD}}{\text{Initial COD}} \times 100 \tag{11.3}$$

The organic compounds upon degradation in MFC are recovered as electricity. Hence, the conversion measure of chemical energy present in organic compounds into electric energy is measured as coulombic efficiency (CE), which is calculated using the equation:

$$C_E = \frac{M_S \int_0^{tb} I\,dt}{F b_{es} v_{an} \Delta c} \tag{11.4}$$

where C_E is the columbic efficiency, M_s is the molecular mass of the substrate, Δc is the change in substrate concentration over time tb, F is the Faraday constant (96,485 C/mol), V_{an} is the volume of the anode, b_{es} is the number of electrons exchanged per mole of oxygen, and I is the current density produced over time dt. However, as in wastewater, the organic matter is present in undefined form; hence, the equation for calculating the unspecified compound is as follows:

$$C_E = \frac{8 \int_0^{tb} I\,dt}{F v_{an} \Delta \text{COD}} \tag{11.5}$$

where 8 is a constant used for COD, based on the molecular weight of O_2, $b_{es} = 4$ for the number of electrons exchanged per mole of oxygen, and ΔCOD is the change in chemical oxygen demand over time tb. The CE in real wastewater treatment is comparatively lesser than in MFC fed with defined substrates such as acetate, lactate, or glucose. This is due to the complexity of the organic substrates present in wastewater, which slows down the microbial metabolism. For example, a study reported by Ref. [3] showed that in an air cathode MFC, 90% of CE was achieved with acetate as substrate, whereas only 22% was observed in a system fed with domestic wastewater.

11.3 FACTORS AFFECTING THE RECOVERY OF ENERGY FROM WASTEWATER IN MFC

11.3.1 Microbial Inoculum

The extent of organic and inorganic contaminants degradation in MFC depends largely on the source of inoculum and microbial community in them. Mixed consortia are always favorable when compared to single-type culture for bioremediation studies due to the presence of a wide variety of microorganisms and symbiosis among themselves. A comparative study performed using mixed and pure (*Shewanella oneidensis* MR 1) culture inoculum showed that mixed consortia outperformed *S. oneidensis* MR 1-fed MFC in COD removal when treating agricultural and domestic wastewater, while the performance with pure culture inoculum was better in food processing wastewater, which is attributed to acidic pH of the medium. The use of mixed inoculum in application studies such as wastewater treatment is advantageous due to the presence of a dense and diverse microbial population [4]. For example, a study on MFC fed with sludge and river sediment showed faster enrichment with sludge inoculum, whereas river sediment was favorable for biofilm development. A similar study reported by Ref. [5] claimed that MFC fed with different inocula (arctic soil and wastewater) had a different microbial community with some common electroactive groups in both. Strategies of pre-treatment for enhancing the performance have also been reported in the literature. Acid pre-treatment of inoculum resulted in approximately seven times higher power density compared to untreated inoculum [6].

11.3.2 Cathode Reaction

Cathode oxygen reduction reaction (ORR) is a key half-cell reaction that plays an important role in enhancing the power density of MFCs. The ORR is a multi-step and sluggish reaction, and the overall power output depends on the faster kinetics of ORR. Generally, Pt-based catalysts are the best and efficient catalysts known for catalyzing ORR. However, due to the high cost of Pt-based catalysts, various other cost-effective, earth-abundant, non-Pt catalysts have been proposed for ORR. A handful of literature is available on cathode catalysts utilized for MFCs.

11.3.3 Separator and Ion Exchange Membrane

The microbes-driven wastewater treatment in MES depends hugely on some operational components of the system, such as electrode spacing, the configuration of the system, and ion exchange membrane [7]. In double-chambered MFCs, membranes are the separator placed between anode and cathode chambers that regulate the selective passage of ions from the anode to cathode compartment based on concentration gradient. However, membraneless (ML) MFCs have also shown effective treatment of wastewater simultaneous to electricity generation. Earlier, the authors of Ref. [8] demonstrated the design of a ML MFC for synthetic wastewater treatment that can reduce COD by up to 526.67 g/m^3. Since then, several reports have suggested designs of efficient ML MFCs for wastewater treatment for long-term operations. Nafion is an excellent membrane for proton conduction and a widely used separator in MFC. However, its high cost and poor mechanical strength make it unsuitable for long-term operations. Hence, researchers have started using alternate cation exchange membranes with better durability and mechanical strength. Another commonly used membrane CEM 7000 is a polymer of polystyrene with divinylbenzene cross-links. It shows equivalent proton conductivity and better durability compared to Nafion and hence is considered a preferred choice for long-term operations. An alternate CEM, disulfonated poly(arylene ether sulfone) showed a higher proton conductivity and is relatively inexpensive to Nafion. Sulfonated poly(ether ether ketone) (SPEEK) membranes have been shown to effectively reduce COD by up to 80% with better coulombic efficiency and maximum power density generation compared to Nafion [9]. A nanocomposite membrane of Fe_3O_4/PES with higher fouling resistance was reported to produce 29% higher power density compared to Nafion membrane [10]. A comparative study on CEM between Zifron and Fumasep showed that Zifron could be a better choice due to its proton-specific and low internal resistance in a single-chambered air cathode MFC treating synthetic wastewater [11]. Microfiltration membrane has shown an excellent effect in treating sludge and wetlands in an air cathode MFC. A twofold increase in maximum power density (878 mW/m^2) compared to MFC (415 mW/m^2) with PEM and over 96% COD removal efficiency were reported [12]. Recently, a polymer of agar membrane and ionic liquid membrane have also been explored for their cost-effective and efficient performance in wastewater treatment.

11.3.4 Design and Configuration of the System

The performance of the MFC depends on the type of configuration and design parameters that play a crucial role in achieving high performance, high pollutant degradation rates, and electricity generation. The MECs are broadly classified as single-chambered and dual-chambered MFCs, which differ from each other by having a PEM separator. Various types of models have also been reported in the literature depending on the objective and goal of the study. A microbial electrochemical snorkel model was introduced by Ref. [13] for effective urban wastewater treatment by monitoring the biofilm growth on the anode. This system was shown to be efficient than MFC in COD removal, by eliminating electricity generation. An annular single-chamber MFC with steel anode showed exceptional result in COD reduction (91%), power density production (20.2 W/m^3), and coulombic efficiency (26.87%) [14]. The steel mesh graphite anode provided good support for bacterial colonization to achieve an OCV of 810 mV in dairy wastewater. A hybrid system of microbial fuel cell–membrane bioreactor (MFC-MBR) constructed using electrical ultrafiltration membrane showed an excellent contaminants removal with 97% of COD and 97% of ammonia-N reduction in wastewater [15]. A submerged MFC with a multi-cathode system showed up to 95% COD removal in synthetic wastewater, but the efficiency was less in domestic wastewater. This is attributed to the microbial community, presence of complex substrates, and low ionic strength in the medium. Similarly, a hybrid system of constructed wetland integrated MFC was claimed to be an excellent system with complete COD removal efficiency [16].

11.3.5 Hydraulic Retention Time

The hydraulic retention time (HRT) is defined as the average length of time that a soluble organic compound remains in the reactor for the microorganism to metabolize it. HRT has a major impact on organic matter degradation and COD removal during wastewater treatment in MFC. For example, a recent study has reported that a HRT of 8–17 h was suitable for the removal of over 90% of COD in a double-chambered MFC [17]. Increasing the HRT from 10 to 24 h increased the COD removal efficiency from 50% to 80% in an osmotic MFC treating domestic wastewater. The HRT is believed to influence the acclimatization and enrichment of microbial communities depending on the type and concentration of substrate used in the system. Song and co-workers reported that an increase in HRT resulted in an increase in microbial diversity and increased electricity production in an up-flow MFC coupled with wetlands [18]. However, due to the dynamic system configuration and different microbial inocula HRT varies with design and operational conditions.

11.4 TREATMENT OF HAZARDOUS POLLUTANTS IN MFC

Some toxic and hazardous species in water bodies cause detrimental effects to the living ecosystem. This is due to the highly reactive and unstable nature of pollutants. The two major toxic pollutants in wastewater are heavy metals and organic dyes.

11.4.1 Reduction and Recovery of Heavy Metals

Water released from steel plants, mining, metal plating, and metal industries contains high concentrations of one or more types of metal residue. Biological reduction of metals in the environment takes place through a process popularly referred to as bioremediation. Chemical treatment involves solvent extraction, purification, precipitation, and filtration that are expensive and complex processes. The presence of metals in water bodies is a serious threat to the ecosystem due to its hazardous effects and resistance to conventional treatment methods. The removal of various metal ions from wastewater in MES is extensively reported in the literature. The authors of Ref. [19] demonstrated the sequential recovery of Cu, Pb, Cd, and Zn ions from a simulated municipal solid waste incineration ash leachate by adjusting the cell potential from −0.66 to +0.51 V. During metals recovery at the cathode, the concentration of individual ions plays a crucial role due to its unique standard reduction potential value. An integrated system of MFC and microbial electrolysis cell (MEC) developed to recover Cu^{2+} and Ni^{2+} showed efficiencies of 99% and 97%, respectively, by adjusting the ions concentration. A similar report on MES–electrolysis reactor (ER) showed removal efficiencies of 98.5%, 95.4%, and 98.1% for Cu^{2+}, Zn^{2+}, and Pb^{2+}, respectively, by adjusting the pH and concentration [20]. Excellent recovery rates of heavy metals from wastewater are reported in the literature for Au (99.89%),

Ag (99.91%), and Cu (99.88%), which proves the efficiency of MES in recovering resources during wastewater treatment.

The ecological habitat and diverse metabolic ability of electroactive microorganisms make them play an important role in biogeochemical cycles in the environment. Due to this, they play an important role in the recovery of non-metals and inorganic molecules during wastewater treatment. The major geochemical cycles by bacteria are carbon, nitrogen, phosphorous, and sulfur. The recovery of nitrogen was first demonstrated via denitrification in a tubular MFC, which showed 0.146 kg NO_3^-–N/m^3 of nitrogen recovery [21]. An MFC–MEC shift system can make 67% of phosphate recovery. A study claimed 72% of sulfate removal simultaneous to 88% of COD reduction in a microbial desalination cell (MDC), which was dominated by *Desulfobulbus*, *Geobacter*, and *Desulfovibrio* genera [22].

11.4.2 Dyes Reduction

Dyes are chemical compounds used for coloring fabrics and textiles in textile industries. Waste effluents from textile industries are another major concern of health hazards due to their toxic and carcinogenic effects on humankind [23]. Dyes are cyclic and heterocyclic aromatic compounds with redox properties due to their conjugated double bonds. Chemical dyes and waste effluents released from textiles industries are one of the major causes of water pollution. Treatment of such polluted water and dyes degradation in MFC have shown promising results. Degradation of dyes is assessed as an extent of decolorization during its metabolism in the reactor. Degradation of azo dye (Acid Orange 7) in an integrated MFC–aerobic bioreactor showed an excellent result with over 90% decolorization and COD removal efficiency [24]. The integrated system of constructed wetland and MFC (CW-MFC) demonstrated that more than 91% of decolorization was achieved, promoted by plants grown in the cathode chamber. The metabolites released from azo dyes degradation act as excellent electron mediators, thereby improving the performance by enhancing power density generation in MFC.

The efficacy of MES in degrading a wide variety of contaminants from water bodies is extensively investigated across the world. Other than these, antibiotics, heavy metals, toxic pollutants and xenobiotics are also widely reported. Apart from treating wastewater, MDC, a type of bio-electrochemical cell, is employed for softening the hard water. In MDC, the cell potential generated from the system is applied to attract the oppositely charged ions in the desalination chamber.

11.5 USE OF MODIFIED ELECTRODES FOR PERFORMANCE IMPROVEMENT

Carbonaceous materials are the most commonly preferred electrode material due to their large surface area, and non-corrosive and non-toxic properties. The low output and slow treatment process are some of the drawbacks of carbon electrodes. The use of electrocatalysts and modified electrodes have shown some promising results in wastewater treatment in MES [25]. Different forms of carbon in native form or in modified form showed excellent results. For example, the treatment of carbon black (CB) with nitric acid resulted in a maximum current density of 1,115 mA/m^2 in a single-chambered MFC treating wastewater [26]. Such treatment resulted in surface modification of carbon material, thereby enhancing the charge transfer phenomenon. In a similar study by Ref. [27], modification of CB with nitrogen and fluorine resulted in maximum current density (672 mA/cm^2) production, higher than that with commercial Pt/C (572 mA cm^{-2}). The stability and durability of catalysts in the medium are very crucial in long-term operations such as wastewater treatment in MFC. Activated carbon treated with Fe-EDTA showed excellent durability with improved performance compared to Pt cathode [28]. Superior electroactivity of multi-stacked carbon materials such as graphite, graphene, carbon nanotubes, and nanofibers is extensively reported in the literature. Modification of CNTs with polymers such as polyaniline and polypyrrole increased the catalytic activity tremendously. For example, over 80% of COD removal was reported with CNT/PPy nanocomposite and a maximum power density of 1,125.4 mW/m^{-2} with CNT-MnO_2/PPy-modified carbon cloth in wastewater treatment [29]. Similarly, graphene doped with nitrogen resulted in an excellent oxygen reduction reaction (ORR) activity in MFC [30].

Among metals, platinum is a widely employed material for improving ORR in fuel cells, due to its strong reducing nature. But its high cost and low availability made researchers search for alternative materials such as metal oxides, nanocomposites, and other non-precious materials [31]. A higher power density was produced with iron phthalocyanine (634 mW m^{-2}) cathode catalyst compared to that with Pt cathode (593 mW m^{-2}) in a single-chambered MFC. In another study, the use of Cu and Cu–Au electrodes resulted in more than 80% COD removal simultaneous to electricity generation of 2.9 mW m^{-2} [32]. Nanomaterials such as graphene [33] and carbon nanotubes [34] can act as excellent support materials for electrocatalysts in MFCs. Carbon nanotubes-supported MnO_2 cathode demonstrated 84.8% of COD removal efficiency, which was higher compared to Pt-based MFC, and the graphene-supported MnO_2 nanotubes resulted in a COD removal efficiency of 83.7% simultaneous to a maximum power density of 4.68 W/m^3 in a single-chambered MFC. The metal oxide composites are believed to enhance the reaction kinetics by increasing the electron transfer rate due to their large surface area, superior conductivity, and metal-oxide (M-O) framework for active catalytic site. Metal catalysts have been dominating species in fuel cells due to their vacant d-orbital and multiple oxidation states enabling them to readily participate in redox reactions. Other than these, TiO_2, V_2O_5, MgO, WO_3, and MoO_2 are some metal oxides reported in MFC for their superior contaminants removal efficiency during wastewater treatment.

11.6 LARGE-SCALE IMPLICATIONS OF MFC IN WASTEWATER TREATMENT AND ELECTRICITY PRODUCTION

A good number of studies have reported the proven application of MES in wastewater treatment on a laboratory scale from 50 mL to 1 L of volume. However, very few reports have shown successful scale-up studies to treat a large quantity of wastewater with continuous operation for few months to a year. One such pilot-scale MFC (20 L) integrated with multiple cathodes and anodes resulted in 80% removal of contaminants from municipal wastewater. An increase in HRT resulted in improved BOD and COD removal and recovered an electricity producing 500 mW/m^2 of maximum power density [35]. A similar study on a 20 L MFC for brewery wastewater treatment reported a COD removal efficiency of 95%. Further continuation of operation showed a drop in COD removal efficiency with time, resulting in a 48% reduction after 300 days. A 45 L MFC treating anaerobic sludge showed 84% COD removal with 17.63 mW m^2 of maximum power density [36], while a similar study showed high power density (73 mW m^{-2}) with reduced COD removal efficiency [37]. This difference in contaminants removal from sludge could be due to the cathode material employed. The platinum-coated carbon cloth used by Hiegemann et al. showed good power density, but the carbon felt used by Ghadge and co-workers resulted in a higher COD reduction.

The large-scale MES operation carries a drawback of mass transport loss, due to an increase in reactor volume and internal resistance of the system. This can be overcome by connecting multiple reactors to improve the performance by minimizing various losses. Stacked MFCs are commonly employed for large-scale study purposes such as wastewater treatment. An efficient multi-stacked MFC with a volume of 72 L showed 97% COD reduction and a power density of 50.9 ± 1.7 W m^{-3} [38]. The superior contaminants removal proficiency was believed to be due to the reactor design and electrode configuration. A five-stacked MFC module has shown the removal of organic contaminants and suspended solids by up to 87% and 86%, respectively, in brewery wastewater simultaneous to the energy production of 0.097 kWh m^{-3} in a 90 L stacked system [39]. The COD removal efficiency of the system depends largely on the configuration of the reactor, volume of each module, electrode materials employed, operational conditions, and anode inocula. For example, a stack of the largest single MFC module of 250 L each connected to treat 1,000 L of wastewater showed around 80% COD reduction [40]. Other reports on 1,000 L MFCs with wastewater treatment showed higher contaminants removal rates (up to 90%) [41] and recovery of organic matter as gases (CO_2) and volatile fatty acids [42]. Further information on the type of MFCs treating different types of wastewater on a large scale is summarized in Table 11.1. The proven application of MES to treat wastewater in large volumes and prolonged duration indicates it as a potent technique with simultaneous resource recovery and energy production.

11.7 FUTURE PROSPECTIVE AND CONCLUSIONS

The microbial electrochemical system provides a complete solution for treating various types of wastewater and softening hard water through a unique working mechanism. Considering the global scenario on the requirement for alternate energy sources and pollution control, MES offers an excellent solution for wastewater treatment simultaneous to sustainable energy generation. However, the existence of various losses and limiting factors in the system makes it challenging to compete with present wastewater treatment methods. Further research is required to overcome these constraints to make it practically applicable to field-scale applications.

TABLE 11.1
Comparison of Performances of Large-Scale MFC (=/>20 L) in Wastewater Treatment

MFC Configuration	Volume (L)	Duration of Operation (Days)	Anode	Cathode	Wastewater	Hydraulic Retention Time (H)	COD Removal Efficiency (%)	Power Density Produced (mW/m²)	Reference
Multi-cathode/anode MFC	20	100	Graphite rods	Carbon cloth coated with MnO_2	Domestic wastewater	20	80	500	[35]
2 DC stacked MFC	20	325	Carbon fiber cloth	Carbon fiber cloth	Brewery wastewater	313	94.6	1.61	[43]
4 SC membraneless MFCs	45	82	Graphite fiber brush	Carbon cloth coated with Pt	Sludge from wastewater treatment plant	44	67	73	[37]
SC MFC	45	90	Carbon cloth	Carbon felt	Sludge from septic tank	16	84	17.63	[36]
Stacked MFC with multi-cathode and anode	72	60	Activated carbon granules	Activated carbon granules	Synthetic wastewater	1.25	97	51	[38]
Stack of 5 MFC modules	90	90	Activated carbon	Carbon brushes	Brewery wastewater	144	84.7	171	[39]
Modularized system with 96 MFC modules	200	300	Carbon brush	N-doped carbon cloth	Effluent from wastewater treatment plant	6	75	200	[44]
Stack of 96 2L MFCs	200	NR	Carbon brush	N-doped carbon cloth	Municipal wastewater	6	75	130	[45]
4 MFC modules of 250 L each	1,000	365	Pt-coated carbon mesh	Carbon brush	Local wastewater	144	80	57	[40]
Stack of 24 MFC modules	1,000	65	Graphite fiber brush	Stainless steel mesh	Winery wastewater	24	62	[a]188 A/m³	[42]
Stack of 50 MFC modules	1,000	200	Granular activated carbon	Granular activated carbon	Municipal wastewater	2	90	125	[41]

NR, not reported; SC, single-chambered; DC, dual-chambered; MFC, microbial fuel cell; COD, chemical oxygen demand.

[a] Reported in volume.

REFERENCES

1. B. Sen Thapa, T.S. Chandra, Kluyvera georgiana MCC 3673: A novel electrogen enriched in microbial fuel cell fed with oilseed cake, *Curr. Microbiol.* 76 (5) (2019) 650–657. https://doi.org/10.1007/s00284-019-01673–0.
2. C. Santoro, C. Arbizzani, B. Erable, I. Ieropoulos, Microbial fuel cells: From fundamentals to applications. A review, *J. Power Sources.* 356 (2017) 225–244. https://doi.org/10.1016/j.jpowsour.2017.03.109.
3. X. Zhang, W. He, L. Ren, J. Stager, P.J. Evans, B.E. Logan, COD removal characteristics in air-cathode microbial fuel cells, *Bioresour. Technol.* 176 (2015) 23–31. https://doi.org/10.1016/j.biortech.2014.11.001.
4. A.S. Mathuriya, Inoculum selection to enhance performance of a microbial fuel cell for electricity generation during wastewater treatment, *Environ. Technol.* 34 (2013) 1957–1964. https://doi.org/10.1080/09593330.2013.808674.
5. E.S. Heidrich, J. Dolfing, M.J. Wade, W.T. Sloan, C. Quince, T.P. Curtis, Temperature, inocula and substrate: Contrasting electroactive consortia, diversity and performance in microbial fuel cells, *Bioelectrochemistry.* 119 (2018) 43–50. https://doi.org/10.1016/j.bioelechem.2017.07.006.
6. B.R. Tiwari, M.M. Ghangrekar, Enhancing electrogenesis by pretreatment of mixed anaerobic sludge to be used as inoculum in microbial fuel cells, *Energy and Fuels.* 29 (2015) 3518–3524. https://doi.org/10.1021/ef5028197.
7. S.E. Oh, B.E. Logan, Proton exchange membrane and electrode surface areas as factors that affect power generation in microbial fuel cells, *Appl. Microbiol. Biotechnol.* 70 (2006) 162–169. https://doi.org/10.1007/s00253-005-0066–y.
8. J.K. Jang, T.H. Pham, I.S. Chang, K.H. Kang, H. Moon, K.S. Cho, B.H. Kim, Construction and operation of a novel mediator- and membrane-less microbial fuel cell, *Process Biochem.* 39 (2004) 1007–1012. https://doi.org/10.1016/S0032–9592(03)00203–6.
9. A. Mayahi, H. Ilbeygi, A.F. Ismail, J. Jaafar, W.R.W. Daud, D. Emadzadeh, E. Shamsaei, D. Martin, M. Rahbari-Sisakht, M. Ghasemi, J. Zaidi, SPEEK/cSMM membrane for simultaneous electricity generation and wastewater treatment in microbial fuel cell, *J. Chem. Technol. Biotechnol.* 90 (2015) 641–647. https://doi.org/10.1002/jctb.4622.
10. M. Rahimnejad, M. Ghasemi, G.D. Najafpour, M. Ismail, A.W. Mohammad, A.A. Ghoreyshi, S.H.A. Hassan, Synthesis, characterization and application studies of self-made Fe_3O_4/PES nanocomposite membranes in microbial fuel cell, *Electrochim. Acta.* 85 (2012) 700–706. https://doi.org/https://doi.org/10.1016/j.electacta.2011.08.036.
11. S. Sevda, X. Dominguez-Benetton, K. Vanbroekhoven, T.R. Sreekrishnan, D. Pant, Characterization and comparison of the performance of two different separator types in air-cathode microbial fuel cell treating synthetic wastewater, *Chem. Eng. J.* 228 (2013) 1–11. https://doi.org/10.1016/j.cej.2013.05.014.
12. J. Sun, Y. Hu, Z. Bi, Y. Cao, Improved performance of air-cathode single-chamber microbial fuel cell for wastewater treatment using microfiltration membranes and multiple sludge inoculation, *J. Power Sources.* 187 (2009) 471–479. https://doi.org/10.1016/j.jpowsour.2008.11.022.
13. B. Erable, L. Etcheverry, A. Bergel, From microbial fuel cell (MFC) to microbial electrochemical snorkel (MES): Maximizing chemical oxygen demand (COD) removal from wastewater, *Biofouling.* 27 (2011) 319–326. https://doi.org/10.1080/08927014.2011.564615.
14. M. Mahdi Mardanpour, M. Nasr Esfahany, T. Behzad, R. Sedaqatvand, Single chamber microbial fuel cell with spiral anode for dairy wastewater treatment, *Biosens. Bioelectron.* 38 (2012) 264–269. https://doi.org/10.1016/j.bios.2012.05.046.

15. L. Malaeb, K.P. Katuri, B.E. Logan, H. Maab, S.P. Nunes, P.E. Saikaly, A hybrid microbial fuel cell membrane bioreactor with a conductive ultrafiltration membrane biocathode for wastewater treatment, *Environ. Sci. Technol.* 47 (2013) 11821–11828. https://doi.org/10.1021/es4030113.
16. Y.L. Oon, S.A. Ong, L.N. Ho, Y.S. Wong, Y.S. Oon, H.K. Lehl, W.E. Thung, Hybrid system up-flow constructed wetland integrated with microbial fuel cell for simultaneous wastewater treatment and electricity generation, *Bioresour. Technol.* 186 (2015) 270–275. https://doi.org/10.1016/j.biortech.2015.03.014.
17. Y. Ye, H.H. Ngo, W. Guo, S.W. Chang, D.D. Nguyen, X. Zhang, S. Zhang, G. Luo, Y. Liu, Impacts of hydraulic retention time on a continuous flow mode dual-chamber microbial fuel cell for recovering nutrients from municipal wastewater, *Sci. Total Environ.* 734 (2020) 139220. https://doi.org/10.1016/j.scitotenv.2020.139220.
18. H.L. Song, H. Li, S. Zhang, Y.L. Yang, L.M. Zhang, H. Xu, X.L. Yang, Fate of sulfadiazine and its corresponding resistance genes in up-flow microbial fuel cell coupled constructed wetlands: Effects of circuit operation mode and hydraulic retention time, *Chem. Eng. J.* 350 (2018) 920–929. https://doi.org/10.1016/j.cej.2018.06.035.
19. O. Modin, X. Wang, X. Wu, S. Rauch, K.K. Fedje, Bioelectrochemical recovery of Cu, Pb, Cd, and Zn from dilute solutions, *J. Hazard. Mater.* 235–236 (2012) 291–297. https://doi.org/10.1016/j.jhazmat.2012.07.058.
20. H.C. Tao, T. Lei, G. Shi, X.N. Sun, X.Y. Wei, L.J. Zhang, W.M. Wu, Removal of heavy metals from fly ash leachate using combined bioelectrochemical systems and electrolysis, *J. Hazard. Mater.* 264 (2014) 1–7. https://doi.org/10.1016/j.jhazmat.2013.10.057.
21. P. Clauwaert, K. Rabaey, P. Aelterman, L. De Schamphelaire, T.H. Pham, P. Boeckx, N. Boon, W. Verstraete, Biological denitrification in microbial fuel cells, *Environ. Sci. Technol.* 41 (2007) 3354–3360. https://doi.org/10.1021/es062580r.
22. T. Jafary, W.R.W. Daud, S.A. Aljlil, A.F. Ismail, A. Al-Mamun, M.S. Baawain, M. Ghasemi, Simultaneous organics, sulphate and salt removal in a microbial desalination cell with an insight into microbial communities, *Desalination.* 445 (2018) 204–212. https://doi.org/https://doi.org/10.1016/j.desal.2018.08.010.
23. R. Kant, Textile dyeing industry an environmental hazard, *Nat. Sci.* 04 (2012) 22–26. https://doi.org/10.4236/ns.2012.41004.
24. E. Fernando, T. Keshavarz, G. Kyazze, Complete degradation of the azo dye Acid Orange-7 and bioelectricity generation in an integrated microbial fuel cell, aerobic two-stage bioreactor system in continuous flow mode at ambient temperature, *Bioresour. Technol.* 156 (2014) 155–162. https://doi.org/https://doi.org/10.1016/j.biortech.2014.01.036.
25. H. Yuan, Y. Hou, I.M. Abu-Reesh, J. Chen, Z. He, Oxygen reduction reaction catalysts used in microbial fuel cells for energy-efficient wastewater treatment: A review, *Mater. Horizons.* 3 (2016) 382–401. https://doi.org/10.1039/c6mh00093b.
26. N. Duteanu, B. Erable, S.M. Senthil Kumar, M.M. Ghangrekar, K. Scott, Effect of chemically modified Vulcan XC-72R on the performance of air-breathing cathode in a single-chamber microbial fuel cell, *Bioresour. Technol.* 101 (2010) 5250–5255. https://doi.org/10.1016/j.biortech.2010.01.120.
27. K. Meng, Q. Liu, Y. Huang, Y. Wang, Facile synthesis of nitrogen and fluorine co-doped carbon materials as efficient electrocatalysts for oxygen reduction reactions in air-cathode microbial fuel cells, *J. Mater. Chem. A.* 3 (2015) 6873–6877. https://doi.org/10.1039/C4TA06500J.
28. X. Xia, F. Zhang, X. Zhang, P. Liang, X. Huang, B.E. Logan, Use of pyrolyzed iron ethylenediaminetetraacetic acid modified activated carbon as air-cathode catalyst in microbial fuel cells, *ACS Appl. Mater. Interfaces.* 5 (2013) 7862–7866. https://doi.org/10.1021/am4018225.

29. P. Mishra, R. Jain, Electrochemical deposition of MWCNT-MnO_2/PPy nano-composite application for microbial fuel cells, *Int. J. Hydrogen Energy.* 41 (2016) 22394–22405. https://doi.org/https://doi.org/10.1016/j.ijhydene.2016.09.020.
30. Y. Liu, Z. Liu, H. Liu, M. Liao, Novel porous nitrogen doped graphene/carbon black composites as efficient oxygen reduction reaction electrocatalyst for power generation in microbial fuel cell, *Nanomater. (Basel, Switzerland).* 9 (2019). https://doi.org/10.3390/nano9060836.
31. K. Ben Liew, W.R.W. Daud, M. Ghasemi, J.X. Leong, S. Su Lim, M. Ismail, Non-Pt catalyst as oxygen reduction reaction in microbial fuel cells: A review, *Int. J. Hydrogen Energy.* 39 (2014) 4870–4883. https://doi.org/https://doi.org/10.1016/j.ijhydene.2014.01.062.
32. F. Kargi, S. Eker, Electricity generation with simultaneous wastewater treatment by a microbial fuel cell (MFC) with Cu and Cu–Au electrodes, *J. Chem. Technol. Biotechnol.* 82 (2007) 658–662. https://doi.org/10.1002/jctb.1723.
33. B. Sen Thapa, S. Seetharaman, R. Chetty, T.S. Chandra, Xerogel based catalyst for improved cathode performance in microbial fuel cells, *Enzyme Microb. Technol.* 124 (2019) 1–8. https://doi.org/10.1016/j.enzmictec.2019.01.007.
34. T. Sharma, A.L. Mohana Reddy, T.S. Chandra, S. Ramaprabhu, Development of carbon nanotubes and nanofluids based microbial fuel cell, *Int. J. Hydrogen Energy.* 33 (2008) 6749–6754. https://doi.org/10.1016/j.ijhydene.2008.05.112.
35. D. Jiang, M. Curtis, E. Troop, K. Scheible, J. Mcgrath, B. Hu, S. Suib, D. Raymond, B. Li, A pilot-scale study on utilizing multi-anode / cathode microbial fuel cells (MAC MFCs) to enhance the power production in wastewater treatment, *Int. J. Hydrogen Energy.* 36 (2010) 876–884. https://doi.org/10.1016/j.ijhydene.2010.08.074.
36. A.N. Ghadge, D.A. Jadhav, M.M. Ghangrekar, Wastewater treatment in pilot-scale microbial fuel cell using multielectrode assembly with ceramic separator suitable for field applications, *Environ. Prog. Sustain. Energy.* 35 (2016) 1809–1817. https://doi.org/10.1002/ep.12403.
37. H. Hiegemann, D. Herzer, E. Nettmann, M. Lübken, P. Schulte, K.G. Schmelz, S. Gredigk-Hoffmann, M. Wichern, An integrated 45 L pilot microbial fuel cell system at a full-scale wastewater treatment plant, *Bioresour. Technol.* 218 (2016) 115–122. https://doi.org/10.1016/j.biortech.2016.06.052.
38. S. Wu, H. Li, X. Zhou, P. Liang, X. Zhang, Y. Jiang, X. Huang, A novel pilot-scale stacked microbial fuel cell for efficient electricity generation and wastewater treatment, *Water Res.* 98 (2016) 396–403. https://doi.org/10.1016/j.watres.2016.04.043.
39. Y. Dong, Y. Qu, W. He, Y. Du, J. Liu, X. Han, Y. Feng, A 90-liter stackable baffled microbial fuel cell for brewery wastewater treatment based on energy self-sufficient mode, *Bioresour. Technol.* 195 (2015) 66–72. https://doi.org/10.1016/j.biortech.2015.06.026.
40. Y. Feng, W. He, J. Liu, X. Wang, Y. Qu, N. Ren, A horizontal plug flow and stackable pilot microbial fuel cell for municipal wastewater treatment, *Bioresour. Technol.* 156 (2014) 132–138. https://doi.org/10.1016/j.biortech.2013.12.104.
41. P. Liang, R. Duan, Y. Jiang, X. Zhang, Y. Qiu, X. Huang, One-year operation of 1000-L modularized microbial fuel cell for municipal wastewater treatment, *Water Res.* 141 (2018) 1–8. https://doi.org/10.1016/j.watres.2018.04.066.
42. R.D. Cusick, B. Bryan, D.S. Parker, M.D. Merrill, M. Mehanna, P.D. Kiely, G. Liu, B.E. Logan, Performance of a pilot-scale continuous flow microbial electrolysis cell fed winery wastewater, *Appl. Microbiol. Biotechnol.* 89 (2011) 2053–2063. https://doi.org/10.1007/s00253-011-3130–9.
43. M. Lu, S. Chen, S. Babanova, S. Phadke, M. Salvacion, A. Mirhosseini, S. Chan, K. Carpenter, R. Cortese, O. Bretschger, Long-term performance of a 20-L continuous flow microbial fuel cell for treatment of brewery wastewater, *J. Power Sources.* 356 (2017) 274–287. https://doi.org/10.1016/j.jpowsour.2017.03.132.

44. Z. Ge, Z. He, Long-term performance of a 200 liter modularized microbial fuel cell system treating municipal wastewater: Treatment, energy, and cost, *Environ. Sci. Water Res. Technol.* 2 (2016) 274–281. https://doi.org/10.1039/c6ew00020g.
45. Z. Ge, L. Wu, F. Zhang, Z. He, Energy extraction from a large-scale microbial fuel cell system treating municipal wastewater, *J. Power Sources*. 297 (2015) 260–264. https://doi.org/10.1016/j.jpowsour.2015.07.105.

12 Biowaste-Based Microbial Fuel Cells

Nidhi Chauhan, Utkarsh Jain, and Kirti Saxena
Amity University Uttar Pradesh

CONTENTS

12.1 INTRODUCTION

Energy sources scarcity is a serious challenge throughout the world. Climate changes and population increase are the two crucial factors that pose serious issues such as water resources scarcity, increased energy demand, waste treatment, and land use challenges. The survival of all living organisms including humans depends upon energy in various aspects (Pant et al. 2012b). As the demand for global energy increases day by day, non-renewable energy resources are diminishing at an alarming rate (Shockey et al. 2010). It is now necessary to develop some new sources for renewable energy, so researchers turned the direction of their study to develop alternative methods for energy generation (Chandrasekhar et al. 2015). Microbial fuel cell (MFC) technology emerges as the newest approach for the generation of bioelectricity or green energy from biomass using bacteria. MFC is one of the most popular techniques for wastewater treatment and green energy production (Pant et al. 2012a).

DOI: 10.1201/9781003178354-15

MFCs are bio-electrochemical devices that oxidize organic wastes/matter by degrading them into smaller molecules and release electrons as well as protons, thereby producing electricity. MFCs consist of an anode compartment/chamber which is anaerobic and a cathode compartment/chamber which is aerobic, and both are separated by a proton exchange membrane (PEM) or anion exchange membrane (AEM). MFC typically utilizes microbes for substrate oxidization in an anodic chamber; afterward, electrons released from microbes are transferred to the cathode through an external wire. MFCs are able to convert chemical energy into electrical energy through the utilization of microorganisms or enzymatic catalysts in an electrochemical reaction.

Generally, MFCs are of two types: single-chamber and two-chamber systems. A single-chamber system can be developed very easily by putting an anode electrode inside a tube and exposing the cathode to air, i.e., at the open end of the tube. In this way, the oxygen or atmospheric air supplied to the cathode electrode surface will act as a catholyte and participate in an oxygen reduction reaction (Ringeisen et al. 2006; Chaudhuri & Lovley 2003) (Figure 12.1).

On the contrary, in the two-chamber system, the major chamber is the anodic chamber where all the activity takes place, i.e., biocatalyst and electron generation, but cathode reaction is also an equally crucial part of the whole mechanism (Luimstra et al. 2013). A wide diversity of biowaste, for example organic wastes, inorganic materials, and soil sediments, can be utilized as a source of fuel generation. MFCs can perform energy generation processes at atmospheric pressure and ambient temperature unlike traditional methods (Larrosa-Guerrero et al. 2010; Du et al. 2007). Additionally, MFCs can provide electrical facilities in a location that lacks electricity and can outperform other technologies such as aerated lagoon and anaerobic disaster (Logan 2008).

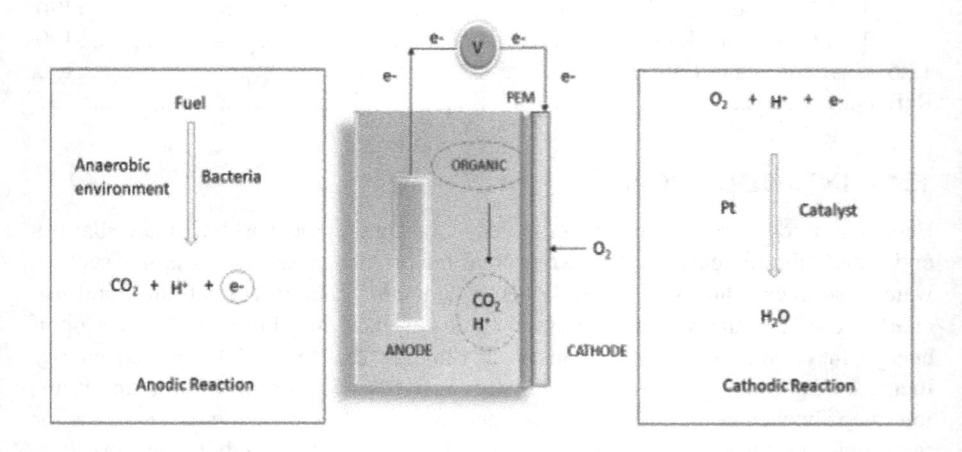

FIGURE 12.1 Generalized structure of a single-chamber microbial fuel cell, displaying a cathodic and anodic reaction with a catalyst.

12.2 DIFFERENT TYPES OF BIOWASTE EXPLOITED AS SUBSTRATE

12.2.1 Food or Kitchen Waste

Food waste or kitchen waste is generally an organic waste, which includes waste from households, cafeterias, restaurants, and municipal solid waste as well. The studies report that one-third of the total food production for human consumption in the world is wasted. If the food waste is not treated well, it would cause health and environmental issues because of the huge number of organic materials, high moisture, and salinity. Therefore, due to its high biodegradability, food waste can be an attractive substrate to be exploited as an alternative source for energy production in MFCs (Kiran et al. 2014; FAO 2012). A study investigated the characteristics of food waste's organic matter before and after MFC treatment. The study described the role of organic material to be biodegraded and transformed in an MFC system. A power density of 5.6 W/m^3 and an average output voltage of 0.51 V were obtained (Li et al. 2016). In another study, kitchen garbage along with bamboo waste was utilized for the generation of bioelectricity in the MFC system. The by-products from this study can be utilized as a soil conditioner (Moqsud et al. 2014). A number of studies have been performed with kitchen and food wastes to be utilized as a substrate for energy generation using MFCs (Xin et al. 2018; Manjerkar et al. 2018; Meicong et al. 2020; Du et al. 2020; Kayode & Hart 2019).

12.2.2 Paper Industry Waste

Paper and pulp industries are major polluters of the environment. The pulping process is used to break the bonds present in the structure of wood. Various chemical and mechanical processes are involved in the conversion of wood into pulp. Several pollutants are released in this process as effluents. The management of these types of organic matter-rich industrial wastes is a problem due to their odor and threat to the environment (aquatic and terrestrial) if not discarded properly. So, these wastes can be exploited directly for energy production. An MFC system has been developed to reduce aquatic Cr(VI) to Cr(III), a non-toxic waste, by utilizing industrial paper waste organic material. In this study, biochar material developed by industrial paper sludge was used to fabricate cathode electrodes of MFC. This biochar composite material showed increased electron transfer capacity (Zhou et al. 2020).

12.2.3 Lignocellulosic Biomaterials

The main sources of lignocellulosic material are either forest wood, natural biomass, or forest wood waste processing industries. These industries are more likely to produce or process plywood, timber, and particle board. A huge amount is generally produced during road or other constructions, the manufacture of households such as furniture, plantation, thinning, etc. Although the degradation of these complex lignocellulosic compounds was slow, they may be used as a direct substrate in MFC for feasible production of energy (Mohd Zaini Makhtar & Tajarudin 2020).

Lignocellulosic biowaste consists of lignin, cellulose, and hemicellulose, and their pretreatment is required to convert them into an easily degradable form. The pretreatment strategy is based on the composition of materials (Bhatia et al. 2017a, b). Several reports described the methods for the degradation of biomass to be used as a suitable substrate in MFC, such as biological, physicochemical, chemical, and physical methods (ElMekawy et al. 2014; Bhatia et al. 2016; Amin et al. 2017; Mulakhudair et al. 2017; Procentese et al. 2017).

12.2.4 Animal Waste

Animal wastes such as manure and by-products from meat processing are produced at a large scale worldwide, especially from dairy and poultry farms. Animal wastes produced from meat processing, such as hair, feathers, animal fat waste, bone, and skin, are very rich in organic matter. Their proper disposal is essential for avoiding foul odor and growth of pathogenic microorganisms. Usually, animal wastes such as cow dung is stored to produce manure by livestock producers, which ultimately releases methane gas into the environment which is a hazardous and more dangerous gas as compared to carbon dioxide. The animal waste can be converted into biofuel by using it as a substrate in MFC, but it requires high maintenance and cost, so its utilization is quite limited (Gebrezgabher et al. 2010). The pretreatment of animal fat waste is required before use in MFC as substrate. Various procedures are reported for its pretreatment (Gooding & Meeker 2016; Dias et al. 2009; Taher et al. 2011; Banković-Ilić et al. 2014; Kara et al. 2018).

12.2.5 Municipal Solid Waste

Due to urbanization, population increase, and economic development, municipal solid waste increases very rapidly at a high rate. Approximately 100–400 kg of municipal solid waste is produced annually by a person in a developing country. The management of this type of waste is a global problem that negatively affects public health, quality of life, and our environment (Islam 2016; Cheng & Hu 2010). Generally, landfills, thermal and biological treatment, and recycling methods are used to manage municipal solid waste. This could be a great option to be used as a substrate in the MFC system because its organic contents can be an attractive source of renewable energy. The two-chambered MFC system utilized municipal solid waste organics to convert it into bioenergy through a number of operational conditions. In this study, the municipal solid waste was first treated with alkali hydrolysis to produce maximum power (Chiu et al. 2016).

12.3 BIOWASTE TO BIOENERGY CONVERSION

The organic matter present in biowaste can be utilized for the production of bioenergy through various biological methods such as fermentation, anaerobic digestion, transesterification, and MFC. Biogas, biodiesel, and bioelectricity are different forms of bioenergy and produced from the exploitation of several biological methods. Biogas is produced from the anaerobic digestion of biological materials by microorganisms. The combination of gases such as methane, CO_2, and traces of other gases is known as biogas. This process is also used to maintain biological waste materials. Biofuels such as biodiesel can also be produced from biological wastes by the process of transesterification. Animal fat waste, vegetable oil waste, or microbial oil is used to produce biodiesel. Sewage sludge is also employed for the production of biodiesel, which is produced during wastewater treatment. Alcohol as a fuel is in higher demand nowadays. Bioethanol can be produced through the fermentation process. So, the biowaste can be utilized for the production of various types of biofuels and bioenergy to fulfill the high requirements of renewable energy. These are eco-friendly and cost-effective.

12.4 BIOWASTE-BASED MFC

Microbial fuel cells emerge as the most attractive technique for the production of clean energy and reduction in pollution. MFC utilizes solid wastes with high organic content, for example biomass, biowaste, sewage sludge, etc., and produces energy without any harmful by-products. Generally, biowaste includes food waste, garden waste, and dung, which produce daily organic matter that is harmful to the environment and humans. Conversely, this biowaste is an alternative resource to produce bioethanol and biodiesel, i.e., bioenergy through MFC. The recently developed biowaste-based MFCs are summarized in Table 12.1.

Biochar is a carbon-rich material developed through thermal decomposition of biomass in anaerobic conditions of low-temperature pyrolysis. Biochar is extensively used in soil amendment which enhances the availability of nutrients and improves soil quality. Biowaste such as seagrass residues, vineyard prunings, olive kernels, and sewage sludge was utilized for biochar production by pyrolysis method at two different temperatures, i.e., 250 °C and 500 °C. Biochar is used to enhance biofuel production and biomass recycling. S. cerevisiae and K. marxianus were used as microbial catalysts for bioethanol production by biochar (Kyriakou et al. 2019).

In MFC, topsoil microbial communities which are able to generate electricity are incorporated to produce biofuels. These types of bacteria are abundant in nature. In an experiment, Himalayan topsoil from Uttarakhand along with kitchen wastes and acetate was incorporated in MFC, and the results revealed that the voltage across the terminals rose up to 1.54 V and the peak power was 0.99 mW. This type of MFC-generated system can be utilized to produce electricity at a small scale for powering lamps in remote rural areas (Bose & Bose 2017). This is a very low in cost and easy-to-fabricate technique.

Another most important application of MFC is the production of hydrogen from biowaste through the disintegration of microbes. Hydrogen can serve as a good alternative option to fuel energy. MFC is able to convert hydrogen into electricity directly, with the production of water as the only by-product. For hydrogen production, extremely thermophilic bacteria have been selected rather than mesophilic bacteria because thermophilic bacteria are known for high hydrogen production efficiency. For this work, crop wastes (rich in sugar), domestic organic wastes, potato peels, and paper sludge were used as biowaste. Various researches have been carried out for the production of hydrogen with several types of microorganisms, for example

TABLE 12.1
Summary of Recently Developed Biowaste-Based MFCs

Biowaste Types	Pretreatment Method	Microbes Employed	Uses	References
Biochar (carbon-rich)	Pyrolysis	*S. cerevisiae* and *K. marxianus*	Improvement in soil quality, biofuel production, bioethanol	Kyriakou et al. (2019)
Kitchen waste	Not reported	Top soil microbes	Electricity production	Bose and Bose (2017)
Crop wastes (sugar-rich), potato peels, paper sludge, domestic organic wastes	Steam acid pretreatment and enzymatic hydrolysis	Thermophilic bacteria (*Thermotoga neapolitana, Caldicellulosiruptor saccharolyticus*, and *Thermotoga elfii*)	Biohydrogen production	Sudhanshu et al. (2013)
Tannery effluents (potassium dichromate), sweet lime waste, cow dung	Not reported	*Geobacter metallireducens*	Heavy metals compound removal, electricity generation	Sindhuja et al. (2018)
Crustacean shells	Freeze and vacuum (lipolyzed)	*Arenibacter palladensis YHY2*	Power generation, hydrolysis of chitin	Gurav et al. (2018)
Sewage and contaminated ground water	Hydrolytic–acidogenic method	Exoelectrogens	Electricity production	Al-Mamun et al. (2020)
Ficus religiosa leaves	Thermochemical treatment	*B. subtilis* (NCIM NO. 5433)	Decolorization of Congo red dye	Prajapati and Yelamarthi (2020)
Lactic acid, ethanol, methanol, glycerol, urea, and glucose	Not reported	Not reported	Biowaste decomposition and generation of electric current	Lui et al. (2017)
Household wastes, glucose-based synthetic wastewater, and anaerobic sludge	Not reported	Not reported	Energy production	Tremouli et al. (2019)
Coffee waste	Thermochemical	*E. coli*	Energy production	Hung et al. (2019)

Thermotoga neapolitana, Caldicellulosiruptor saccharolyticus, and *Thermotoga elfii* (Sudhanshu et al. 2013). Microbes are utilized to produce hydrogen in some specific ways other than MFC, known as photofermentation, dark fermentation, hybrid system, and biophotolysis (Jadhav et al. 2017).

A two-chambered MFC exploits potassium dichromate and wastewater from tannery effluent. This MFC was fabricated mainly for the reduction of toxic chromium compound present in sweet lime waste inoculated from cow dung (anolyte), and a graphite electrode was utilized to reduce chromium. Apart from the reduction of chromium, MFC also generates power simultaneously. In this study, Cr(VI) was reduced to Cr_2O_3 in the cathode chamber within 24 h. Cr(VI) present in tannery effluents was completely reduced within 10 days through MFC. *Geobacter metallireducens* present in a mixed culture of bacteria in an anodic anaerobic chamber played an important role in the production of energy. The power density obtained in this experiment was noticed as 396.7 mW/m^2. Electrochemical studies were done through the EIS technique (Sindhuja et al. 2018).

Marine bacteria are also exploited in MFC to generate power frequently. *Arenibacter palladensis* YHY2 is a novel marine bacterium isolated from the eastern sea of South Korea. The bacteria are able to hydrolyze crab chitin in the MFC system. In a study, Arenibacter palladensis YHY2 was co-cultivated with *Ralstonia eutropha* H16. Co-cultivation increased the production of chitinase surprisingly. The maximum current output density was calculated at the initial stage as 15.15 and 10.72 μA/cm^2 at 204 h. Other metabolites present in MFC system were butyrate, lactate, propionate, and acetate. This report provides a novel method for the hydrolysis of chitin into GlcNAc. The method was also called one-pot hydrolysis. Several techniques were employed for the characterization of MFC, such as SEM, TEM, FT-IR, and XRD (Gurav et al. 2018).

MFC can be employed for the treatment of wastewater from sewage and contaminated groundwater. This is a cost-effective way to oxidize organic compounds from wastewater through exoelectrogens on an anode electrode to produce electricity, whereas the cathode utilized this electricity to reduce nitrogen of wastewater. In a recent study, an anaerobic sequencing batch reactor was integrated to achieve the conversion of complex organics by hydrolytic–acidogenic method, to boost power recovery, and to remove nitrogen and carbon from sewage and other contaminated water, because exoelectrogens were not able to oxidize complex insoluble organics. The result revealed that for pretreated sewage, optimum power and current generation were 7.1 W/m^3 and 45.88 A/m^3 (Al-Mamun et al. 2020).

Kitchen waste is also used as a substrate for the generation of bioelectricity through an MFC. Several studies have been carried out to demonstrate the effect of substrate and mediators to enhance the production through MFC. Mediators are known to enhance the activity of MFC, such as $K_3[Fe(CN)_6]$, methylene blue, and neutral red potassium permanganate. The research revealed that the use of mediators effectively increases the percentage of energy contribution and power density. The highly moisturized contents of kitchen waste substrate can facilitate the production of better electron mobile solution and boost up the movement of electrons toward cathode in MFC. It was also reported that 10% higher content of water in MFC can increase voltage output threefold (Wang et al. 2014; Adebule et al. 2018).

In an interesting study, MFC was also utilized for the decolorization of Congo red dye. In this study, "*Ficus religiosa* leaves" biowaste was exploited for the development of bioelectrode, which is used to improve hydrophilicity, surface area, and roughness to produce biofilms. The dye concentration, hydraulic retention time, glucose concentration, and effect of dye decolorization were also studied. This study aimed to develop an anode-based biowaste-derived MFC for the decolorization of Congo red dye. The results showed a maximum decolorization of "80.95 ± 2.08%" and a COD reduction of 73.96 ± 1.76% (Prajapati & Yelamarthi 2020).

Very recently, a wearable power generator has been synthesized using biowaste sources as fuel, such as ethanol, methanol, glycerol, glucose, lactic acid, and urea. The photocatalytic fuel cell exploited light irradiation along with biowaste decomposition to generate electric current. It was employed as a sweat band that helps to generate power from human sweat (Lui et al. 2017).

A study described an MFC that contains four air cathodes and a single chamber without membrane for the production of energy using household wastes. In this research work, MnO_2 was employed as a cathode catalyst. All the setups of this MFC provide a maximum output power calculated as 3.2 mW in 120 mL volume of anolyte. EIS characterization was used for analytical and numerical calculations on internal resistance (Tremouli et al. 2019). In another research, "coffee-waste-derived" activated carbons were utilized as an anode electrode material in an *E. coli*-based MFC system. A higher power density was obtained (3,927 Mw/m^2), which led to the long-term performance of MFC, high conductivity, and fast electron transfer (Hung et al. 2019).

12.5 APPLICATIONS

The world's economy, as well as industrial development, depends upon the accessibility of energy resources. So to overcome the scarcity of energy, there is an urgent need to develop an alternative such as MFC. MFC was developed as a green technology for electricity generation and treatment of various pollutants as well. Several works have been carried out to improve and evaluate the performance of MFC. Over the decades, the capacity of MFC has increased from μL to few liters. Design, performance, capacity, and cost-effectiveness have improved over time. MFC systems exhibit a variety of applications (Figure 12.2). Some of the prominent applications of MFCs are given as follows:

12.5.1 Bioelectricity Production

Electricity generation is the prime and most studied application of MFC technology. The most important component of MFCs is the substrate which can be oxidized completely to generate electrons to achieve higher power output and coulombic efficiency. The most important microorganism utilized in MFC is Geobacter sulfurreducens, which is of great interest among the researchers at present time. It is known for the complete reduction of acetate to generate electrons as well as protons (Reguera et al. 2005).

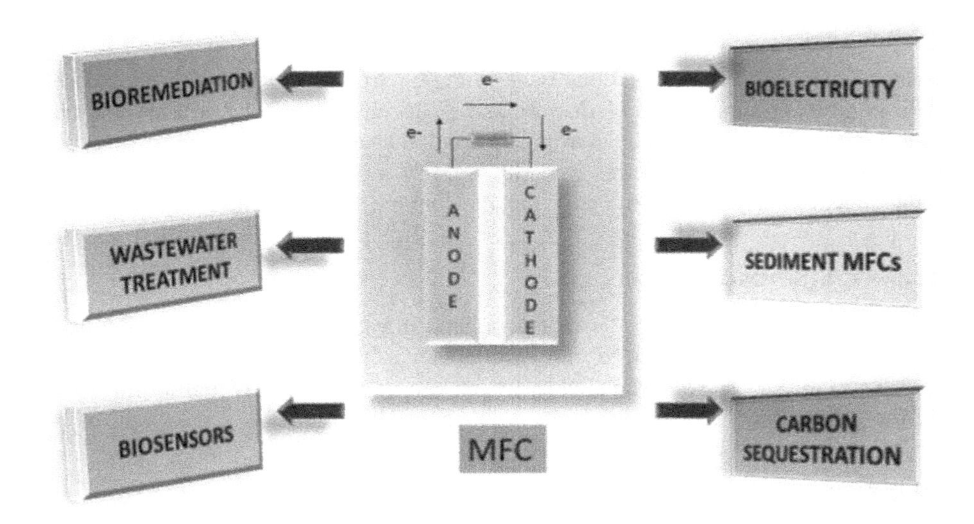

FIGURE 12.2 Applications of microbial fuel cells.

12.5.2 Wastewater Treatment

MFCs can potentially treat the wastewater from industries and urban or domestic sources (Oh et al. 2010; Rhoads et al. 2005). MFC is able to remove up to 98% COD (Oh et al. 2010). It is a more sensitive method for wastewater treatment over conventional methods, such as COD, BOD, and total solids (Dong et al. 2015). Figure 12.3 shows the mechanism involved in wastewater treatment through microbial fuel cells.

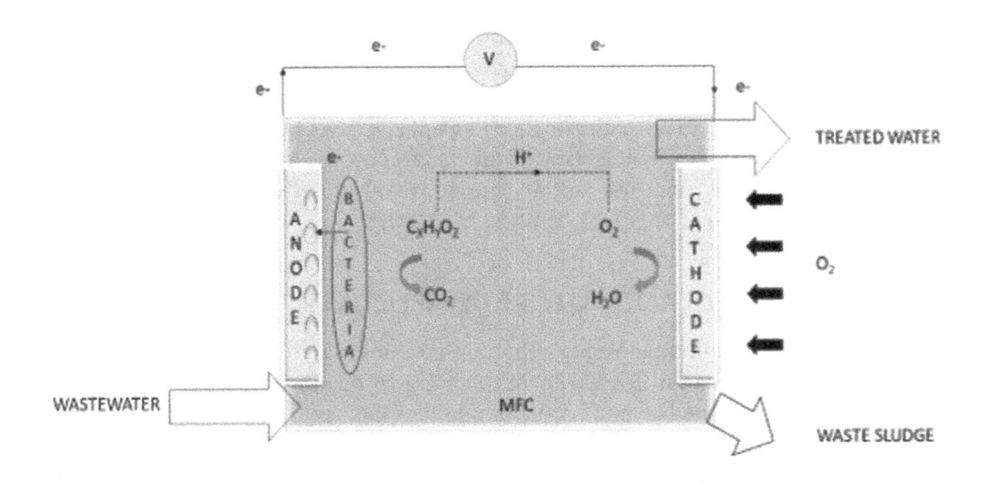

FIGURE 12.3 Schematic representation of MFC-based wastewater treatment plant.

12.5.3 Removal/Recovery of Heavy Metals

Heavy metals cause a very serious issue to the environment and health because of their non-biodegradability, toxic nature, and bioaccumulation. The primary sources of heavy metals release in the environment are domestic wastewater, industrial effluents, and medical wastes (Sotres et al. 2015). MFC system is highly investigated for removal as well as recovery of heavy metals from the environment.

12.5.4 Biohydrogen Production

Biohydrogen production is one of the popular applications of MFC. A typical two-chamber MFC can be employed in microbial electrolysis cells to produce hydrogen for several purposes. The produced hydrogen can be stored easily and used further for the generation of electricity (Rorke & Gueguim 2016).

12.5.5 Biosensor Fabrication

Biosensors are analytical devices that consist of a bio-recognition element to detect an analyte and a transducer surface to convert the response into measurable signals (Su et al. 2011; D'Souza 2001). MFC technology is utilized to develop a biosensor for the detection of pollutants in water bodies. MFC can be used as a BOD biosensor because there is a linear relationship between wastewater strength and coulombic yield of MFC. These biosensors are cost-effective, are time efficient, and require zero maintenance. So BOD biosensors are more reliable and stable biosensors (Kim et al. 2003).

12.5.6 Bioremediation

It is a process to remove pollutants, oil spills, toxins, and contaminants from water and soil by employing living organisms. Several different configurations have been exploited in MFC for simultaneous treatment of water and bioremediation along with the production of energy. Electroactive bacteria, a unique bacterial species, have been employed in these systems, which donate or accept electrons from their surroundings by cell surface. Mostly oxygen is used as an electron acceptor by these bacteria, but other than oxygen they can use sulfate, nitrate, manganese, etc., if conditions are anaerobic (Sherafatmand & Ng 2015; Wang et al. 2015).

12.6 CHALLENGES AND FUTURE PERSPECTIVES

Energy production and output require thorough rethinking and a potential improvement in the development of MFC. There is an urgent need to increase energy generation by scaling up the bacterial ecology and development of power management. The use of pure culture and development of air cathode enhance the applications of mediators, leading to an increase in the applied value and kinetics limitations of the MFC system. However, the research and development of MFC technology noticeably expanded with time and the performance has also been improved exponentially.

MFC can provide cheap and green renewable energy sources and thus received much attraction from researchers. MFC system is able to recover valuable products such as heavy metals, industrial chemicals, nutrients, and gaseous fuels in addition to serving the main purposes such as electricity generation and wastewater treatment. The actual power production limit and optimum size of MFC need to be determined as they are still an unknown fact. There is a need to better understand the microbial population which metabolizes organic matter more efficiently and to have knowledge of electron transfer mechanism. It also to be get modeled that how electrogens physiology and microbial ecology develop over time. However, the MFC technology develops from laboratory to field-scale trials, which show the potential of this technology to overpower other conventional renewable energy techniques. Moreover, this technology is the most promising technology that can resolve the future energy crisis by keeping our environment safe and free from pollutants. It is a cost-effective technique that simultaneously treats wastewater and produces energy as well. To resolve economic challenges and for a greener future, this technology will emerge as an integrated part of our waste management system in the future time. Further advancement will be much needed to develop more procedures such as the incorporation of nanomaterial engineering, materials science, and bio-electrochemistry for the development of more efficient MFCs. Despite some limitations, MFC is a very fascinating research subject to produce energy.

REFERENCES

Adebule, A., Aderiye, J., & Adebayo, A. (2018). Improving bioelectricity generation of microbial fuel cell (MFC) with mediators using kitchen waste as substrate. *Annals of Applied Microbiology & Biotechnology Journal*, 2, 1–5.

Al-Mamun, A., Jafary, T., Baawain, M. S., Rahman, S., Choudhury, M. R., Tabatabaei, M., Lam, S. S. (2020). Energy recovery and carbon/nitrogen removal from sewage and contaminated groundwater in a coupled hydrolytic-acidogenic sequencing batch reactor and denitrifying biocathode microbial fuel cell. *Environmental Research*, 183, 109273.

Amin, F. R., Khalid, H., Zhang, H., u Rahman, S., Zhang, R., Liu, G., & Chen, C. (2017). Pretreatment methods of lignocellulosic biomass for anaerobic digestion. *Amb Express*, 7(1), 72.

Banković-Ilić, I. B., Stojković, I. J., Stamenković, O. S., Veljkovic, V. B., & Hung, Y. T. (2014). Waste animal fats as feedstocks for biodiesel production. *Renewable and Sustainable Energy Reviews*, 32, 238–254.

Bhatia, S. K., et al. (2017b). Microbial biodiesel production from oil palm biomass hydrolysate using marine Rhodococcus sp. YHY01. *Bioresource Technology*, 233, 99–109.

Bhatia, S. K., Kim, S. H., Yoon, J. J., & Yang, Y. H. (2017a). Current status and strategies for second generation biofuel production using microbial systems. *Energy Conversion and Management*, 148, 1142–1156.

Bhatia, S. K., Lee, B. R., Sathiyanarayanan, G., Song, H. S., Kim, J., & Jeon, J. M. (2016). Biomassderived molecules modulate the behavior of Streptomyces coelicolor for antibiotic production. *Biotech*, 6, 223.

Bose, D., & Bose, A. (2017). Electrical power generation with Himalayan mud soil using microbial fuel cell. *Nature Environment and Pollution Technology*, 16, 433–439.

Chandrasekhar, K., Amulya, K., & Venkata Mohan, S. (2015b). Solid phase bioelectrofermentation of food waste to harvest value-added products associated with waste remediation. *Waste Management*, 45, 57–65.

Chaudhuri, S. K., & Lovley, D. R. (2003). Electricity generation by direct oxidation of glucose in mediatorless microbial fuel cells. *Nature Biotechnology*, 21, 1229–1232.

Cheng, H., & Hu, Y. (2010). Municipal solid waste (MSW) as a renewable source of energy: Current and future practices in China. *Bioresource Technology*, 101(11), 3816–3824.

Chiu, H. Y., Pai, T. Y., Liu, M. H., Chang, C. A., Lo, F. C., Chang, T. C., & Lo, S. W. (2016). Electricity production from municipal solid waste using microbial fuel cells. *Waste Management & Research*, 34(7), 619–629.

Dias, J. M., Alvim-Ferraz, M. C., & Almeida, M. F. (2009). Production of biodiesel from acid waste lard. *Bioresource Technology*, 100(24), 6355–6361.

Dong, Y., Qu, Y., He, W., Du, Y., Liu, J., Han, X., & Feng, Y. (2015). A 90-litre stackable baffled microbial fuel cell for brewery wastewater treatment based on energy self-sufficient mode. *Bioresource Technology*, 195, 66–72.

D'Souza, S. F. (2001). Microbial biosensors. *Biosensors & Bioelectronics*, 16, 337–353.

Du, Z., Li, H., & Gu, T. (2007). A state of the art review on microbial fuel cells: A promising technology for wastewater treatment and bioenergy. *Biotechnology Advances*, 25, 464–482.

Du, H., Wu, Y., & Wu, H. (2020). Dissolved organic matter and bacterial population changes during the treatment of solid potato waste in a microbial fuel cell. *Water Science and Technology*, 82(10), 1982–1994.

ElMekawy, A., Diels, L., Bertin, L., De Wever, H., & Pant, D. (2014). Potential biovalorization techniques for olive mill biorefinery wastewater. *Biofuels, Bioproducts and Biorefining*, 8(2), 283–293.

FAO. (2012). Towards the Future We Want: End hunger and make the transition to sustainable agricultural and food systems.

Gebrezgabher, S. A., Meuwissen, M. P., Prins, B. A., & Lansink, A. G. O. (2010). Economic analysis of anaerobic digestion—A case of Green power biogas plant in The Netherlands. *NJAS-Wageningen Journal of Life Sciences*, 57(2), 109–115.

Gooding, C. H., & Meeker, D. L. (2016). Comparison of 3 alternatives for large-scale processing of animal carcasses and meat by-products. *The Professional Animal Scientist*, 32(3), 259–270.

Gurav, R., Bhatia, S., Moon, Y-M., Choi, T-R., Jung, H-R., Yang, S-Y., Song, H., Jong-Min, J., Yoon, J-J., Kim, Y-G., & Yang, Y-H. (2018). One-pot exploitation of chitin biomass for simultaneous production of electricity, n-acetylglucosamine and polyhydroxyalkanoates in microbial fuel cell using novel marine bacterium Arenibacter palladensis YHY2. *Journal of Cleaner Production*, 209, 324–332.

Hung, Y. H., Liu, T. Y., & Chen, H. Y. (2019). Renewable coffee waste-derived porous carbons as anode materials for high-performance sustainable microbial fuel cells. *ACS Sustainable Chemistry & Engineering*, 7(20), 16991–16999.

Islam, K. M. (2016). Municipal solid waste to energy generation in Bangladesh: Possible scenarios to generate renewable electricity in Dhaka and Chittagong city. *Journal of Renewable Energy*, 2016.

Jadhav, P., Sarkar, J., Patil, R., & Vinchurkar, S. (2017). Microbial disintegration of bio-waste for hydrogen generation for application in fuel cell. *International Research Journal of Engineering and Technology (IRJET)*, 4, 229–232.

Kara, K., Ouanji, F., Lotfi, E. M., El Mahi, M., Kacimi, M., & Ziyad, M. (2018). Biodiesel production from waste fish oil with high free fatty acid content from Moroccan fish-processing industries. *Egyptian Journal of Petroleum*, 27(2), 249–255.

Kayode, B., & Hart, A. (2019). An overview of transesterification methods for producing biodiesel from waste vegetable oils. *Biofuels*, 10(3), 419–437.

Kim, B. H., Chang, I. S., Gil, G. C., Park, H. S., & Kim, H. J. (2003). Novel BOD (biological oxygen demand) sensor using mediator-less microbial fuel cell. *Biotechnology Letters*, 25, 541–545.

Kiran, E. U., Trzcinski, A. P., Ng, W. J., & Liu, Y. (2014). Bioconversion of food waste to energy: A review. *Fuel*, 134, 389–399.

Kyriakou, M., & Chatziiona, V. K., Costa, C. N., Kallis, M., Koutsokeras, L., Constantinides, G., & Koutinas, M. (2019). Biowaste-based biochar: A new strategy for fermentative bioethanol overproduction via whole-cell immobilization. *Applied Energy*, Elsevier, 242(C), 480–491.

Larrosa-Guerrero, A., Scott, K., Head, I. M., Mateo, F., Ginesta. A., & Godinez, C. (2010). Effect of temperature on the performance of MFCs. *Fuel*, 89(12), 3985–3994.

Li, H., Tian, Y., Zuo, W., Zhang, J., Pan, X., Li, L., & Su, X. (2016). Electricity generation from food wastes and characteristics of organic matters in microbial fuel cell. *Bioresource Technology*, 205, 104–110.

Logan, B. E. (2008). *Microbial Fuel Cells* (1st ed.). Hoboken: Wiley-Interscience. https://www.engr.psu.edu/ce/enve/logan/publications/2010-Logan-AMB.pdf.

Lui, G., Jiang, G., Lenos, J., Lin, E., Fowler, M., Yu, A., & Chen, Z. (2017). Advanced biowaste-based flexible photocatalytic fuel cell as a green wearable power generator. *Advanced Materials Technologies*, 2(3), 1600191.

Luimstra, V. M., et al. (2013). A cost-effective microbial fuel cell to detect and select for photosynthetic electrogenic activity in algae and cyanobacteria. *Journal of Applied Phycology*. https://doi.org/10.1007/s10811-013-0051-2.

Manjerkar, Y., Kakkar, S., & Durve-Gupta, A. (2018). Bio-electricity generation using kitchen waste and molasses powered MFC.

Meicong, W., Zinuo, W., Fei, H., Liping, F., & Xuejun, Z. (2020). Polyelectrolytes/α-Fe_2O_3 modification of carbon cloth anode for dealing with food wastewater in microbial fuel cell. *Carbon Resources Conversion*, 3, 76–81.

Mohd Zaini Makhtar, M., & Tajarudin, H. A. (2020). Electricity generation using membraneless microbial fuel cell powered by sludge supplemented with lignocellulosic waste. *International Journal of Energy Research*, 44(4), 3260–3265.

Moqsud, M. A., Omine, K., Yasufuku, N., Bushra, Q. S., Hyodo, M., & Nakata, Y. (2014). Bioelectricity from kitchen and bamboo waste in a microbial fuel cell. *Waste Management & Research*, 32(2), 124–130.

Mulakhudair, A. R., Hanotu, J., & Zimmerman, W. (2017). Exploiting ozonolysis-microbe synergy for biomass processing: Application in lignocellulosic biomass pretreatment. *Biomass and Bioenergy*, 105, 147–154.

Oh, S. T., Kim, J. R., Premier, G. C., Lee, T. H., Kim, C., & Sloan, W. T. (2010). Sustainable wastewater treatment: How might MFCs contribute. *Biotechnology Advances*, 28, 871–881.

Pant, D., et al. (2012b). Bioelectrochemical systems (BES) for sustainable energy production and product recovery from organic wastes and industrial wastewaters. *RSC Advances*, 2, 1248–1263.

Pant, D., Singh, A., Van Bogaert, G., Singh Nigam, P., Diels, L., & Vanbroekhoven, K. (2012a). Bioelectrochemical systems (BES) for sustainable energy production and product recovery from organic wastes and industrial wastewaters. *RSC Advances*, 2, 1248–1263.

Pawar, S. S., Nkemka, V. N., Zeidan, A. A., Murto, M., & van Niel, Ed W. J. (2013). Biohydrogen production from wheat straw hydrolysate using Caldicellulosiruptor saccharolyticus followed by biogas production in a two-step uncoupled process. *International Journal of Hydrogen Energy*, 38(22), 9121–9130.

Prajapati, S., & Yelamarthi, P. S. (2020). Microbial fuel cell-assisted Congo red dye decolorization using biowaste-derived anode material. *Asia-Pacific Journal of Chemical Engineering*, 15(5), e2558.

Procentese, A., Raganati, F., Olivieri, G., Russo, M. E., Rehmann, L., & Marzocchella, A. (2017). Low-energy biomass pretreatment with deep eutectic solvents for bio-butanol production. *Bioresource Technology*, 243, 464–473.

Reguera, G., McCarthy, K. D., Mehta, T., Nicoll, J. S., Tuominen, M. T., & Lovley, D. R. (2005). Extracellular electron transfer via microbial nanowires. *Nature*, 435, 1098–1101.

Rhoads, A., Beyenal, H., & Lewandowski, Z. (2005). Microbial fuel cell using anaerobic respiration as an anodic reaction and biomineralized manganese as a cathodic reactant. *Environmental Science & Technology*, 39, 4666–4671.

Ringeisen, B. R., Henderson, E., Wu, P. K., Pietron, J., Ray, R., Little, B., Biffinger, J. C., & JonesMeehan, J. M. (2006). High power density from a miniature microbial fuel cell using Shewanella oneidensis DSP10. *Environmental Science & Technology*, 40, 2629–2634.

Rorke, D. C. S., & Gueguim Kana, E. B. (2016). Biohydrogen process development on waste sorghum (Sorghum bicolor) leaves: Optimization of saccharification, hydrogen production and preliminary scale up. *International Journal of Hydrogen Energy*, 41(30), 12941–12952.

Sherafatmand, M., & Ng, H. Y. (2015). Using sediment microbial fuel cells (SMFCs) for bioremediation of polycyclic aromatic hydrocarbons (PAHs). *Bioresource Technology*, 195, 122–130.

Shockey, I., Kinicki, R., Delorey, J., Arruda, K., Baldwin, S., & Quinn, A. (2010). Feasibility study of alternative energy sources and conservation techniques for implementation in El Yunque National Forest. Available at http://web.cs.wpi.edu/~rek/Projects/USF_Proposal. Accessed 2 Dec 2015.

Sindhuja, M., Harinipriya, S., Bala, A. C., & Ray, A. K. (2018). Environmentally available biowastes as substrate in microbial fuel cell for efficient chromium reduction. *Journal of Hazardous Materials*, 355, 197–205.

Sotres, A., Cerrillo, M., Vias, M., & Bonmat, A. (2015) Nitrogen recovery from pig slurry in a two-chambered bioelectrochemical system. *Bioresource Technology*, 194, 373–382. https://doi.org/10.1016/j.biortech.2015.07.036.

Su, L., Jia, W., Hou, C., & Lei, Y. (2011). Microbial biosensors: A review. *Biosensors & Bioelectronics*, 26(5), 1788–1799.

Taher, H., Al-Zuhair, S., AlMarzouqui, A., & Hashim, I. (2011). Extracted fat from lamb meat by supercritical CO_2 as feedstock for biodiesel production. *Biochemical Engineering Journal*, 55(1), 23–31.

Tremouli, A., Karydogiannis, I., Pandis, P. K., Papadopoulou, K., Argirusis, C., Stathopoulos, V. N., & Lyberatos, G. (2019). Bioelectricity production from fermentable household waste extract using a single chamber microbial fuel cell. *Energy Procedia*, 161, 2–9.

Wang, L., et al. (2015). Efficient degradation of sulfamethoxazole and the response of microbial communities in MFCs. *RSC Advances*, 5, 56430–56437.

Wang, X., Tang, J., Cui, J., Liu, Q., & Markus, H. (2014). Synergy of electricity generation and waste disposal in solid-state microbial fuel cell (MFC) of cow manure composting. *International Journal of Electrochemical Science*, 9, 3144–3157.

Xin, X., Ma, Y., & Liu, Y. (2018). Electric energy production from food waste: Microbial fuel cells versus anaerobic digestion. *Bioresource Technology*, 255, 281–287.

Zhou, S., Zhang, B., Liao, Z., Zhou, L., & Yuan, Y. (2020). Autochthonous N-doped carbon nanotube/activated carbon composites derived from industrial paper sludge for chromate (VI) reduction in microbial fuel cells. *Science of the Total Environment*, 712, 136513.

13 Recent Development in Microbial Fuel Cells Using Biowaste

Abhinay Thakur, Shveta Sharma, and Ashish Kumar
Lovely Professional University

CONTENTS

13.1 INTRODUCTION

In recent years, an increase in energy usage has been recorded in accordance with population growth. Approximately six billion people are living on earth, with an estimated population of 9.4 billion by 2050 [1–3]. Based on their origins and consistency, energy sources have been categorized as nuclear, fossil fuel, and renewable energy sources. Non-renewable energy sources include nuclear and fossil fuels, which account for the majority of the energy production. Unearthing and burning

DOI: 10.1201/9781003178354-16

fossil fuels releases CO_2 frozen under the earth's layers for years, adding significantly to global warming and environmental pollution. Although fossil fuels often aided a country's development and economic growth over the last century, it's indeed apparent that they have been insufficient to support a global economic output. Oil would hardly run up for another 100 years; although its massive utilization anticipates to outstrip its supply from existing and projected oil reserves, rendering its depletion before 2025 [4,5]. This may seem far away to many customers and businesses who rarely plan upward of 3–5 years ahead, but for society, it's a very limited period. By 2030, energy demand is projected to rise by nearly 44%, owing primarily to increased demand from emerging economies such as India and China. Non-renewable energy sources such as petroleum, natural gas, coal, and nuclear currently meet over 90% of global energy demand (500 quadrillions Btu) [6]. Renewable energy sources now provide just 7% of the overall energy demand (100 quadrillions Btu) in the USA [7,8]. Our firm reliance on non-renewable energy sources has several unavoidable implications, including negative effects on economic growth, national security, and the local and global climate.

To address the global energy crisis and the challenges posed by fossil fuels, several nations are looking for alternative renewable energy sources. Fuel cell technology uses precious metal catalysts to produce electricity. Fuel cells possess numerous advantages over other existing energy generation methods, including the absence of toxic inorganic oxides (CO_2, SO_2, NO, and CO) and higher performance. However, operating costs are the main drawbacks restraining the commercialization of fuel cell technology. Energy production, consumption, and storage are the topics that are becoming more common in the modern research fields with a potency toward global interest and significance [9]. Microbial fuel cells (MFCs) are one possible renewable energy source. MFC technologies are one of the most recent approaches toward electric current production from biomass using bacteria. This chapter will cover the whole essential information regarding the mechanism and usage of MFCs in various fields owing to its advantages and disadvantages.

13.2 MICROBIAL FUEL CELLS

Instead of relying on electricity, MFCs use the biocatalytic capacities of viable microorganisms and can use a variety of organic fuel sources by transforming the energy contained in chemical bonds into an electrical current. Microorganisms, such as bacteria, can produce energy from biodegradable substrates and organic matter and biodegradable products such as municipal wastewater. Microorganisms dissolve (oxidize) organic matter in an MFC, releasing electrons that pass through the cell's respiratory enzymes and producing energy in ATP form. After that, the electrons are transmitted to a terminal electron acceptor (TEA), which receives the e^- and then reduces itself [10]. The reduction of oxygen to water is catalyzed by an electron–proton process. Most TEAs, such as nitrate, oxygen, and sulfate, disperse easily into cells, where they accept electrons and then produce compounds that can propagate. But some bacteria can extracellularly move e^- (i.e., outside the cell), including metal oxides such as iron oxide, to TEA. Such bacteria, known as exoelectrogens, can transmit electrons exogenously that could be utilized to produce

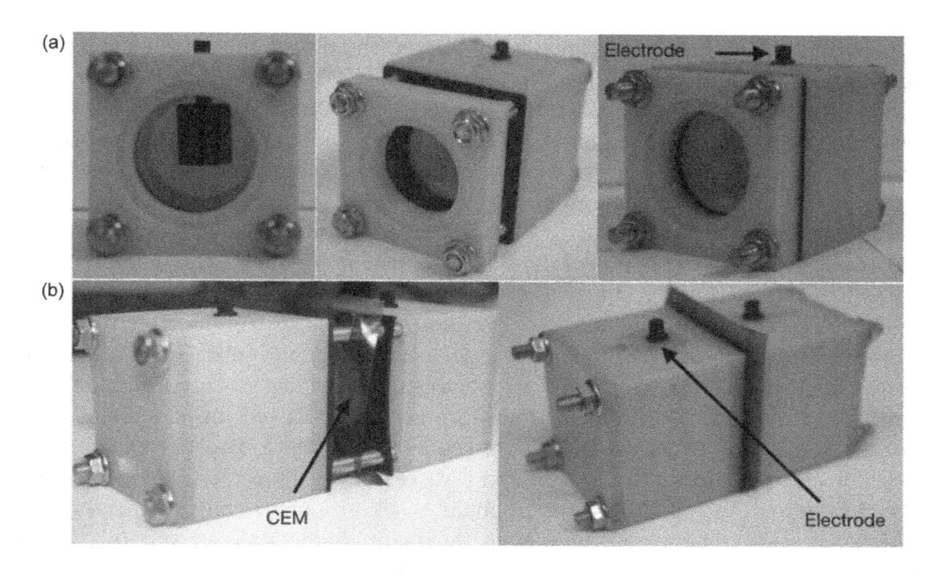

FIGURE 13.1 Exploded and compact views of (a) a single-chamber MFC and (b) a double-chamber MFC. (Reproduced from Ref. [11]. Copyright (2018) Elsevier. The article is printed under a CC-BY license.)

power in the MFC. Methanogenesis, methanogens, and anaerobic digesters are the terms used to classify the processes, microorganisms, and reactors that produce methane. Since MFCs are so recent, little effort has been put into developing functional architectures that use low-cost materials. MFCs are a new technology that generates energy from waste by using bacteria. Bacteria in an MFC could break down food and bodily wastes, extracting energy from products that would otherwise be discarded. The generation of clean, renewable energy could be at a low cost by tapping into this unexploited source of electricity. MFCs should be designed for use in the industry that would most likely generate the most revenue. MFCs are particularly useful because they can be used in numerous ways to mitigate emissions and suppress water treatment expenditure in a safe, sustainable, friendly manner. Figure 13.1 shows the exploded and compact views of (i) a single-chamber MFC and (ii) a double-chamber MFC [11].

13.2.1 Structural Configurations

MFCs come in various structural configurations, including single- and two-chamber configurations, with the presence or absence of PEM. The MFC is comprised of an anodic chamber, a cathode chamber, and a half-cell separator. Separators enable protons to travel to the cathode because of a possible gradient, thus preventing O_2 (or e-acceptor used in cathodic chamber) from diffusing to anode, in which the presented bacteria could be damaged [12]. The compartments' structural design can differ drastically to boost MFC's power performance. Two-compartment MFCs generally utilized in the batch mode having a specific medium (such as acetate or glucose).

Miniature MFCs are attracting a lot of interest in applied and fundamental studies, owning to their insightful and positive aspects. At the milliliter to microliter scale, miniature MFCs can generate electricity. Miniature MFCs have proven efficacy as energy sources for autonomous sensors since they can absorb nutrients from atmosphere and the electrogenic biofilm can get retained at anode. In remote locations, miniature MFC configurations might be particularly useful. These results suggest that the diffusion range via the PEM affects MFC power output more often than electrode size, resulting in potent power deliverance in miniaturized configurations.

Upflow mode MFC configurations gained a lot of attention, owing to their enhanced productibility for use in wastewater treatment, as well as effective usage in the scaling up to required industrial aspects as shown in Figure 13.2 [8]. Min and Logan designed the flat-plate MFC to resemble the specifications used in traditional hydrogen fuel cells, in which electrodes are generally packed inside a single strip split with PEM, allowing them to be pressed close together to boost proton conduction between two electrodes. Moreover, PEMs permeable to oxygen, such as Nafion, are commonly used in MFCs, which could be hazardous to obligate anaerobes if utilized as selected bacteria to be used in the anodic chamber. As a result, this prototype was sorely tested to see if it could withstand the possibility of oxygen penetration into the anode. The results demonstrated a peak power density of 71.9 mW/m^2 while residential wastewater has been used as a source of fuel, which is a 2.8 times improvement in power production over a single-chambered MFC design examined by similar research team. Numerous structural configurations are often used in MFC approaches, with designs such as tubular topologies proving to be helpful in terms of improving power outputs.

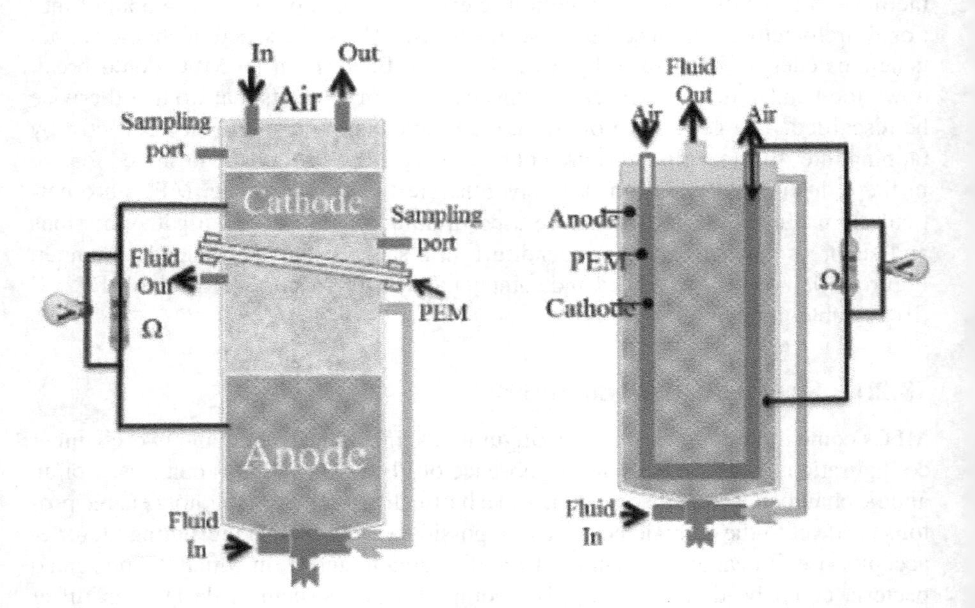

FIGURE 13.2 Upflow microbial fuel cell configuration model. (Reproduced from Ref. [8]. Copyright (2019) MDPI. The article is printed under a CC-BY license.)

Sediment microbial fuel cells (SMFCs) have indeed been explored extensively for energy production through biological sediment, including a primary emphasis on wastewater treatment alternatives. SMFCs generate electricity from organic debris in sediment by using bacterial metabolism. The presence of anoxic environments at the anode distinguishes SMFCs from conventional MFCs, as does its membrane-less composition, which employs the sediment–water interaction as membrane. Despite this, SMFC research has recently been accelerated owing to the dual functionalities, which allow for both electrical energy production and wastewater treatment by removing unique contaminants. Bose et al. [6], at Jaipur, Rajasthan, India, used an SMFC to test the bioelectricity generation capacity of dark flocculent sediment samples. Six graphite-based anodes were installed in sediments and arranged in series to the cathode to improve device output with the inclusion of a glucose-rich substrate medium. After a week of testing, current density with a relatively low resistance and power density at a peak of 2,122 80 mW/m^2 and 4.31 mA/m^2 was attained. The generation of bioelectricity results from the electrochemical oxidation of deposited inorganic and organic compounds that exist in soil, which is the first step in utilizing SMFCs as a vital technique for bioremediation. An FTIR study of soil also showed that the functional components are abundant in sulfur, nitrogen, silicon, aluminum, and copper, as well as other inorganic and organic compounds easily disintegrated by the metabolic processes of soil bacteria, as a result rendering them a valuable asset for bioremediation and bioelectricity production.

13.3 TYPES OF MFCs ON THE BASIS OF COMMERCIALIZATION

13.3.1 Low-Cost MFCs

An important step in enhancing MFC efficiency is the effective transport of electrons through bacteria to electrodes. Artificial additives serve as e- shuttles, transporting e- through the bacteria's internal compounds and respiration chain to the electrode; redox mediators have increasingly been used in the control of electron transfer rate. The electron transfer is aided by the dissipation of reduced mediator that disperses through bacterial cells oxidized in the electrode. Helinando et al. [13] explored the effect of various variables on the capacity of energy generation by *E. coli*-provided MFC to ensure the MFC is cost-effective. A blend of chemicals in the anodic and cathode chambers in an MFC results in adequate e- transfer through bacteria to electrode and cation transfer, together with a membrane separator, resulting in optimal situations for electricity generation. As a consequence, when E. coli is used as an active agent in MFC, the processing potential is enhanced in terms of days. In the parameters outlined, PMMA could be a feasible alternative to traditional charge transfer membranes for implementing inexpensive and effective energy production in MFCs [14]. Similarly, Shrok et al. [15] used the low-cost membrane to clean effluent from petroleum refineries in dual-chamber MFCs. They used $KMnO_4$ for cathodic electron acceptors throughout the cathode compartment of MFC via an inexpensive proton exchange membrane rather than the relatively expensive Nafion to tackle PRW from Al-Dura refinery. $KMnO_4$ was employed as a cathodic electron acceptor to enhance MFC energy efficacy. As a result, researchers looked at the impacts of potassium permanganate content on MFC production and PRW intervention effects. After 48 h, the optimum COD reduction effectiveness was 71.24% with a maximum output of 1.0032 W/m^2 and a permanganate content of 0.125 g/L [15].

13.3.2 Compost-Based MFCs

MFCs generate electricity by breaking apart fuel with bacteria or released enzymes. Bacteria and enzyme serve as biocatalysts for MFCs, allowing electrochemical reactions in compost soil matrices to create electricity. Almost all liquid MFCs provide a potential hazard since they deteriorate quickly owing to ammonia volatilization, resulting in toxic effects, instability, and spillage. Compost soil is chosen over liquid soil for decreasing hazards, using lower fuel, and creating greater effort and energy. Furthermore, reliable sources are generally less expensive and do not cause volatilization; additionally, maintaining a solid-state pH level in the soil is easy. Urea is one of the most abundant sources of renewable energy. Urea is a suitable fuel for MFCs among all of the existing sources of renewable energy. Advantageously, compost soil microbes participate in natural nitrification and denitrification processes in the nitrogen cycle during their development.

Bibiana et al. [16] utilized a compost inoculum to produce microbial bioanodes having potential properties. At 0.4 V/SCE, the reduced currents were attributable to the lower interfacial electron transfer rate as well as usual electrochemical kinetics at that potential. The three major bacterial groups Anaerophaga, Geobacter, and Pelobacter were similar for all bioanodes and did not depend on polarization potential, according to DNA- and RNA-based DGGE. The only CVs that revealed changes in the biofilms' redox capabilities were non-turnover CVs, which had the greatest potential for promoting numerous electron transfer pathways.

13.4 FUNDAMENTAL BIOELECTRICITY GENERATION IN MFCs

The generation of biopotential in an anodic chamber (resulting in reduction processes) and e- acceptor parameters at the cathode are responsible for the generation of bioelectricity in MFCs (that are segregated with a membrane). Electrochemically active microorganisms in an anodic chamber may donate e- to anode that is released by oxidizing organic/inorganic waste, resulting in an energy source. A single-chamber MFC with a bacterial biofilm on anode and diffusion layers at the cathode was studied [17].

An oxidation reaction involving electrochemically active bacteria in an anodic chamber and acetate as a fuel source is as follows:

$$CH_3COO^- + 4H_2O \rightarrow 2HCO_3^- + 9H^+ + 8e^-$$

Protons formed by electrochemically active bacteria in an anode disperse into the cathodic chamber via a proton exchange membrane (PEM). Due to high reduction potential and quantity, O_2 is used as the primary oxidant in the cathodic compartment. Even in the case of significant kinetics and overpotential, the oxygen reduction process is still a constraint that prevents it from being optimized for better use and hence improving MFC designs. Metal oxidants, such as copper, cadmium, and chromium, have been used in the cathodic compartment in other experiments. Once protons have entered the cathode through the PEM, they combine with the available oxygen to produce H_2O through the oxygen reduction reaction:

$$O_2 + 4H^+ + 4e^- \rightarrow 2H_2O$$

A system must be able to have a reliable fuel source, which could be oxidized at the substrate–anode interface and refilled continuously or intermittently, to be labeled as an MFC; otherwise, the design is referred to as biobattery rather than MFC. Anaerobic systems are used in the majority of MFC configurations. This is because anaerobic conditions are currently regarded as the "gold standard" in terms of electron transfer characteristics for bacterial organisms. MFCs are commonly used as closed-system instruments, with an anodic chamber that is anaerobically sealed. It is essential to promote the growth of anaerobic bacteria to transfer electrons, e.g., G. sulfurreducens. So far, marine sediment soil, soil, wastewater, activated sludge, and freshwater sediment have all been established as fuel sources with bacteria capable of electron transfer. Bacteria can generate electrons in such systems by oxidizing substrates isolated in the anodic compartment. The mechanism segment has gone over the entire procedure. Based on the effluent strength, MFC can generate 1.43–1.8 kWh/m^3. However, it only absorbs 0.024 kW, approximately one order of magnitude less than the 0.3 kW involved in the anaerobic digestion. The results show that MFC consumes about 10% less energy than the activated sludge process, indicating that it has a better capacity for the production of renewable energy and cost-effective wastewater treatment. MFC technology can effectively treat wastewater while simultaneously producing electricity; it still faces challenges in moving beyond the laboratory for potential application or monetization [2,18]. MFC technology in direct field applications is limited by a variety of factors, including the cost of electrode materials, the need for precious metal catalysts, poor efficiency, low power densities, and expensive PEMs. The major benefit of MFC technologies is that they are less reliant on geographical position and seasonal change than existing renewable energy models.

13.5 PROGRESS IN THE DEVELOPMENT OF COST-EFFECTIVE ELECTRODE MATERIALS FOR MFCs

13.5.1 Electrode Materials

In MFCs, e- are generated at the anode by the anaerobic oxidation of fuel with the action of microbes and transferred to the cathode and the oxidizing agent is reduced, via the external circuits. The simultaneous phenomenon of reduction and oxidation processes results in an uninterrupted flow of electrons, resulting in the generation of electricity. The efficiency of an MFC is thought to be influenced by several factors including electrode configuration, electrode materials, membranes, mediators, and biocatalysts. The quality of the electrode materials has a substantial influence over the other factors mentioned because variations in e- transportation between electrodes and microbes lead to constrained power output. The procurement of electrode components is a critical step that must be handled with extreme caution to

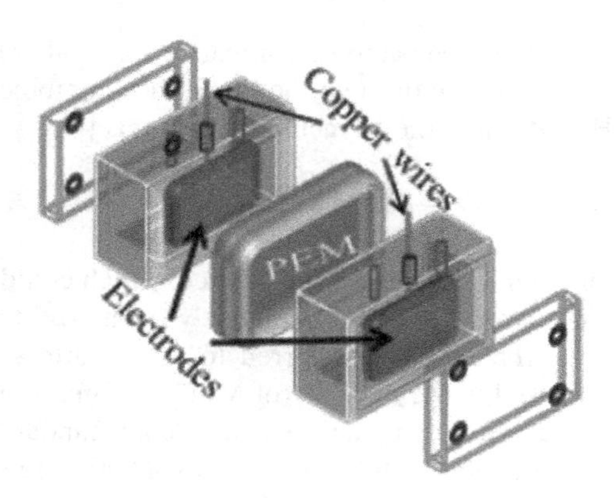

FIGURE 13.3 A single-electrode/PEM assembly used in a traditional chemical fuel cell. (Reproduced from Ref. [8]. Copyright (2019) MDPI. The article is printed under a CC-BY license.)

avoid corrosion [19,20]. Due to galvanic corrosion, copper appears to have higher power densities when utilized as electrodes. Copper ions, on the other hand, have been discovered to have antibacterial properties. Similarly, the use of stainless-steel electrodes is determined by the alloy's chromium content. Figure 13.3 shows a mini square-shaped compact flat-plate MFC, identical to a conventional chemical fuel cell with PEM assembly, wherein the cathode is hot-compressed to a Nafion PEM and has been in conjunction via an anode to create a PEM/electrode framework [8].

13.5.2 Anode Materials

An MFC's efficiency is largely determined by its anode performance, even though this is a platform for important bio-electrochemical reactions and a route for electron transmission from exoelectrogens to electrodes. Chemical resistivity, durability, surface area, and electrical conductivity are all anodic parameters that have a major impact on MFC efficiency. Several studies have been carried out to increase the performance of MFCs with minimal resources by utilizing inexpensive electrode materials in a variety of situations [21]. To improve MFC performance, carbon cloth anodes are equipped with ammonia gas and phosphate buffer. The intervention resulted in a significant boost in efficiency due to increased electrode surface charge and solution conductivity. When compared to MFC with an untreated electrode, the combined effect of these two treatments culminated in a 48% improved power generation. This novel approach resulted in a 1,970 mW/m^2 boost in power density. As an anode material, a low-cost carbon grid was investigated. The increased N/C atomic ratio on mesh electrode substrate was discovered to be the cause of the improved power densities. According to the research, carbon mesh treated with ammonia gas may be a cheaper alternative to carbon fabric and paper electrodes. Three different methods were used on carbon fiber brushes: acid soaking, heating, or its mixture. The carbon mesh anode with a mix of acid and heat treatment increased power generation by up to 34%. Carbon cloth anodes equipped with heat and ammonia gas outperformed anodes treated with a combination of acid soaking and heat treatment.

13.5.3 Cathode Materials

The cathode chamber in an MFC is assumed to be an electron trap, where O_2 is reduced to H_2O. At the three-phase contact of liquid, air, and solid in the cathode chamber, oxygen is decreased. A conventional MFC cathode consists of an electrode support, air diffusion layer, and the catalyst. While anode materials are often used as cathodes, a potential cathode may have higher electrical conductivity, catalytic activity, and mechanical properties. Hereby, the rate of O_2 reduction is quite slow, resulting in overpotential and limiting MFC performance. In addition to their high cost, they foul when using low-quality water. There has been a lot of research into finding less expensive substitutes to Pt without affecting the performance of cathodic catalysts [22,23]. In an experiment, a base layer of carbon powder and polytetrafluoroethylene was applied to one side of the carbon fabric, followed by several layers of polytetrafluoroethylene as a diffusion layer. A platinum catalyst was employed in the carbon fabric which was facing the water. In a single-chamber MFC, the fabricated carbon cloth was used as the cathode. The MFC outperforms carbon cloth with merely a base layer in terms of power density (42.4%) and coulombic efficiency (200%). As per studies, four polytetrafluoroethylene coatings reduced water losses through the cathode, resulting in the highest power density. At neutral pH, a cathode based on iron phthalocyanine demonstrated higher oxygen reduction rates than a platinum catalyst.

In the fabrication of traditional catalysts, generally costly, non-environmentally friendly, and complex materials are often used, which leads to the production of biocathodes, wherein microorganisms serve as a catalyst. There are two types of biocathodes: aerobic and anaerobic biocathodes. O_2 serves as a junction for e- acceptor in aerobic biocathodes, while hydrogen peroxide is an intermediate. Between the electrode and oxygen, transition metals including manganese and iron serve as electron mediators. Electrons produced from the biofilm reduce Fe(III) to Fe(II), which is then oxidized by O_2. The e- move from the cathode to the terminal e- acceptor. Sulfates and nitrates can serve as terminal electron acceptors in anaerobic environments, where oxygen is not present. Nitrate, manganese, and iron have cathodic potentials that are equivalent to oxygen. Furthermore, an anaerobic biocathode has the benefit of preventing electron loss via oxygen, which could otherwise diffuse to the anode through PEM [17].

13.6 FACTORS AFFECTING THE MFC's EFFICIENCY

13.6.1 pH Buffer and Electrolyte

The reaction rate of electrons, protons, and oxygen at cathode corresponds to the generation rates of protons at anode when a functional MFC has no buffer solution, resulting in an apparent pH discrepancy between the cathodic and anodic chambers[24]. The driving force of proton diffusion from the anode to the cathode chamber is increased by the pH difference, and a dynamic equilibrium is formed.

Seyed et al. [25] investigated the activity of a chamber MFC at various pH values utilizing purified terephthalic acid wastewater. To investigate the pH impact, the MFC is regularly supplied tenfold diluted wastewater at pH values of 5.4, 7.0, and

8.5. The overall power density was found to be 4.2, 12.4, and 7.3 mW/m^2 for 5.4, 7.0, and 8.5 [25]. It could be inferred from the pattern of power production at various values that an alkaline environment is optimal for the growth of electrogenic bacteria.

13.6.2 Effect of Temperature

An MFC transforms biomass into electric current electrochemically via the catalytic action of bacteria known as exoelectrogen or anode-respiring bacteria, which can pass electrons beyond their outer membrane. Hao Ren et al. [26] with a Geobacter sulfurreducens-dominated mixed inoculum examined the impact of temperature on a miniaturized MFC. The experiment was conducted in the optimum temperature range of 322–326 K for miniaturized MFC. The current density increases by 282% as the temperature rises from 294 to 322 K, from 2.2 to 6.2 Am^2. They miniaturized the MFC by installing it in a gravity convection furnace with a closed-loop temperature regulator. The miniaturized MFC had a chamber capacity of only 50 L because of its less thermal mass and regulation of the temperature of MFC inside the furnace [26].

13.7 APPLICATIONS OF MFCs

The MFC is currently and potentially used in domestic wastewater treatment, hydrogen production, remote sensing, desalination, pollution remediation, among other applications. Numerous applications are currently practiced, and they could become widely used in the future [27–29]. Using a pre-treatment procedure, lignocelluloses could be broken down into reducing sugars, which could be used as a fuel for EAMs to produce electricity [12].

13.7.1 Biobattery

Like MFC, the bacteria-powered battery (or biobattery) transfers biological energy to electrical energy by retaining its chemical reactions within and not refilling or removing them. As a result, a biobattery can power battery-powered devices that merely need a little quantity of electricity for a matter of seconds. Biobattery has the advantage of being able to create energy from a nearly limitless number of ecological or microbiological liquids that appear to be freely available, even in resource-scarce situations [30,31]. The biobatteries are designed to be used as a primary cell that can be removed once the organic fuel collected inside is depleted. Microorganisms may be implemented as a biocatalyst during service or pre-loaded in the battery before being used. By transporting electrons through the cell membrane to external electrodes in the ambient liquid of a lake, sea, or pond, a variety of microorganisms can create electricity. Any liquid that is readily available in the surrounding environment should be used to fuel the bacteria-powered battery. The battery would be unable to operate on-site in the desert and arid circumstances due to the lack of liquids. As a result, the pre-loading technique of properly storing highly electroactive anodophiles as a self-contained system prior to operation would be extremely powerful and reliable for battery-type MFC devices. The electroactive bacteria used as a biocatalyst are immobilized in an anodic chamber once the system is created. The device is then tightly enclosed, trapping the bacterial cells inside and preventing them from being released into the air while remaining safe from outside contamination.

13.7.2 Wastewater Treatment

The activated sludge method has been the most effective and commonly used biological technology for wastewater treatment so far. Pumping and aeration are the primary energy consumers in the process; for example, pumping consumes 21% of total treatment energy demand and aeration consumes 30%–55%. Various approaches to wastewater treatment are preferred due to the high cost of service and high energy usage. Liu et al. [32] presented the very first example of MFCs employing household wastewater as the foundation. They utilized single-chamber MFCs that didn't need any oxygen aeration in the cathode chamber, and they were able to remove up to 80% of COD from domestic wastewater.

Similarly, Yajing Guo et al. [14] presented an overview on the MFC-based energy harvesting and wastewater treatment study, as well as several biocatalysts employed in MFCs and related mechanisms underlying pollutant reduction and electricity generation through wastewater. While significant progress has been made in terms of pollutants removal efficiency and energy recovery, they claim that research on nutrient separation and restoration from wastewater has got the least attention. MFCs have the potential to convert a wide range of contaminants into beneficial resources. Alternative wastewater treatment approaches that collect and reuse resources simultaneously must be investigated in order to contribute to zero carbon emissions. Other factors, such as high capital costs, low power harvesting performance, and poor long-term system reliability, are also preventing MFCs from being used in wastewater treatment in the real world [33].

13.7.3 Remote Biosensors

MFCs can be used to control low-power sensors that collect information in remote locations. In the Palouse River in Washington, scientists have substituted a conventional wireless thermometer with the one operated by an MFC. This MFC was incorporated into the riverbed for research purposes. This was a basic MFC with a metal wire connecting the cathode and anode. The anode is placed in the anaerobic sediment of a stream or river, while the cathode is placed in the aerobic water directly over the sediment to create a current. Anaerobic bacteria in the sediment produce a low current that could be utilized to power a capacitor and reserve energy till the sensor needs it. One of the most significant benefits of employing an MFC instead of a conventional battery in remote sensing is the replication of bacteria, offering MFC a better life span than that of conventional batteries. As a result, the sensor could be left in a remote location without having to be serviced for many years.

Bombelli et al. [34] used Physcomitrella patens (spreading earth moss), a bryophyte model to combine PMFCs (non-vascular seedless plant). They discovered that PMFCs using non-sterile moss generated a peak power output of 6.7 mW/m^2. The electrical outputs produced by Physcomitrella patens are adequate to control a radio or an environmental sensor, as shown in Figure 13.4 [34].

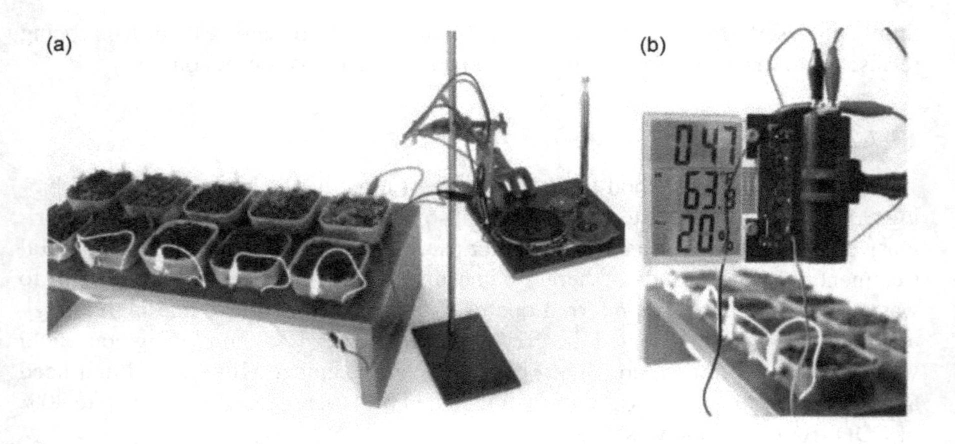

FIGURE 13.4 Ten Physcomitrella patens-containing plastic containers are enough to power (a) a commercial radio receiver and (b) an environmental sensor. (Reproduced from Ref. [34]. Copyright (2016) Royal Society Open Science. The article is printed under a CC-BY license.)

13.8 CONCLUSIONS

Microbial fuel cells, which were once only seen as a trend at workshops, are now a growing fact with a lot of room for development in cleaning techniques and power generation phenomena. These bacteria produce electricity by breaking down food waste and sewage. They are used to generate energy in MFC and are also converted into less toxic metabolites, demonstrating yet another possible application in managing waste and controlling pollution. To date, a wide range of microbes and substrates which include xenobiotics and waste have been utilized to generate electricity. Furthermore, MFCs are highly efficient, do not need fossil fuels for electricity, and can operate efficiently on waste and sewage. They contribute to the world's transformation into a more prosperous and environmentally friendly climate. MFC has the ability to be efficient in the transition away from fossil fuel-based technology and toward sustainable energy sources. This, in combination with extensive research into the behavior of exoelectrogenic bacteria, their biological processes, and the effect of crucial pathways such as electron transfer processes, could result in power outputs that are relatively close to theoretical goals while also advancing the concept of electromicrobiology. The development of fuel cells with enhanced designs and electron transfer mechanisms to electrodes could support current generation, biosensing output, bioremediation, bio-fouling reduction, biogenesis, and biobatteries on biodegradable fuels. Bio-electrochemical technology holds promise for generating renewable energy and reducing emission risks to humans. Several applications of bio-electrochemical technology are being explored, but more applications are likely to emerge in the coming years. Although this field is getting advanced, far more work remains to be done before MFC technologies are widely adopted in enterprise and environment. With future progress, MFCs will attain the efficiency to produce hydrogen to be used for fuel cells as well as deliver renewable energy options for rural regions.

REFERENCES

1. Narayanaswamy Venkatesan P, Dharmalingam S (2017) Characterization and performance study of phase inversed Sulfonated Poly Ether Ether Ketone – Silico tungstic composite membrane as an electrolyte for microbial fuel cell applications. *Renew Energy* 102:77–86.
2. Ayyaru S, Mahalingam S, Ahn YH (2019) A non-noble V_2O_5 nanorods as an alternative cathode catalyst for microbial fuel cell applications. *Int J Hydrogen Energy* 44:4974–4984.
3. Cui Y, Lai B (2019) Microbial fuel cell-based biosensors. *Biosensors* 9:92.
4. Das I, Das S, Dixit R, Ghangrekar MM (2020) Goethite supplemented natural clay ceramic as an alternative proton exchange membrane and its application in microbial fuel cell. *Ionics (Kiel)* 26:3061–3072.
5. Rashid T, Sher F, Hazafa A, Hashmi RQ, Zafar A, Rasheed T, Hussain S (2021) Design and feasibility study of novel paraboloid graphite based microbial fuel cell for bioelectrogenesis and pharmaceutical wastewater treatment. *J Environ Chem Eng* 9:104502.
6. Bose D, Santra M, Sanka RVSP, Krishnakumar B (2021) Bioremediation analysis of sediment-microbial fuel cells for energy recovery from microbial activity in soil. *Int J Energy Res* 45:6436–6445.
7. Papiya F, Nandy A, Mondal S, Kundu PP (2017) Co/Al_2O_3-rGO nanocomposite as cathode electrocatalyst for superior oxygen reduction in microbial fuel cell applications: The effect of nanocomposite composition. *Electrochim Acta* 254:1–13.
8. Flimban SGA, Ismail IMI, Kim T, Oh S-E (2019) Review: Overview of recent advancements in the microbial fuel cell from fundamentals to applications: Design, major elements, and scalability. *Energies* 12:1–20.
9. Santoro C, Arbizzani C, Erable B, Ieropoulos I (2017) Microbial fuel cells: From fundamentals to applications. A review. *J Power Sources* 356:225–244.
10. Santoro C, Serov A, Gokhale R, Rojas-Carbonell S, Stariha L, Gordon J, Artyushkova K, Atanassov P (2017) A family of Fe-N-C oxygen reduction electrocatalysts for microbial fuel cell (MFC) application: Relationships between surface chemistry and performances. *Appl Catal B Environ* 205:24–33.
11. Jannelli E, Di Trolio P, Flagiello F, Minutillo M (2018) Development and performance analysis of biowaste based microbial fuel cells fabricated employing additive manufacturing technologies. *Energy Procedia* 148:1135–1142.
12. Moradian JM, Fang Z, Yong YC (2021) Recent advances on biomass-fueled microbial fuel cell. *Bioresour Bioprocess* 8:1–13.
13. De Oliveira HP, Da Costa MM, De Oliveira AHP, Nascimento MLF (2015) Development of low cost microbial fuel cell based on Escherichia coli. *IEEE-NANO 2015-15th Int Conf Nanotechnol*: 940–942.
14. Guo Y, Wang J, Shinde S, Wang X, Li Y, Dai Y, Ren J, Zhang P, Liu X (2020) Simultaneous wastewater treatment and energy harvesting in microbial fuel cells: An update on the biocatalysts. *RSC Adv* 10:25874–25887.
15. Allami S, Hasan B, Redah M, Hamody H, Abd Ali ZD (2018) Using low cost membrane in dual-chamber microbial fuel cells (MFCs) for petroleum refinery wastewater treatment. *J Phys Conf Ser.* https://doi.org/10.1088/1742–6596/1032/1/012061.
16. Cercado B, Byrne N, Bertrand M, Pocaznoi D, Rimboud M, Achouak W, Bergel A (2013) Garden compost inoculum leads to microbial bioanodes with potential-independent characteristics. *Bioresour Technol* 134:276–284.
17. Mustakeem (2015) Electrode materials for microbial fuel cells: Nanomaterial approach. *Mater Renew Sustain Energy* 4:1–11.
18. Rojas Flores S, Naveda RN, Paredes EA, Orbegoso JA, Céspedes TC, Salvatierra AR, Rodríguez MS (2020) Agricultural wastes for electricity generation using microbial fuel cells. *Open Biotechnol J* 14:52–58.

19. Cao Y, Mu H, Liu W, Zhang R, Guo J, Xian M, Liu H (2019) Electricigens in the anode of microbial fuel cells: Pure cultures versus mixed communities. *Microb Cell Fact* 18:1–14.
20. Ezziat L, Elabed A, Ibnsouda S, El Abed S (2019) Challenges of microbial fuel cell architecture on heavy metal recovery and removal from wastewater. *Front Energy Res* 7:1–13.
21. Gao Y, Mohammadifar M, Choi S (2019) From microbial fuel cells to biobatteries: Moving toward on-demand micropower generation for small-scale single-use applications. *Adv Mater Technol.* https://doi.org/10.1002/admt.201900079.
22. Gajda I, Greenman J, Santoro C, Serov A, Melhuish C, Atanassov P, Ieropoulos IA (2018) Improved power and long term performance of microbial fuel cell with Fe-N-C catalyst in air-breathing cathode. *Energy* 144:1073–1079.
23. Kumar AG, Singh A, Komber H, Voit B, Tiwari BR, Noori MT, Ghangrekar MM, Banerjee S (2018) Novel sulfonated co-poly(ether imide)s containing trifluoromethyl, fluorenyl and hydroxyl groups for enhanced proton exchange membrane properties: Application in microbial fuel cell. *ACS Appl Mater Interfaces* 10:14803–14817.
24. Wang CT, Lee YC, Liao FY (2015) Effect of composting parameters on the power performance of solid microbial fuel cells. *Sustain* 7:12634–12643.
25. Kamran S, Marashi F, Kariminia H (2015) Environmental health performance of a single chamber microbial fuel cell at different organic loads and pH values using purified terephthalic acid wastewater. *J Environ Heal Sci Eng* 1–6.
26. Ren H, Jiang C, Chae J (2017) Effect of temperature on a miniaturized microbial fuel cell (MFC). *Micro Nano Syst Lett* 3–9.
27. ElMekawy A, Hegab HM, Pant D, Saint CP (2018) Bio-analytical applications of microbial fuel cell–based biosensors for onsite water quality monitoring. *J Appl Microbiol* 124:302–313.
28. Kumar R, Singh L, Zularisam AW, Hai FI (2018) Microbial fuel cell is emerging as a versatile technology: A review on its possible applications, challenges and strategies to improve the performances. *Int J Energy Res* 42:369–394.
29. Kodali M, Herrera S, Kabir S, Serov A, Santoro C, Ieropoulos I, Atanassov P (2018) Enhancement of microbial fuel cell performance by introducing a nano-composite cathode catalyst. *Electrochim Acta* 265:56–64.
30. Chatzikonstantinou D, Tremouli A, Papadopoulou K, Kanellos G, Lampropoulos I, Lyberatos G (2018) Bioelectricity production from fermentable household waste in a dual-chamber microbial fuel cell. *Waste Manag Res* 36:1037–1042.
31. Ma H, Peng C, Jia Y, Wang Q, Tu M, Gao M (2018) Effect of fermentation stillage of food waste on bioelectricity production and microbial community structure in microbial fuel cells. *R Soc Open Sci* 5:180457.
32. Liu G, Yates MD, Cheng S, Call DF, Sun D, Logan BE (2011) Examination of microbial fuel cell start-up times with domestic wastewater and additional amendments. *Bioresour Technol* 102:7301–7306.
33. Liu W, Cheng S (2014) Microbial fuel cells for energy production from wastewaters: The way toward practical application. *Jr Zhe Univ Sci A* 15:841–861.
34. Bombelli P, Dennis RJ, Felder F, Cooper MB, Iyer DMR, Royles J, Harrison STL, Smith AG, Jill Harrison C, Howe CJ (2016) Electrical output of bryophyte microbial fuel cell systems is sufficient to power a radio or an environmental sensor. *R Soc Open Sci* 3:160249.

14 Waste-Derived Carbon Materials for Hydrogen Storage

Mohamed Aboughaly and Hossam A. Gabbar
University of Ontario Institute of Technology

CONTENTS

14.1 INTRODUCTION

Hydrogen is considered a promising energy source, an abundant element, an excellent energy carrier, and a clean alternative fuel that can help reduce reliance on fossil fuels and meet the current industrial and commercial energy demands as well as help reduce greenhouse gases emissions by more than 80% [1–4]. Due to the high industrial demand for hydrogen as a reactant and its use for sulfur removal in oil refineries, the annual global demand for hydrogen production is expected to reach more than 70 million tonnes by 2050 [5]. The current challenge facing the economic expansion of hydrogen fuel economy is the availability of efficient engineering systems that allows safe storage, safe transportation, and practical usage of hydrogen gas [6,7]. Hydrogen is considered the next-generation convenient fuel due to the rapid depletion of fossil fuels, many routes for the production of hydrogen, highly efficient combustion

DOI: 10.1201/9781003178354-17

properties, and zero to low carbon emissions [8]. Hydrogen adsorption plays a vital role as a key enabling technology for the advancement of hydrogen and fuel cell technologies in power and transportation applications [9,10]. To be able to use hydrogen effectively in useful applications, the hydrogen adsorption mechanism has to be developed to operate in industrial applications between 5 and 50 MPa and 273–300 K [11,12]. Hydrogen gas is a promising renewable fuel with the highest energy density among non-nuclear fuel sources at 120 MJ/kg [13]. However, utilization of hydrogen is limited by technological challenges such as safe storage, high storage density, and safe transportation [14,15].

The development of carbon-based materials derived from biowaste feedstock could enhance hydrogen adsorption technologies and provide affordable and economical materials and structures for the hydrogen adsorption process for large-scale production and usage of hydrogen as a reliable and safe renewable fuel [16]. Hydrogen is also an important reactant in petroleum, petrochemical, and fertilizer industries where it can be used for reduction of metal oxides, oil refining, production of ammonia, metal ore reduction, hydrochloric ore reduction, welding, and as a cooling agent. The current global production of hydrogen exceeds 50 million tonnes per annum, mostly by steam reforming, methane reforming, and electrolysis [17]. However, technical challenges of safe storage and transportation of hydrogen still possess unacceptable risks and limit the mass production of hydrogen as a fuel [18]. The three current challenges of the hydrogen economy are high production, safe storage, and transportation. Based on several investigations, storage of hydrogen in solid form overcomes storage and transportation challenges [19].

The most common and economical chemical processes used for hydrogen production are steam reforming processes using natural gas, methane-steam cracking, and coal gasification which is widely used in petroleum refineries and chemical plants to generate hydrogen for chemical reactions [20]. Alternative methods for hydrogen production are steam-methane reforming, biowaste gasification, and water electrolysis. In chemical plants, hydrogen gas is required in catalytic cracking chemical reactions for conversion of heavy petroleum fractions to lighter ones using hydrocracking and other processes such as hydrodesulfurization and ammonia production [17]. Hydrogen gas is an important reactant in the chemical industry and is used in several chemical reactions in ammonia chemical plants, refining, and methanol production units as shown in Figure 14.1.

The current challenges facing the economic expansion of hydrogen production is large operating cost, hazardous and flammable nature, and high safety engineering costs despite the abundance and availability of industrial processes for mass production [21]. Not to mention, hydrogen is an important primary reactant in chemical and petrochemical industries such as petroleum refining, petrochemicals, mining, and fertilizer industries. Thus, advancements in hydrogen storage and transportation technologies are required with the increase in industrial production rates in various industries, the development of safe transportation methods, and less reliance on fossil fuels [16,17]. The widespread use of hydrogen in chemical and petrochemical technologies are still limited by the current challenges faced in its storage and mass-scale production which are associated with energy efficiency, durability, refueling time, life cycle cost, production, and safe storage [22]. Hydrogen storage via

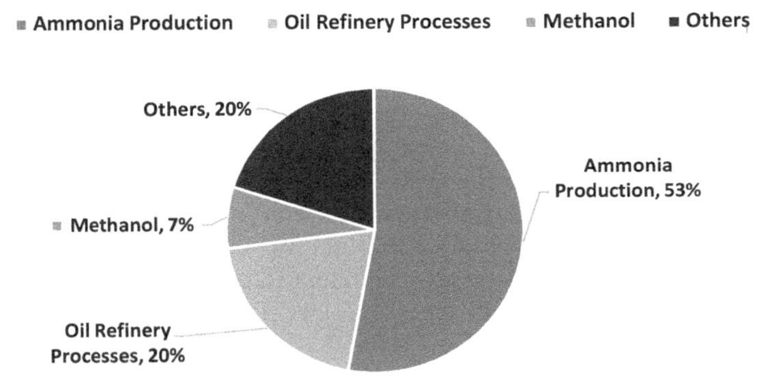

FIGURE 14.1 Hydrogen usage in chemical industries.

adsorption provides a safe and reliable method that prevents the hazardous combination of hydrogen gas and provides easy methods for fast adsorption and desorption processes for the utilization of hydrogen.

Novel hydrogen storage technologies should be able to achieve up to 1,500–2,000 refuel cycles, provide safe storage, and increase process reliability at lower operational costs [20,22]. Prior to large-scale implementation of the hydrogen economy, viable hydrogen storage technologies should be developed to reduce operational costs and allow scale-up of hydrogen economy to primary sectors such as transportation and energy [23]. The current commercial hydrogen storage technologies are the following: (i) high-pressure metal cylinders (350–700 bar), (ii) high-pressure cryogenic cylinders below −253 °C and 70 MPa, (iii) electrochemical hydrogen storage using alkaline fuel cells, and (iv) physisorption using porous carbon and nanostructured materials such as biomass, activated carbon, zeolites, and MOFs [23].

Hydrogen storage is the primary technology for the advancement of hydrogen transportation and for fuel cell technologies that can provide energy for different engineering applications such as stationary power, portable power, fuel, and transportation [1]. Not to mention, hydrogen storage technologies could be used as a secondary medium for energy storage. Renewable power sources such as wind, solar, hydro, and other renewable energy sources that can be used as a supplementary source of energy production during periods of high energy demand as well as saving operational costs and reduction in greenhouse gaseous emissions [7,22].

Moreover, hydrogen plays a vital role as a sustainable solution to global warming and clean fuel for energy generation without producing by-products such as greenhouse gases emissions from fossil fuels [24]. The current challenges facing the fuel storage and technological expansion of hydrogen storage technology are associated with the production, storage, transportation, and safety of hydrogen. Prior to realizing a large-scale implementation of the hydrogen economy and shifting from fossil fuels, viable hydrogen storage materials must be developed to allow safe transportation and storage of hydrogen in a stable non-reactive form [16].

Based on several investigations reported in the literature, it is observed that hydrogen storage using solid molecular sieves is the safest and most economical solution to overcome the current challenges including storage and transportation [20].

Biomass-based carbon materials could be activated to store hydrogen by adjusting the reaction parameters that enhance the adsorption process and achieving control over the morphology and texture prepared through the carbonization and activation process of biomass waste [6]. Different types of carbon-based materials such as borohydrides, metal–organic frameworks (MOFs), and porous materials are used to store hydrogen. The most desirable features are large surface area, high porous concentration and volume, lightweight, affordable manufacturing cost, and high recyclability [25]. The wide textural properties and large porous surface area and microporosity increase hydrogen adsorption rates [26]. Different solid-state hydrogen storage technologies are shown in Figure 14.2.

14.1.1 Hydrogen Physical Storage Practices

Hydrogen can be stored physically in the liquid or gaseous state or on the surface of solid materials in a mass transfer process known as surface adsorption at very high pressures (350–700 bar) [16]. Hydrogen is typically stored in the gaseous or liquid form either using high-pressure storage tanks or at cryogenic temperatures due to low boiling point at atmospheric pressure, typically lower than −252.8 °C [17,22]. Hydrogen gas has a low density (i.e., 0.0708 g/L) at room temperature, which makes it uneconomic for storage in the gaseous state. Porous carbon-based materials offer a viable solution to increase its storage density and safety [22]. Alternative storage practices such as high-pressure adsorption are required to reduce volume size and make the process more economical and improve its industrial applicability. Hydrogen gas at room temperature has a low density, and 1 kg of hydrogen gas requires 11 m^3 for storage in a gaseous state at room temperature [27]. Therefore, alternative technologies such as porous carbon-based materials are able to reduce storage volume [16,22].

14.1.2 Carbon-Based Porous Materials for Hydrogen Storage

Physisorption using porous activated carbon materials with high surface area is a promising strategy due to the ability to control hydrogen storage capacity using manipulation of physical properties such as surface area, reactive porosity, and high volume of porous material [1]. Among several approaches to hydrogen storage, the molecular sorption of H_2 in porous solid-state materials is preferred due to fast kinetics, excellent process cyclability, durability, and high adsorption capacity [28]. The pore geometry also affects the hydrogen adsorption rates, showing that porous geometries with higher accessible volume have shown better adsorption than non-accessible pore structures [6,22]. Three-dimensional carbon-based hydrogen storage materials are zeolites, MOFs, and covalent organic frameworks (COFs). The MOFs have a high surface area and high porosity, which allow them to have optimum adsorption rates. The adsorption rate decreases with high temperature and can be maximized using catalysts [29]. The most important parameter in the adsorption of hydrogen on carbon-based porous materials is the adsorption enthalpy, which ranges from 20 to 160 kJ/mol, as well as the internal surface area of microporous surfaces [7].

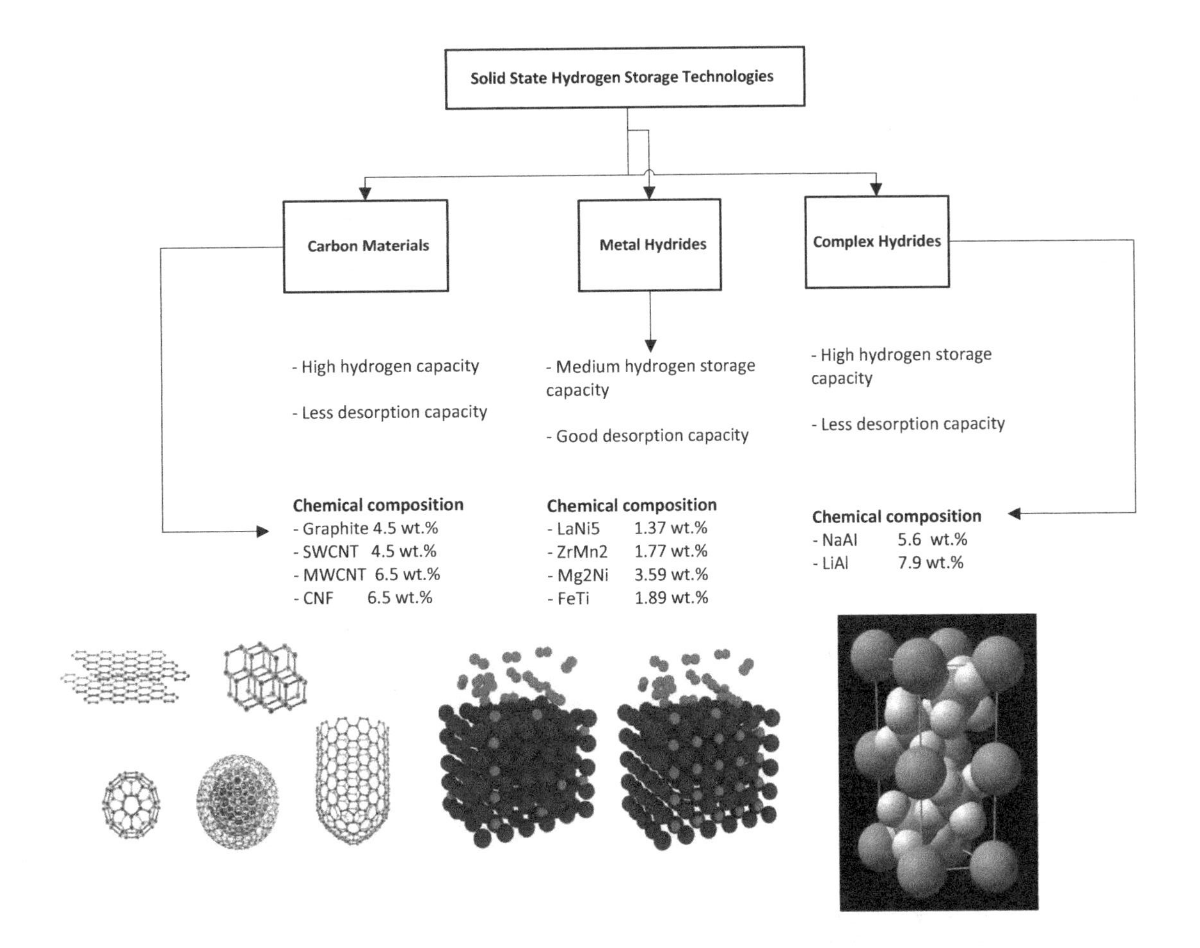

FIGURE 14.2 Solid-state hydrogen storage technologies.

This method of hydrogen storage is more recommended than alternative methods such as cryogenic storage due to less operating cost and storage at room temperature [12]. This storage method has shown high reliability, and the most recommended operating conditions are high operating pressures up to 70 MPa, which increases adsorption rate [12]. The influencing factors are pore size, packing density, operating pressure, and porous material [30]. The recommended operating conditions for the adsorption process of activated carbon are 80–100 bar at 298 K [8]. The reported storage capacity at different operating temperatures and pressures ranges from 4.0 to 6.5 wt.% [8,31].

The microporous carbon structures derived from biomass exhibit high surface area of up to 3,100 m^2/g, large pore structures, and high volumetric hydrogen uptake of up to 5 wt.%. The highest reported hydrogen storage capacity of 7.03 wt.% was reported at cryogenic temperature of 77 K and high pressure of 20 bar [32]. Based on the experimental conditions, mesoporous and microporous carbons can be produced with excellent textural properties and a high surface area ranging from 886 to 2,314 m^2/g [33]. At the highest surface area, the maximum adsorption capacity is 1.78 wt.% at 77 k and ambient pressure. The rate of H_2 adsorption capacity was directly proportional to the surface area [34]. It is reported that the surface area plays a vital role in the rate of hydrogen adsorption on porous carbon materials. It is also reported that at cryogenic temperatures, the amount of hydrogen uptake increases from 0.3 to up to 7.8 wt.% [34].

14.1.3 Carbon Nanostructures Derived from Biomass Waste for Hydrogen Storage

The biomass waste is converted into microporous carbon structures with controlled morphologies and textures for hydrogen storage using inexpensive carbonaceous materials [40]. Microporous carbons with controlled morphologies and textures can be developed using low-cost biomass waste deposits. The synthesized carbon materials possess high adsorption surface area, large adsorption volumes, and large pore sizes especially at cryogenics conditions, which increases hydrogen uptake [41]. This chapter focuses on illustrating the possible routes for the conversion of biomass waste to carbon-based porous structures with high adsorption capacities. Examples of carbon materials that possess high surface area are porous carbons, MOFs, and COFs, which have shown excellent adsorption properties for hydrogen storage, such as large surface area, pore volume, and pore size [42]. Below are the steps for the transformation process of biomass into microporous carbon structure for hydrogen adsorption as shown in Figure 14.3.

The operating conditions for the optimum hydrogen adsorption capacity are controlled by the physisorption conditions that affect the H_2 uptake, such as the porous material surface area, micropore volume, and pore size. The mechanism of hydrogen adsorption occurs by the van der Waals forces between activated carbon (AC) and H_2 molecules. The weak adsorption van der Waals forces allow only single layers of H_2 atom packing onto carbon nanostructures [36]. Since the H_2 and activated carbon are a monolayer structure, a large adsorption surface area is essential for higher hydrogen uptake [43].

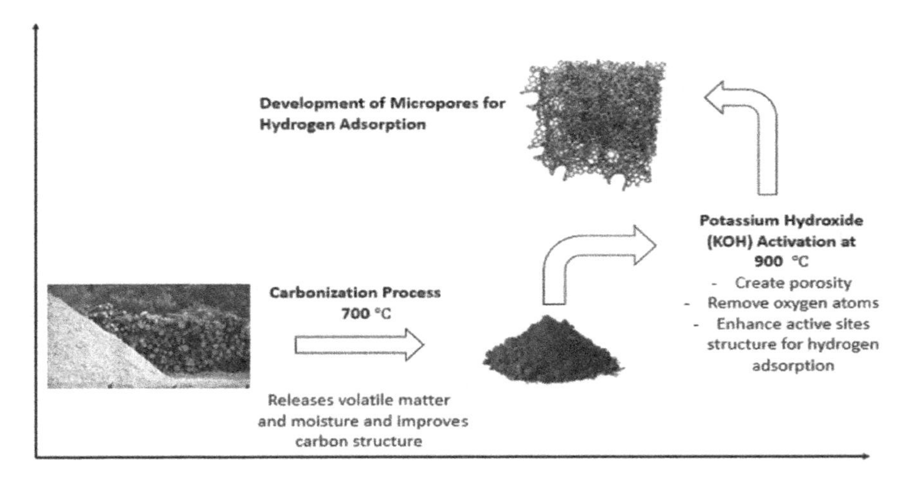

FIGURE 14.3 Conversion of biomass to microporous carbon structure using KOH for hydrogen adsorption.

14.2 MECHANISM OF HYDROGEN ADSORPTION AND STORAGE USING POROUS MATERIALS

Adsorption is the adhesion of atoms, ions, or molecules from a gas or liquid to the surface of a solid via weak van der Waals molecular forces [44]. The mechanism of isothermal adsorption of hydrogen on porous materials such as activated carbons and MOFs is determined by different assessment methods, most commonly the comparison of the stored hydrogen with a specified volume of adsorbents against the same volume of hydrogen without adsorbents [45]. The fluid (the adsorbate) is attached to the solid molecule via weak molecular forces. Metal sites have shown stronger molecular bonds than non-metal or ring sites [46].

The adsorption process of the hydrogen molecule is recommended since it doesn't react and therefore doesn't accumulate impurities or poison the catalyst. Also, the mentioned process is fast compared to solid-state diffusion or chemical dissociation. The process is fully reversible, which makes it a preferred method for storage, separation, and dissociation of gaseous products [44]. Below are the required adsorption enthalpies for different adsorbent materials and optimal operating temperatures, as shown below in Table 14.1 [16].

TABLE 14.1
Optimal Conditions for Hydrogen Uptake Using Adsorbent Materials

Adsorbent Material	Average Adsorption Enthalpy (KJ/mol)	Optimal Operating Temperature (K)	Usable Capacity at T_{opt} (wt.%)
AX-21-33 (activated carbon)	5.2–5.4	85	3.25%
MOF-177	5.0	95	7.45%
Magnesium formate	7.2	112	0.78%

In order to fully understand the hydrogen physisorption on porous materials, the amount adsorbed at a certain pressure has to be recorded. Also, the measurement of amount adsorbed at different temperature profiles as well as the adsorption enthalpy helps determine the adsorption curve [38,47]. Below are the influential factors that affect the adsorption process on molecular sieves.

14.2.1 Molecular Potential

The hydrogen storage capacity is calculated using a thermodynamic model of the molecular interaction between hydrogen and activated carbon sites [48]. The interaction potential is either planar, cylindrical, or spherical and can be formulated as a sum of individual interaction energies of each carbon atom C_i at the surface of a hydrogen molecule.

$$V(s) = \sum U(ri, s) \tag{14.1}$$

The type of interaction potential between a hydrogen atom and carbon surface is of either planar, cylindrical, or spherical shape and is calculated by the addition of individual interaction atomic energies influenced by molecular pore size and diameter [49].

14.2.2 Physical Adsorption Rate

The hydrogen molecule is dissociated during adsorption on the surface of the carbon atom and is stored in the carbon lattice structure filling the interstitial sites such as tetrahedral and octahedral pores [34]. The complex structure of the carbon atom affects the adsorption rate and the energy levels for adsorption [35,37,39]. The shape and size of the carbon atom pores as well as the mechanical structure have a direct influence on the storage capacity [48]. The adsorption rate is affected by the operating temperature and pressure. It is reported that under cryogenic conditions, the adsorption capacity increases up to 20 wt.% [38].

Physical adsorption is defined as the dissociation of a gaseous molecule on the surface of an adsorbent with weak intermolecular forces between the solute and the adsorbent molecule that is greater than the required dissociation energy. The adsorption process is an exothermic reaction, and the enthalpy change of the physical adsorption is represented by the following equation:

$$\Delta H = \Delta G + T\Delta S \tag{14.2}$$

14.2.3 Modeling Equations of Physical Adsorption of Hydrogen on Carbon Porous Materials

The hydrogen adsorption capacity on activated carbon is around 0.5 wt.% at 60 bar and room temperature. At the same pressure and cryogenic conditions, the adsorption capacity increases to 5 wt.% [50].

The modeling equation used for hydrogen adsorption on activated carbons is "Dubinin–Astakhov (D–A) model":

$$n = n_o \exp\left[-\left(\frac{A}{E}\right)^b\right] \tag{14.3}$$

(i) The virial equation which permits to determine the isosteric heats at zero coverage [10,11], and (ii) the Dubinin–Astakhov (D–A) equation that makes use of volume filling of micropores (TVFM) concept and some structural or energetic parameters derived from best fits with measured isotherms.

The equation below calculates the absolute amount of adsorbed hydrogen [mol/kg] using the following expression:

$$n_{\text{abs}} = n_{\text{exc}} + \rho_g V_a \tag{14.4}$$

where n_{abs} is the absolute hydrogen adsorbed, n_{exc} is the difference between the hydrogen uptake on the surface of the adsorbent at a specific temperature and pressure, ρ_g is the hydrogen bulk phase density, and V_a is the volume of the adsorbed phase. The mass and heat transfer of the hydrogen adsorption process using activated carbon can be modelled using mass conservation, momentum, and energy balance equations [51]. The time-dependent mass flow rate of hydrogen in the porous carbon adsorbent bed is expressed using the following equation:

$$\frac{\delta \rho_g}{\delta t} + \nabla\left(\rho_g v\right) = S_m \tag{14.5}$$

Momentum conservation equations model the gaseous adsorption in an adsorbent bed by Ergun equation which models the fluid flow in porous media. The Langmuir equation also models the adsorption isotherms for individual gaseous components and is expressed in the below equation [52]:

$$\frac{-\partial p}{-\partial z} = \frac{150\left(1-\varepsilon_b\right)^2}{2\left(R_p\right)^2} \tag{14.6}$$

where ∂p is the differential pressure, z is the axial position in the adsorbent bed, ε_b is the bed porosity, and R_p is the particle radius, m.

14.3 CURRENT CHALLENGES OF HYDROGEN STORAGE USING CARBON-BASED MATERIALS

In order to implement hydrogen gas as an energy source for transportation, storage of large volumes of hydrogen in limited space remains a challenge. The constraints for storage systems are weight, volume, storage efficiency, and cost. For example, the implementation of hydrogen storage systems for a conventional driving range (>300 km) requires a storage capacity of 423.3 kg of hydrogen gas [22,53]. For large-scale implementation of hydrogen storage technologies, durability and performance assessments have to be verified and validated and the development of refueling times suitable for commercial use has to be done [1,23]. The main challenge in hydrogen storage via adsorption is the development of new adsorbents that have high hydrogen storage capabilities for practical use at a reasonable operational cost. This includes low operating pressures and temperatures since high operating pressure increases both the capital and operational costs [54].

Also, the physical interaction of hydrogen gas with most solid adsorbents is weak due to weak dipole bonds and low polarizability. In addition to that, at low pressure the H_2 molecule has weak molecular bonds leading to adsorption capacities lower than 8.6 wt.% [54]. Below are the current physical challenges faced in hydrogen storage:

- **Weight and Volume:** The hydrogen gas at ambient temperature has a very low weight and large volume, resulting in large storage requirements which increases both the capital and operational costs.
- **Storage System Cost:** The cost of hydrogen storage systems is significantly higher than that of fossil fuel storage systems, which makes it economically unfeasible. Low-cost hydrogen storage technologies, as well as high-volume manufacturing technologies, are required for large-scale implementation of hydrogen.
- **Energy Efficiency:** Low energy efficiency remains a challenge for the advancement of hydrogen storage systems. The high energy requirements for adsorption and desorption of carbon-based adsorption storage systems increase the operating cost for hydrogen storage. The energy management for charging and releasing hydrogen in carbon-based storage systems needs energy optimization to increase system reliability and energy efficiency.
- **Durability and Reliability:** The carbon-based hydrogen storage systems require improvement and verification for porous materials design and containment vessels that enable hydrogen storage with acceptable lifetimes and higher adsorption efficiency. The current durability of hydrogen storage systems is inadequate, and the current carbon-based adsorbent materials cannot withstand 1,500 recharging cycles.
- **Refueling Time:** The current processing times for adsorption and desorption are too long, which makes the technology inapplicable for industrial use due to long refueling times and low adsorption efficiency.
- **Capital Cost:** The current cost of onboard hydrogen storage systems is high in comparison with other storage systems such as petroleum liquid storage systems. The development of low-cost hydrogen storage technologies such as carbon-based hydrogen storage as well as new manufacturing methods for high volume production can achieve product development and implementation of the technology on a large scale.
- **Life Cycle and Efficiency Analysis:** The lack of full life cycle analysis, life cycle cost analysis, and feasibility studies hinders the development of the technology.

Some carbon structures have shown high storage capacities in comparison with other structures. Carbon nanotubes, activated carbons, mesoporous carbons, and MOFs have shown high hydrogen storage capacities [55]. Some carbon-based structures such as graphene have shown very high desorption rates, but slow adsorption rates, which shows the unique hydrogen storage properties of graphene compared to other carbon structures [55].

14.4 CONCEPTS FOR IMPROVEMENT OF HYDROGEN ADSORPTION ON NANOPOROUS ADSORBENT MATERIALS

The maximum achieved adsorption capacity of hydrogen on nanoporous materials exceeds 10 wt.% at low temperatures around 77 K [56]. However, low adsorption volumes at room temperatures make the technology economically impractical at ambient temperature [56]. Promising results using machine learning have shown that new materials formed by a combination of different materials including MOFs have shown improved adsorption characteristics and achieve higher hydrogen storage capacities [56].

Hydrogen storage at ambient temperature requires the development of materials with higher volumetric and gravimetric capacities to be economically feasible. Higher gravimetric capacities reduce the volume of hydrogen storage tanks and operational cost [27]. The adsorption enthalpy increases the strength of the interaction between the hydrogen gas and the adsorbent material, leading to better molecular bondage and storage [54]. Novel concepts such as modification of adsorbent molecular architecture and structure have shown better absorption rates [54]. Activated carbon adsorbents have shown a reduction in volume storage capacity by 23 wt.% compared to gas compression at 20 MPa and room temperature [49].

Another approach for the improvement of adsorption rate is the formation of monoliths that increase the bulk density of the adsorbent and the overall volumetric capacity. Also, interpenetration or catenation improves the geometry of the porous materials and increases the rate of adsorption [56]. New adsorption materials such as MOFs have shown higher adsorption enthalpy due to the presence of more active metal sites and higher pore density that result in higher adsorption rates of hydrogen [57]. Also, the control of pore mechanical structures for selective gas separation for hydrogen and the development of precise pore size and volume improve the selectivity of the porous materials and hydrogen adsorption rate.

14.5 PREPARATION AND ACTIVATION OF HIERARCHAL POROUS CARBON

The preparation of biomass into a hierarchal porous carbon (HPC) exhibits high performance in various applications such as hydrogen adsorption. Biomass feedstock such as wood, grass, and nutshell has widely been used due to their hierarchical porosity and ease of preparation and activation using simple pyrolysis reactions. For non-structured feedstock such as sucrose and plastics, novel technologies such as hydrothermal carbonization, chemical vapor deposition, spray pyrolysis, and autogenic pressure carbonization are used for activation [9].

14.6 HYDROGEN ADSORPTION RATES OF DIFFERENT CARBON-BASED POROUS MATERIALS

By definition, the adsorbent or loading capacity is the working or deliverable capacity which refers to the amount of adsorbate taken up by the adsorbent per unit mass in a specified time unit. Hydrogen storage capacities are reported as absolute or uptake pressure of adsorbents. The most important parameter during the adsorption process is the working or deliverable capacity [56].

For carbon-based porous materials such as single-walled carbon nanotubes (SWCNTs), hydrogen adsorption reached 1.5 wt.% at 0.1 MPa [58].

REFERENCES

1. D. P. Broom, C. J. Webb, G. S. Fanourgakis, G. E. Froudakis, P. N. Trikalitis, and M. Hirscher, "Concepts for improving hydrogen storage in nanoporous materials," *International Journal of Hydrogen Energy*, vol. 44, no. 15, 2019, doi: 10.1016/j.ijhydene.2019.01.224.
2. S. A. Montzka, E. J. Dlugokencky, and J. H. Butler, "Non-CO_2 greenhouse gases and climate change," *Nature*, vol. 476, no. 7358, 2011, doi: 10.1038/nature10322.
3. S. J. Yang, H. Jung, T. Kim, and C. R. Park, "Recent advances in hydrogen storage technologies based on nanoporous carbon materials," *Progress in Natural Science: Materials International*, vol. 22, no. 6, 2012, doi: 10.1016/j.pnsc.2012.11.006.
4. E. Rivard, M. Trudeau, and K. Zaghib, "Hydrogen storage for mobility: A review," *Materials*, vol. 12, no. 12, 2019, doi: 10.3390/ma12121973.
5. M. Noussan, P. P. Raimondi, R. Scita, and M. Hafner, "The role of green and blue hydrogen in the energy transition—A technological and geopolitical perspective," *Sustainability (Switzerland)*, vol. 13, no. 1, 2021, doi: 10.3390/su13010298.
6. Nick, Kanellopoulos, and N. Kanellopoulos, *Nanoporous Materials: Advanced Techniques for Characterization, Modeling, and Processing*, 2012, CRC Press, New York.
7. G. Sdanghi, R. L. S. Canevesi, A. Celzard, M. Thommes, and V. Fierro, "Characterization of carbon materials for hydrogen storage and compression," *C — Journal of Carbon Research*, vol. 6, no. 3, 2020, doi: 10.3390/c6030046.
8. R. Ströbel, J. Garche, P. T. Moseley, L. Jörissen, and G. Wolf, "Hydrogen storage by carbon materials," *Journal of Power Sources*, vol. 159, no. 2, 2006, doi: 10.1016/j.jpowsour.2006.03.047.
9. J. Ren, S. Gao, S. Tan, L. Dong, A. Scipioni, and A. Mazzi, "Role prioritization of hydrogen production technologies for promoting hydrogen economy in the current state of China," *Renewable and Sustainable Energy Reviews*, vol. 41, 2015, doi: 10.1016/j.rser.2014.09.028.
10. M. McPherson, N. Johnson, and M. Strubegger, "The role of electricity storage and hydrogen technologies in enabling global low-carbon energy transitions," *Applied Energy*, vol. 216, 2018, doi: 10.1016/j.apenergy.2018.02.110.
11. O. K. Alekseeva, I. v. Pushkareva, A. S. Pushkarev, and V. N. Fateev, "Graphene and graphene-like materials for hydrogen energy," *Nanotechnologies in Russia*, vol. 15, no. 3–6, 2020, doi: 10.1134/S1995078020030027.
12. M. A. de La Casa-Lillo, F. Lamari-Darkrim, D. Cazorla-Amorós, and A. Linares-Solano, "Hydrogen storage in activated carbons and activated carbon fibers," *Journal of Physical Chemistry B*, vol. 106, no. 42, 2002, doi: 10.1021/jp014543m.
13. I. Staffell et al., "The role of hydrogen and fuel cells in the global energy system," *Energy and Environmental Science*, vol. 12, no. 2, 2019, doi: 10.1039/c8ee01157e.

14. I. Dincer and C. Acar, "Innovation in hydrogen production," *International Journal of Hydrogen Energy*, vol. 42, no. 22, 2017, doi: 10.1016/j.ijhydene.2017.04.107.
15. F. Dawood, M. Anda, and G. M. Shafiullah, "Hydrogen production for energy: An overview," *International Journal of Hydrogen Energy*, vol. 45, no. 7, 2020, doi: 10.1016/j.ijhydene.2019.12.059.
16. Y. Kojima, "Hydrogen storage materials for hydrogen and energy carriers," *International Journal of Hydrogen Energy*, vol. 44, no. 33, 2019, doi: 10.1016/j.ijhydene.2019.05.119.
17. S. A. Treese, P. R. Pujadó, and D. S. J. Jones, *Handbook of Petroleum Processing*, vol. 1, 2015, doi: 10.1007/978-3-319-14529-7.
18. L. van Hoecke, L. Laffineur, R. Campe, P. Perreault, S. W. Verbruggen, and S. Lenaerts, "Challenges in the use of hydrogen for maritime applications," *Energy and Environmental Science*, vol. 14, no. 2, 2021, doi: 10.1039/d0ee01545h.
19. S. Rostami, A. N. Pour, and M. Izadyar, "A review on modified carbon materials as promising agents for hydrogen storage," *Science Progress*, vol. 101, no. 2, 2018. doi: 10.3184/003685018X15173975498956.
20. R. Moradi and K. M. Groth, "Hydrogen storage and delivery: Review of the state of the art technologies and risk and reliability analysis," *International Journal of Hydrogen Energy*, vol. 44, no. 23, 2019, doi: 10.1016/j.ijhydene.2019.03.041.
21. T. Covert, M. Greenstone, and C. R. Knittel, "Will we ever stop using fossil fuels?" *Journal of Economic Perspectives*, vol. 30, no. 1, 2016, doi: 10.1257/jep.30.1.117.
22. J. Andersson and S. Grönkvist, "Large-scale storage of hydrogen," *International Journal of Hydrogen Energy*, vol. 44, no. 23, 2019, doi: 10.1016/j.ijhydene.2019.03.063.
23. S. J. Yang, H. Jung, T. Kim, and C. R. Park, "Recent advances in hydrogen storage technologies based on nanoporous carbon materials," *Progress in Natural Science: Materials International*, vol. 22, no. 6, 2012, doi: 10.1016/j.pnsc.2012.11.006.
24. N. Abdoulmoumine, S. Adhikari, A. Kulkarni, and S. Chattanathan, "A review on biomass gasification syngas cleanup," *Applied Energy*, 2015, doi: 10.1016/j.apenergy.2015.05.095.
25. A. K. Dalai, N. Batta, I. Eswaramoorthi, and G. J. Schoenau, "Gasification of refuse derived fuel in a fixed bed reactor for syngas production," *Waste Management*, 2009, doi: 10.1016/j.wasman.2008.02.009.
26. Y. Xia, Z. Yang, and Y. Zhu, "Porous carbon-based materials for hydrogen storage: Advancement and challenges," *Journal of Materials Chemistry A*, vol. 1, no. 33, 2013, doi: 10.1039/c3ta10583k.
27. N. Stetson and M. Wieliczko, "Hydrogen technologies for energy storage: A perspective," *MRS Energy & Sustainability*, vol. 7, no. 1, 2020, doi: 10.1557/mre.2020.43.
28. A. Ariharan and B. Viswanathan, "Porous activated carbon material derived from sustainable bio-resource of peanut shell for H_2 and CO_2 storage applications," *Indian Journal of Chemical Technology*, vol. 25, no. 2, 2018.
29. H. W. Langmi, J. Ren, B. North, M. Mathe, and D. Bessarabov, "Hydrogen storage in metal-organic frameworks: A review," *Electrochimica Acta*, vol. 128, 2014, doi: 10.1016/j.electacta.2013.10.190.
30. S. Tedds, A. Walton, D. P. Broom, and D. Book, "Characterisation of porous hydrogen storage materials: Carbons, zeolites, MOFs and PIMs," *Faraday Discussions*, vol. 151, 2011, doi: 10.1039/c0fd00022a.
31. Y. Li, D. Zhao, Y. Wang, R. Xue, Z. Shen, and X. Li, "The mechanism of hydrogen storage in carbon materials," *International Journal of Hydrogen Energy*, vol. 32, no. 13, 2007, doi: 10.1016/j.ijhydene.2006.11.010.
32. S. S. Samantaray, S. R. Mangisetti, and S. Ramaprabhu, "Investigation of room temperature hydrogen storage in biomass derived activated carbon," *Journal of Alloys and Compounds*, vol. 789, 2019, doi: 10.1016/j.jallcom.2019.03.110.

33. M.I. Maulana Kusdhany, S.M. Lyth, New insights into hydrogen uptake on porous carbon materials via explainable machine learning, Carbon N. Y. 179 (2021) 190–201. https://doi.org/https://doi.org/10.1016/j.carbon.2021.04.036.
34. M. Hirscher et al., "Materials for hydrogen-based energy storage – past, recent progress and future outlook," *Journal of Alloys and Compounds*, vol. 827, 2020, doi: 10.1016/j.jallcom.2019.153548.
35. D. Silambarasan, V. J. Surya, V. Vasu, and K. Iyakutti, "Reversible and reproducible hydrogen storage in single- walled carbon nanotubes functionalized with borane," in *Carbon Nanotubes – Recent Progress*, 2018, doi: 10.5772/intechopen.75763.
36. T. Zhao, X. Ji, W. Jin, W. Yang, and T. Li, "Hydrogen storage capacity of single-walled carbon nanotube prepared by a modified arc discharge," *Fullerenes Nanotubes and Carbon Nanostructures*, vol. 25, no. 6, 2017, doi: 10.1080/1536383X.2017.1305358.
37. M. Marella and M. Tomaselli, "Synthesis of carbon nanofibers and measurements of hydrogen storage," *Carbon*, vol. 44, no. 8, 2006, doi: 10.1016/j.carbon.2005.11.020.
38. "Hydrogen adsorption and storage on porous materials," in *Handbook of Hydrogen Energy*, 2020, doi: 10.1201/b17226-28.
39. D. J. Browning, M. L. Gerrard, J. B. Lakeman, I. M. Mellor, R. J. Mortimer, and M. C. Turpin, "Studies into the storage of hydrogen in carbon nanofibers: Proposal of a possible reaction mechanism," *Nano Letters*, vol. 2, no. 3, 2002, doi: 10.1021/nl015576g.
40. J. Deng, M. Li, and Y. Wang, "Biomass-derived carbon: Synthesis and applications in energy storage and conversion," *Green Chemistry*, vol. 18, no. 18, 2016, doi: 10.1039/c6gc01172a.
41. M. I. Maulana Kusdhany and S. M. Lyth, "New insights into hydrogen uptake on porous carbon materials via explainable machine learning," *Carbon*, vol. 179, 2021, doi: 10.1016/j.carbon.2021.04.036.
42. T. S. Blankenship, N. Balahmar, and R. Mokaya, "Oxygen-rich microporous carbons with exceptional hydrogen storage capacity," *Nature Communications*, vol. 8, no. 1, 2017, doi: 10.1038/s41467-017-01633-x.
43. A. I. Sultana, N. Saha, and M. Toufiq Reza, "Synopsis of factors affecting hydrogen storage in biomass-derived activated carbons," *Sustainability (Switzerland)*, vol. 13, no. 4, 2021, doi: 10.3390/su13041947.
44. K. Vasanth Kumar, A. Salih, L. Lu, E. A. Müller, and F. Rodríguez-Reinoso, "Molecular simulation of hydrogen physisorption and chemisorption in nanoporous carbon structures," *Adsorption Science and Technology*, vol. 29, no. 8, 2011, doi: 10.1260/0263-6174.29.8.799.
45. J. Juan-Juan, J. P. Marco-Lozar, F. Suárez-García, D. Cazorla-Amorós, and A. Linares-Solano, "A comparison of hydrogen storage in activated carbons and a metal-organic framework (MOF-5)," *Carbon*, vol. 48, no. 10, 2010, doi: 10.1016/j.carbon.2010.04.025.
46. T. B. Lee et al., "Understanding the mechanism of hydrogen adsorption into metal organic frameworks," *Catalysis Today*, vol. 120, no. 3–4 Spec. iss., 2007, doi: 10.1016/j.cattod.2006.09.030.
47. R. Roszak, L. Firlej, S. Roszak, P. Pfeifer, and B. Kuchta, "Hydrogen storage by adsorption in porous materials: Is it possible?" *Colloids and Surfaces A: Physicochemical and Engineering Aspects*, vol. 496, 2016, doi: 10.1016/j.colsurfa.2015.10.046.
48. I. Cabria, M. J. López, and J. A. Alonso, "Simulation of the hydrogen storage in nanoporous carbons with different pore shapes," *International Journal of Hydrogen Energy*, vol. 36, no. 17, 2011, doi: 10.1016/j.ijhydene.2011.05.125.
49. P. Ramirez-Vidal et al., "A step forward in understanding the hydrogen adsorption and compression on activated carbons," *ACS Applied Materials and Interfaces*, vol. 13, no. 10, 2021, doi: 10.1021/acsami.0c22192.
50. R. Ströbel et al., "Hydrogen adsorption on carbon materials," *Journal of Power Sources*, vol. 84, no. 2, 1999, doi: 10.1016/S0378-7753(99)00320-1.

51. G. Hermosilla-Lara, G. Momen, P. H. Marty, B. le Neindre, and K. Hassouni, "Hydrogen storage by adsorption on activated carbon: Investigation of the thermal effects during the charging process," *International Journal of Hydrogen Energy*, vol. 32, no. 10–11, 2007, doi: 10.1016/j.ijhydene.2006.10.048.
52. J. Xiao, A. Mei, W. Tao, S. Ma, P. Bénard, and R. Chahine, "Hydrogen purification performance optimization of vacuum pressure swing adsorption on different activated carbons," *Energies*, vol. 14, no. 9, 2021, doi: 10.3390/en14092450.
53. E. Rivard, M. Trudeau, and K. Zaghib, "Hydrogen storage for mobility: A review," *Materials*, vol. 12, no. 12, 2019, doi: 10.3390/ma12121973.
54. D. P. Broom et al., "Outlook and challenges for hydrogen storage in nanoporous materials," *Applied Physics A: Materials Science and Processing*, vol. 122, no. 3, 2016, doi: 10.1007/s00339-016-9651-4.
55. L. Wang, N. R. Stuckert, and R. T. Yang, "Unique hydrogen adsorption properties of graphene," *AIChE Journal*, vol. 57, no. 10, 2011, doi: 10.1002/aic.12470.
56. D. P. Broom, C. J. Webb, G. S. Fanourgakis, G. E. Froudakis, P. N. Trikalitis, and M. Hirscher, "Concepts for improving hydrogen storage in nanoporous materials," *International Journal of Hydrogen Energy*, vol. 44, no. 15, 2019, doi: 10.1016/j.ijhydene.2019.01.224.
57. D. P. Broom et al., "Outlook and challenges for hydrogen storage in nanoporous materials," *Applied Physics A: Materials Science and Processing*, vol. 122, no. 3, 2016, doi: 10.1007/s00339-016-9651-4.
58. H. Takagi, H. Hatori, Y. Soneda, N. Yoshizawa, and Y. Yamada, "Adsorptive hydrogen storage in carbon and porous materials," *Materials Science and Engineering B: Solid-State Materials for Advanced Technology*, vol. 108, no. 1–2, 2004, doi: 10.1016/j.mseb.2003.10.095.

15 Organic Waste for Hydrogen Production

Yassine Slimani and Essia Hannachi
Imam Abdulrahman Bin Faisal University

CONTENTS

15.1 INTRODUCTION

The interest in energy, especially in clean energy, green energy, and renewable energy, has increased amazingly during the last few years [1,2]. The different recent statistics indicates that the reserves of fossil fuels will finish soon in the upcoming decades. Hence, the search for alternative sources of renewable energy has become an urgent interest for researchers worldwide. Till now, many energy sources including biogases (with an energy content of 9.6 MJ/kg), biodiesel (~45 MJ/kg), bioethanol (~29 MJ/kg), wind energy, solar energy, and hydrogen (~120 MJ/kg) have been exploited as alternative sources [3]. They could be extracted from renewable sources (natural), could be converted to electricity or heat, are non-polluting, and display elevated energy content. Hydrogen could be exploited as a fuel for transportation and as a raw material for chemical purposes. Hydrogen is not available naturally like other sources of energy such as wind and the sun. This source of energy is frequently being generated using thermochemical processes such as coal gasification, plasma technology, partial oxidation of fossil fuels, and pyrolysis. Nowadays, the use of these processes has resulted in extensive emissions of carbon dioxide, which is harmful to

DOI: 10.1201/9781003178354-18

the environment. For that reason, researchers aim to develop eco-friendly and cost-effective processes for producing hydrogen. The alternate methods generally involve biological processes such as microbial electrolysis cells, photofermentation, dark fermentation, and indirect or direct photolysis. In fact, these processes are versatile in exploiting diverse raw materials and can derive energy from organic wastes.

Organic wastes are wastes rich in organic substances that decompose into methane, water, carbon dioxide (CO_2), etc. The increase in the living level leads to the rise in wastes. For example, the municipal solid wastes in the USA raised from about 88 million tons in the 1960s to about 259 million tons in 2014. Starting from 2013 and during the next 6 years, the dry solid waste-activated sludges in China raised by approximately two times (from 3.2 to 6.2 million tons) [4]. The rise in the quantity of organic wastes is being a severe issue. Indeed, the throw of organic wastes causes pollution of air and water and hence weakens the quality of human life. The production of energy from organic wastes is one of the solutions to reducing the amount of organic wastes. This attracts the interest of researchers around the world. The organic wastes employed in the production of hydrogen through the dark fermentation method principally involves organic wastewater, food waste, cellulose-based biomass, algae biomass, and waste-activated sludge.

The use of organic wastes in producing biohydrogen is an encouraging route to generate sustainable energy. Accordingly, the current chapter delivers an overview on the up-to-date advances and current situation of fermentative generation of hydrogen by exploiting organic wastes, thus assisting the upcoming research works.

15.2 ORGANIC WASTES: TYPES AND COMPONENTS

Different organic wastes such as wastewater, food wastes, starch-based biomasses, cellulose-based biomasses, algal biomasses, and waste-activated sludges have been investigated as feedstocks to produce hydrogen by fermentation.

- **Waste-activated sludges:** One could categorize the components of waste-activated sludge into four [5]: (i) sources of non-toxic carbon, which principally constitute extracellular polymeric matters and microbial cells. It represents about 60%–70% of the total dry weight. (ii) Inorganic compounds such as components containing magnesium, calcium, and silicates. (iii) Toxic organic and inorganic pollutants such as pathogenic microorganisms, estrogens, pesticides, and heavy metals. (iv) Inorganic nutrients that involve compounds comprising of phosphorous and nitrogen. The large content of water and the existence of toxic matters are making treating and discarding waste-activated sludges hard and expensive. Taking into consideration that they are rich in organic matters, waste-activated sludges are getting more interest owing to their potential applications to produce renewable fuels.
- **Algal biomasses:** They are simple organisms that contain chlorophyll. They possess elevated photosynthetic proficiencies to convert atmospheric carbon dioxide to a vast range of organic matters such as lipids, polysaccharides, and proteins. Algae could be classified into microalgae (such as *Anabaena cylindrica*, *Spirulina* sp., *Chlamydomonas* sp., *Arthrospira* sp., *Nannochloropsis* sp.,

Scenedesmus sp., and *Chlorella* sp.) and macroalgae (such as *Gelidium amansii* and *Laminaria japonica*). Macroalgae contain a high amount of carbohydrates, minerals, and water. The microalgae constituents differ based on their classes and the growing environment. The components of microalgae cells are nucleic acids up to 10%, lipids up to 30%, carbohydrates up to 60%, and proteins between 40% and 60% [6]. Figure 15.1 presents the different compositions of microalgae and their potential applications.

- **Cellulose-based biomasses:** Cellulose-based biomasses principally include municipal and agricultural wastes such as wood chips, grass, yard clippings, forest residues, stalks, and straws. They constitute three major biopolymers: lignin (10%–25%), hemicelluloses (15%–30%), and cellulose (30%–70%). The lignin and cellulose fibers are linked with hemicellulose. The hemicelluloses are covalently linked with lignin via the complex of carbohydrates and lignin. Lignin is considered the most obstructing component for producing hydrogen from cellulose-based biomasses.
- **Starch-based biomasses:** These signify biomasses that are rich in starch, such as potatoes, cassava, rice, corn, and wheat. Starch is a type of polysaccharide that plants produce for energy storage. It has been shown that starch-based biomasses demonstrate more biodegradability compared to cellulose-based biomasses. They could simply hydrolyze into simple sugars (principally involving maltose and glucose) via acid or enzymatic saccharification.

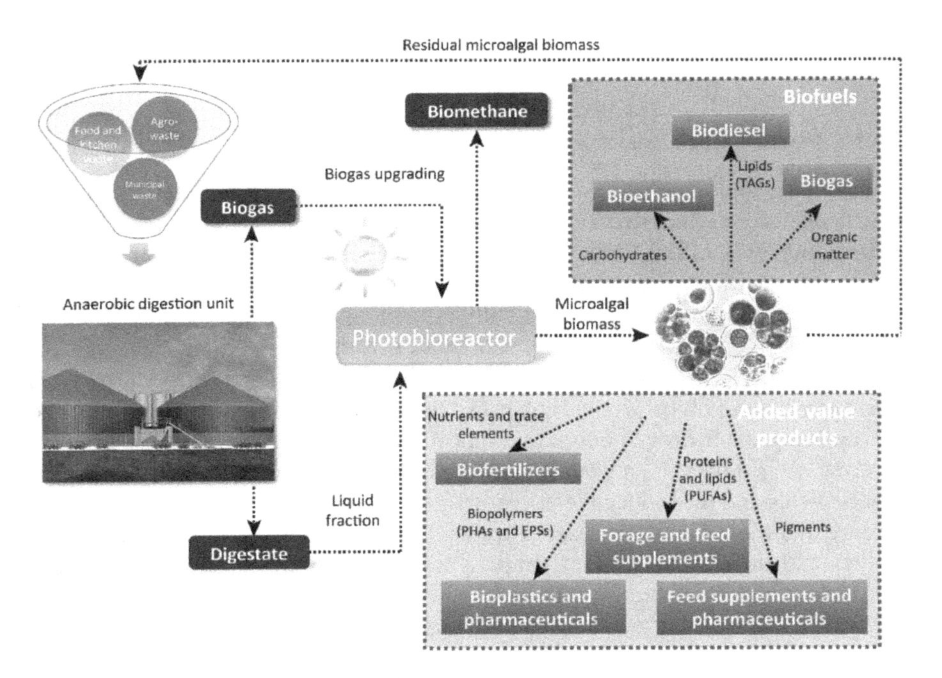

FIGURE 15.1 Different compositions of microalgae and their potential applications. (Adapted with permission from Ref. [7]. Copyright (2018) Elsevier.)

- **Food wastes:** Food wastes include fruits, meats, vegetables, rice, and grain flours. Depending on the components, they could be classified principally into four categories including organic acids, lipids, proteins, and carbohydrates (hemicelluloses, cellulose, starch, and sugars).
- **Wastewater:** Wastewater rich in organic substances is usually utilized for recovering bioenergy. Biofuels including biodiesel, alcohols, hydrogen, and methane have been obtained from wastewater rich in the organic matter. For the generation of hydrogen through dark fermentation, wastewater rich in organic substances, such as dairy wastewater, palm oil mill effluent, beverage wastewater, and distillery wastewater, has been exploited as substrates.

15.3 PRETREATMENTS OF ORGANIC WASTES

Generally, a pretreatment process is needed to hydrolyze the biomasses because of the low enzymatic activities with the anaerobic methods. This will enhance the efficiency of hydrogen generation. Throughout the process of pretreatment, the confined constituents are liberated via the lysis of the cell walls and the removal of lignocellulosic biomasses. Consequently, the process of pretreatment could demolish the structure of macromolecular constituents, releasing their degree of polymerization. Thus, a high amount of rapidly fermentable matter becomes available to microorganisms. The pretreatment approaches could be classified into four groups: biological methods, chemical methods, physical methods, and a combination of diverse methods. The usually employed pretreatment techniques to treat biomasses are reported in Figure 15.2.

15.3.1 Physical Treatment Methods

Physical pretreatments of biomasses involve irradiation (ionizing radiation and microwave), temperature effect (freeze and thaw, and heat), and mechanical (ultrasonication, grinding, and milling) treatments. For instance, irradiating the organic wastes via ionizing radiations with high energy could alter the chemical or physical characteristics of the target substances via both indirect and direct actions (Figure 15.3). In the direct process, the molecules' atoms absorb the energy that is enough for removing electrons from atoms, causing a break of bonds. The indirect process is mediated by the radiolysis products of H_2O. On the other hand, microwaves generate heat, rupturing the hydrogen bonds and modifying the hydration area. The efficacy of microwave irradiations in destroying cells is influenced by the power of electromagnetic fields and the induced temperature.

Freeze and thaw treatment process consists of freezing of sludges at severe temperatures and thawing at ambient temperature numerous times. Throughout these strong temperature fluctuations, cells are swelled due to the formation of intracellular ice crystals. This will cause damage and disrupt cell aggregates.

Heat treatment is the extensively applied process to solubilize biomasses for producing fermentative hydrogen. Elevated temperatures could provoke a disruption in the chemical bonds of the membrane and wall of cells, which leads to solubilizing

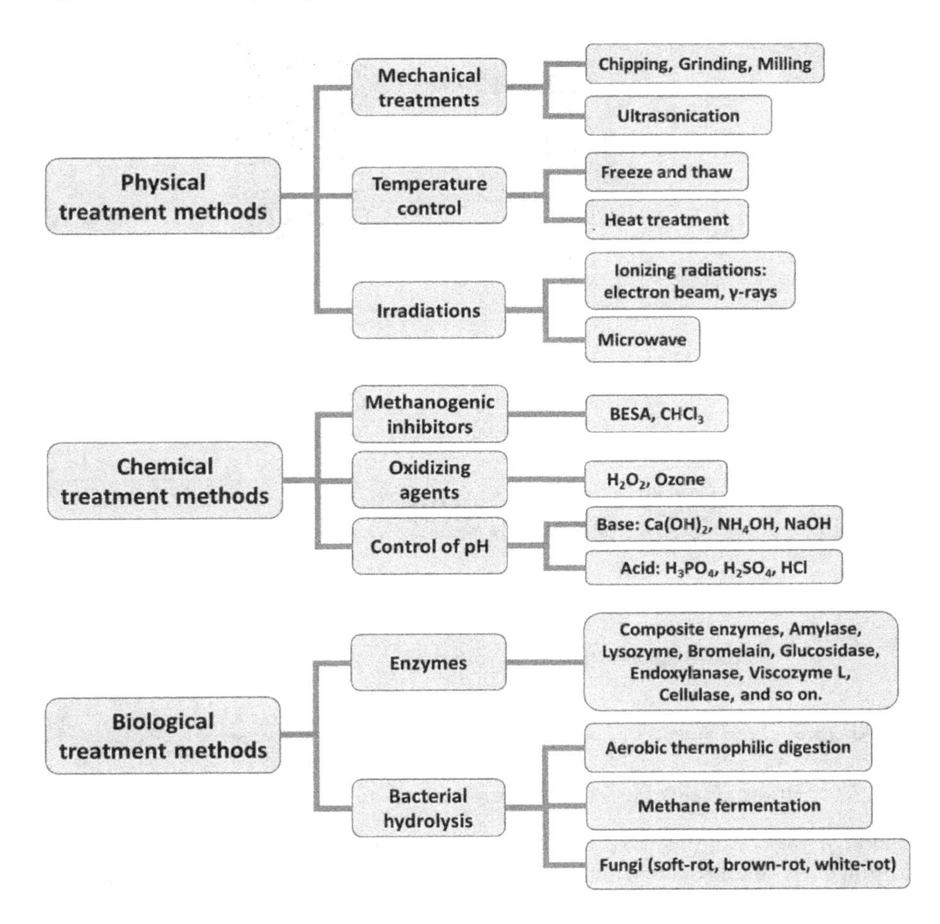

FIGURE 15.2 Pretreatment techniques used to treat biomasses.

cell constituents and deteriorating the microbial proteins. In addition to its great effectiveness for disintegrating biomasses, the broad applications of heat pretreatment are also attributed to its easy control and simple operation. Here, one should note that it is important to select a suitable temperature in heat pretreatment.

Throughout the ultrasonication treatment, extremely high pressure (180 MPa), elevated localized temperature (5,000 K), shear forces, and highly activated radicals are produced in the medium, which results in disruption of the cells and solubilization of the particles.

The pretreatment by employing mechanical techniques (such as chipping, grinding, and milling) is mostly utilized as an initial stage for reducing the size of particles, creating a larger accessible surface zone, and decreasing the crystallinity. Typically, mechanical pretreatment is linked with another treatment process (biological, chemical, and physical treatment methods).

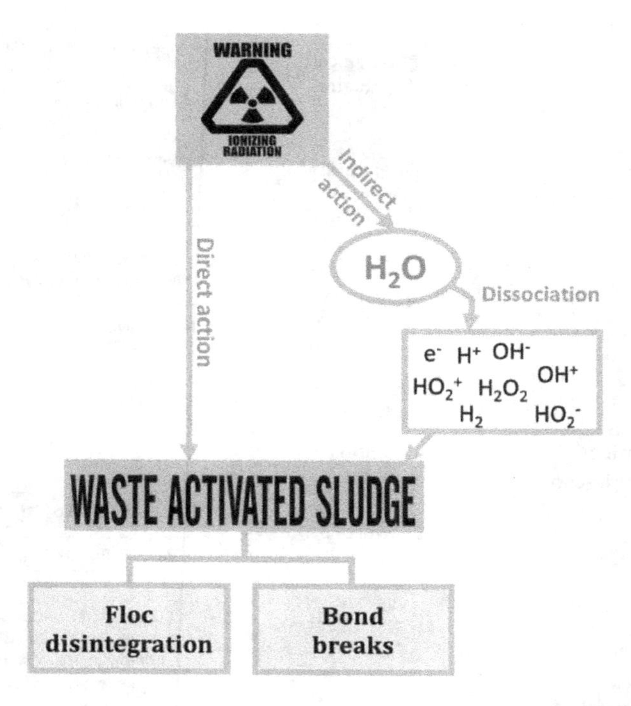

FIGURE 15.3 Mechanisms of ionizing radiations to treat organic wastes.

15.3.2 Chemical Treatment Methods

Generally, these approaches include the addition of methanogenic inhibitors, the addition of oxidizing agents, and the control of pH (Figure 15.2). Base and acid treatments are usually the utmost utilized chemical treatment approaches for a suitable operation and superior efficiencies.

15.3.3 Biological Treatment Methods

The treatment with enzymes is also exploited to treat algae biomasses and cellulose-based biomasses. However, it is typically performed after chemical or physical treatments. Different enzymes have been utilized for this process (Figure 15.2), where cellulase is the most exploited one. The effect of dissolution of enzymatic treatment is influenced by hydrolyzing time, the dose of enzymes, characteristics of substrates, and enzyme varieties.

15.4 PRODUCTION OF HYDROGEN FROM ORGANIC WASTES

15.4.1 Waste-Activated Sludges for Hydrogen Production

Some recent studies have been performed on waste-activated sludges treated by dark fermentation to produce hydrogen [8–19]. The pretreatment process of waste-activated sludges is a crucial stage before hydrogen generation. The pretreatment process does not only disintegrate sludges, but also inhibits non-hydrogen producers existing in waste-activated sludges. Accordingly, the reported investigations are categorized in accordance with the treatment approaches employed.

The concentrations of total solids of sludges vary considerably with the diverse sources and time of retention. Inocula are not important when waste-activated sludges are utilized as substrates because they are abundant in microorganisms. Nevertheless, in some investigations, producers of hydrogen, such as *Enterobacter* sp. and *Clostridium* sp., have been injected into the system of fermentation to improve the efficiency of hydrogen generation [20]. Primary pH values in the ranges of 5–11 have been utilized, which is very diverse. This is due to the dissimilar treatment methods employed. For the alkali or the acid treatment process, alkaline or acidic conditions have been implemented to avoid the costs for adjusting pH. Both thermal (55 °C–78 °C) and mesophilic (30 °C–37 °C) fermentations have been investigated. The mesophilic conditions are more usually utilized because of their lower operational costs.

15.4.2 Algae Biomasses for Hydrogen Production

Fermentative hydrogen generation from diverse algae biomasses has been investigated recently [21–28]. Among microalgae species, *Chlorella* sp. is broadly explored. Roy and co-workers [21] revealed that the treatment of *Chlorella sorokiniana* biomasses by combining acid and heat effects leads to a high yield of hydrogen of about 960 mL H_2/g VS. *Scenedesmus* sp. has also been explored widely due to its elevated amount of carbohydrates where a maximum hydrogen yield of about 113 mL H_2/g VS has been reached after heat pretreatment [29]. Among macroalgae classes, *Laminaria japonica* is the largest examined one owing to its low requirements of pretreatment and high amount of carbohydrates [30]. Other macroalgae classes have rarely been examined. Generally, the majority of reports have been performed at mesophilic conditions (29 °C–38 °C), the initial pH is in the range of 6–8, the content of substrates varies between 5 and 117 g TS/L, and the *Clostridium butyricum* and anaerobic sludge species are the most usually utilized inoculum.

15.4.3 Cellulose-Based Biomasses for Hydrogen Production

Biomasses based on celluloses, involving leaves, grasses, bagasse, cob, corn stover, stalks (such as sunflower stalk and corn stalk), and straws (e.g., sugarcane straw, corn straw, wheat straw, and rice straw), have broadly been utilized as feedstocks for fermentative hydrogen generation. Some of the studies performed on treated cellulose-based materials for hydrogen production can be found in the references [31–39]. The concentration of cellulose ranging between 0.1 and 10 g/L has been investigated. In addition to the investigations performed without pretreatment, some of the usually utilized pretreatments including combined acid and heat treatment, bacterial hydrolysis, and enzyme treatment have been reported. Pure strains have largely been employed as inoculum compared to others. The species of thermophilic bacteria are particularly the preferable ones among the pure species. Consequently, operating temperatures range between 55 °C and 70 °C. A temperature of 37 °C has been considered for the investigations performed at mesophilic conditions. In addition to operating temperature, the pH value in the range of 5.5–8 has been implemented. Hydrogen yields in the interval of about 0.6–19 mmol H_2/g celluloses have been attained, and superior yields have been achieved from thermophilic fermentation systems. Similarly, straws have also been employed broadly for hydrogen generation. In general, 10–340 g TS/L straws has been utilized as substrates after pretreatments. For straws, temperature $T = 35$ °C–80 °C and pH = 4–8 have largely been employed. A temperature of 35 °C and pH value of 6–7 are the most applied. Hydrogen yields gained from treated straws are considerably greater compared to untreated ones. In fact, the obtained hydrogen yields range between 3.5 and 93.5 mL H_2/g TS.

15.4.4 Starch-Based Biomasses for Hydrogen Production

Like biomasses based on cellulose, starch-based biomasses are largely available and non-expensive. However, biomasses based on starch could be hydrolyzed into simple sugars more easily than biomasses reported above. Therefore, they offer particular benefits for hydrogen generation. Some examples have been reported in the following studies [40–42]. Clearly, pretreatment processes are ignored in the majority of investigations, and the conditions used to treat starch-based biomasses are greatly softer compared to the cases of cellulose-based biomasses. Both pure and mixed cultures have been employed as inoculum. The most usually employed conditions are mesophilic conditions, whereas some reports revealed that thermophilic fermentation could also lead to elevated hydrogen yields [43]. The commonly applied pH values range between 5.5 and 7.2. Hydrogen contents are in the intervals of 0.2–3.5 mol H_2/mol hexose, which are influenced by operational conditions, inoculum, and starch sources.

15.4.5 Food Wastes for Hydrogen Production

Food wastes used for hydrogen generation have widely been reported during the recent years [44–46]. There are investigations that rarely involve pretreatment process before fermentation because most of the food wastes are already cooked. Due to the complexity in the components of food wastes, anaerobic cultures are preferred over pure ones. The operating pH values and temperatures are in the intervals of 4–7.2 and 34 °C–39 °C, respectively. It is clear that acidic environments are most extensively preferred. Hydrogen yields are influenced by both operating conditions and food wastes content.

15.4.6 Wastewater for Hydrogen Production

Numerous types of wastewater include high amounts of organic substances, which generally need specific pretreatment. Due to the rapid growth of interest in environmental issues as well as on clean energy, researchers are focusing on the treatment of wastewaters to shift from only controlling pollution to recovering resources. Therefore, diverse organic wastewaters are utilized as substrates to produce hydrogen. Some examples of treated organic wastewaters are listed in references [47–52].

For treated distillery wastewaters, COD content varies between 3 and 96 g/L. In certain conditions, sucrose is added to improve the content of the carbon source. As distillery wastewaters are commonly abundant in microorganisms, limited reports have inoculated pure cultures to produce hydrogen. The most extensively employed inoculum is anaerobic sludges. The effects of thermophilic and mesophilic conditions have been explored. The operating temperatures and pH values are in the intervals of 28 °C–70 °C and 5–7, respectively. The largest employed temperatures for thermophilic and mesophilic conditions are about 55 °C and 28 °C. Researchers reached hydrogen yields of about 0.1–2.8 mol H_2/mol hexose.

Proteins-based wastewaters such as textile wastewaters, dairy wastewaters, and cheese whey wastewaters are also utilized as substrates to produce hydrogen. In these cases, COD content varies between 4.5 and 88 g/L. The most widely preferred inoculum is anaerobic sludges. Also, some pure cultures such as strains *Thermotoga neapolitana* and genes-modified *Escherichia coli* are employed. For proteins-based wastewaters, the pH values and operating temperatures are in the ranges of 5.5–8.5 and 29 °C–77 °C, respectively. The commonly utilized pH values and temperatures are 6–7 and 35 °C–37 °C, respectively. Hydrogen yields of up to 2.3 mol H_2/mol hexose are reached. It should be noted that the hydrogen yields have considerably been influenced by the properties of proteins-based wastewaters.

Moreover, sugar-based wastewaters are promising substrates to produce hydrogen. Beverage wastewaters are abundant in mixed micromolecule sugars such as sucrose and glucose. In these cases, the content of carbohydrates ranges between 5 and 40 g/L. Likewise, anaerobic sludges are the largely utilized inoculum. The most employed temperatures are about 35 °C–37 °C. Also, the most applied pH value is about 5 and can reach 9 in some cases. Hydrogen yields reached about 3.6 mol H_2/mol hexose.

Also, some industrial wastewaters, biodiesel wastewaters, and oil mill wastewaters are utilized as substrates to produce hydrogen. Biodiesel wastewaters are abundant in glycerol, which is a type of biodegradable carbohydrate. Oil mill effluents contain elevated content of proteins, minerals, lipids, carbohydrates, and organic acids, which could help as growth nutrients for the microorganisms. Therefore, both biodiesel and oil mill wastewaters are appropriate substrates to produce hydrogen. In these cases, the COD content of oil mill effluents varies between 3 and 80 g/L. Most of the investigations have been performed at pH values of 5.5–5.8 and operating temperatures of 35 °C–38 °C. Hydrogen yields vary between 0.12 and 2.88 mol H_2/mol hexose. In the case of biodiesel wastewaters, the content of glycerol ranges between 1 and 25 g/L. Here, the majority of investigations on treated biodiesel wastewaters were carried out with pure cultures, neutral conditions (pH around 7), and operating temperatures varying between 35 °C and 80 °C. Hydrogen yields achieved values of 3.4 mol H_2/mol hexose, and maximum yields were achieved at thermophilic conditions.

15.5 CONCLUSIONS

The production of hydrogen from fossil fuels is currently dominant due to its high efficiency and applicable technologies. However, the significant rise in energy consumption and polluting residues has greatly reduced the environmental benefits of the hydrogen produced. Therefore, processes of sustainable hydrogen generation are needed to be developed. By using organic wastes, dark fermentative hydrogen generation reaches dual advantages of wastes treatment and hydrogen production, which create a map for an auspicious future for the sustained community. Diverse kinds of organic wastes have been reported in the current chapter, including wastewaters, food wastes, starch-based biomasses, cellulose-based biomasses, algae biomasses, and waste-activated sludges. Cellulose-based biomasses are the most broadly explored due to their rich sources. Starch-based biomasses as well as food wastes are rich in nutrients and have a high amount of carbohydrates, making them the best substrates to produce hydrogen. Wastewaters generally do not require a pretreatment process and could be simply applied in the fermentation process. Waste-activated sludges are being a serious environmental issue. Algal biomasses have attracted interest during the recent years because of their CO_2 fixation and ease of cultivation. Several investigations have revealed the viability of dark fermentative hydrogen generation by exploiting organic wastes. Nevertheless, the technology looks to be still in its infancy. The rate of substrate disintegration and the low hydrogen yields have impeded its intensive applications. Therefore, additional investigations are required to evaluate the possibility of implementing the fermentable hydrogen generation process.

Pretreatment is a crucial procedure for the high efficiency of hydrogen generation from organic wastes. The performed investigations showed that the heating treatment process is the largest utilized one, and the combinations of diverse pretreatment processes are gradually examined. Diverse pretreatment processes reported in this chapter showed their efficiency to treat organic wastes. However, some kind of wastes, such as cellulose-based biomasses, needs harder treatment processes. Further investigations are needed to realize a more economical and efficient process. This will lead to choosing the appropriate pretreatment process to the characteristics of biomasses and their components. On the other hand, the biological methods greatly depend on the fermentation conditions such as hydrogen partial pressure, constituents of inorganic substances, pH value, and temperature. These operating parameters influence not only the hydrogen yields, but also the formation of by-products.

All of the reported methods in this chapter are still in their beginnings and should be further investigated. Due to all the above-reported reasons, extensive efforts should be made to further explore, examine, and test them from different sides (practical and technical features) to attain a wide range of applications of hydrogen generation from organic wastes.

REFERENCES

1. Y. Slimani, E. Hannachi, Green chemistry and sustainable nanotechnological developments: Principles, designs, applications, and efficiency, in: *Green Polym. Chem. Compos.*, Apple Academic Press, 2021: pp. 1–18. https://doi.org/10.1201/9781003083917–1.
2. S. Noreen, K. Khalid, M. Iqbal, H.B. Baghdadi, N. Nisar, U.H. Siddiqua, J. Nisar, Y. Slimani, M.I. Khan, A. Nazir, Eco-benign approach to produce biodiesel from neem oil using heterogeneous nano-catalysts and process optimization, *Environ. Technol. Innov.* 22 (2021) 101430. https://doi.org/10.1016/j.eti.2021.101430.

3. S. Satyapal, J. Petrovic, C. Read, G. Thomas, G. Ordaz, The U.S. Department of Energy's National Hydrogen Storage Project: Progress towards meeting hydrogen-powered vehicle requirements, *Catal. Today.* 120 (2007) 246–256. https://doi.org/10.1016/j.cattod.2006.09.022.
4. G. Yang, G. Zhang, H. Wang, Current state of sludge production, management, treatment and disposal in China, *Water Res.* 78 (2015) 60–73. https://doi.org/10.1016/j.watres.2015.04.002.
5. W. Rulkens, Sewage sludge as a biomass resource for the production of energy: Overview and assessment of the various options, *Energy and Fuels.* 22 (2008) 9–15. https://doi.org/10.1021/ef700267m.
6. E. Uggetti, B. Sialve, E. Trably, J.P. Steyer, Integrating microalgae production with anaerobic digestion: A biorefinery approach, *Biofuels, Bioprod. Biorefining.* 8 (2014) 516–529. https://doi.org/10.1002/bbb.1469.
7. E. Koutra, C.N. Economou, P. Tsafrakidou, M. Kornaros, Bio-based products from microalgae cultivated in digestates, *Trends Biotechnol.* 36 (2018) 819–833. https://doi.org/10.1016/j.tibtech.2018.02.015.
8. T. Assawamongkholsiri, A. Reungsang, S. Pattra, Effect of acid, heat and combined acid-heat pretreatments of anaerobic sludge on hydrogen production by anaerobic mixed cultures, in: *Int. J. Hydrogen Energy*, Pergamon, 2013: pp. 6146–6153. https://doi.org/10.1016/j.ijhydene.2012.12.138.
9. C.C. Wang, C.W. Chang, C.P. Chu, D.J. Lee, B.V. Chang, C.S. Liao, Producing hydrogen from wastewater sludge by Clostridium bifermentans, *J. Biotechnol.* 102 (2003) 83–92. https://doi.org/10.1016/S0168–1656(03)00007–5.
10. Z. Wang, S. Shao, C. Zhang, D. Lu, H. Ma, X. Ren, Pretreatment of vinegar residue and anaerobic sludge for enhanced hydrogen and methane production in the two-stage anaerobic system, *Int. J. Hydrogen Energy.* 40 (2015) 4494–4501. https://doi.org/10.1016/j.ijhydene.2015.02.029.
11. S.M. Kotay, D. Das, Novel dark fermentation involving bioaugmentation with constructed bacterial consortium for enhanced biohydrogen production from pretreated sewage sludge, *Int. J. Hydrogen Energy.* 34 (2009) 7489–7496. https://doi.org/10.1016/j.ijhydene.2009.05.109.
12. L. Guo, M. Lu, Q. Li, J. Zhang, Z. She, A comparison of different pretreatments on hydrogen fermentation from waste sludge by fluorescence excitation-emission matrix with regional integration analysis, *Int. J. Hydrogen Energy.* 40 (2015) 197–208. https://doi.org/10.1016/j.ijhydene.2014.10.141.
13. C.C. Chen, Y.S. Chuang, C.Y. Lin, C.H. Lay, B. Sen, Thermophilic dark fermentation of untreated rice straw using mixed cultures for hydrogen production, in: *Int. J. Hydrogen Energy*, Pergamon, 2012: pp. 15540–15546. https://doi.org/10.1016/j.ijhydene.2012.01.036.
14. S.S. Yang, W.Q. Guo, G.L. Cao, H.S. Zheng, N.Q. Ren, Simultaneous waste activated sludge disintegration and biological hydrogen production using an ozone/ultrasound pretreatment, *Bioresour. Technol.* 124 (2012) 347–354. https://doi.org/10.1016/j.biortech.2012.08.007.
15. M. Cai, J. Liu, Y. Wei, Enhanced biohydrogen production from sewage sludge with alkaline pretreatment, *Environ. Sci. Technol.* 38 (2004) 3195–3202. https://doi.org/10.1021/es0349204.
16. Y. Yin, J. Wang, Gamma irradiation induced disintegration of waste activated sludge for biological hydrogen production, *Radiat. Phys. Chem.* 121 (2016) 110–114. https://doi.org/10.1016/j.radphyschem.2016.01.009.
17. J. Massanet-Nicolau, A. Guwy, R. Dinsdale, G. Premier, S. Esteves, Production of hydrogen from sewage biosolids in a continuously fed bioreactor: Effect of hydraulic retention time and sparging, *Int. J. Hydrogen Energy.* 35 (2010) 469–478. https://doi.org/10.1016/j.ijhydene.2009.10.076.

18. N.H.M. Yasin, M. Fukuzaki, T. Maeda, T. Miyazaki, C.M.H.C. Maail, H. Ariffin, T.K. Wood, Biohydrogen production from oil palm frond juice and sewage sludge by a metabolically engineered Escherichia coli strain, *Int. J. Hydrogen Energy.* 38 (2013) 10277–10283. https://doi.org/10.1016/j.ijhydene.2013.06.065.
19. J.H. Kang, D. Kim, T.J. Lee, Hydrogen production and microbial diversity in sewage sludge fermentation preceded by heat and alkaline treatment, *Bioresour. Technol.* 109 (2012) 239–243. https://doi.org/10.1016/j.biortech.2012.01.048.
20. M.S. Miah, C. Tada, S. Sawayama, Enhancement of biogas production from sewage sludge with the addition of Geobacillus sp. strain AT1 culture, *Japanese J. Water Treat. Biol.* 40 (2004) 97–104. https://doi.org/10.2521/jswtb.40.97.
21. S. Roy, K. Kumar, S. Ghosh, D. Das, Thermophilic biohydrogen production using pre-treated algal biomass as substrate, *Biomass and Bioenergy.* 61 (2014) 157–166. https://doi.org/10.1016/j.biombioe.2013.12.006.
22. P. Sinha, A. Pandey, Biohydrogen production from various feedstocks by Bacillus firmus NMBL-03, *Int. J. Hydrogen Energy.* 39 (2014) 7518–7525. https://doi.org/10.1016/j.ijhydene.2013.08.134.
23. E.N. Efremenko, A.B. Nikolskaya, I.V. Lyagin, O.V. Senko, T.A. Makhlis, N.A. Stepanov, O.V. Maslova, F. Mamedova, S.D. Varfolomeev, Production of biofuels from pretreated microalgae biomass by anaerobic fermentation with immobilized Clostridium acetobutylicum cells, *Bioresour. Technol.* 114 (2012) 342–348. https://doi.org/10.1016/j.biortech.2012.03.049.
24. B.P. Nobre, F. Villalobos, B.E. Barragán, A.C. Oliveira, A.P. Batista, P.A.S.S. Marques, R.L. Mendes, H. Sovová, A.F. Palavra, L. Gouveia, A biorefinery from Nannochloropsis sp. microalga – Extraction of oils and pigments. Production of biohydrogen from the leftover biomass, *Bioresour. Technol.* 135 (2013) 128–136. https://doi.org/10.1016/j.biortech.2012.11.084.
25. Z. Yang, R. Guo, X. Xu, X. Fan, S. Luo, Fermentative hydrogen production from lipid-extracted microalgal biomass residues, *Appl. Energy.* 88 (2011) 3468–3472. https://doi.org/10.1016/j.apenergy.2010.09.009.
26. A. Xia, J. Cheng, L. Ding, R. Lin, W. Song, J. Zhou, K. Cen, Enhancement of energy production efficiency from mixed biomass of Chlorella pyrenoidosa and cassava starch through combined hydrogen fermentation and methanogenesis, *Appl. Energy.* 120 (2014) 23–30. https://doi.org/10.1016/j.apenergy.2014.01.045.
27. N. Wieczorek, M.A. Kucuker, K. Kuchta, Fermentative hydrogen and methane production from microalgal biomass (Chlorella vulgaris) in a two-stage combined process, *Appl. Energy.* 132 (2014) 108–117. https://doi.org/10.1016/j.apenergy.2014.07.003.
28. K.W. Jung, D.H. Kim, H.W. Kim, H.S. Shin, Optimization of combined (acid+thermal) pretreatment for fermentative hydrogen production from Laminaria japonica using response surface methodology (RSM), *Int. J. Hydrogen Energy.* 36 (2011) 9626–9631. https://doi.org/10.1016/j.ijhydene.2011.05.050.
29. A.P. Batista, P. Moura, P.A.S.S. Marques, J. Ortigueira, L. Alves, L. Gouveia, Scenedesmus obliquus as feedstock for biohydrogen production by Enterobacter aerogenes and Clostridium butyricum, *Fuel.* 117 (2014) 537–543. https://doi.org/10.1016/j.fuel.2013.09.077.
30. J. Il Park, J. Lee, S.J. Sim, J.H. Lee, Production of hydrogen from marine macro-algae biomass using anaerobic sewage sludge microflora, *Biotechnol. Bioprocess Eng.* 14 (2009) 307–315. https://doi.org/10.1007/s12257-008-0241–y.
31. H. Jiang, S.I. Gadow, Y. Tanaka, J. Cheng, Y.Y. Li, Improved cellulose conversion to bio-hydrogen with thermophilic bacteria and characterization of microbial community in continuous bioreactor, *Biomass and Bioenergy.* 75 (2015) 57–64. https://doi.org/10.1016/j.biombioe.2015.02.010.

32. S.I. Gadow, Y.Y. Li, Y. Liu, Effect of temperature on continuous hydrogen production of cellulose, in: *Int. J. Hydrogen Energy*, Pergamon, 2012: pp. 15465–15472. https://doi.org/10.1016/j.ijhydene.2012.04.128.
33. J. Wu, S. Upreti, F. Ein-Mozaffari, Ozone pretreatment of wheat straw for enhanced biohydrogen production, *Int. J. Hydrogen Energy*. 38 (2013) 10270–10276. https://doi.org/10.1016/j.ijhydene.2013.06.063.
34. P. Kongjan, S. O. Thong, M. Kotay, B. Min, I. Angelidaki, Biohydrogen production from wheat straw hydrolysate by dark fermentation using extreme thermophilic mixed culture, *Biotechnol. Bioeng.* 105 (2010) 899–908. https://doi.org/10.1002/bit.22616.
35. J. Cheng, H. Su, J. Zhou, W. Song, K. Cen, Microwave-assisted alkali pretreatment of rice straw to promote enzymatic hydrolysis and hydrogen production in dark- and photo-fermentation, *Int. J. Hydrogen Energy*. 36 (2011) 2093–2101. https://doi.org/10.1016/j.ijhydene.2010.11.021.
36. S. Talluri, S.M. Raj, L.P. Christopher, Consolidated bioprocessing of untreated switchgrass to hydrogen by the extreme thermophile Caldicellulosiruptor saccharolyticus DSM 8903, *Bioresour. Technol.* 139 (2013) 272–279. https://doi.org/10.1016/j.biortech.2013.04.005.
37. M. Cui, J. Shen, Effects of acid and alkaline pretreatments on the biohydrogen production from grass by anaerobic dark fermentation, *Int. J. Hydrogen Energy*. 37 (2012) 1120–1124. https://doi.org/10.1016/j.ijhydene.2011.02.078.
38. L. Ozkan, T.H. Erguder, G.N. Demirer, Effects of pretreatment methods on solubilization of beet-pulp and bio-hydrogen production yield, *Int. J. Hydrogen Energy*. 36 (2011) 382–389. https://doi.org/10.1016/j.ijhydene.2010.10.006.
39. A.F. Saripan, A. Reungsang, Simultaneous saccharification and fermentation of cellulose for bio-hydrogen production by anaerobic mixed cultures in elephant dung, *Int. J. Hydrogen Energy*. 39 (2014) 9028–9035. https://doi.org/10.1016/j.ijhydene.2014.04.066.
40. C.Y. Chu, B. Sen, C.H. Lay, Y.C. Lin, C.Y. Lin, Direct fermentation of sweet potato to produce maximal hydrogen and ethanol, *Appl. Energy*. 100 (2012) 10–18. https://doi.org/10.1016/j.apenergy.2012.06.023.
41. T. Sreethawong, S. Chatsiriwatana, P. Rangsunvigit, S. Chavadej, Hydrogen production from cassava wastewater using an anaerobic sequencing batch reactor: Effects of operational parameters, COD:N ratio, and organic acid composition, *Int. J. Hydrogen Energy*. 35 (2010) 4092–4102. https://doi.org/10.1016/j.ijhydene.2010.02.030.
42. S. Wang, T. Zhang, H. Su, Enhanced hydrogen production from corn starch wastewater as nitrogen source by mixed cultures, *Renew. Energy*. 96 (2016) 1135–1141. https://doi.org/10.1016/j.renene.2015.11.072.
43. A. Cakr, S. Ozmihci, F. Kargi, Comparison of bio-hydrogen production from hydrolyzed wheat starch by mesophilic and thermophilic dark fermentation, *Int. J. Hydrogen Energy*. 35 (2010) 13214–13218. https://doi.org/10.1016/j.ijhydene.2010.09.029.
44. Y. Lin, J. Liang, S. Wu, B. Wang, Was pretreatment beneficial for more biogas in any process? Chemical pretreatment effect on hydrogen-methane co-production in a two-stage process, *J. Ind. Eng. Chem.* 19 (2013) 316–321. https://doi.org/10.1016/j.jiec.2012.08.018.
45. G. Cappai, G. De Gioannis, M. Friargiu, E. Massi, A. Muntoni, A. Polettini, R. Pomi, D. Spiga, An experimental study on fermentative H_2 production from food waste as affected by pH, *Waste Manag.* 34 (2014) 1510–1519. https://doi.org/10.1016/j.wasman.2014.04.014.
46. C. Sreela-Or, T. Imai, P. Plangklang, A. Reungsang, Optimization of key factors affecting hydrogen production from food waste by anaerobic mixed cultures, in: *Int. J. Hydrogen Energy*, Pergamon, 2011: pp. 14120–14133. https://doi.org/10.1016/j.ijhydene.2011.04.136.

47. S. Singh, M. Khwairakpam, C.N. Tripathi, A comparative study between composting and vermicomposting for recycling food wastes, *Int. J. Environ. Waste Manag.* 12 (2013) 231–242. https://doi.org/10.1504/IJEWM.2013.056119.
48. H. Hafez, G. Nakhla, H. El Naggar, An integrated system for hydrogen and methane production during landfill leachate treatment, *Int. J. Hydrogen Energy.* 35 (2010) 5010–5014. https://doi.org/10.1016/j.ijhydene.2009.08.050.
49. G.Z. Kyzas, M.P. Symeonidou, K.A. Matis, Technologies of winery wastewater treatment: A critical approach, *Desalin. Water Treat.* 57 (2016) 3372–3386. https://doi.org/10.1080/19443994.2014.986535.
50. E.B. Sydney, C. Larroche, A.C. Novak, R. Nouaille, S.J. Sarma, S.K. Brar, L.A.J. Letti, V.T. Soccol, C.R. Soccol, Economic process to produce biohydrogen and volatile fatty acids by a mixed culture using vinasse from sugarcane ethanol industry as nutrient source, *Bioresour. Technol.* 159 (2014) 380–386. https://doi.org/10.1016/j.biortech.2014.02.042.
51. B.B. Romão, F.R.X. Batista, J.S. Ferreira, H.C.B. Costa, M.M. Resende, V.L. Cardoso, Biohydrogen production through dark fermentation by a microbial consortium using whey permeate as substrate, *Appl. Biochem. Biotechnol.* 172 (2014) 3670–3685. https://doi.org/10.1007/s12010-014-0778–5.
52. S. Venkata Mohan, G. Mohanakrishna, P.N. Sarma, Integration of acidogenic and methanogenic processes for simultaneous production of biohydrogen and methane from wastewater treatment, *Int. J. Hydrogen Energy.* 33 (2008) 2156–2166. https://doi.org/10.1016/j.ijhydene.2008.01.055.

16 Recycling E-Waste for Hydrogen Energy Production and Replacement as Building Construction Materials

Ramji Kalidoss
Bharath Institute of Higher Education and Research

Radhakrishnan Kothalam and Ganesh Vattikondala
SRM Institute of Science and Technology

CONTENTS

16.1 INTRODUCTION

One of the most important factors of economic growth of the world and the backbone of human society is the usage of electrical and electronic components, which indeed relies on industrial growth and the global economy. The imagination of survival of mankind without the usage of electrical and electronic components or devices

DOI: 10.1201/9781003178354-19

is obsolete. In other words, it is also most convenient to say that they are the most important aspect of contemporary lifestyle, luxury, adequacy, and even growth in developing countries as well as developed countries. For example, nowadays, the modern world is talking about digital transactions, online payments, wallet pay, etc., which are completely dependent on various electronic devices, mostly mobile phones and personal computers. On the other hand, the usage of electronic and electrical components can largely be seen in house appliances such as fridges, television, microwave oven, alarms, etc. Therefore, it is the responsibility of every human being to discard electrical and electronic devices properly without exploiting the environment, which may also be called e-waste management.

Over a certain period, every electrical and electronic device or equipment becomes e-waste. E-waste is defined as follows as per the European Union (EU) WEEE Directive:

> the waste from electrical and electronic equipment (WEEE) that includes all the components of electronic equipment made with several components like diodes, transistors, integrated circuits and soldering or connections between the components using various metals, its subassemblies, and consumables which are the part of the product at the time it is discarded [1].

Old, flawed, and undesirable electronic materials and devices that are not functioning are also e-waste. The utilization of electronic devices is rapidly emerging due to e-waste, which is the fastest developing zone in the concurrent universe economy and dependent on various factors such as technological developments, the rate at which older forms of technologies are being replaced or discard with newer forms of technologies, fast budgetary growth, urbanization, industrialization, and increased demand [2]. Because of the rapid development in the manufacturing technologies of the EE components and increased dependencies on the usage of electronic instruments, the demand for such equipment is increasing every year. People have started investing hugely in electronic instruments than ever before. To idealize the amount of e-waste generated, Figure 16.1 shows the increase in obsolete computers per year globally in the period of 2019–2030 [3]. Some of the electronic devices directly or indirectly use heavy metals such as Pb, Hg, Cd, polychlorinated biphenyls (PCBs), and brominated flame retardants (BFRs), which are toxic in nature. On the other hand, outdoor burning of circuit boards/wires and/or materials containing both chlorine and bromine can release hazardous by-products including multiple chlorinated and brominated dioxin compounds with halogenated dibenzo-*p*-dioxins/dibenzofurans [4]. Therefore, it is essential to recycle e-waste in order to eliminate the stress on the environment in addition to the pollution already made during the manufacturing of these goods.

On the contrary, the shrinking finite resources of fossil fuels for energy generation and the carbon emission causes motivate the research community to generate alternative fuels. Therefore, the generation of hydrogen using e-waste opens up an avenue to effectively manage e-waste and simultaneously generate renewable energy sources, leading to a circular economy, as shown in Figure 16.2. Moreover, e-wastes are effectively utilized as an alternative concrete mixture for building construction that claims to increase the structural strength in comparison with the conventional materials. These applications of e-waste indicate the possibility of recycling it into

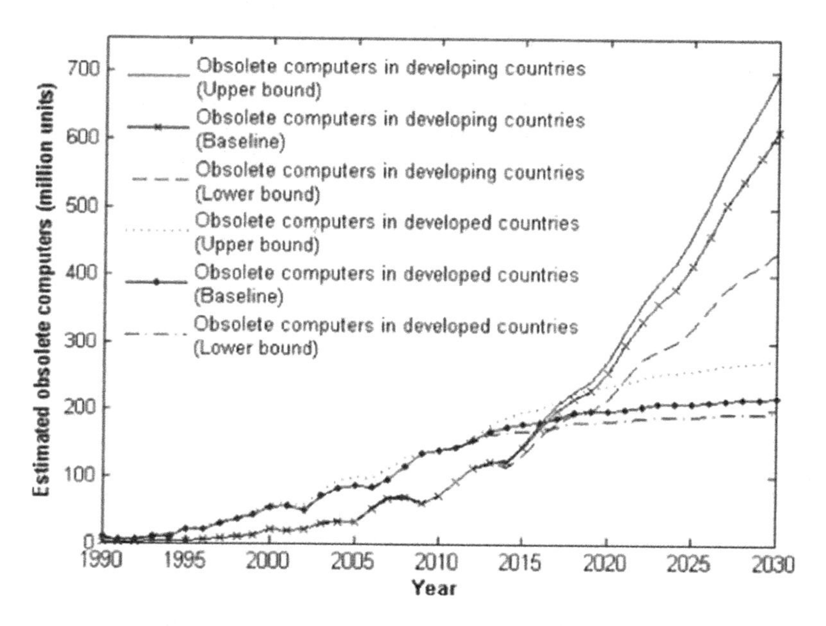

FIGURE 16.1 E-waste prediction for the period 2019–2030. (Adapted with permission from Ref. [3].)

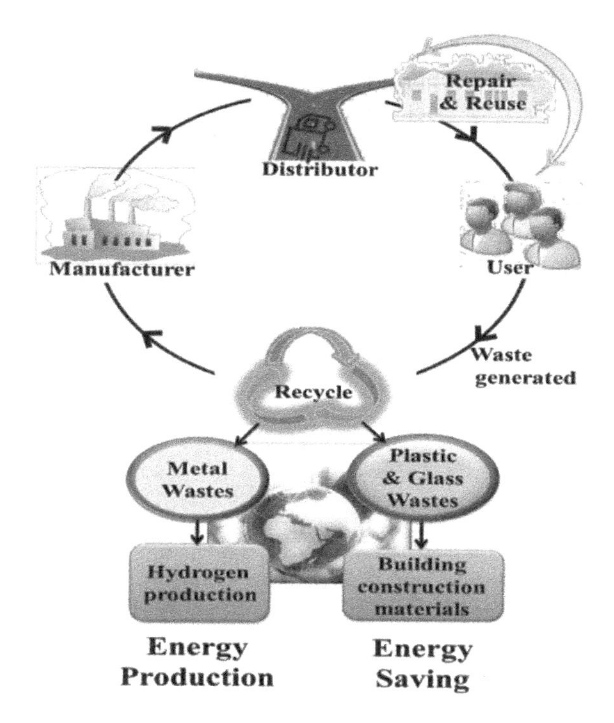

FIGURE 16.2 Circular economy of e-waste.

useful products rather than simply destroying it without causing harm to the environment. Hence, this chapter is focused on the availability of resources in e-waste from different equipment, their processing techniques, and recycling applications of energy generation and building construction materials.

16.2 E-WASTE COMPOSITION

The waste generated from these products can be categorized as toxic heavy metals, precious metals, and various types of plastics. The components of e-waste are mostly dependent on several factors such as model, device type, company, date of manufacturer, and the age of the scrap. A scrap of several IT and telecommunication devices contains precious metals such as silver, gold, and palladium in large quantities compared to household appliances [5]. For example, base metals such as copper (Cu) and tin (Sn) and special metals such as lithium (Li), cobalt (Co), indium (In), lead (Pb), and antimony (Sb) can be found in mobile phones. Several other toxic materials such as arsenic, cadmium, led, chromium, and mercury are used in circuit boards, which are the backbone of many electronic devices. Typically, PCBs are treated with soldering of tin–lead metals with a quantity of 50 g per square meter. Outdated refrigerators, freezers, and air conditioning units contain ozone-depleting chlorofluorocarbons (CFCs). Electronic equipment is manufactured using various components of electronic parts, which include PCBs in which precious and semi-precious metals are used as contact materials due to their high chemical stability along with good conductivity. The more the electric charge transfer, the more the required mobility of the material; gold-plated contacts in earphone jacks and integrated circuit connectors in smartphones, laptops, sensors, and other electronic circuits where several parts of electrical and electronic components are connected are some examples. Among the other inert metals, platinum group metals are the mostly used in integrated circuits (platinum silicide ohmic and Schottky contacts), displays, sensors (MOSFETs and thermocouples) as they have high temperature coefficient or long-term stability. Figure 16.3 shows the quantitative analysis of precious and semi-precious metals along with other common e-waste components [6]. Special treatment

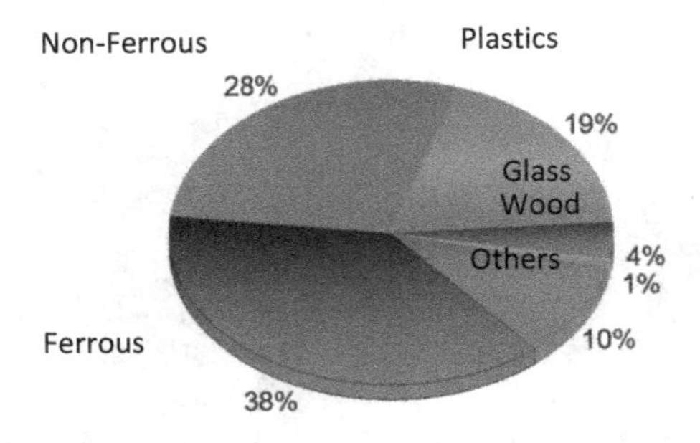

FIGURE 16.3 Approximate quantification of material composition in e-waste. (Adapted from Ref. [6].)

of e-waste should be considered to prevent wasting valuable materials and rare elements. In general, instead of mining for precious metals such as gold and palladium from ores, these metals can also be sourced more effectively from urban mines, e-waste. Urban mines provide secondary resources for refurbishment, remanufacture, and recycling. These resources can be appropriately processed to extract secondary resources for numerous other applications [6].

16.3 E-WASTE PROCESSING TECHNIQUES

E-waste processing is an important process to ensure the environment safe from harmful chemicals and toxic metals that are being used in the electrical and electronic industry. One way to process e-waste in developing countries is dismantling the different instruments and breaking them into smaller-sized components. For example, computers, TVs, mobile phones, etc., are broken down into frames, circuit boards, batteries, and plastic casings. Mostly, the above-mentioned activities are done by hand through automatic shredding of equipment. However, this process is mostly followed in developing countries. But this process increases the exposure of workers to toxic chemicals. Humans are also able to effectively dismantle the appliances into various parts, which cannot be done through automation. However, such manual handling must be conducted under strict health and hygiene standards.

Developed and developing nations process e-waste differently; for instance, the calorific content of the waste is converted into electricity in developed nations, whereas e-waste is recycled manually from the garbage and waste is reduced in developing countries. However, e-waste processing can be categorized into the following techniques.

16.3.1 Landfill

It refers to the burial of e-waste in a dedicated space in the landmass. Landfills are one of the most traditional, inexpensive, and oldest methods of waste management. However, the toxic heavy metals such as cadmium, mercury, and lead in EE components possess leaching properties, which may end up in the soil and groundwater contamination entering the food chain of living organisms and thus posing a threat to the ecosystem. Moreover, they cause devastating health effects including mental disorders, tumors, and cancers [7]. Among the EE components, Personal Computers (PCs) and CRTs in televisions contribute to the significant concentration of lead in the e-waste stream. A study by the US Environmental Protection Agency using toxicity characteristic leaching procedure (TCLP) estimated 18.5 mg/L of lead leaches from CRTs of televisions and monitors, which exceeds the prescribed limit of 5.0 mg/L [8]. Also, an alarming level of lead (150–500 mg/L) was observed in TCLP extracts from the printed wire boards such as disk drives, motherboards, and power supply [9]. A 2-year study to identify the effect of toxic metals leaching with a combination of solid and e-wastes showed that the lead can readily leach from e-waste and is strongly absorbed by the solid waste around it [10]. However, efforts to divert e-waste ending in landfills are necessary to protect the environment in the long run. Similarly, related research on the effect of brominated flame retardants and plastics from e-wastes indicated the exploration of chemical degradation kinetics of different constituents of e-waste and the adoption of policies for their safe disposal [11].

16.3.2 Thermochemical Combustion Techniques

The production of energy through thermal treatment of various constituents of e-waste has three broad pathways (Figure 16.4).

- Incineration – combustion of e-wastes at high temperature
- Pyrolysis – thermal decomposition at inert atmosphere
- Gasification – pyrolysis and subsequent oxidation in an oxygen-rich environment.

A comparison of the working conditions of the above-mentioned thermochemical combustion techniques is given in Table 16.1.

Incineration: It is the process of burning the e-waste in open land or smelters and refineries. Incineration is a controlled process and involves the complete combustion of e-waste at 900 °C–1,000 °C [12]. Incineration significantly reduces the volume of e-waste and converts hazardous organic waste into less hazardous compounds, leading to detoxification and sterilization. Yet, the demand for cleaning the incinerators using flue gas and emission of residues after combustion contributes to air pollution. The concentration of heavy metals is higher in e-wastes compared to other solid and medical wastes. This diffusion causes health effects to the workers at incinerators

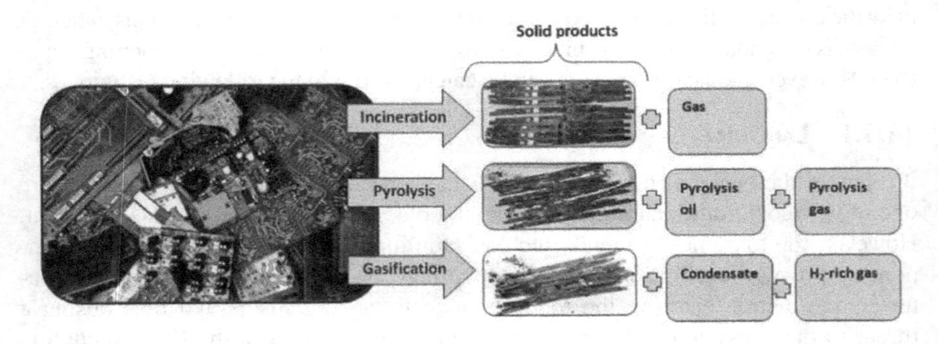

FIGURE 16.4 Schematic illustration of energy generation from e-wastes through thermochemical techniques. (Adapted with permission from Ref. [24].)

TABLE 16.1
Comparison of Thermochemical Conversion of E-Waste to Energy (Adapted from Ref. [9])

Technique	Temperature (°C)	Pressure (MPa)	Catalyst	Drying
Incineration	700–1,400	>0.1	Not required	Not essential, but may help
Pyrolysis	380–530	0.1–0.5	Not required	Necessary
Gasification	500–1,300	>0.1	Not essential, but may help	Necessary

and people living in the vicinity through inhalation, ingestion, and dermal contact. These claims were substantiated by the revelation of phosphate flame retarders in the urinary metabolites of workers in incinerators and significantly correlated with the occupational exposure time [13,14].

Pyrolysis and gasification: Pyrolysis refers to the thermal decomposition of e-waste in an inert atmosphere, i.e., with little or no oxygen [15]. Meanwhile, gasification involves pyrolysis and subsequently partial oxidation in a well-controlled oxidizing environment [16]. In general, e-waste is treated at high temperatures and pressures. The waste is converted to solid, liquid, and non-condensable gases. The solids are refined to various forms of carbon, and the liquid is converted to a high-quality oil which can be used as an energy source in combustion engines for power and heat generation [15]. However, the quality of the oil depends on the temperature at which the combustion is performed. The pyrolyzed oil from plastics used in EE components at different temperatures exhibited its capacity to drive an internal combustion engine with better performance and emission in comparison with mineral diesel [17]. In addition, organic and metallic fractions from pyrolysis residue can be easily separated, which facilitates material recovery and source for energy generation [18]. Moreover, thermochemical decomposition methods limit the diffusion of toxic heavy metals into the atmosphere due to their lower operating temperature in comparison with the incinerators [19]. However, a trade-off between the operating temperature and toxic metals diffusion should be maintained for higher pyrolysis conversion rates [20]. Yet, some studies have shown that pyrolysis oil consists of aromatic compounds due to the combustion of flame retardants present in e-waste. Hence, additional steps are necessary to upgrade the oil, including catalytic hydrothermal treatment, shock cooling steps, and mechanochemical treatments, resulting in the gasification process and production of hydrogen energy [21,22]. Hence, plastics from e-waste can be used to source hydrogen production and recover precious metals to be used as electrodes for fuel cells [23].

16.4 HYDROGEN ENERGY PRODUCTION FROM E-WASTES

Electronic wastes are a rising issue, and they create a business opportunity with increasing significance. A huge volume of toxic and expensive resources of e-waste is generated. The question is how effectively one can recycle or use it in another way to produce energy. One such process involved is pyrolysis and gasification. As mentioned earlier, e-waste contains several components and some among them are metals such as Al, Cu, Fe, Pb, In, and Zn as well as noble metals such as Ag, Au, Pt, and Pd. A process involving energy generation, simultaneous reduction of e-waste, and potential material recovery is a way for natural energy resources relief, energy conservation, and waste management. The e-waste contains nearly 40% of metals; a typical metal scrap consists of copper (20%), iron (8%), tin (4%), nickel (2%), lead (2%), zinc (1%), silver (0.2%), gold (0.1%), and palladium (0.005%) [6]. One way to recover precious metals from e-waste is through high-temperature processes such as pyrometallurgical processing, hydrometallurgical processing, and biometallurgical processing [25]. A system is developed by a team of researchers from the University of the Basque Country to produce hydrogen from e-waste (discarded electronic boards). The discarded plastic is treated not as a waste product, but is used in the gasification process. By treating the components with steam, the metals act as a catalyst and gaseous hydrogen is released [23]. Hydrogen is one of the most renewable energies among several and is one of the future energy sources to the globe.

Why hydrogen as a fuel: Hydrogen is a clean fuel that can be used in fuel cells to generate electricity or power and heat. A fuel cell produces only water, electricity, and heat. Moreover, the highly conductive precious metals recovered from e-waste can be used as electrode materials in fuel cells [23]. The use of hydrogen fuel cells potentially reduces the greenhouse gas emissions and shows high efficiency in many applications. The potential uses of fuel cells are in all sectors such as transportation, commercial, industrial, and residential. Hydrogen and fuel cells can provide energy for use in diverse applications, including distributed or combined heat-and-power; backup power; systems for storing and enabling renewable energy; portable power; auxiliary power for trucks, aircraft, rail, and ships; specialty vehicles such as forklifts; and passenger and freight vehicles including cars, trucks, and buses [26].

Several methods have been used to produce hydrogen as follows:

- Natural gas reforming
- Electrolytic process
- Solar-driven water splitting
- Biological processing.

16.4.1 Natural Gas Reforming

Today, it is an essential technology to produce hydrogen in industrial plants and has a larger pipeline deliverance structure. The thermal processes, including partial oxidation and steam methane reformation process, are used to produce hydrogen from natural gas.

Steam methane reformation: The steam methane reforming has explored the hydrogen production at high temperatures in the range of 700 °C–1,000 °C. The methane sources react with high-temperature steam in the presence of a catalyst to generate hydrogen and other by-products such as carbon monoxide and carbon dioxide. In this endothermic process, heat energy is provided to progress the reaction. The water–gas shift process is carried out by the reaction of steam with carbon monoxide in the presence of a catalyst to generate hydrogen and carbon dioxide. The pressure swing adsorption process is the final stage of pure hydrogen production, where carbon dioxide and other impurities are separated [27]. The noble metals recovered from e-waste are referred to be used as catalytic materials, and metal wastes are used for pipeline construction.

Partial oxidation: The natural gas that reacts with an inadequate level of oxygen entirely oxidizes the methane to carbon monoxide and hydrogen. Accordingly, the water–gas shift process is carried out by the reaction of carbon with water to generate more hydrogen and carbon dioxide. In this exothermic process, heat energy is liberated as the reaction progresses in a shorter time in a smaller reactor [28].

Therefore, hydrogen production of the partial oxidation process is more efficient than the steam reforming process. Natural gas reforming is cost-effective and provides hydrogen fuel for electronic vehicles and other applications. The production of hydrogen gas is compared with nuclear, renewable, other low-carbon, and domestic energy resources, and coal. The toxic emissions of natural gas reforming are lower for an internal combustion engine.

16.4.2 Electrolytic Process

Electrolysis is a hopeful preference for splitting water into hydrogen and oxygen by using electricity in a smaller unit called an electrolyzer. The electrolytic system consists of an anode and cathode compartment. The functions of the electrolyzer differ slightly with the choice of electrolyte material used.

Polymer membrane electrolyzer: The polymer membrane consists of solid electrolytic plastic matter and operates at the temperature of 70 °C–90 °C. In the anodic compartment, water is split into oxygen and hydrogen ions. In this process, hydrogen ions move through the polymer membrane to reach the cathode, while electron reaches the external circuits. The cathode reactions organize the combination of hydrogen ions with electrons [29].

Alkaline electrolyzers: Alkaline electrolyzer carries the hydroxide ions toward the anode through an electrolyte, while hydrogen gas was produced at the cathode at an operating temperature of 100 °C–150 °C. The hydroxide salt of sodium or potassium was used as an electrolyte. Recently, the solid alkaline membrane has been used as an electrolyte for promising hydrogen gas generation on a laboratory scale [30].

Solid oxide electrolyzers: The ceramic material behaves as an electrolyte at a high temperature of 700–800 °C, and it specifically conducts oxygen ions toward anode to support hydrogen production. The cathode reactions organize the combination of water with electron to produce hydrogen gas and oxygen ions. The negatively charged oxygen ion reaches the anode through ceramic membranes to produce oxygen gas, and the electron moves to the external circuits. Hydrogen gas generation by this cyclic process operates with a low consumption of electricity in the electrolytic process [31].

16.4.3 Solar-Driven Water Splitting

The solar-driven water splitting was performed using semiconductor materials and sunlight. In this photo-electrochemical reaction, sunlight was directly used to split water into hydrogen and oxygen gases. The semiconducting electrolyte arrangement was similar to that of a photovoltaic solar cell to produce electricity, but in solar-driven water splitting, the electrolyte was immersed in water to generate hydrogen gas [26]. The photo-electrochemical reactor was assembled with a panel of an electrode, and this system has broadly been studied due to its advantage over slurry-based systems and challenges. To date, electronic wastes have been examined to substitute electrode systems. The solar panels consist of gallium, cadmium telluride, copper, and lead, which are extracted individually. The separation and recovery of these elements for solar panels involves tedious processes. Auspiciously, recycling tools have offered a maximum recovery efficiency. The solar panel materials are recycled and reused at the end of their life.

16.5 E-WASTE AS AN ALTERNATIVE TO THE CONCRETE MIXTURE FOR BUILDING CONSTRUCTION

As e-waste can be used for energy production, it can also be used for saving energy by avoiding mining of mountains and rivers in the search for building construction materials. Moreover, as the urban infrastructure is expanding along with the population growth, the availability of raw materials is depleting and alternate sources are necessary. The utilization of e-waste plastics and glasses in various forms such as concrete, bricks, glass ceramics, and cement mortar solves the dual purpose of e-waste management and increasing the structural stability of the buildings. This approach also minimizes the economic and environmental implications. For instance, the glasses in e-waste that are used as displays possess low water absorption capability and considerable intrinsic strength. Similarly, plastics in e-waste can be used as aggregate in a concrete mixture, which tends to increase the tensile and structural strength.

16.5.1 E-Waste in Concrete and Cement Pastes

Concrete mixtures usually consist of 1:2:3 parts of cement, sand, and gravel, respectively. Among them, sand and gravel or powdered stones are known as aggregates and constitute about 70% of the concrete mix. The aggregates are categorized into fine and coarse aggregates, where e-waste finds its application. Partial replacement of coarse aggregate with e-waste plastics tends to increase the quality of concrete in terms of compressive and tensile strengths [32]. The compressive strength is the ability of the concrete to resist the amount of load which tends to compress in contrast to tensile strength where the concrete can withstand the load that tends to elongate the mixture. However, it is necessary to ensure the following properties of e-waste-replaced concrete, which are vital for building construction materials [33].

Workability and setting time: The ability of mixed concrete essentially put into operation such as they surround the reinforcement and other structures in a beam are called workabilities. The time required for the concrete made of e-waste to harden and attain the shape of the mold is settling time.

Density: It is a measure of concrete's solidity. The density of the concrete varies depending on numerous factors including the concentration of e-waste, cement–water ratio, and the type of cement. A higher or lower density of concrete can be modified by the process of mixing the aggregates for specified applications such as urban infrastructures and residential homes.

Leaching: Leaching is the loss of essential materials from concrete in the long run due to contact with water. Usually, calcium will gradually elude from the structures. However, the usage of e-waste causes toxic metals to leach.

Alkali–silica reaction (ASR): It is a swelling action that occurs due to the reaction between highly alkaline cement paste and silica found commonly in all aggregates. The reaction results in a gel, and it expands when it absorbs sufficient moisture due to the internal pressure. And the glasses in e-waste are rich in silica. ASR is also called concrete cancer.

Shrinkage: Shrinkage is the decrease in the volume of concrete material resulting from the changes in moisture content or chemical reaction. It is classified into plastic, drying, autogenous, and carbonation shrinkages based on what causes it.

Radiation shielding: It is the property of the concrete to shield alpha, beta, gamma, and X-rays. Concrete with high density and large water content tends to reduce the rays penetrating the structure.

Excessive mining of river sand and other natural resources for the synthesis of concrete causes serious environmental problems. Hence, glasses from e-waste can be replaced as they possess a similar chemical structure to the fine aggregates used in a concrete mixture. An increase in fluidity was reported to cause better workability and molding based on the cast. This was attributed to the smoothness of the glass surface and lower water absorption capability [34]. However, the absorption capability increases when the glass particle sizes are much finer due to the trapping of atmospheric air by the large surface area [3]. Moreover, optimal glass sizes showed improved workability rather than smaller sizes. Even though the replacement of aggregates with e-waste glasses exhibited better properties, flexural strength was not significant compared to the traditional mixture owing to the smooth surface of the glass and hence weak bonding with the cement. Yet, finer particle size increases the flexural strength. However, an increase in glass size increased the compressive strength due to the uniform particle size distribution [34]. Similarly, ASR values increase with the increase in particle size and lead leaching decreases with an increase in particle size [35].

Similarly, the volume of e-waste glasses as a replacement in aggregates showed different concrete properties. An increase in workability and compressive strength was observed with an increase in the volume of e-waste glasses, which is attributed to the acceleration in cement hydration [36]. A similar pattern was observed for drying shrinkage owing to a reduction in the porosity of the concrete [37]. The higher the volume of e-waste glasses, the higher the possibility of lead leaching. Yet, studies have reported leaching values less than the prescribed limits [38]. Moreover, casting methods and aggregate-to-cement ratio also influence the levels of lead leaching from the concrete [39]. Hence, a trade-off between concrete performance and the particle size, volume, casting methods, and other factors of the e-waste glass is necessary to synthesize a desirable concrete aggregate.

16.5.2 E-Waste in Bricks

E-waste glasses are partially replaced with clay and cement or concrete to manufacture bricks used as building materials as they possess large quantities of silica similar to the composition of sand. The manufacturing processes usually involve the preparation of clay/cement with a mixture of e-waste glass and molding it to the desired shape. After molding, the bricks are dried to eliminate the resident moisture and subsequently burnt at 1,100 °C to gain strength and hardness. Lee et al. utilized CRT glasses as an ingredient for clay bricks with 40% replacement and 2% for cement bricks. They found that the bending strength, dimension deformation, and water absorption rate of the bricks are better compared to the bricks manufactured with traditional ingredients [40]. Similarly, these e-waste glasses can be used as a partial replacement for other building construction materials including glass tiles and ceramic glazes. These replacements not only improve the performance of the construction material in all discussed aspects, but also tend to be environmentally friendly by eliminating the need to excavate river sand and by reducing the emission of greenhouse gas, and economic by reducing the production cost as well as the management of e-waste effectively without dumping in a wasteland.

16.6 CONCLUSIONS

Economic growth demands the utilization of electronic goods. The rapid technological advancements for improving standards of life have made a consequent surge in the volume of e-waste globally. Management of e-waste without environmental damage is a serious concern. On the other hand, emission of greenhouse gases from the conventional energy production techniques and mining of construction materials owing to the large-scale infrastructural construction lead to the exploitation of depleting natural resources. Both these scenarios possess a serious threat to the environment and hence drive the research community to explore the utilization of e-wastes for hydrogen energy production and in building construction materials. These approaches synergistically reduce harm to the environment and create a potential for a circular economy. However, these technologies are still in their infancy on a laboratory scale and further understanding is required for the realization of energy production/saving. The efficiency of energy conversion, the performance of replacement of precious metals from e-waste as electrodes and catalysts, adverse effects in terms of secondary pollution, the performance of e-waste materials as construction materials, and the optimal percentage of e-waste mixture in conventional concrete mixture require further exploration. Moreover, receptive government policies to encourage the recycling of e-waste for alternative fuel production applications and building construction materials would produce and save energy, respectively.

REFERENCES

1. Directive 2012/19/Eu of the European Parliament and of the Council. Available online: https://eur-lex.europa.eu/legal ontent/EN/TXT/?uri=CELEX%3A32012L0019 (accessed on 20 May 2021).
2. Balakrishnan Ramesh B, Anand Kuber P, Chiya Ahmed B (2007) Electrical and electronic waste: A global environmental problem. *Waste Management & Research* 25: 307–318.
3. Yu J, Williams E, Ju M, Yang Y (2010) Forecasting global generation of obsolete personal computers. *Environmental Science & Technology* 44: 3232–3237.
4. Tue NM, Takahashi S, Subramanian A, Sakai S, Tanabe S (2013) Environmental contamination and human exposure to dioxin-related compounds in e-waste recycling sites of developing countries. *Environmental Science: Processes* 15: 1326.
5. Ghosh M, Sur D, Basu S, Banerjee PS (2018) Metallic Materials From E-Waste. Reference Module in Materials Science and Materials Engineering.
6. Ghosh M, Sur D, Basu S, Banerjee PS (2020) Metallic materials from e-waste. *Encyclopedia of Renewable and Sustainable Materials* 1: 438–455.

7. Premalatha M, Abbasi T, Abbasi SA (2014) The generation, impact, and management of e-waste: State of the art. *Critical Reviews in Environmental Science and Technology* 44: 1577–1678.
8. Musson SE, Jang Y-C, Townsend TG, Chung I-H (2000) Characterization of lead leachability from cathode ray tubes using the toxicity characteristic leaching procedure. *Environmental Science & Technology* 34: 4376–4381.
9. Li Y, Richardson JB, Walker AK, Yuan P-C (2006) TCLP heavy metal leaching of personal computer components. *Journal of Environmental Engineering* 132: 497–504.
10. Li Y, Richardson JB, Mark Bricka R, Niu X, Yang H, Li L, Jimenez A (2009) Leaching of heavy metals from E-waste in simulated landfill columns. *Waste Management* 29: 2147–2150.
11. Danon-Schaffer MN, Mahecha-Botero A, Grace JR, Ikonomou M (2013) Mass balance evaluation of polybrominated diphenyl ethers in landfill leachate and potential for transfer from e-waste. *Science of the Total Environment* 461–462: 290–301.
12. Jayapradha A (2015) Occupational health hazards related to informal recycling of E-waste in India: An overview. *Scenarios for Water* 19: 61–65.
13. Yan X, Zheng X, Wang M, Zheng J, Xu R, Zhuang X, Lin Y, Ren M (2018) Urinary metabolites of phosphate flame retardants in workers occupied with e-waste recycling and incineration. *Chemosphere* 200: 569–575.
14. Li T-Y, Zhou J-F, Wu C-C, Bao L-J, Shi L, Zeng EY (2018) Characteristics of polybrominated diphenyl ethers released from thermal treatment and open burning of e-waste. *Environmental Science & Technology* 52: 4650–4657.
15. Mandal S, Kunhikrishnan A, Bolan NS, Wijesekara H, Naidu R (2016) Application of biochar produced from biowaste materials for environmental protection and sustainable agriculture production. *Environmental Materials and Waste*: 73–89.
16. Kang JJ, Lee JS, Yang WS, Park SW, Alam MT, Back SK, Choi HS, Seo YC, Yun YS, Gu JH, Saravanakumar A, Kumar, KV (2016) A study on environmental assessment of residue from gasification of polyurethane waste in e-waste recycling process. *Procedia Environmental Sciences* 35: 639–642.
17. Kalargaris I, Tian G, Gu S (2017) The utilisation of oils produced from plastic waste at different pyrolysis temperatures in a DI diesel engine. *Energy* 131: 179–185.
18. Estrada-Ruiz RH, Flores-Campos R, Gámez-Altamirano HA, Velarde-Sánchez EJ (2016) Separation of the metallic and non-metallic fraction from printed circuit boards employing green technology. *Journal of Hazardous Materials* 311: 91–99.
19. Dimitrakakis E, Janz A, Bilitewski B, Gidarakos E (2009) Small WEEE: Determining recyclables and hazardous substances in plastics. *Journal of Hazardous Materials* 161: 913–919.
20. Kantarelis E, Yang W, Blasiak W, Forsgren C, Zabaniotou A (2011) Thermochemical treatment of E-waste from small household appliances using highly pre-heated nitrogen-thermogravimetric investigation and pyrolysis kinetics. *Applied Energy* 88: 922–929.
21. Vasile C, Brebu MA, Totolin M, Yanik J, Karayildirim T, Darie H (2008) Feedstock recycling from the printed circuit boards of used computers. *Energy & Fuels* 22: 1658–1665.
22. Tongamp W, Zhang Q, Shoko M, Saito F (2009) Generation of hydrogen from polyvinyl chloride by milling and heating with CaO and $Ni(OH)_2$. *Journal of Hazardous Materials* 167: 1002–1006.
23. Gangadharan P, Nambi IM, Senthilnathan J (2015) Liquid crystal polaroid glass electrode from e-waste for synchronized removal/recovery of Cr^{+6} from wastewater by microbial fuel cell. *Bioresource Technology* 195: 96–101.
24. Gurgul A, Szczepaniak W, Zablocka-Malicka M (2018) Incineration and pyrolysis vs. steam gasification of electronic waste. *Science of the Total Environment* 15: 1119–1124.

25. Flandinet L, Tedjar F, Ghetta V, Fouletier J (2012) Metals recovering from waste printed circuit boards (WPCBs) using molten salts. *Journal of Hazardous Materials* 213–214: 485–490.
26. Ganesh V, Pandikumar A, Alizadeh M, Kalidoss R, Baskar K (2020) Rational design and fabrication of surface tailored low dimensional Indium Gallium Nitride for photoelectrochemical water cleavage. *International Journal of Hydrogen Energy* 45: 8198–8222.
27. Mosinska M, Szynkowska MI, Mierczynski P (2020) Oxy-steam reforming of natural gas on Ni catalysts—A minireview. *Catalysts* 10: 896.
28. Qi J, Liao M, Wang C, Jiang Z, Chen Y, Liang B, Song Q (2020) Hydrogen production via catalytic propane partial oxidation over $Ce_{1-x}M_xNiO_{3-\lambda}$ (M=Al, Ti and Ca) towards solid oxide fuel cell (SOFC) applications. *International Journal of Hydrogen Energy* 45: 8941–8954.
29. Sapountzi FM, Orlova ED, Sousa JPS, Salonen LM, Lebedev OI, Zafeiropoulos G, Tsampas MN, Niemantsverdriet HJW, Kolen'ko YV (2020) FeP nanocatalyst with preferential [010] orientation boosts the hydrogen evolution reaction in polymer-electrolyte membrane electrolyzer. *Energy Fuels* 34: 6423–6429.
30. Bashir SM, Nadeem MA, Al-Oufi M, Al-Hakami M, Isimjan T T, Idriss H (2020) Sixteen percent solar-to-hydrogen efficiency using a power-matched alkaline electrolyzer and a high concentrated solar cell: Effect of operating parameters. *ACS Omega* 5: 10510–10518.
31. Mehrpooya M, Karimi M (2020) Hydrogen production using solid oxide electrolyzer integrated with linear Fresnel collector, Rankine cycle and thermochemical energy storage tank. *Energy Conversion and Management* 224: 113359.
32. Yao Z, Ling TC, Sarker PK, Su W, Liu J, Wu W, Tang J (2018) Recycling difficult-to-treat e-waste cathode-ray-tube glass as construction and building materials: A critical review. *Renewable and Sustainable Energy Reviews* 81: 595–604.
33. Ling TC, Poon CS (2012) Effects of particle size of treated CRT funnel glass on properties of cement mortar. *Materials and Structures* 46: 25–34.
34. Ling TC, Poon CS (2011) Properties of architectural mortar prepared with recycled glass with different particle sizes. *Materials & Designs* 32: 2675–2684.
35. Lee G, Liang TC, Wong YL, Poon CS (2011) Effects of crushed glass cullet sizes, casting methods and pozzolanic materials on ASR of concrete blocks. *Construction and Building Materials* 25: 2611–2618.
36. Park SB, Lee BC, Kim JH (2004) Studies on mechanical properties of concrete containing waste glass aggregate. *Cement and Concrete Research* 34: 2181–2189.
37. Rougelot T, Skoczylas F, Burlion N (2009) Water desorption and shrinkage in mortars and cement pastes: Experimental study and poromechanical model. *Cement and Concrete Research* 39: 36–44.
38. Moncea AM, Badanoiu A, Georgescu M, Stoleriu S (2013) Cementitious composites with glass waste from recycling of cathode ray tubes. *Materials and Structures* 46: 2135–2144.
39. Ling TC, Poon CS (2014) Use of CRT funnel glass in concrete blocks prepared with different aggregate-to-cement ratios. *Green Materials* 2: 43–51.
40. Lee JS, Yoo HM, Park SW, Cho SJ, Seo YC (2016) Recycling of cathode ray tube panel glasses as aggregates of concrete blocks and clay bricks. *Journal of Material Cycles and Waste Management* 18: 552–562.

Part 4

Waste for Advanced Energy Devices

17 Biowaste-Derived Carbon for Solar Cells

Fahmeeda Kausar
Shanghai Jiao Tong University

Jazib Ali
University of Rome Tor Vergata

Ghulam Abbas Ashraf
Zhejiang Normal University

Muhammad Bilal
Huaiyin Institute of Technology

CONTENTS

17.1 INTRODUCTION

At present time, many complications that play a major role in affecting the environment, such as excessive usage of non-renewable supplies, scarcity of energy, and insufficient provision, are encountered [1,2]. Therefore, endeavors have been made to manufacture new and valuable functional compounds from environment-friendly, reproducible, healthy assets to cope with the revolution of environment contamination and global warming [3]. Owing to the outstanding characteristics such as durability, minor toxicity, modifiable architecture, and larger surface area, carbon-rooted compounds are of substantial interest out of the reported modern functional materials with fascinating practical applications [4]. For over 3,000 years, carbon-structured compounds have been prepared and utilized [5]. Carbon-based materials science and engineering has gained remarkable attention after the introduction of

DOI: 10.1201/9781003178354-21

fullerenes in the 1980s [6]. Cosmetics, cell biology, fuel cells, electrodes, carbon fuel cells, catalytic support, storage of gas, and fixation of carbon are the noticeable areas for the prominent implication of compounds with carbon structure [7]. Compounds with various morphologies that are crystalline, amorphous, and carbonaceous, were prepared using numerous methodologies, for example hydrothermal carbonization, electric current with high voltage arc, removal through laser, carbonization, and thermal decomposition [8].

Usually, petroleum coke [9] and coal [10] are considered customary non-renewable resources for synthesizing carbon-rooted compounds. On the other hand, the demand for manufacturing compounds with carbon architecture from reproducible unprocessed materials has increased due to the quick reduction in non-renewable assets. Because of the greater carbon constituent, i.e., 45–55 wt.%, biomass residues such as solid sanitation residues, leftover of consumed eatables, and farming residues can be used to synthesize excellent carbon materials [11]. Charcoal is the material obtained with surplus carbon after the decomposition and consequent elimination of carbohydrates [11]. A major content of biomass residue (70%–80%) is dumped as junk pile [12], and only 20%–30% [12] are reprocessed, burned, and converted into fertilizer from the total content of biomass constituting carbon [13]. Currently, many significant implications such as biomedicine, sorbents, and preservation of hydrogen have been unveiled by carbon-based compounds prepared with biowaste [14]. Nonetheless, for proficient execution and usefulness, further endeavors are needed in the industrialization of carbon compounds obtained from biomass.

For transforming biomass into varied products, numerous methodologies such as hydrothermal carbonization (HTC), liquefaction, gasification, and pyrolysis have been in practice for many decades, which have also been a fascinating subject for scientists [15]. For achieving biochar from biomass, HTC and pyrolysis have been implemented [16]. Generally, charcoal is obtained via slow pyrolysis technique. Slow pyrolysis is performed in the following conditions: 400 °C–800 °C [17] while combusting at 1 °C–30 °C per minute [18] with 1–5 h reaction time [19]. Conversely, ca. >100 °C/min is the applied rate for rapid pyrolysis [19] that is executed to get liquid compound [20]. Water facilitates in decaying of biomass thermochemically called HTC and the expected water and biomass ratios are 5:1–75:1 [21], and applied as prospective implications with a new methodology of transforming biomass to carbon compounds at high temperature pressure [22]. HTC has been divided into two prominent groups depending upon experimental parameters and reaction mechanism. For preparing CNTs, graphite, and activated carbon [23], elevated pressure and temperature (400 °C–800 °C) were employed in high-temperature HTC reactions. Alternatively, carbonaceous compounds have been manufactured at low temperatures (200 °C–300 °C) with diverse surface characteristics and morphology through low-temperature HTC reactions [24]. Thus, the HTC reaction exhibits friendliness with the environment at decreased temperature and exerts numerous chemical modifications [25]. Environment, biology, sensors, catalysis, and energy production are the noticeable realms for applying carbonaceous compounds having varying characteristics and morphology, which are produced following the HTC pathway. In addition, NPs of noble metals such as gold or platinum have been considered important constituents for improving the physical and chemical characteristics of carbon materials. For quite a long time, the avidity to prepare zero-dimensional (0D) LKs by transforming carbonaceous compounds [26] has increased enormously for investigating the edge effects and quantum detention. A single-layered micrometric sheet of carbon atoms decorated hexagonally is graphene. Quantum dots could be acquired from these sheets when their size would be reduced to 20 nm. Therefore, 1–20 nm sized graphene is given the name of graphene quantum dots (GQDs). Bandgap makes the distinction between graphene quantum dots and graphene. Because of edge factors, GQDs produce bandgap; however, graphene exhibits no bandgap. These GQDs show good crystallinity and are constituted of carbons with sp2 hybridization [27]. Armchair brinks, zigzag pattern, smoothness quality, and enclosure of quantum make GQDs fluorescent. In comparison with semiconductor quantum dots (SQDs), GQDs have been analyzed, estimated, and evaluated with effective edge factors and enclosure of quantum. SQDs are not able to display captivating facts as compared to GQDs based on this characteristic [28].

Additionally, nano-sized (2–10 nm) particles composed of different proportions of turbostratic carbon and graphite having quasi-spherical shape are referred to as carbon quantum dots or simply CDs [29]. These show amorphous nature and hybridization of sp3 carbon atoms. Graphite component and amorphous nature were responsible for the dual behavior at 18.2° and 23.8° (2θ), which was confirmed with XRD spectra. Contrary to GQDs, a lesser quantity of carbon and a higher portion of oxygen constitute CDs [30]. The name CNDs has been assigned to the earlier described compounds because of a greater number of oxygen atoms. Because of the presence of functional groups and exterior deformities of CDs, FL emerges that leads to cognition of quenching in the selective environment [29].

Up till now, graphene has the structure of carbon; that's why GQDs are often contemplated as CDs because only a little difference is there between the two groups. Hence, carbon being a similar agent is supposed to form different groups of quantum dots. However, the cadmium selenide, cadmium sulfide, etc, semiconductor quantum dots are nicely apprehended for quantum confinement effect in comparison with the GQDs and C-dots [28].

Many productive applications of GQDs have been reported, such as LEDs, photovoltaic devices, supercapacitors, batteries, and bioimaging due to dispersion in many solvents, non-poisonous nature, environment-friendliness, and proficient light-emitting capacity [31]. Thermal remedy, lithography using an electron beam, microwaves, electrochemical reactions, and hydrothermal synthesis are the diverse approaches used to prepare GQDs [28]. However, graphene, GO and its modified compounds, and CNTs are likely non-producible precursors involved in given strategies. Moreover, the manufacturing of GQDs has been lowered due to less product output because of reaction with acid at elevated temperature and pressure for obtaining durable and good quality smaller-sized pieces while dispersion [32]. Although the output is deficient for practical implications, environment-friendly methodologies (e.g., electrochemical remedy) are applied to carbon structures such as starch, glucose, and C60 graphene compounds [32].

GQDs could be proficiently manufactured applying biomass as a reproducible, durable, cheap, naturally occurring, and organic unprocessed material. However, other raw materials, e.g., $C_6H_{12}O_6$, citric acid, CNTs, carbon fibers, and graphite, can also be used, but biomass is inexpensive and profoundly available for generating GQDs. The precise price has not been mentioned yet in any report. Relating the yield with graphene-rooted exorbitant forerunner material, charcoal from wood, grounded coffee, husks of rice, grassy, and other diverse leaves was utilized as a vast source of biomass to prepare GQDs with proportional effect [33]. Fascinatingly, graphene derivatives demonstrated lower quantum yield (QY) values when correlated with GQDs obtained from biomass [34]. Green synthesis of GQDs is possible using plants leaves, excluding all organic solvents, oxidizers, reducers, and passivating agents [34]. In contrast to carbon nanotubes or effectively arranged thick unicomponent graphene, facile dispersion is favored because of different functional groups constituting biomass. Furthermore, collocated neighboring species arranged in the disorganized pattern have emerged through remediation of biomass via pyrolysis residing nano-sized graphene resembling carbon structure with finely divided smaller pores. Thus, it has been given the name turbostratic carbon [33]. The large-scale preparation of GQDs was thermally possible using disorganized carbon at a lower temperature (90 °C), making low-cost industrial-scale manufacturing easy. Contrarily, GQDs particles were observed with larger size (i.e., 5–7 nm) that are extracted from biomass and hardly tackle the issue of standard size control. Control of size is crucially important on the ground of quantum confinement phenomenon of GQDs, which in turn is related to size. Therefore, scalable biomass transformation to GQDs necessitates additional endeavors. Currently, by way of feasible transformation to GQDs, with greater carbon proportion, durability, and cost-effectiveness, biomass has been engrossing interest being a valuable raw material. However, the various controversies encountered should be handled cautiously for economical industrial manufacturing. Thus, a summary of controllable synthesis of GQDs from biomass is cardinally required in the form of chapter.

The description of this chapter explains reprocessing issues of biowaste and the challenge to fulfill the energy needs in the future through implementing GQDs and carbogenic compounds obtained from biomass conversion in diverse potential appliances, e.g., energy reposition and transformation. The preparation of GQDs using activated carbon (AC) formed after the conversion of biowaste follows various pathways. Furthermore, energy-transforming appliances, e.g., photovoltaic, were estimated based on GQDs' working.

17.2 BRIEF HISTORY

The year 1839 was the beginning of PV history when the origin of electronic currents from light-induced chemical reactions was witnessed by a physicist named Alexandre-Edmond Becquerel [5], and few years later, other researchers also analyzed the same effects in Se, a solid. Nevertheless, the progress of foremost solid-state appliances led to the industrial establishment of primary Si solar cells having 6% efficiency until the late 1940s. The ultimate step for the progression of solar technologies was the establishment of the first Si solar cell, so some potential applications were implied for characterizing the power conversion unit of the PV system. Modules were furnished with silicon cells instead of using them individually. The thirst for research and advancement is continuously expanding and improving the energy accumulator, whereas on the commercial scale, there are different kinds of solar cells used in the present era.

The development of structure and materials always influences the progress of this type of automation having the objective of least cost extreme power. The solar cells contain the basic constituents of a semiconductor for a PV panel to produce desired current and voltage levels, whatever their structure is, either series combination or parallel combination. To increase the desired absorption along with reflection prior to absorption, the power rating of solar cells confirms that the efficiency can be maximized. A current source having a parallel diode was demonstrated as a model for the primary solar cell equivalent circuit. Globally, the problems regarding the reduction in the emission of greenhouse gases and global warming have become a matter of agitation. Additionally, meeting the increase in the energy demand and compensating for the depletion of primary sources that are relevant to fossil fuels are the main goals of present energy schemes. Particularly, solar PV and, generally, renewable energy reservoirs are growing at a fast rate. The photovoltaic effect is the combination of all the solar energy harvesting technologies for direct electric current production. Moreover, nanotechnology, compound semiconductors, thin films, and crystalline forms are the subdivisions of this technology.

17.3 SYNTHESIS TECHNIQUES

GQDs could be synthesized using several starting reagents through the following diverse methodologies, for example decaying biomass matter. "Top-down" and "bottom-up" are the two vital strategies that provide the basis for the above-mentioned methodologies. Carbonization approach along with intermolecular coupling was employed to organic moieties to fabricate carbon domains (sp2) (the bond angle between carbons attached consecutively in two-dimensional (2D) disposition is 120) having a desirable appearance and adjustable dimension, which is a "bottom-up" process. Nevertheless, the technologies described here are intricate and cumbersome. Mechanical shearing and oxidizing methods, e.g., cutting/etching, are indicated as the majorly used techniques in the "top-down" approach. Enhanced production of GQDs is achieved through mechanical shearing or chemical oxidation of the vast carbon (sp2) volume to a narrower range, which constitutes a "top-down" approach [35]. Although such approaches indicate efficient performance and are facile and time-effective, initial reagents have been insufficient for larger volumes of compounds containing double bond (sp2) carbon (C), for instance, C-black, C-fiber, and graphene [35].

17.4 TOP-DOWN APPROACH

Mechanical shearing or oxidative reactions are used to convert 2D sheets or 3D massive compounds of bigger carbon (sp2) domains into zero-dimensional (0D) quantum dots. Disparate synthesis mechanisms are evolved to gain such objectives. Solvothermal/hydrothermal cleaving, chemical etching, electrochemical scissoring are the introduced techniques for oxidizing bigger carbon domains (sp2). Categorization of major oxidizing procedures involves molecular oxidation mainly utilizing sulfuric acid, nitric acid, perchloric acid, and potassium permanganate [36], and free radical oxidation techniques applying electrochemical scissoring and hydrogen peroxide. Currently, numerous scientists furnish comprehensive reports [37] for preparing graphene quantum dots applying electrochemical scissoring [38], chemical etching [39], hydrothermal/solvothermal cleaving [40], ball milling [41], and ultrasonic remediation [42]. Surface configuration, appearance, volume arrangement, and shape of GQDs are greatly influenced by temperature, duration of reaction, capacity to heat, capability to fill, and starting reagents that are well explained in logical researches [28]. The following discussion will describe the implementation of various synthetic methodologies to fabricate GQDs.

17.5 BOTTOM-UP TECHNIQUE

Gathering simple organic constituting units to chemically fabricate GQDs shows the significance of "bottom-up" methods. Carbonization and intermolecular interactions are implemented in the working methodology of such processes. Interestingly, the regulated shape and size of GQDs are the significance of intermolecular coupling. Even though aggregation resulting from π–π coupling and substandard yield were a great hindrance, GQDs could also be fabricated with greater accuracy, tracking the regular schematic procedure involved in organic synthesis [43]. Simple organic configurations have been applied to develop GQDs via the hydrothermal carbonization process [44]. Production of bigger-sized GQDs and CDs is possible with nucleation and condensation of organic small molecules by treating them with a temperature higher than their (melting point) m.p. Several initiating reagents were employed in the approaches, such as organic-based salts [45], grounded coffee leaves [46], 2-hydroxypropane-1,2,3-tricarboxylic acid [47], $C_6H_8O_6$ (ascorbic acid), and propane-1,2,3-triol. Noteworthily, biomass could be applied proficiently as a raw material for the generation of GQDs utilizing the "bottom-up" method, furnishing controllable size. Rice grains dispense an excellent reservoir for tunable yield of GQDs and were introduced by Kalita and co-workers [48]. Different sizes of GQDs, i.e., 2–6.5 nm, were attained when heated for contrasting time duration in a baking pot. Oligomers of $C_6H_{12}O_6$ were generated from thermally decaying starch of powdered rice. GQDs are also achieved through the pyrolysis of the product of nucleation of oligomers at 200°C, as shown in Figure 17.1a. GQDs showed a controllable size arrangement in the consequence of regulating yield via heating duration. Figure 17.1b displays the emulsion-patterned process put into practice for obtaining GQDs, while carbohydrate raw material was supplied from honey [49].

Fascinatingly, initiating reagents provide effective heteroatomic doping features using "bottom-up" methods. To modulate the characteristics utilizing diverse methods, $CO(NH_2)_2$, ethylenediamine, and $HOCH_2CH_2NH_2$ are used a source of GQDs, while carbon is equipped from citric acid to synthesize nitrogen-doped GQDs implementing hydrothermal methods. Nonetheless, the exact working strategies have still not been thoroughly apprehended. Profound studies are compulsory to have a better perception of the reaction mechanism that leads to an enhanced link between configuration and characteristics for potential implication. Previously, bigger-sized GQDs and CDs were procured from nucleation and condensation

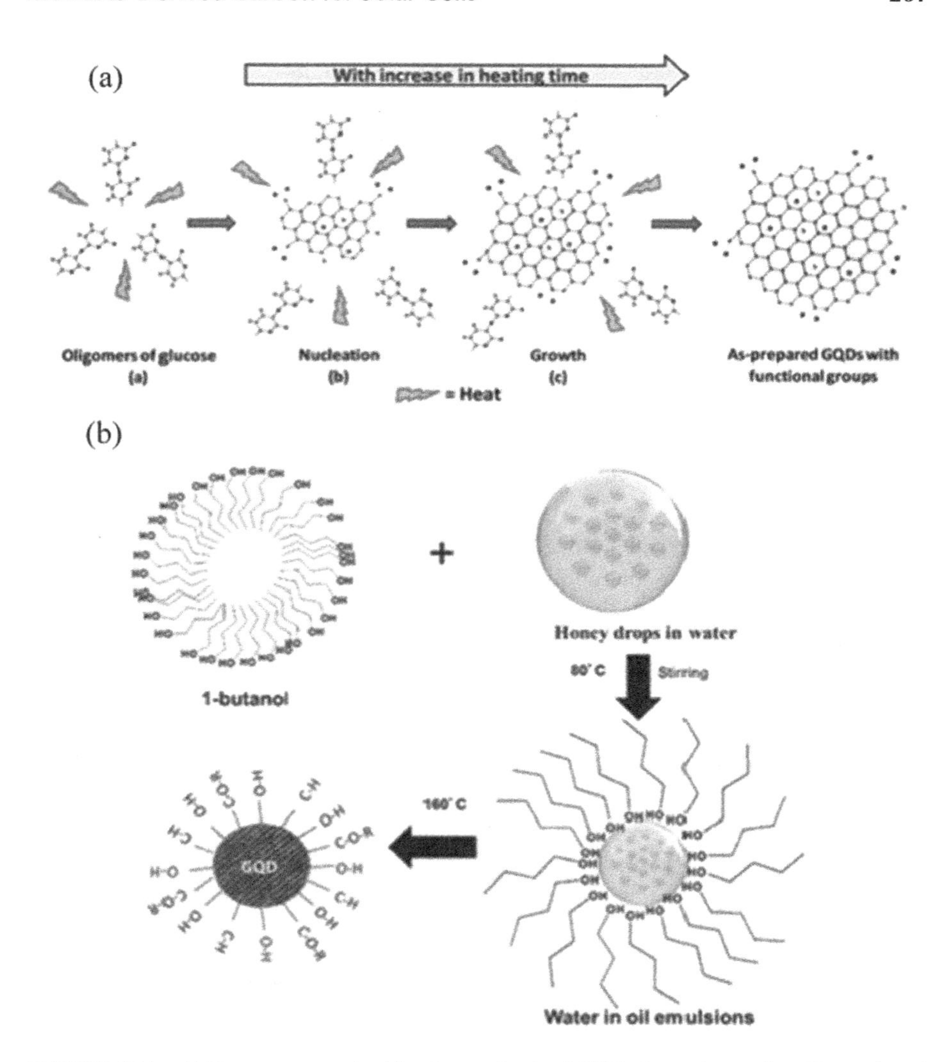

FIGURE 17.1 (a) Bottom-up method for the synthesis of GQDs from rice grains. (Reproduced with permission [48]. Copyright 2016, Royal Society of Chemistry.) (b) Emulsion template carbonization of honey for the synthesis of GQDs. (Reproduced with permission [49]. Copyright 2016, Wiley-VCH.)

resulting from treating organic small molecules at temperatures higher than optimal melting point. Eventually, the "bottom-up" process has been classified into the following major categories: (i) carbonization and (ii) dehydrogenation. Applying solvothermal/hydrothermal constraints, GQDs could also be obtained with carbonizing and dehydrogenating small molecules. However, the given mechanism still has many controversial logics to be debated. Instead of lattice doping, the exterior area gets incorporated with heteroatoms as a consequence of complicated carbonizing and dehydrogenating reactions.

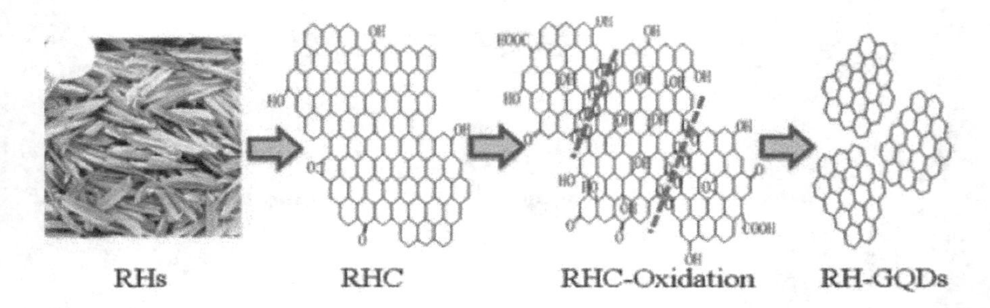

FIGURE 17.2 Schematic diagram of the synthetic strategy of RH-GQDs from RHs. (Reproduced with permission [50]. Copyright 2016, American Chemical Society.)

17.6 TOP-DOWN COLLECTIVE TECHNIQUE

Currently, numerous reacting agents are used to prepare GQDs, such as waste biomass, through collective induction of bottom-up and top-down processes [50]. Figure 17.2 demonstrates the application of the top-down process for the production of carbon (sp2) domains through hydrothermal cleavage of carbon flakes, which were acquired via conversion of decaying biomass with the bottom-up process [50]. A greater amount of GQDs as a product is possible with this method because of many advantages such as controllability, facility, and trouble-free nature. Fascinatingly, by applying this technique, GQDs could be acquired having different characteristics that would be useful in various practical implications. Bioimaging, for instance, has been perceived with vivid blue luminescence when rice husk-based GQDs (RHGQDs) dispersed after exciting them with 365 nm UV light.

17.7 PHOTOVOLTAICS

Miscellaneous instrumental pieces, e.g., catalytic agents, active layers, activating agents, sensitizing agents, and photovoltaics, imply GQDs as opposite/parallel electrodes due to outstanding optical and electrical characteristics. For the fabrication of silicone/GQDs heterojunctional solar cell, Gao and team members applied GQDs as an active layer [51]. Figure 17.3a shows the detachment of electron–hole induced by subduing charge reconnection through the active layer of GQDs. Sheets of graphene oxide showed a photoconversion efficiency (PCE) of 3.99%, while silicone-based cell exhibited 2.26% PCE; on the other hand, GQDs was detected with even higher PCE of 6.6%. Sole zinc oxide nanowire-supported cell evidenced 75% lower current density (J_{sc}) when compared to the solar cell of heterojunctional GQDs/zinc oxide prepared by Dutta and co-workers [52]. However, reduced hole-assembling proficiency has been observed owing to a lesser chance of hole transport layer (HTL) to get in contact with GQDs, resulting in decreased PCE. Tavakoli et al. prepared a hybrid of lead sulfide-QDs/GQDs shell/core by applying the hot injection method as given in Figure 17.3b [53]. Charge bearers could be swiftly extracted with a decrement in trap sites by furnishing imperfect surface inertness of sulfanylidenelead-QDs, which has been achieved through an extremely thin layer of

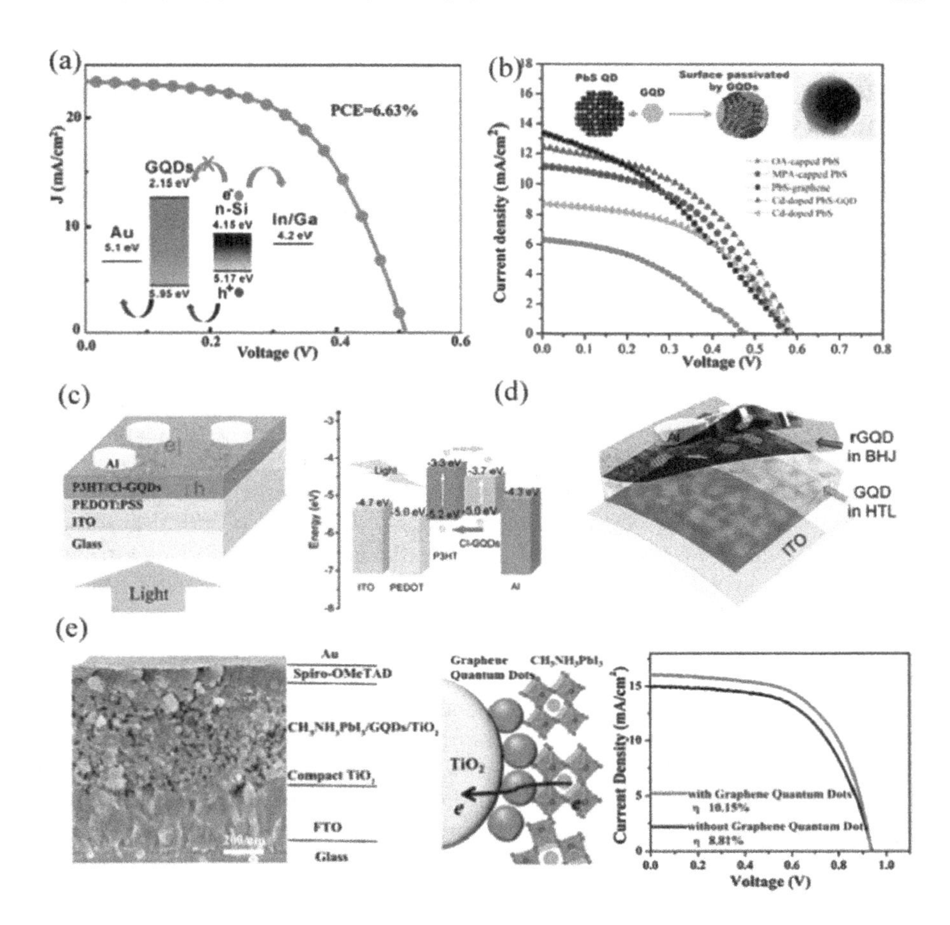

FIGURE 17.3 (a) The J–V curve of a CH3eSI/GQD heterojunction solar cell at Am1.5G. (Reproduced with permission [51]. Copyright 2014, American Chemical Society.) (b) A schematic diagram of PbS QDs surface passivated by GQDs and PbS based solar cells' J–V curves under Am1.5G. (Reproduced with permission [53]. Copyright 2015, American Chemical Society.) (c) Schematic diagram of a photovoltaic device based on GQD-Cl hybrid and its working mechanism. (Reproduced with permission [54]. Copyright 2015, Royal Society of Chemistry.) (d) Illustration of a photovoltaic device based on GQDs in hole transport layer (HTL) and hydrothermally reduced GQDs in the BHJ layer. (Reproduced with permission [57]. Copyright 2015, Springer Nature.) (e) An SEM image of a perovskite solar cell is shown on the left-hand side, the illustration of its working mechanism is in the middle, and the J–V curves of the cell without or with GQD are on right-hand side. (Reproduced with permission [63]. Copyright 2014, American Chemical Society.)

GQDs, contrary to customarily implied capping molecules. Instead of utilizing lead sulfide-QDs with a capping of organic compounds, PbS-QDs/GQDs active-layered solar cell provides an enhanced 0.58 V open voltage, 3.6% PCE, and 13.4 mA cm^2 J_{sc}, as given in Figure 17.3b [53]. Electron transfer and isolation of photo-produced excitons were boosted through suffice generation of the p-n-junctions interior of active layer of P3HT:GQDs. Preferably, GQDs addition enhanced the voltage of an open circuit, i.e., 0.5–0.8 V. 0.67 V V_{oc} and 6.3 mA cm^2 J_{sc} have been estimated using a real device demonstrating 1.28% PCE. Figure 17.3c illustrates the lamination of P3HT with GQDs after chlorine doping, which ameliorated the polymer device by enhancing 30% charge bearer amount [54]. Layer arrangement of ITO/PEDOT:PSS/P3HT:ANI-GQDs/lithium fluoride/aluminum was adopted by Gupta et al. for fabricating solar cell and P3HT impregnated with aniline-modified GQDs [55]. In comparison with an activated layer of P3HT:ANI-graphene-sheet of control device showing 0.65% PCE, this device exhibited a higher PCE of 1.14%. Substandard working was observed with an exceptional extent of diffusing excitons by way of greater phase dissociation encouraged by the uneven surface of B3HT:ANI-graphene owing to large-area sheets of graphene [56]. Diverse oxygen-containing compounds incorporated with GQDs were implemented with an active sheet of PTB7 and PC71BM for fabricating solar appliances by Kim and co-workers [56]. The fill factors (FFs) capability of GQDs was refined via implementing lesser oxygen atoms and their high-quality conduction proficiently detached chare carriers, while GQDs' surface decorated with oxygen groups intensified not only J_{sc}, but also optical absorption property. A PCE of 7.6% was noticed owing to the equilibrium of given attributes. An improved current density (15.6–17.3 mA cm^2) for BHJ solar cells has been detected when the hole transferring layer (PEDOT:PSS) was laminated with GQDs of hydrophilic nature. PEDOT is positively charged, while GQDs carry a negative charge, and a noticeable binding has been observed between them bestowing ITO/PEDOT:PSS a boosted conductivity for charge bearers [57]. PTB7:PC71BMBHJ layer subsumed with hydrophobic rGQDs after hydrothermal reduction treatment, while PEDOT:PSS HTL was integrated with hydrophilic GQDs for apprehending harmonious phenomena, as shown in Figure 17.3d, and attained a remarkably improved PCE of 6.87%. In solar cells, Yan et al. applied GQDs lacking chemical groups for sensitization of titanium oxide photo-anode [44]. Nevertheless, owing to low-quality bonding attraction between titanium oxide and GQDs, dye-sensitized solar cells (DSSCs) showed a twofold higher J_{sc} than that of a cell. Titanium oxide communicated incredibly with GQDs, resulting in noticeably enhanced working when oxygen moieties were incorporated accompanying GQDs [58]. An increase in J_{sc} (9.72–14.07 mA cm^2) and PCE (4.9%–6.1%) was noticed for titanium oxide photo-anodes with a combination of GQDs and N719 being utilized as auxiliary and major sensitizing agents [59]. N_3 Ru-dye accompanying N-GQDs acting as an auxiliary sensitizing agent for DSSC with elevated PCE and J_{sc} values was demonstrated by Mihalache et al. [60]. Titanium oxide was laminated over GQDs for obtaining photoluminescence upconversion and elevated PCE (7.28%–9.2%), and consequently, DSSCs introduced by Lee et al. absorbed greater radiations [61]. To achieve improved PCE from DSSCs, a combination of polypyrrole and GQDs as parallel electrode having fine pores was presented by Chen et al. [62]. Recently, an increase in J_{sc} (17.07–15.34 mA cm^2) and PCE (8.81%–10.15%) has been estimated when $CaTiO_3$/GQDs/titanium oxide arrangement was implemented via furnishing $CaTiO_3$ with an extremely thin layer of GQDs, which is displayed in Figure 17.3e [63]. The photoluminescence of $CaTiO_3$-TiO_2 film was quenched by 75%, which assures an increase in the production of current from photons, covering a visible region that corresponds to proficient electron withdrawal.

The units of the layer utilized, small bandgap (0.3–1.5 eV), greater hole movement, and bipolar properties make phosphorene an interesting candidate for photoelectronic applications [64]. For acquiring a p–n diode demonstrating a PCE of 0.3% for photovoltaic energy preservation, a heterojunctional combination was developed by Deng and co-workers that consisted of multilayered p-type phosphorene and monolayered n-type molybdenum disulfide [65]. Two-layered phosphorene was speculated to improve the PCE of p–n diode by 18% [64]. Buscema and team members synthesized a heterojunctional p–n diode and employed dielectric gates (2 h-BN) for biasing in two distinctive areas and applied a single sheet of phosphorene based on a bipolar property of multilayered phosphorene [66]. Consequently, NIR region was also expected to be covered through photovoltaic effect. Phosphorene was applied to fill the electrical bandgap existing between TMDs and graphene ends. Therefore, phosphorene quantum dots were preferred for administering the "implication gap" of TMD QDs and graphene QDs.

17.8 CONCLUSIONS

Biowaste is a sustainable and ironic means of carbon production. With the increasing consumption of carbon resources, it is essential to recognize and understand the knowledge of carbon chemistry and features that affect their properties. This chapter emphasized the different scientific researches on biowaste-derived GQDs and their application in the field of photovoltaic. The materials derived from biowaste are obtaining growing interest as zero-dimensional materials which exhibit electrochemical, chemical, catalytic, electronic, and optical, etc., properties and advantages compared to 2D counterparts. These tunable properties have sensitivity by size, defects, edge configurations, chemical functionalities, thickness, or heteroatom dopants. These materials exhibit encouraging applications in the field of fuel cells, drug delivery, catalysis, carbon fixation, bioimaging, and gas sensors.

REFERENCES

1. Franco, S., V.R. Mandla, and K.R.M. Rao, Urbanization, energy consumption and emissions in the Indian context: A review. *Renewable and Sustainable Energy Reviews*, 2017. **71**: pp. 898–907.
2. Alam, H. and S. Ramakrishna, A review on the enhancement of figure of merit from bulk to nano-thermoelectric materials. *Nano Energy*, 2013. **2**(2): pp. 190–212.
3. Ali, J., et al., Modalities for conversion of waste to energy—Challenges and perspectives. *Science of the Total Environment*, 2020. **727**: p. 138610.
4. Ye, R., et al., Coal as an abundant source of graphene quantum dots. *Nature Communications*, 2013. **4**(1): pp. 1–7.
5. Suh, W.H., et al., Nanotechnology, nanotoxicology, and neuroscience. *Progress in Neurobiology*, 2009. **87**(3): pp. 133–170.
6. Kroto, H.W., et al., C 60: Buckminsterfullerene. *Nature*, 1985. **318**(6042): pp. 162–163.
7. Titirici, M.M., A. Thomas, and M. Antonietti, Replication and coating of silica templates by hydrothermal carbonization. *Advanced Functional Materials*, 2007. **17**(6): pp. 1010–1018.
8. Su, F., et al., Pt nanoparticles supported on nitrogen-doped porous carbon nanospheres as an electrocatalyst for fuel cells. *Chemistry of Materials*, 2010. **22**(3): pp. 832–839.

9. Kawano, T., et al., Preparation of activated carbon from petroleum coke by KOH chemical activation for adsorption heat pump. *Applied Thermal Engineering*, 2008. **28**(8–9): pp. 865–871.
10. Chingombe, P., B. Saha, and R. Wakeman, Surface modification and characterisation of a coal-based activated carbon. *Carbon*, 2005. **43**(15): pp. 3132–3143.
11. Xie, X. and B. Goodell, Thermal degradation and conversion of plant biomass into high value carbon products. Deterioration and Protection of Sustainable Biomaterials, 2014. ACS Publications: pp. 147–158.
12. Themelis, N.J. and L. Arsova, Calculating tons to composting in the US. *Bio Cycle*, 2015. **56**: p. 27.
13. McKendry, P., Energy production from biomass (part 1): Overview of biomass. *Bioresource Technology*, 2002. **83**(1): pp. 37–46.
14. Schmidt, L.D. and P.J. Dauenhauer, Hybrid routes to biofuels. *Nature*, 2007. **447**(7147): pp. 914–915.
15. Jain, A., R. Balasubramanian, and M. Srinivasan, Hydrothermal conversion of biomass waste to activated carbon with high porosity: A review. *Chemical Engineering Journal*, 2016. **283**: pp. 789–805.
16. Libra, J.A., et al., Hydrothermal carbonization of biomass residuals: A comparative review of the chemistry, processes and applications of wet and dry pyrolysis. *Biofuels*, 2011. **2**(1): pp. 71–106.
17. Williams, P.T. and S. Besler, The influence of temperature and heating rate on the slow pyrolysis of biomass. *Renewable Energy*, 1996. **7**(3): pp. 233–250.
18. Ronsse, F., et al., Production and characterization of slow pyrolysis biochar: Influence of feedstock type and pyrolysis conditions. *Gcb Bioenergy*, 2013. **5**(2): pp. 104–115.
19. Gani, A. and I. Naruse, Effect of cellulose and lignin content on pyrolysis and combustion characteristics for several types of biomass. *Renewable Energy*, 2007. **32**(4): pp. 649–661.
20. Lin, Y.-C., et al., Kinetics and mechanism of cellulose pyrolysis. *The Journal of Physical Chemistry C*, 2009. **113**(46): pp. 20097–20107.
21. Toufiq Reza, M., et al., Hydrothermal carbonization (HTC) of cow manure: Carbon and nitrogen distributions in HTC products. *Environmental Progress & Sustainable Energy*, 2016. **35**(4): pp. 1002–1011.
22. Funke, A. and F. Ziegler, Hydrothermal carbonization of biomass: A summary and discussion of chemical mechanisms for process engineering. *Biofuels, Bioproducts and Biorefining*, 2010. **4**(2): pp. 160–177.
23. Jia, B., L. Gao, and J. Sun, Self-assembly of magnetite beads along multiwalled carbon nanotubes via a simple hydrothermal process. *Carbon*, 2007. **45**(7): pp. 1476–1481.
24. Shen, Y., et al., Hydrothermal carbonization of medical wastes and lignocellulosic biomass for solid fuel production from lab-scale to pilot-scale. *Energy*, 2017. **118**: pp. 312–323.
25. Owsianiak, M., et al., Environmental performance of hydrothermal carbonization of four wet biomass waste streams at industry-relevant scales. *ACS Sustainable Chemistry & Engineering*, 2016. **4**(12): pp. 6783–6791.
26. Nirala, N.R., et al., One step electro-oxidative preparation of graphene quantum dots from wood charcoal as a peroxidase mimetic. *Talanta*, 2017. **173**: pp. 36–43.
27. Jang, M.-H., et al., Origin of extraordinary luminescence shift in graphene quantum dots with varying excitation energy: An experimental evidence of localized sp2 carbon subdomain. *Carbon*, 2017. **118**: pp. 524–530.
28. Wang, X., et al., Quantum dots derived from two-dimensional materials and their applications for catalysis and energy. *Chemical Society Reviews*, 2016. **45**(8): pp. 2239–2262.
29. Georgakilas, V., et al., Broad family of carbon nanoallotropes: Classification, chemistry, and applications of fullerenes, carbon dots, nanotubes, graphene, nanodiamonds, and combined superstructures. *Chemical Reviews*, 2015. **115**(11): pp. 4744–4822.

30. Baker, S.N. and G.A. Baker, Luminescent carbon nanodots: Emergent nanolights. *Angewandte Chemie International Edition*, 2010. **49**(38): pp. 6726–6744.
31. Ananthanarayanan, A., et al., Nitrogen and phosphorus co-doped graphene quantum dots: Synthesis from adenosine triphosphate, optical properties, and cellular imaging. *Nanoscale*, 2015. **7**(17): pp. 8159–8165.
32. Zheng, X.T., et al., Glowing graphene quantum dots and carbon dots: Properties, syntheses, and biological applications. *Small*, 2015. **11**(14): pp. 1620–1636.
33. Suryawanshi, A., et al., Large scale synthesis of graphene quantum dots (GQDs) from waste biomass and their use as an efficient and selective photoluminescence on–off–on probe for Ag+ ions. *Nanoscale*, 2014. **6**(20): pp. 11664–11670.
34. Roy, P., et al., Plant leaf-derived graphene quantum dots and applications for white LEDs. *New Journal of Chemistry*, 2014. **38**(10): pp. 4946–4951.
35. Pan, D., et al., Hydrothermal route for cutting graphene sheets into blue-luminescent graphene quantum dots. *Advanced Materials*, 2010. **22**(6): pp. 734–738.
36. Ritter, K.A. and J.W. Lyding, The influence of edge structure on the electronic properties of graphene quantum dots and nanoribbons. *Nature Materials*, 2009. **8**(3): pp. 235–242.
37. Li, L., et al., Focusing on luminescent graphene quantum dots: Current status and future perspectives. *Nanoscale*, 2013. **5**(10): pp. 4015–4039.
38. Ananthanarayanan, A., et al., Facile synthesis of graphene quantum dots from 3D graphene and their application for Fe^{3+} sensing. *Advanced Functional Materials*, 2014. **24**(20): pp. 3021–3026.
39. Dong, Y., et al., One-step and high yield simultaneous preparation of single-and multi-layer graphene quantum dots from CX-72 carbon black. *Journal of Materials Chemistry*, 2012. **22**(18): pp. 8764–8766.
40. Park, H., et al., Large scale synthesis and light emitting fibers of tailor-made graphene quantum dots. *Scientific Reports*, 2015. **5**(1): pp. 1–9.
41. Wang, L., et al., Carbon quantum dots displaying dual-wavelength photoluminescence and electrochemiluminescence prepared by high-energy ball milling. *Carbon*, 2015. **94**: pp. 472–478.
42. Zhuo, S., M. Shao, and S.-T. Lee, Upconversion and downconversion fluorescent graphene quantum dots: Ultrasonic preparation and photocatalysis. *ACS Nano*, 2012. **6**(2): pp. 1059–1064.
43. Yan, X., et al., Large, solution-processable graphene quantum dots as light absorbers for photovoltaics. *Nano Letters*, 2010. **10**(5): pp. 1869–1873.
44. Wang, L., et al., Gram-scale synthesis of single-crystalline graphene quantum dots with superior optical properties. *Nature Communications*, 2014. **5**(1): pp. 1–9.
45. Bourlinos, A.B., et al., Surface functionalized carbogenic quantum dots. *Small*, 2008. **4**(4): pp. 455–458.
46. Deng, Y., et al., Long lifetime pure organic phosphorescence based on water soluble carbon dots. *Chemical Communications*, 2013. **49**(51): pp. 5751–5753.
47. Dong, Y., et al., Blue luminescent graphene quantum dots and graphene oxide prepared by tuning the carbonization degree of citric acid. *Carbon*, 2012. **50**(12): pp. 4738–4743.
48. Kalita, H., et al., Efficient synthesis of rice based graphene quantum dots and their fluorescent properties. *RSC Advances*, 2016. **6**(28): pp. 23518–23524.
49. Mahesh, S., et al., Simple and cost-effective synthesis of fluorescent graphene quantum dots from honey: Application as stable security ink and white-light emission. *Particle & Particle Systems Characterization*, 2016. **33**(2): pp. 70–74.
50. Wang, Z., et al., Large-scale and controllable synthesis of graphene quantum dots from rice husk biomass: A comprehensive utilization strategy. *ACS Applied Materials & Interfaces*, 2016. **8**(2): pp. 1434–1439.
51. Gao, P., et al., Crystalline Si/graphene quantum dots heterojunction solar cells. *The Journal of Physical Chemistry C*, 2014. **118**(10): pp. 5164–5171.

52. Dutta, M., et al., ZnO/graphene quantum dot solid-state solar cell. *The Journal of Physical Chemistry C*, 2012. **116**(38): pp. 20127–20131.
53. Tavakoli, M.M., et al., Quasi core/shell lead sulfide/graphene quantum dots for bulk heterojunction solar cells. *The Journal of Physical Chemistry C*, 2015. **119**(33): pp. 18886–18895.
54. Zhao, J., et al., Fabrication and properties of a high-performance chlorine doped graphene quantum dot based photovoltaic detector. *RSC Advances*, 2015. **5**(37): pp. 29222–29229.
55. Gupta, V., et al., Luminscent graphene quantum dots for organic photovoltaic devices. *Journal of the American Chemical Society*, 2011. **133**(26): pp. 9960–9963.
56. Kim, J.K., et al., Balancing light absorptivity and carrier conductivity of graphene quantum dots for high-efficiency bulk heterojunction solar cells. *Acs Nano*, 2013. **7**(8): pp. 7207–7212.
57. Kim, J.K., et al., Surface-engineered graphene quantum dots incorporated into polymer layers for high performance organic photovoltaics. *Scientific Reports*, 2015. **5**(1): pp. 1–10.
58. Long, R., Understanding the electronic structures of graphene quantum dot physisorption and chemisorption onto the TiO_2 (110) surface: A first-principles calculation. *ChemPhysChem*, 2013. **14**(3): pp. 579–582.
59. Fang, X., et al., Graphene quantum dots optimization of dye-sensitized solar cells. *Electrochimica Acta*, 2014. **137**: pp. 634–638.
60. Mihalache, I., et al., Charge and energy transfer interplay in hybrid sensitized solar cells mediated by graphene quantum dots. *Electrochimica Acta*, 2015. **153**: pp. 306–315.
61. Lee, E., J. Ryu, and J. Jang, Fabrication of graphene quantum dots via size-selective precipitation and their application in upconversion-based DSSCs. *Chemical Communications*, 2013. **49**(85): pp. 9995–9997.
62. Chen, L., et al., Graphene quantum-dot-doped polypyrrole counter electrode for high-performance dye-sensitized solar cells. *ACS Applied Materials & Interfaces*, 2013. **5**(6): pp. 2047–2052.
63. Zhu, Z., et al., Efficiency enhancement of perovskite solar cells through fast electron extraction: The role of graphene quantum dots. *Journal of the American Chemical Society*, 2014. **136**(10): pp. 3760–3763.
64. Dai, J. and X.C. Zeng, Bilayer phosphorene: Effect of stacking order on bandgap and its potential applications in thin-film solar cells. *The Journal of Physical Chemistry Letters*, 2014. **5**(7): pp. 1289–1293.
65. Deng, Y., et al., Black phosphorus–monolayer MoS2 van der Waals Heterojunction p–n Diode. *ACS Nano*, 2014. **8**(8): pp. 8292–8299.
66. Buscema, M., et al., Photovoltaic effect in few-layer black phosphorus PN junctions defined by local electrostatic gating. *Nature Communications*, 2014. **5**(1): p. 4651.

18 Biowastes for Metal-Ion Batteries

C. Nithya
PSGR Krishnammal College for Women

CONTENTS

18.1 INTRODUCTION

Recently, the global market has boomed with electric vehicles and portable electronic devices because of the rapid development of energy storage systems, and further expansion of electric vehicles requires inexpensive, high energy density, and safe energy storage devices. Among the state-of-the-art energy storage devices, alkali metal (Li/Na/K)-ion batteries are highlighted with great merits such as high energy density, long calendar life, and safety. However, the production cost of electrode materials of lithium-ion batteries and their subsequent members is quite high, which directly increases the production cost of these energy storage devices. Hence, there is an urgent need to develop low-cost energy storage devices for the long run without compromising energy and power densities. Recently, there has been an increased awareness about conserving non-renewable energy sources (which releases energy upon burning) by the sustainable development of energy storage devices to adopt green chemistry techniques. With the rapid advancement of technologies, these non-renewable energy sources can be easily converted into electrode materials and incorporated into alkali-ion batteries, solar cells, supercapacitors, and fuel cells, which are considered the most promising alternative opportunity to reduce greenhouse gas emissions. In the meantime, electrode materials are derived from earth-abundant, recyclable, and environmentally friendly materials, which is considered one of the significant strategies for the design and development of green technology.

Biowaste/biomass contains a large amount of biologically produced carbon, hydrogen, nitrogen, oxygen, and sulfur, which greatly enhance the energy density of

DOI: 10.1201/9781003178354-22

alkali-ion batteries. Recently, the fabrication of energy storage devices from this kind of feedstock has been considered a promising green energy technology. Interestingly, carbon materials derived from various biowaste materials are used as electrode materials for lithium/sodium/potassium-ion batteries. For instance, in recent years carbon materials derived from rice husk, corn/wheat straw, olive stones, apple waste, mushrooms, sweet potatoes, algal blooms, soybean residues, coir pith, pistachio shells, lignin, grass, and pine pollen. These biowaste-derived carbon materials not only provide low-cost electrode materials, but also offer sustainable structures for alkali-ion insertion/extraction during cycling. The method of deriving anode materials from biowaste is very important to determine the physical and chemical properties of the materials. In this view, several important techniques have been explored in recent days and reported in the production of heteroatom-doped carbons as well as porous micro-/nanostructures. The important methods of deriving electrode materials from biowastes are pyrolysis, hydrothermal method, air expansion method, and emulsion method. This chapter deals with the recent achievements regarding electrode materials derived from waste biomass and their applicability in alkali-ion batteries. In this contribution, how waste biomass-derived materials (e.g., carbon and carbon/silicon/metal oxide or sulfide composites) with improved properties and specific morphologies are highlighted as well as how these electrode materials solve the interfacial and bulk properties of alkali-ion batteries are discussed. This chapter provides a critical analysis of the influence of various heteroatoms present in biowaste-derived carbon materials toward the electrochemical performance of LIBs and beyond LIBs (SIBs and PIBs). In the end, future perspectives and various possible research directions for further development of biowaste-derived electrode materials for alkali-ion batteries are proposed.

18.2 BIOWASTE-DERIVED CARBONS FOR ALKALI-ION BATTERIES

Graphite is an ideal intercalation anode host for lithium-ion batteries, which provides the theoretical capacity of 372 mAh/g. It is a commercially used anode host for lithium-ion batteries; however, it has certain limitations such as solvent co-intercalation, thermal resistance at high temperature, and safety concerns. This shows that graphite anode couldn't be used to meet the ever-growing demand of high energy and high power density next-generation lithium-ion batteries. Moreover, the lithium-ion storage sites (LiC_6) of graphite have imposed poor rate performance (Li^+ ion diffusion coefficient 10^{-8} cm^2/s), which becomes a drawback of utilizing graphite anode for electric vehicles. The utilization of graphite in sodium-ion batteries (SIBs) is strongly limited owing to the large size of Na^+ ion, and the poor thermodynamic stability of sodium-graphite intercalation compounds leads to the poor electrochemical performance of SIBs. Similarly, the intercalation of large size K^+ ion in graphite anode leads to large volume expansion, which results in sluggish electrode kinetics in potassium-ion batteries (PIBs). Furthermore, natural graphite is considered an abundant anode material for LIBs and is cheaper than other carbon-based materials synthesized in the laboratory. In the viewpoint of economic production of carbon materials, biowaste has widely been used to synthesize carbonaceous materials for alkali-ion batteries owing to its low cost, easy accessibility, and eco-friendliness. These uniquely featured carbon materials are derived from various biowastes/biomass such as corn starch, seashells, peanut dregs/shells, ox horns, bamboo chopsticks, lignin, wheat straw, corn starch, sweet potatoes, and corn stalk, etc. In this view, considerable research interest is explored to synthesize cost-effective, high-performance anode materials for LIBs, thanks to the recent development of various biowaste-derived anode materials and their composites for their active exploration. The following section emphasizes the recent advances in biowaste-derived carbon materials and their composites for advanced LIBs and technologies beyond LIBs (SIBs and PIBs). The structural features of biowaste-derived carbonaceous materials are specifically investigated and examined, such as morphology, nanostructure, unique structural properties, and the influence of heteroatoms toward the electrochemical performance of LIBs/SIBs/PIBs.

18.2.1 Non-doped Carbonaceous Materials

The synthesis of carbon anode materials derived from naturally available biowaste materials is considered as one of the significant strategies for their application in electric and hybrid electric vehicles. Yu's group [1] used corn straw as a low-cost raw material for the preparation of cellulose-derived carbon nanospheres without heteroatom doping via hydrothermal carbonization process and used them as anode material for LIBs. By carefully controlling the reaction parameters such as time, temperature, and heating rate, they have obtained carbon nanospheres with cross-linked structures. These cross-linked nanospheres not only enhance the electrochemical performance of LIBs, but also decrease the structural damage caused by lithium-ion intercalation. Significantly, carbon nanospheres produced by 36 h of hydrothermal carbonization offer excellent cycling stability and outstanding specific capacity of 577 mAh/g at 0.2C after 100 cycles. These cellulose-derived carbon nanospheres not only provide more active sites for lithium-ion intercalation, but also increase the specific surface area, resulting in more capacity for LIBs. Parallel to the work of corn straw-derived carbon, Yuan's group [2] carried out microstructure engineering toward the preparation of porous carbon materials derived from peanut dregs. Peanut dregs contain a rich amount of proteins, carbohydrates, and lipids, and one of the by-products of edible oil processing industry, they are considered as a cheap and easily available raw material to produce highly disordered carbon materials (HDCMs) and graphene-like carbon materials (GLCMs). GLCMs are applied as an anode material for LIBs and deliver 286 mAh/g at a current density of 1,000 mAh/g and can retain over 1,000 charge/discharge cycles. The obtained specific capacity and cycling stability are 6.9 times higher than those of graphite (41.1 mAh/g at 1,000 mA/g). This excellent performance is attributed to the following factors: (i) high degree of graphitization, microstructure engineering toward mesoporosity, and large lattice spacing, which results in favorable lithium-ion intercalation/deintercalation.

Recently, it has been found that hard carbon stands out as a promising anode candidate for rechargeable batteries owing to its distinct structural features such as expanded interlayer spacing, short-range lamellar order, and plentiful micropores. In order to reduce greenhouse gas emissions, the development of electrode materials from low-cost natural waste biomass is considered an eco-friendly approach. In this

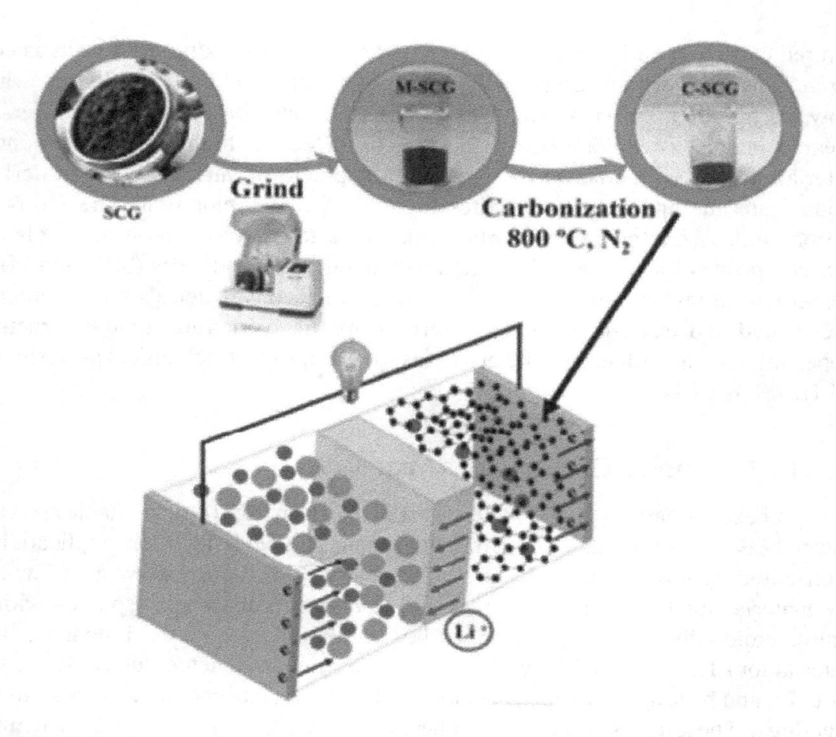

FIGURE 18.1 Spent coffee ground (SCG) is grounded in ball mill (350 rpm for 10 min) and carbonized at 800°C under N_2 atmosphere for 2 h and applied as an anode material for lithium-ion batteries. (Reproduced with permission from Ref. [3]. Copyright (2018) Elsevier.)

view, Luna-Lama's group [3] derived a non-porous carbon material from spent coffee ground (SCG) and used it as a highly sustainable anode for LIBs. The scheme of the synthesis of carbonized SCG (C-SCG) is shown in Figure 18.1. A specific reversible capacity of 285 mAh/g at 0.1 A/g is obtained over 100 cycles when C-SCG is employed as an anode material for LIBs, and the coulombic efficiency reaches ~100%. Inspired by the work of SCG, H. Darjazi's group [4] was involved in the preparation of anodes for LIBs and SIBs from coffee grounds. This group synthesized a hard carbon anode material through chemical acid activation of coffee grounds followed by carbonization at 970°C under argon atmosphere over 6 h. This anode delivers a specific capacity of 339 and 113 mAh/g for LIBs and SIBs, respectively, at the rate of 0.2C with 70%–80% capacity retention after 100 cycles. This excellent electrochemical performance is ascribed to a certain degree of graphitization besides a disordered structure, which results in the high conductivity of hard carbon.

Li et al. [5] synthesized carbon material from sunflower seed husks through physical carbonization and chemical activation using calcium chloride as an activator and pyromellitic acid as a processing assistant. The obtained microporous carbon is employed as an anode material for LIBs and delivers a specific capacity of 590 mAh/g at 2C rate over 450 cycles. The porous structure increase the specific surface area, offers more active sites, and reduces the lithium-ion transport pathway.

Utilizing unburned charcoal (UC), Yu et al. [6] synthesized a hard carbon (UC-x) using acid and heat treatment. The prepared UC-x demonstrated a high specific surface area (57.9 m^2/g) and unique morphology and possessed a large interlayer spacing, which facilitates Li^+ and Na^+ ion intercalation. With the benefit of unique structural features, the delivered specific capacity of UC-1,200 °C was 292.3 and 256 mAh/g for SIBs and LIBs, respectively. Inspired by the unique structural properties of nanocellulose, Li et al. [7] synthesized a binder-free hard carbon film from a nanocellulose precursor via two-step thermal treatment. The synthesized hard carbon film has a hyper-cross-linked framework formed through aromatization and cyclization of small molecular fragments. Significantly, carbon nanofibers served as an interconnected conductive carbon network for fast electron transfer and they act as an interlinked mechanical skeleton network that is favorable for lithium-ion diffusion. This hard carbon exhibits a reversible capacity of 513 mAh/g over 1,000 cycles without any capacity decay at the current density of 50 mA/g. This excellent electrochemical performance is attributed to the abundant micropores with accessible redox-active carbonyl groups beneficial for the accommodation of Li^+ ions.

The next rival for lithium-ion batteries is sodium-ion batteries; however, the larger ionic radius of sodium impedes the reaction kinetics, and also the larger volume expansion of anode materials results in slow kinetics on sodium-ion intercalation/deintercalation in the electrode materials, leading to poor reversible capacity [8]. One of the strategies to solve this issue of SIBs is to synthesize an electrode material to store sodium ions with larger interlayer spacing. Two kinds of carbon materials are employed as an anode material for SIBs, viz. ordered soft carbon and disordered hard carbon. Sodium-ions have a difficulty intercalating into soft carbon since they have a tightly packed structure, whereas a large number of Na^+ ions access more channels in the case of hard carbon. These hard carbons are composed of randomly packed microdomains of larger interlayer spacing with loosely stacked graphene layers than graphite. Sustainable hard carbons are synthesized from available macadamia nutshell biowaste using thermal transformation and are used as an anode material for SIBs [9], which deliver a reversible capacity as high as 220 mAh/g at 10 mA/g. Zhang et al. [10] investigated a honeycomb-structured hard carbon derived from pine pollen precursor and used it as an anode material for SIBs. This carbon has a large surface area of 174 m^2/g with a large interlayer spacing of 0.41 nm and delivers a discharge capacity of 203 mAh/g after 200 cycles at 0.1 A/g with outstanding rate performance.

Various studies have reported the influence of sodium-ion storage on carbonization temperature, but Wang's group [11] investigated the influence of carbonization time on sodium-ion storage using hard carbon derived from mangosteen shells. The hard carbon is obtained by carbonizing the mangosteen shell at 1,500 °C for 2 h, and it exhibited a high reversible capacity of 330 mAh/g at a current density of 20 mA/g with 98% capacity retention. The hard carbon obtained at 1,200 °C for 2 h produced high capacity owing to more defects and nanovoids as compared to other hard carbons obtained at various carbonization temperatures (800 °C–1,100 °C and 1,300 °C–1,600 °C) and time (0.5, 1, 2, 3, and 5 h). Anish Raj et al. [12] reported a bio-derived porous hard carbon that is obtained from Litchi chinensis inedible seeds for sodium-ion battery. This seed contains a rich quantity of polyphenols, crude fibers, and starch, which make it a good precursor material for the synthesis of carbon using low carbonization temperature (>500 °C) [13]. The biocarbon anode is used as an anode for SIBs and delivered a reversible capacity of 146 mAh/g at a current density of 0.2 A/g over 100 cycles. This anode is also used to construct a full-cell prototype for SIBs coupled with vanadium phosphate cathode and delivered a specific capacity of 266 mAh/g over 50 cycles at a current density of 0.1 A/g. The influence of pore size on the waste coffee ground-derived porous carbon was investigated by Chiang et al. [14]. The carbon derived from the ratio of coffee ground to KOH of 1:5 delivered a highest reversible capacity of 206 mAh/g at a current density of 50 mA/g without any capacity decay after 250 cycles than other samples. This excellent performance is attributed to the high surface area and large pore size favorable for sodium-ion diffusion, which results in excellent cycling stability and high rate capability. Rybarczyka et al. [15] reported a hard carbon derived from rice husk and investigated the influence of treatment temperatures because it determines the material morphology, degree of graphitization, and nanovoids formation. The sodium-ion accessibility is significantly influenced by the channels present in the nanovoids. The carbon was derived from rice husk at a temperature of 1,600°C and exhibited a highest specific capacity of 276 mAh/g, which is attributed to the favorable insertion/extraction of sodium ions in nanovoids.

Sepals of palmyra palm fruit calyx don't have any biological importance, so they are discarded as biowaste every year; however, they are chemically significant because they have a hierarchical porous network to transport essential nutrients and water from the roots. The carbon derived from palmyra palm fruit calyx produces a porous three-dimensional network structure, but carbonization temperature is very important to retain the structural features. Therefore, Damodar et al. [16] derived carbons from palmyra palm fruit calyx sepals using carbonization at low temperatures (500 °C, 700 °C, and 900 °C under Ar atmosphere at a rate of heating of 5 °C/min) to retain structural features. This carbon has the morphology of a honeycomb structure and consists of tube-like wood fibers/cells aligned in a longitudinal direction. It has a pore diameter of 2–10 μm and a cell wall thickness of 2 μm. This porous carbon hard carbon is used as an anode material for SIBs and delivers a specific capacity of 280 mAh/g at 30 mA/g with excellent capacity retention. The superior performance of hard carbon derived from palmyra palm fruit calyx sepals is attributed to the porous microstructure with larger interlayer spacing supporting electrolyte infiltration, which enables favorable sodium-ion intercalation/deintercalation and also enhances the surface-dominated capacitive effect. An apple pomace-derived biocarbon anode for SIBs with high aerial capacity was investigated by Fu's group [17]. Aerial capacity is an important factor to consider bio-derived carbon for practical applications. This hard carbon is synthesized from apple pomace using the activation–carbonization method and was investigated for its applicability to SIBs. The shell-like macroporous carbon with a diameter of 500 nm exhibited a reversible coulombic efficiency >99.3% with a gravimetric capacity of 208 mAh/g after 200 cycles. This is mainly attributed to the 3D macroporous carbon network, which provides more active sites for sodium-ion intercalation and accommodate volume expansion. Saavedra Rios et al. [18] investigated the effect of inorganic contents on the electrochemical performance of a bio-derived hard carbon for SIBs. They synthesized different hard carbons from various biowastes such as beech wood, wheat straw, miscanthus, and pine at different pyrolysis temperatures (1,000 °C,

1,200 °C, and 1,400 °C). Pine- and beech wood-derived carbons were of high purity in nature and delivered a reversible specific capacity of 300 mAh/g with 80% coulombic efficiency, but wheat straw- and miscanthus-derived carbons (240–200 mAh/g with 70%–60% coulombic efficiency) were strongly affected by inorganic impurities such as Si, P, and Ca content. These impurities promote intrinsic activation and graphitic domain growth and enhance the formation of the silicon carbide phase. A hard carbon derived from biowaste is considered as an alternative anode for PIBs since graphite shows poor cycling performance when employed as anode material for PIBs. Verma's group [19] investigated an orange peel-derived hard carbon anode material for PIBs. This hard carbon is obtained by carbonization of orange peel at 1,000 °C and exhibited an impressive specific capacity and cycling performance (112 mAh/g after 3,000 cycles at a current density of 500 mA/g). The excellent performance is ascribed to the enlarged interlayer spacing, presence of mesopores, and abundant sp^2 domains that support favorable potassium-ion insertion/deinsertion.

18.2.2 Doped Carbonaceous Materials

It is well known that doped carbonaceous materials improve the electrochemical performance of LIBs, SIBs, and PIBs significantly. These heteroatom-doped carbons produce more active sites, create defects, and effectively modify the structural and electronic characteristics, resulting in improved electrochemical performance of LIBs, SIBs, and PIBs. So far, doping of heteroatoms such as N, S, B, and P has been shown to significantly improve the electrochemical performance of carbonaceous materials. Among heteroatom doping, N-doping significantly improves the electrochemical performance, which is attributed to the enhancement of electronic conductivity after heteroatom doping because electronegativity difference between carbon and nitrogen give rise to polarity difference. Like N-doping, S-doping also enhances the electrochemical performance of LIBs and SIBs because of the presence of lone pair of electrons.

Using eggshell membranes as raw materials, Gao et al. [20] successfully fabricated 3D architectured carbon materials and used them as anode for lithium-ion batteries. The presence of naturally doped nitrogen in a porous carbonized eggshell membrane (CEM) allows a unique hybrid Li storage mechanism without dendrite formation. CEM was employed as an anode and delivered an aerial capacity as high as 10 mAh/cm^2 with an average coulombic efficiency of 97.6% over 1,300 h. This excellent performance is attributed to the presence of natural nitrogen in eggshell membrane and offers more active sites for Li-ion intercalation and high electronic conductivity. To investigate the effect of tap density along with the presence of nitrogen in carbon materials, Cheong et al. [21] transformed the gum waste into a high tap density micron-sized functional carbon (MFC) material for LIBs. They fabricated MFCs from various gum wastes, namely guar gum (GG), xanthan gum (XG), gum arabic (GA), and gum karaya (GK), with high tap densities. Among the various MFCs derived from different gum wastes, gum karaya-derived MFCs exhibited a high tap density of around 1.7 g/cm^3 and an outstanding electrochemical performance for LIBs. It exhibits a volumetric capacity of 175.4 mAh/cm^3 over 5,000 cycles at a current density of 3,000 mA/g, which is attributed to the structural integrity and minimal cell resistance because it contains a substantial amount of nitrogen. In order to investigate the influence of inorganic impurities on biowaste-derived carbon, Beda et al. [22] successfully prepared carbons from different sources such as asparagus, grape, and potato. Washing done before thermal treatment led to fewer changes in the material porosity, and also the oxygen content in carbon material largely improved the electronic conductivity of the carbon. These biocarbons are employed as an electrode material for SIBs and deliver a charge capacity between 215 and 230 mAh/g.

The carbon material doped with two different heteroatoms can offer a high charge density and greater asymmetrical spin, resulting in more defects and electronic conductivity, which produce more electrochemical active sites as compared to single-atom doping. Doping of multiple atoms into carbon derived from biowaste material or biowaste often containing two or more heteroatoms also broaden interlayer distance supports for Li/Na/K-ion intercalation/deintercalation during cycling results improved electrochemical performance. D. Bosubabu et al. [23] selected a biowaste bagasse that contains multiple heteroatoms such as N, S, and O and synthesized a porous biocarbon using hydrothermal treatment. Further, they investigated the influence of heteroatom doping on the electrochemical performance of Li-ion batteries. XPS measurements revealed the presence of S, O, and N, and N is present in three different forms in the biocarbon, viz. graphitic, pyrrolic, and pyridinic nitrogen. When investigated as an anode for LIBs, N, S, and O co-doped carbon sample (NSOC-10) exhibited a reversible capacity of 574 mAh/g even after 1,000 charge/discharge cycles, while pristine bagasse (undoped carbon) retained only 26.6% capacity after 250 cycles. This excellent electrochemical performance is attributed to the presence of pores in carbon inclines to enhance Li^+ percolation, decrease the Li^+ diffusion path distance and enhance the lithium-ion storage capability. Cao's group [24] synthesized a N/S-co-doped disordered carbon with enlarged interlayer distance from *Cirsium setosum* biowaste precursor. When used as an anode material for SIBs, N/S-co-doped carbon exhibited a reversible capacity of 268 mAh/g after 100 cycles at a current density of 100 mA/g. It also exhibited a good cycling stability (198.6 mAh/g after 320 cycles at a current density of 1,000 mA/g) and outstanding rate capability. This excellent performance is ascribed to the enlarged interlayer spacing and the increase in the degree of disorderness due to N/S co-doping. Besides N, O, and S doping of biocarbon, Ding's group [25] synthesized a S/P-co-doped biocarbon from oak seeds using one-step pyrolysis process. The S/P-co-doped hard carbon obtained from oak seeds exhibited a remarkable cycling stability (88% capacity retention after 1,000 cycles at a current density of 1,000 mA/g) for SIBs because the large amount of active sites offered by S/P co-doping results in fast charge transfer kinetics.

18.3 COMPOSITES OF BIOWASTE-DERIVED CARBONACEOUS MATERIALS FOR ALKALI-ION BATTERIES

Various materials such as metals, metal oxides, and metal sulfides are considered as a high-capacity (800–3,000 mAh/g) anode materials for Li/Na/K-ion batteries. However, the aforementioned anode materials possess two serious challenges such as large volume expansion during cycling and side reactions that occur in the interface, liming their practical applications. One of the viable opportunities to overcome the volume expansion is the incorporation of anode materials into a conductive carbon matrix. This not only accommodates the volume expansion, but also makes the perfect electrical contact between electrode materials and carbon matrix, which results in enhanced capacity retention and cycling stability. Recently, with the advancement in technology, many green methods have become available for making composites of metals/metal oxides/metal sulfides with carbon derived from biowaste. Following these green chemistry principles, composites of metals/metal oxides/metal sulfides with carbon derived from biowastes, viz. silicon nanocrystals, are decorated with activated carbon (AC<nc-Si>AC) from brown rice husks (Figure 18.2) [26], nano-/microstructured Si/C from corn starch biowaste (Figure 18.3) [27], hierarchical nanocomposite of Si@C [28], MnO nanoparticle embedded into porous hard carbon

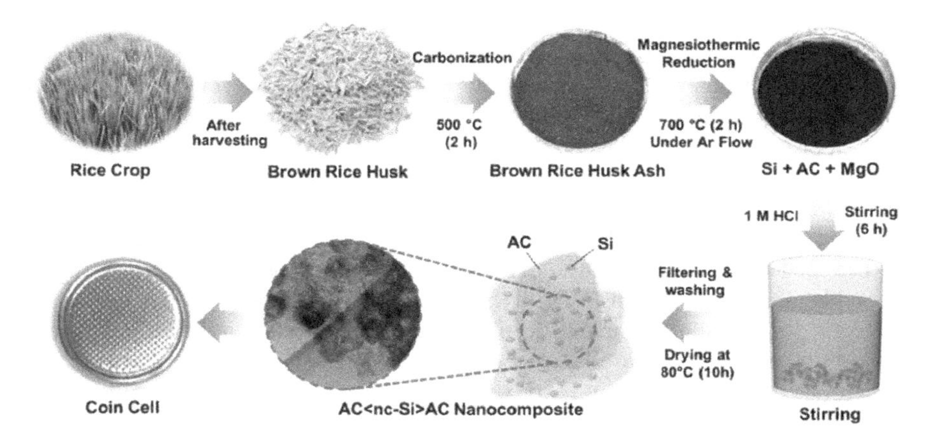

FIGURE 18.2 Stepwise synthesis and simultaneous production of activated carbon-decorated silicon nanocrystal (AC<nc-Si>AC) nanocomposites from brown rice husks (BRHs) through manganesiothermic reduction and schematic representation of AC<nc-Si>AC. (Reproduced with permission from Ref. [26]. Copyright MDPI (2019).)

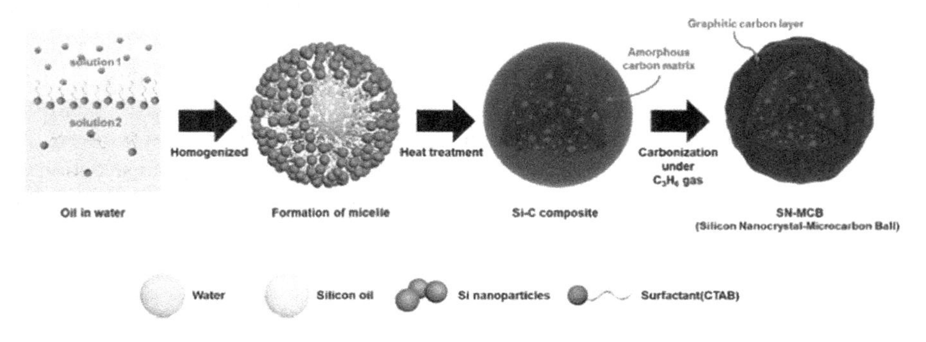

FIGURE 18.3 Oil in water is used as a microemulsion for the preparation of SiC hybrid. The microemulsion is treated with CTAB to form the micelles through homogenization process. Micelle is heated at 320 °C, and further carbonization is carried out under propene gas atmosphere at 800 °C for 24 h. (Reproduced with permission from Ref. [27]. Copyright American Chemical Society (2020).)

[29,30], SiO_x@N-doped carbon microspheres (SiO_x@NC) [31], Sb_2O_4 nanorods@biocarbon (biocarbon derived from Indian blackberry seed waste) [32], FeS nanoparticles into interwoven carbon nanofiber [33], hierarchical nanostructure of SnS_2 with carbon derived from chewed sugarcane [34], MoS_2@carbon nanotube [35], and $MoSe_2$/biocarbon/CNT composite [36].

18.4 SUMMARY AND FUTURE PERSPECTIVES

Waste biomass has been demonstrated as an excellent carbon source for synthesizing various porous/microstructured/nanostructured biocarbons owing to cheap raw material cost, ease of synthesis, abundant resources, eco-friendliness, and accessibility. Sustainable waste biomass-derived electrode materials possess multiple advantages such as versatile nano-/microstructures, varied compositions, abundant functional groups, in-built heteroatom dopants, and eco-friendliness (to protect the environment), and they serve as an efficient energy source for energy storage devices such as LIBs, those beyond LIBs, supercapacitors, and fuel cells. There are two significant reasons why waste biomass-derived carbons and their composites with metal oxides/metal sulfides have been considered as electrode materials for LIBs, SIBs, and PIBs. The first reason is the synthesis of carbonaceous materials with unique structural features can be achieved through the green chemistry principle with low cost. The second reason is the cycling stability of electrode materials and their composites is largely improved as compared to their pristine electrode materials. Researchers working in different parts of the world have rendered their constant efforts to derive biocarbons with unique structural properties such as the presence of in-built heteroatom, large surface area, and porosity from various biowastes.

In this viewpoint, based on the unique properties and various functionalities present in the various waste biomass materials, the structural features such as micro-/nanoporosity and particle size and the influence of the presence of heteroatoms in the biowaste on account of electrochemical characteristics of LIBs, SIBs, and PIBs were emphasized. More significantly, how these unique structural features address the critical issues faced by LIBs, SIBs, and PIBs is discussed. Waste biomass-derived carbonaceous materials and their composites for LIBs (and beyond LIBs) address the following challenges: (i) volume expansion, (ii) ion transport characteristics, (iii) growth of dendrites, (iv) intercalation/deintercalation difficulties, and (v) electrical conductivity. Simultaneously, some other issues need to be addressed while using the biomass, such as (i) the selection of suitable waste biomass, (ii) the selection of suitable synthesis method to convert a maximum amount of biowaste into carbon, and (iii) the choice of waste biomass to synthesize the composites. Even though the biowaste-derived electrode materials addressed various challenges faced by alkali-ion batteries, some important issues still persist in biowaste-derived materials. For example, the presence of impurities in the biowaste that can't be removed easily, leading to a detrimental effect on the electrode fabrication process, and also all the time, one can't expect positive results from the obtained biowaste-derived carbons. On account of the various advances achieved in recent years, some important strategies have been proposed to overcome the challenges of waste biomass materials for their application in LIBs and beyond LIBs. The significant strategies are the following:

i. An appropriate amount of biomass materials must be available locally to produce a maximum amount of carbonaceous materials with a low production cost. The selected waste biomass materials are required to produce a bulk amount of good-quality carbonaceous materials without any impurities after the pre-treatment process. This will influence the structural characteristics, which further leads to the expected electrochemical performance.
ii. Raw biomass consistency is a very important criterion to select the waste biomass for various applications. It is a mixture of different constituents; therefore, it is very hard to control the consistency of derived carbonaceous materials. The qualitative and quantitative techniques are needed to further analyze the structure, morphology, composition (the presence of heteroatoms), and properties of raw biowaste materials for obtaining good-quality carbonaceous materials. In addition to that, the physical and chemical stability of raw biomass materials should not be affected by the synthesis method, which is also considered as an essential factor for further process.
iii. Different battery systems require functional electrode materials of different structures for acquiring excellent electrochemical performance. Therefore, the appropriate structure is required for ion intercalation/deintercalation in which interplanar spacing is a crucial factor to determine the electrochemical performance of LIBs, SIBs, and PIBs. However, the recent research scenario relies on the different experimental approaches, i.e., trial-and-error method, which is a time-consuming process. Therefore, screening of the available waste biomass materials is very hard and also couldn't define the required parameter accurately. In this perspective, efficient screening methods are required to screen the waste biomass materials in terms of structure, morphology, functionality, and surface/interfacial properties.
iv. The selected electrode materials must be compatible with the carbon derived from biowaste for obtaining composites that show high theoretical capacity, cycling stability, and excellent physiochemical stability during cycling.
v. Advanced theoretical models/in situ characterization techniques are also required and needed to be developed for the rational designing of biowaste-derived carbons and their composites to determine the properties such as efficient ion insertion/deinsertion process, surface-oriented properties, and molecular chemistry/interactions. This will be helpful for researchers to improve the capacity and cycle life, minimize the volume expansion, and control the properties of the SEI layer.

To conclude, carbonaceous materials are important constituents of every battery system; more significantly, they not only serve as a host material for Li/Na/K-ion batteries, but also function as a conductive additive to enhance the conductivity of different electrode materials. Using biowaste as a precursor for synthesizing carbonaceous materials is much helpful to society in terms of making our environment green and acts as an efficient energy source for various battery systems. It also addresses the various critical issues of alkali-ion batteries; consequently, it improves the energy and power density of LIBs and beyond LIBs. Therefore, future research efforts move forward to explore efficient screening of biomass precursor, mechanistic analysis of structure-oriented properties, and cost-effective synthesis strategies for LIBs, SIBs, and PIBs. In this view, I hope this chapter provides an inspiration for the design aspects of functional biowaste materials toward bio-derived next-generation alkali-ion batteries.

ACKNOWLEDGMENT

Dr. C. Nithya wishes to thank the Department of Science and Technology, India, for the DST-Women Scientist Award (SR/WOS-A/CS-20/2017).

REFERENCES

1. Yu K, Wang J, Song K, Wang X, Liang C, Dou Y (2020) Hydrothermal synthesis of cellulose-derived carbon nanospheres from corn straw as anode materials for lithium ion batteries. *Nanomater.* 9: 93–106.
2. Yuan G, Li H, Hu H, Xie Y, Xiao Y, Dong H, Liang Y, Liu Y, Mingtao Z (2019) Microstructure engineering towards porous carbon materials derived from one biowaste precursor for multiple energy storage applications. *Electrochim. Acta.* 326: 134974.
3. Luna-Lama F, Rodríguez-Padrón D, Puente-Santiago AR, Muñoz-Batista MJ, Caballero A, Balu AM, Romero AA, Luque R (2019) Non-porous carbonaceous materials derived from coffee waste grounds as highly sustainable anodes for lithium-ion batteries. *J. Clear Prod.* 207: 411–417.
4. Darjazi H, Staffolani A, Sbrascini L, Bottoni L, Tossici R, Nobili F (2020) Sustainable anodes for lithium- and sodium-ion batteries based on coffee ground-derived hard carbon and green binders. *Energies.* 13: 6216–6235.
5. Li Y, Shi H, Liang C, Yu K (2021) Turning waste into treasure: biomass carbon derived from sunflower seed husks used as anode for lithium-ion batteries. *Ionics.* 27: 1025–1039.
6. Yu HY, Liang HJ, Gu ZY, Menga YF, Yang M, Yu MX, Zhao CD, Wu XL (2020) Waste-to-wealth: low-cost hard carbon anode derived from unburned charcoal with high capacity and long cycle life for sodium-ion/lithium-ion batteries. *Electrochim. Acta.* 361: 137041–137049.
7. Li Y, Du YF, Sun GH, Cheng JY, Song G, Song MX, Su FY, Yang F, Xie LJ, Chen CM. (2020) Self-standing hard carbon anode derived from hyper-linked nanocellulose with high cycling stability for lithium-ion batteries. *EcoMat.* 2: 1–14.
8. Ge P, Fouletier M (1988) Electrochemical intercalation of sodium in graphite. *Solid State Ionics.* 28–30: 1172–1175.
9. Kumar U, Wu J, Sharma N, Sahajwalla V (2021) Biomass derived high areal and specific capacity hard carbon anodes for sodium-ion batteries. *Energy Fuels.* 35: 1820–1830.
10. Zhang Y, Li X, Dong P, Wu G, Xiao J, Zeng X, Zhang Y, Sun X (2018) Honeycomb-like hard carbon derived from pine pollen as high performance anode material for sodium-ion batteries. *ACS Appl. Mater. Interfaces.* 10: 42796–42803.
11. Wang K, Jin Y, Sun S, Huang Y, Peng J, Luo J, Zhang Q, Qiu Y, Fang C, Han J (2017) Low-cost and high-performance hard carbon anode materials for sodium-ion batteries. *ACS Omega.* 2: 1687–1695.
12. Anish Raj K, Panda MR, Dutta DP, Mitra S (2018) Bio-derived mesoporous disordered carbon: an excellent anode in sodium-ion battery and full-cell lab prototype. *Carbon.* 143: 402–412.
13. Ding G. (1999) The research progress of Litchi chinensis, Lishizhen Med. *Mater. Medica Res.* 10: 145–146.

14. Peng-Hsuan C, Shih-Fu L, Yu-Hsuan H, Hsin T, Chun-Han G, Han-Yi C (2020) Coffee-Ground-derived nanoporous carbon anodes for sodium-ion batteries with high rate performance and cyclic stability. *Energy Fuels*. 34: 7666–7675.
15. Rybarczyka MK, Li Y, Qiaoc M, Hub YS, Titirici MM, Lieder M (2019) Hard carbon derived from rice husk as low cost negative electrodes in Na-ion batteries. *J. Energy Chem*. 29: 17–22.
16. Damodar D, Ghosh S, Rani MU, Martha SK, Deshpande AS (2019) Hard carbon derived from sepals of Palmyra palm fruit calyx as an anode for sodium-ion batteries. *J. Power Sources*. 438 (2019) 227008.
17. Fu H, Xu Z, Li R, Guan W, Yao K, Huang J, Yang J, Shen X (2018) Network carbon with macropores from apple pomace for stable and high areal capacity of sodium storage. *ACS Sustainable Chem. Eng*. 6: 14751–14758.
18. Saavedra Rios CM, Simonin L, Geyer A, Ghimbeu CM, Dupont C (2020) Unraveling the properties of biomass-derived hard carbons upon thermal treatment for a practical application in Na-ion batteries. *Energies*. 13: 3513–3538.
19. Verma R, Singhbabu YN, Didwal PN, Nguyen AG, Kim J, Park CJ (2020) Biowaste orange peel-derived mesoporous carbon as a cost-effective anode material with ultra-stable cyclability for potassium-ion batteries. *Batteries & Supercaps*. 3: 1099–1111.
20. Gao S, Jiang Q, Shi Y, Kim H, Busnaina A, Jung HY, Jung YJ (2020) High-performance lithium battery driven by hybrid lithium storage mechanism in 3D architectured carbonized eggshell membrane anode. *Carbon* 166: 26–35.
21. Cheong JY, Venkateshaiah A, Yun TG, Shin SH, Cerník M, Padil MVT, Kim ID, Varma RS (2021) Transforming gum wastes into high tap density micron-sized carbon with ultra-stable high-rate Li storage. *Electrochim. Acta* 367: 137419.
22. Beda A, Meins JML, Taberna PL, Simon P, Ghimbeu CM (2020) Impact of biomass inorganic impurities on hard carbon properties and performance in Na-ion batteries. *Sustainable Mater. Technol*. 26: 227.
23. Bosubabu D, Sampathkumar R, Karkera G, Ramesha K (2021) Facile approach to prepare multiple heteroatom-doped carbon material from bagasse and its applications toward lithium-ion and lithium–sulfur batteries. *Energy Fuels*. 35: 8286–8294.
24. Cao L, Wang Y, Hu H, Huang J, Kou L, Xu Z, Li J (2019) A N/S codoped disordered carbon with enlarged interlayer distance derived from cirsium setosum as high performance anode for sodium ion batteries. *J. Mater. Sci: Mater. Electron*. 30: 21323–21331.
25. Ding J, Zhang Y, Huang Y, Wang X, Sun Y, Guo Y, Jia D, Tang X (2021) Sulfur and phosphorus co-doped hard carbon derived from oak seeds enabled reversible sodium spheres filling and plating for ultra-stable sodium storage. *J. Alloys Compd*. 851: 156791.
26. Sekar S, Ahmed ATA, Inamdar AI, Lee Y, Im H, Kim DY, Lee S (2019) Activated carbon-decorated spherical silicon nanocrystal composites synchronously-derived from rice husks for anodic source of lithium-ion battery. *Nanomater*. 9: 1055.
27. Kwon HJ, Hwang JY, Shin HJ, Jeong MG, Chung KY, Sun YK, Jung HG (2020) Nano/microstructured silicon–carbon hybrid composite particles fabricated with corn starch biowaste as anode materials for li-ion batteries. *Nano Lett*. 20: 625–635.
28. Meščeriakovas A, Murashko K, Alatalo SM, Karhunen T, Leskinen JTT, Jokiniemi J, Lahde A (2020) Influence of induction-annealing temperature on the morphology of barley-straw-derived Si@C and SiC@graphite for potential application in Li-ion batteries. *Nanotechnology*. 31: 335709–335721.
29. Zhan D, Luo W, Kraatz HB, Fehse M, Li Y, Xiao Z, Brougham DF, Simpson A, Wu B (2019) Facile approach for synthesizing high-performance MnO/C electrodes from rice husk. *ACS Omega*. 4: 18908–18917.
30. Zhu L, Qu Y, Huang X, Luo J, Zhang H, Zhang Z, Yang Z (2020) Novel agaric-derived olive-like yolk-shell structured MnO@C composites for superior lithium storage. *Chem. Commun*. 56: 13201–13204.

31. Liu D, Jiang Z, Zhang W, Ma J, Xie J (2020) Micron-sized SiOx/N-doped carbon composite spheres fabricated with biomass chitosan for high-performance lithium-ion battery anodes. *RSC Adv.* 10: 38524.
32. Dutta DP (2020) Composites of Sb_2O_4 and biomass-derived mesoporous disordered carbon as versatile anodes for sodium-ion batteries. *Chem. Select.* 5: 1846–1857.
33. Zhao J, Syed JA, Wen X, Lu H, Meng X (2019) Green synthesis of FeS anchored carbon fibers using eggshell membrane as a biotemplate for energy storage application. *J. Alloys Compd.* 777: 974–981.
34. Li R, Miao C, Zhang M, Xiao W (2019) Novel hierarchical structural SnS_2 composite supported by biochar carbonized from chewed sugarcane as enhanced anodes for lithium ion batteries. *Ionics.* 26: 1239–1247.
35. Zhang Y, Liu W, Wang T, Du Y, Cui Y, Liu S, Wang H, Liu S, Chen M, Zhou J (2020) Space-confined fabrication of MoS_2@carbon tubes with semienclosed architecture achieving superior cycling capability for sodium ion storage. *Adv. Mater. Interfaces.* 7: 2000953.
36. Su C, Ru Q, Gao Y, Shi Z, Zheng M, Chen F, Ling FCC, Wei L (2021) Biowaste-sustained $MoSe_2$ composite as an efficient anode for sodium/potassium storage applications. *J. Alloys Compd.* 850: 156770.

19 $NaFePO_4$ Regenerated from Failed Commercial Li-Ion Batteries for Na-Ion Battery Applications

Dona Susan Baji, Anjali V. Nair, Shantikumar Nair, and Dhamodaran Santhanagopalan
Amrita Vishwa Vidyapeetham

CONTENTS

19.1 INTRODUCTION

Electrical and electronic equipment has become an essential part of modern civilization. Over several decades, energy storage has become a major global attraction, yet the overpriced fossil fuels and their environmental impacts have become a global concern.

DOI: 10.1201/9781003178354-23

Conventional fuels are the major contributors to greenhouse gas emissions, which necessitates sustainable energy harvest and storage [1,2]. The second half of the 20th century witnessed more lithium-ion battery (LIB) based electric vehicles (EVs) [3]. As the world is becoming more electrified, it is anticipated that by 2040, more than half of the vehicles will be powered by batteries [4]. Considering the life span of EV batteries, which is predicted to be 9–10 years when its capacity reduces to 80%, the productive way to make use of the remaining energy is to reuse it in stationary applications along with other renewable energy sources before recycling [5,6]. The EV global sale was 2 million units in 2018, it could reach about 10.79 million by 2025, and it is estimated to be 140 million by 2030. As the production and sales increase, it is said that 11 million tons of batteries will be spent and discarded by 2030, yet global LIB recycling is less [3]. As modern society demands an enormous number in the production of batteries from small to large scale, implementation of recycling has become a key element for sustainability [7].

The EVs were initially introduced in the 1800s; later, the demand hyped in the 1900s due to the increased fuel shortage and environmental crisis. But later in the 1930s, the gasoline and oil have become available faster and cheaper and the demand for EVs reduced. The crisis in the environmental condition exacerbates people to look forward to a better sustainable transport system. The demand for battery electric vehicle (BEV) is higher compared to plug-in hybrid electric vehicles (PHEVs), which builds up a foundation of dominating the EVs in the future [8]. Perhaps electric vehicles have become an ideogram of green transportation; however, EVs still have some problems such as range anxiety, safety, and less reliability [9]. Despite all these advantages, lithium-ion batteries possess aspects such as environmental and energy efficiency. The indispensable cost of rare elements as well as the limited resources of lithium has turned the focus of researchers into a super-alternative technology by introducing sodium-ion battery technology due to the similarity of the electrochemistry and the abundance of material that will dominate the future [10]. LIB technology also basically uses less abundant and toxic materials such as lithium and cobalt. As the production of lithium-ion batteries increases, the extra burden is taken to find the resources for lithium, cobalt, and nickel, which in reality turns around to be having the high-cost raw materials [11,12]. These points highlight the importance of recycling spent LIB materials, which may provide techno-economic advantages. In this work, we describe spent lithium iron phosphate-based commercial LIB cathode recycling. In particular, optimization of the regeneration process of converting lithium iron phosphate into sodium iron phosphate. Process parameters are optimized by correlating structure, morphology, and electrochemical performance of sodium iron phosphate as a sodium-ion battery cathode.

19.2 LITERATURE SURVEY

19.2.1 Brief Note on Recycling Methods

The aim of all the recycling methods is to retrieve the resources by eco-friendly, safe, and environmentally feasible processes [13]. Active material extraction methods are classified as hydrometallurgy, pyrometallurgy, direct recycling, and biometallurgy. The primary one involves the most well-known and long chemical process in the field of valuable metal recovery from spent LIBs. This process depends on solid-to-liquid ratio, leachant and reductant species, leaching time, temperature, concentration, etc. [4,14]. Pyrometallurgy, also called smelting, reduces valuable metals using high temperatures. This is a three-step process: pyrolysis or thermal degradation of spent LIBs, metal reduction at 1,500 °C using proper reductant, and finally, gas incineration at 1,000 °C to prevent dioxins liberation. Although a simple technique, it is not eco-friendly due to the release of toxic gases, high energy consumption, and secondary pollution [14,15]. Direct recycling involves non-destructive methods, which include mechanical, electrochemical, cathode-to-cathode, and cathode healing techniques without decomposition of cathode material. The main challenges in this process are the isolation of electrodes and their purification and toxic gases such as HF liberation [16]. Biometallurgy is known to be the most relevant technique due to its low cost and mild reaction conditions. The main challenge during the recycling process is the identification of battery type. If the chemistry of battery is identified, then the process will be very simple and efficient [13,17].

19.2.2 Brief Note on Commercialized Lithium-Ion Batteries

Currently, LIB technology is the prominent storage technology used in the electric vehicle industry due to its rigid milepost in terms of energy density, power, and safety. Electric vehicles have application-specific stringent requirements, namely reliability, safety, cost, low toxicity, high power, and availability [18]. The most used LIB cathodes in electric automobiles are of spinel lithium manganese oxide ($LiMn_2O_4$, LMO), lithium cobalt oxide ($LiCoO_2$, LCO), lithium iron phosphate ($LiFePO_4$, LFP), lithium nickel manganese cobalt oxide ($LiNi_xMn_yCo_zO_2$, NMC), and lithium nickel cobalt aluminum ($LiNi_xCo_yAl_zO_2$, NCA) [19,20]. The first-generation commercial cathode LCO is capable of delivering 274 mAh/g, and practically, it could deliver a capacity of 165 mAh/g if its voltage is limited to 4.35 V. The spinel LMO could also deliver practically a capacity of 120 mAh/g with an operating voltage of 4.0 V. NCA cathode-based batteries possess a high energy density, while NMC cathodes are of interest due to its energy density and cycle life. Although LFP-based batteries provide less energy density than NMC and NCA, the low cost and thermal stability are major advantages [21,22]. Even though LFP has a theoretical capacity of 170 mAh/g, it could only practically deliver a capacity of 120–160 mAh/g; irrespective of its lower energy density, it has widely been used in electric forklifts, sightseeing ships, etc. Apart from better safety, LFP also uses abundant materials such as iron and phosphorous rather than more expensive Ni and Co even though they are widely studied throughout. On the other hand, almost 60% of the anode market is shared by graphite. The transition metal oxide lithium titanate (LTO) is also achieving importance in grid storage and electric power train [18,23].

19.2.3 Brief Note on Opportunities and Challenges in Reuse and Recycling

Conceptually, the beginning of EV batteries starts from the mining of raw materials followed by manufacturing in the industry and first life followed by end-of-life discarding. Considering the implementation of EV battery second use (b2U) to grid storage, increases its life cycle and subsequent recycling would enable circular economy [24]. EV batteries meet their end of life (EOL) when they can no longer meet the essential requirement of vehicles, say reduced range and acceleration which cannot be employed for the transportation facility. It is said that the battery will still retain a capacity of 70% at the end of its first life [25,26]. Researchers have put efforts to reuse the EOL battery in stationary applications rather than recycling immediately. This indeed provides better opportunities to lessen the scarcity in resources as well as provide new instinct in economic aspects.

The surplus opportunities have been categorized into three, which primarily include a stationary application which again can be classified into on-grid and off-grid solutions such as powering short-term production plants and storage for private households. Secondarily, the application comes under quasi-stationary application, which supplies decentralized energy from major events. Finally, the application of second life can be implemented in mobile applications in industrial as well as private sectors [27,28]. An economic and environmental resource balance will be the major advantage of b2U. Nevertheless, the major challenges in reusing and repurposing EOL batteries are unpredictability of safety, lifetime, and quality of the batteries. As the circular economy demands, the next favorable opportunity is to recycle the battery, which again technically has challenges. The necessity of recycling has been recognized by scientific communities. This recycling of spent LIBs has been a hot topic for a decade; however, the large-scale implementation of the recycling of spent batteries is still in its infancy [17,29]. The major reason behind this is the consumer electronic batteries are challenging to recycle due to the smaller size of the battery and the least economic value. However, EV LIB recycling also has many challenges such as reliable product quality and supplier reliability. Ideally, spent batteries should be reformed into their original quality, which is a challenge. Recycling costs and environmental impacts are major obstructions in the recycling process. Sorting of battery chemistries before unpacking is still a challenge. A good balance is necessary for recycling LIBs, which majorly includes affordability, efficiency, and environment-friendliness [17,30,31].

19.2.4 $NaFePO_4$ as Cathode for Sodium-Ion Battery

The olivine $LiFePO_4$ has proved its success as a cathode material for LIBs. Ease of synthesis, stability, and abundance make it a much demanding cathode. However, the preparation of isostructural $NaFePO_4$ is challenging mainly because its polymorph marcite is thermodynamically more stable, but lacks electrochemical activity [32,33]. Researchers have found a cation exchange of $LiFePO_4$ method that seems to be successful in the production of $NaFePO_4$ [34]. The olivine $NaFePO_4$ is electrochemically active, and the sodium-ion could easily intercalate through the *b*-axis. It is also reported that the ionic mismatch during the conversion of LFP to NFP ends up in a cyclic inability and capacity fading [35]. Ali et al. [36] reported NFP wrapped in polythiophene to enhance the electronic conductivity. They used commercially available $LiFePO_4$ for delithiation and further sodiation under argon reflux; the bare electrode delivered a discharge capacity of 108 mAh/g initially, and at the end of 100 cycles, it deteriorated to 59 mAh/g at 10 mA/g. Similarly, a NFP/MWCNT hybrid electrode was by Karthik et al. [37], which delivered a discharge capacity of 90 mAh/g at 0.1C, whereas the bare NFP could only reach 64 mAh/g. Seung et al. [32] synthesized LFP through a chemical co-precipitation route followed by electrochemical sodiation, and the group reported a capacity of 125 mAh/g at C/20 ratio for up to only 50 cycles.

In view of all these tedious processes, our effort is focused on a scalable regeneration of LFP cathode from the spent LIB as an NFP cathode for sodium-ion battery. We have particularly explored the possibilities of LFP cathode regeneration as a sodium-ion battery cathode. We have implemented rapid microwave oxidation for retrieving the $FePO_4$ from the LFP battery and subsequently chemically sodiate FP to form NFP. Since lithium extractions have been demonstrated in most of the literature, we did not focus on that aspect. Hence, the study will focus on a complete optimization and exploration of LFP to NFP regeneration technology.

19.3 REGENERATING SPENT $LiFePO_4$ TO $NaFePO_4$

19.3.1 Delithiation of Spent $LiFePO_4$

In the first instance, a 3.2 V (32,650) commercial LFP/graphite full cell has been cycled for 1,000 cycles at 5 A current and, at the end of cycling, it was used for recycling as we have reported recently [38]. The obtained LFP samples directly went through a delithiation process before sodiation process in an Anton Paar Multiwave Pro system. This expeditious microwave delithiation has been determined to circumvent the difficulty of synthesis as well as to make sure a high-yield, scalable, fast delithiation technique. The pH before and after microwave delithiation was noted, where the pH after the reaction turns to 1, while the solution pH was equivalent to the deionized water pH before undergoing the reaction. The transformation of the pH was due to the formation of by-product lithium sulfate and potassium sulfate whose pH is acidic, which could be inferred to note a successful delithiation process. Thus, the obtained material was thoroughly washed few times to neutralize the solution to the pH of deionized water to clear out the by-products formed after the reaction. The spent delithiated $FePO_4$ was been obtained after the evaporation of water at 70 °C overnight.

19.3.2 Regeneration of $LiFePO_4$ to $NaFePO_4$

Herein, we have structured a regeneration strategy for $FePO_4$ by a reducing agent sodium iodide (NaI) in an acetonitrile medium as well as other solvents (acetonitrile, tetrahydrofuran – THF, and isopropyl alcohol – IPA) at a low temperature and short time. Such an experiment using sodiation of $FePO_4$ in acetonitrile has recently been reported by our group [38]. The examination of sodiation potency has effectively been structured into several parts, namely the effect of sodiation time, NaI stoichiometry, solvents, and sodiation temperature. These samples were named according to the concentration of NaI used along with sodiation time, solvent, or temperature used. The stoichiometric calculations were performed according to the equation given in our previous report [38]. As stoichiometric NaI led to lesser sodiation, an excessive amount of NaI concentration was used for the optimization of sodiation process as described here, so the excess amount was represented along with the sample names. For example, NFP-2.5–6 min represents the sodiation of $FePO_4$ with an excessive concentration of 2.5 times NaI from the stoichiometry and 6 min of sodiation process. Hereafter, all the samples will be named according to this definition. Typically, delithiated $FePO_4$ was taken in a closed bottle along with acetonitrile (or the respective solvents used in the effect of solvent experiments) and the temperature was increased to 80 °C (which was typically the fixed temperature for all the other experiments, except for the effect of temperature experiments). The excess amount from the stoichiometric sodium iodide was added to the $FePO_4$ dispersion under stirring to a particular time below 10 min. The immediate color change of $FePO_4$ to brownish red visibly authenticated the completion of sodiation process. The obtained samples were washed immediately after the sodiation process with the solvent medium served for sodiation. The sodiated sample was henceforth generally named as Na_xFePO_4 (NFP). For the evaluation of the sodiation accuracy, electrochemical investigations were conducted by a half-cell assembly inside an argon-filled glove box, for which the slurry was cast on to an aluminum foil in a ratio of 80:10:10 Na_xFePO_4:carbon black (Timical super C65, MTI, the USA):PVDF (Sigma-Aldrich) and dried at 80 °C overnight. An in-house prepared 1 M $NaClO_4$ in PC:VC (5 wt.%) was utilized as electrolyte for the electrochemical studies. Further, the fabricated Swageloks has been undergone galvanostatic charge/discharge cycles in Biologic Instruments (the USA) in a voltage window of 2.0–4.0 V. Scheme 19.1 illustrates the recycling process, which begins initially by disassembling a failed battery, followed by separation of cathode material from aluminum foil by sonication. Further a rapid microwave process was carried out for delithiation and effective chemical sodiation followed by regeneration of $FePO_4$ to $NaFePO_4$ as sodium-ion battery cathode.

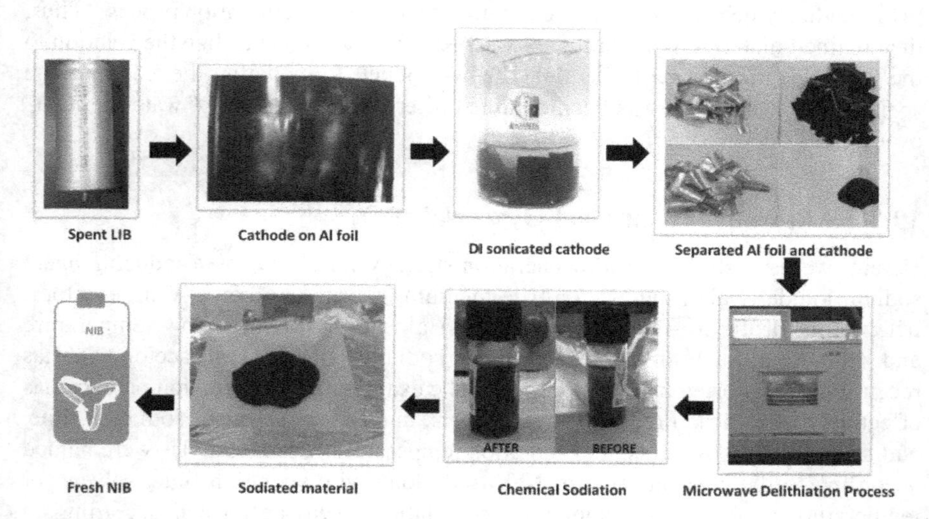

SCHEME 19.1 Illustration of the recycling and regeneration process.

19.3.3 Recycling of $LiFePO_4$

The directly extracted powder from the aluminum current collector was named as spent $LiFePO_4$ (SLFP); further, the extracted cathode material (SLFP) was washed with NMP for a time period of 10 min followed by drying overnight, and this sample was labeled as SLFP-washed. A rapid microwave-assisted delithiation was conducted on the washed sample, and the obtained sample was named FP-MW, which indicates $FePO_4$ microwave. The structural characterization of the spent LFP and FP-MW ($FePO_4$) was conducted and depicted in our previous work [38]. The XRD profile matches with the olivine structure with a slight Li loss, and the efficiency of the rapid microwave delithiation technique was confirmed by the formation of fully delithiated SLFP to $FePO_4$. Figure 19.1c shows the TGA; from the data, we can identify the similarity in the weight loss percentage among all the samples, which is approximately less than 10 wt.%. This loss percentage can be attributed to the residual carbon content in the spent samples. Figure 19.1d–f represents the FESEM image, and it is obvious from the figure that the SLFP-washed and FP-MW samples are almost similar with an average particle size approximately in a range of 250 nm, which indicates the delithiation process didn't affect the particle morphology. However, in Figure 19.1d SLFP-unwashed sample looks like crumbled and agglomerated particles, this is possibly due to the presence of additives.

19.3.4 Effect of Sodiation Time

The microwave delithiated sample FP-MW was chemically sodiated by NaI, which was explained in the previous section. Herein, an excess amount of 1.5 moles was taken apart from the stoichiometric amount and considered as a standard for the

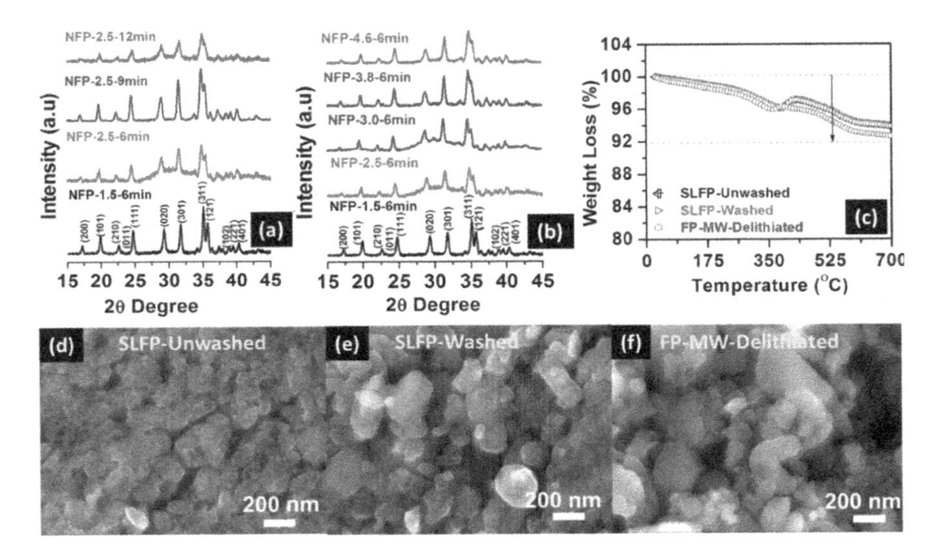

FIGURE 19.1 (a) and (b) XRD analysis of "effect of sodiation time" and "effect of NaI stoichiometry"; (c) TGA; (d), (e), and (f) FESEM images as per the legends.

evaluation purpose (as stoichiometric amount leads to poor sodiation). This section is structured to study in detail the sodiation efficiency according to the time duration of sodiation. For this assessment, four kinds of samples were chosen and the solvent and the temperature were fixed for this experiment, which is acetonitrile at a temperature of 80 °C. In this section, hereafter the sodiated $NaFePO_4$ samples will be abbreviated as NFP along with the excess sodium iodide concentration and sodiation time. Figure 19.1a represents the XRD analysis of all the samples taken for the study of "effect of sodiation time". The peak evolution of (301) represents the formation of Na_xFePO_4. The peaks represent the olivine structure of $NaFePO_4$ with the Pnma space group, which is in accordance with the literature. The chemical sodiation efficiency has been revealed from the XRD data, which have been shown earlier also by our group [38]. Figure 19.2a implies the rate performance study of NFP sample at 2.5 times excess concentration of sodium iodide at various sodiation times in an acetonitrile medium, and for comparison, NFP-1.5–6 min results are included. The first charge–discharge capacity is depicted in Figure 19.3a. From the rate performance, NFP-1.5–6 min and NFP-2.5–6 min have a similar capacity, whereas the capacity decreased for the NFP-2.5–9 min and NFP-2.5–12 min at higher current densities. The samples showed excellent retention values of 76%, 75%, 69%, and 70% for NFP-1.5–6 min, NFP-2.5–6 min, NFP-2.5–9 min, and NFP-2.5–12 min samples, respectively. Although the retention and performance of NFP-1.5–6 min were comparable, the initial charge capacity reveals that the sodiation was relatively less than other samples. It is noted that this sample only had a first charge capacity of 73 mAh/g at 25 mA/g. However, the initial discharge capacity of the sample was 104 mAh/g,

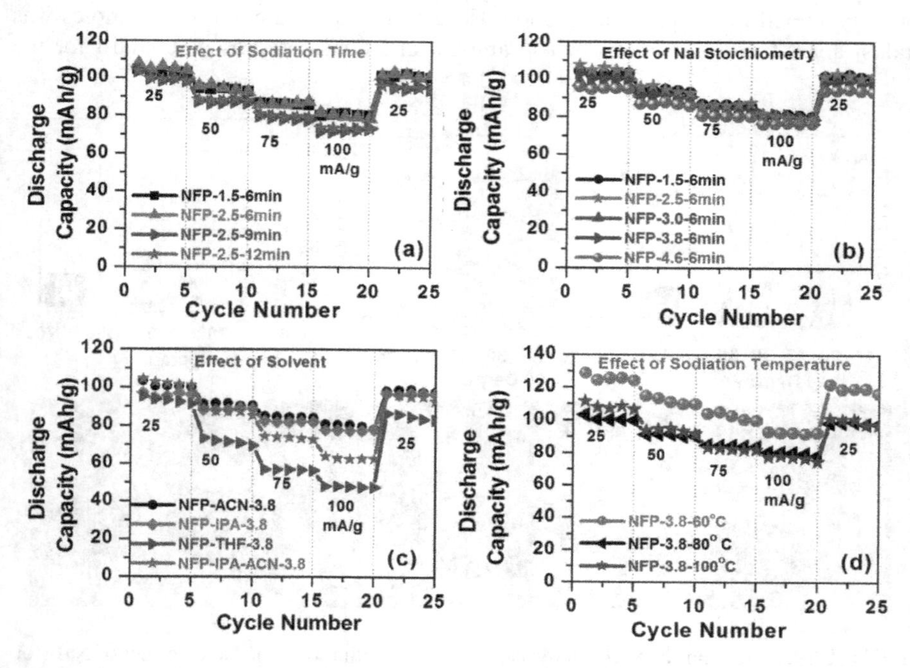

FIGURE 19.2 Electrochemical rate test of all the set of samples as per the legends.

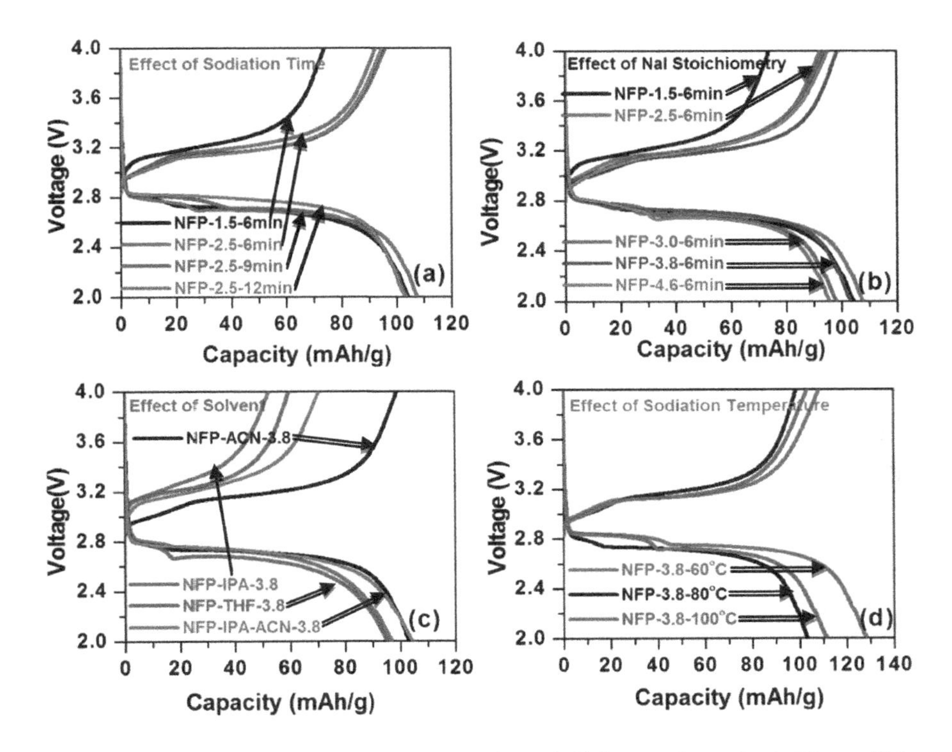

FIGURE 19.3 The first charge–discharge profiles of all the set of samples in Figure 19.2.

which indicates an excessive electrochemical sodiation during the first discharge that is higher than the chemical sodiation of the as-prepared NFP samples. The initial charge capacity of NFP-2.5–6 min, NFP-2.5–9 min, and NFP-2.5–12 min were 92, 96, and 94 mAh/g respectively. It is evident that although the NFP-1.5–6 min sample could perform well in the rate study, it was a sodium-deficient cathode to start with. Nevertheless, NFP-2.5 with various sodiation times showed better sodiation properties. Considering the rate performances, NFP-2.5–6 min sample has retained good capacity at all cycle numbers and current densities. Figure 19.4a depicts the cyclic performance of all the effects of sodiation time samples at 50 mA/g for the initial two cycles and, from the third cycle onward, at 100 mA/g for 100 cycles. The data reveal that the discharge capacity of all the samples initially was similar for the first few cycles, but in the later stage, the least chemically sodiated sample NFP-1.5–6 min merely showed a retention of 44% at the end of 100 cycles, whereas NFP-2.5–6 min, NFP-2.5–9 min and NFP-2.5–12 min showed a retention of 76%, 84%, and 68%, respectively, calculated with respect to their third cycle.

19.3.5 Effect of NaI Stoichiometry

The second approach of our study was to optimize the "effect of NaI stoichiometry" at a fixed time of 6 min in an acetonitrile solvent at a temperature of 80 °C. Here, the temperature, solvent, and time were kept fixed by altering the

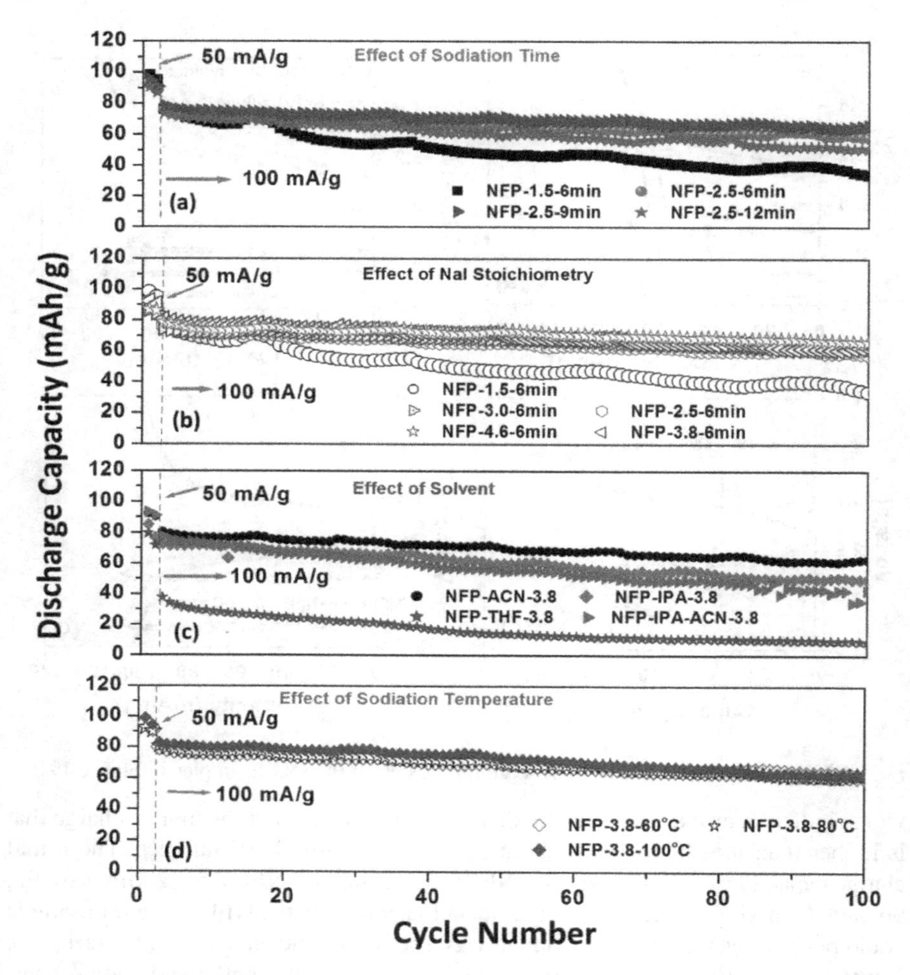

FIGURE 19.4 Cycling performance of all set of samples.

concentration in excess of stoichiometry, which is represented along with the sample IDs. The sodiated NFP samples were labeled as mentioned previously according to the concentration multiple taken from the sample and sodiation time. The data of NFP-1.5–6 min and NFP-2.5–6 min were recorded already and included here for comparison by incorporating more stoichiometry of NaI. Similar to the effect of sodiation time samples, all the analyses were performed for the effect of NaI stoichiometry samples. Figure 19.1b displays the XRD results confirming the formation of $NaFePO_4$ olivine structure. Figure 19.2b presents the rate performance of the NFP sample at varying NaI concentrations and fixed time. The first cycle charge/discharge profiles are depicted in Figure 19.3b. The initial charge capacity of NFP-1.5–6 min, NFP-2.5–6 min, NFP-3.0–6 min, NFP-3.8–6 min, and NFP-4.6–6 min was 73, 92, 94, 98, and 95 mAh/g, respectively. The discharge capacity of the samples was 104, 107, 98, 103, and 95, respectively. From these results, it may be noted that NFP-3.8–6 min delivered a better charge capacity and comparable discharge capacity, indicating proper chemical sodiation. Even though the charge and discharge values of NFP-4.5–6 min overlap, the capacity it delivered was lesser than that of NFP-3.8–6 min sample. From the rate performance (Figure 19.2b), these samples showed a retention of 76%, 75%, 79%, 78%, and 79% for NFP-1.5–6 min, NFP-2.5–6 min, NFP-3.0–6 min, NFP-3.8–6 min, and NFP-4.6–6 min, respectively. The samples when cycled back to 25 mA/g retained a capacity similar to the initial cycle. Nonetheless, the sample named NFP-3.8–6 min had a better chemical sodiation capability and retained higher capacity as well. The long cycling performance of the effect of NaI stoichiometry samples was performed and is depicted in Figure 19.4b. The first discharge capacity of NFP-1.5–6 min, NFP-2.5–6 min, NFP-3.0–6 min, NFP-3.8–6 min, and NFP-4.6–6 min electrodes was 98, 91, 86, 92 and 86, respectively, at 50 mA/g; further, the corresponding samples' retention noted from the third cycle were 44.1%, 76.4%, 82.9%, 78%, and 83.8%. The figure implies a conclusion that the less chemically sodiated sample showed a capacity decay after long cycles. All other samples retained a capacity higher than 75% of their initial capacity at the end of 100 cycles.

19.3.6 Effect of Solvents

The study of the effect of solvent was conducted by varying the solvent with a fixed concentration of 3.8 at a temperature of 80°C and a time of 6 min. The abbreviation ACN, IPA, THF, and IPA-ACN represent acetonitrile, isopropyl alcohol (Merck), tetrahydrofuran (Sigma-Aldrich), and mixed isopropyl alcohol and acetonitrile at 1:1 volume ratio. The sample denoted as NFP-ACN-3.8 in this section is NFP-3.8–6 min discussed in the previous sections, which is taken for comparison here. This section identifies the effect of different solvents on sodiation and differentiates the results. Figure 19.2c shows the electrochemical rate performance of the samples with different solvents. Comparatively, the capacity of NFP-ACN-3.8 was better. Even though the discharge capacity of NFP-IPA-3.8 was comparable with the acetonitrile sample from Figure 19.3c, it is prominent that the capacity achieved by the IPA sample was by electrochemical sodiation than by chemical sodiation. The retention percentage of NFP-ACN-3.8, NFP-IPA-3.8, NFP-THF-3.8, and NFP-IPA-ACN-3.8 were 78%, 80%, 61%, and 50%, respectively. All the samples retained almost 90% of its initial capacity when cycled back again to 25 mA/g. Initially, the mixed sample NFP-IPA-ACN-3.8 showed a comparable discharge capacity similar to other samples; with increasing current density, the capacity started to decrease. Among all the samples, NFP-ACN-3.8 indicated a higher chemical sodiation followed by NFP-IPA-ACN-3.8, NFP-THF-3.8, and NFP-IPA-3.8. From the solvent effect study, NFP-IPA-3.8 sample displayed the least chemical sodiation based on first charge capacity represented in Figure 19.3c. Figure 19.4c displays the long cycle performance of "effect of solvent". The performance portrays among all the solvents, acetonitrile based sample had better capacity and retention. This revealed that completely sodiated sample performed well in the rate test which can be correlated from the rate performance data in figure 19.2c. The samples NFP-ACN-3.8, NFP-IPA-3.8, NFP-IPA-ACN-3.8, and NFP-THF-3.8 showed a retention capacity of 77%, 67%, 55%, and 24%, respectively.

19.3.7 Effect of Sodiation Temperature

Similar to other experiments, the effect of sodiation temperature on the sodiation process was also investigated by altering the temperature to 60 °C, 80 °C, and 100 °C. It should be noted that in all other experiments, the temperature for sodiation was fixed to 80 °C. Here, we fix the concentration, solvent, and time parameters to 3.8, acetonitrile, and 6 min, respectively. For comparison, we take the sample NFP-3.8–6 min sample included in the previous section and label it as NFP-3.8–80 °C. In correspondence to the previous sections, electrochemical tests at different rates were performed varying the temperature of sodiation. The NFP-3.8–60 °C showed both the higher chemical sodiation and capacity retention (Figures 19.2 and 19.3d). The discharge capacities delivered were 128, 103, and 111 mAh/g for 60 °C, 80 °C, and 100°C samples, respectively; from these data, it is implied that NFP-3.8–60°C sample outperformed all the other samples in terms of charge capacity. The capacity retention of the samples was recorded as 72%, 78%, and 69% for NFP-3.8–60 °C, NFP-3.8–80 °C, and NFP-3.8–100 °C, but when evaluating the reversibility in terms of the first cycle, NFP-3.8–80 °C sample performed better. Figure 19.4d depicts the cycling performance of the samples studied for the "effect of sodiation temperature". This shows that the initial discharge capacity was 95, 92, and 99 for NFP-3.8–60 °C, NFP-3.8–80 °C, and NFP-3.8–100 °C electrodes, respectively. The capacity retention of all the samples was almost similar at the end of 100 cycles with respect to the third cycle capacity calculated to be 77%, 77%, and 75% for the three different temperatures of sodiation.

19.4 CONCLUSIONS

In summary, the study was conducted toward optimizing the process parameters in regenerating failed commercial $LiFePO_4$ cathode to $NaFePO_4$. Sodiation efficiency as a function of different process parameters is optimized and evaluated through electrochemical performance of $NaFePO_4$ as a sodium-ion battery cathode. It is concluded that an excess concentration of sodium iodide is needed for the higher sodiation of $FePO_4$. Chemical sodiation of $FePO_4$ resulted in olivine $NaFePO_4$ as confirmed by the XRD, and the process did not influence the morphology or particle size as evident from FESEM image analysis. Based on the process optimization, it is realized that sodiation time of 6 min, with NaI stoichiometry of 3.8, in acetonitrile solvent at a sodiation temperature of 60°C provides the best-performing $NaFePO_4$. Elemental analysis by inductively coupled plasma spectrometry indicates, a maximum sodiation of 90% was achieved. Moreover, this study is an opportunity for economic balance (or for circular economy) and can be scaled up. This process enabled regeneration of spent $LiFePO_4$ to $NaFePO_4$ cathode for sodium-ion battery applications. The advantages of chemical processing, low temperature, and rapid processing make this a scalable technique. With the number of automobile batteries that would end up at the recycling units in the next decade, a reliable and economically viable process such as this is required.

ACKNOWLEDGMENTS

The authors are thankful to the DST, India (Ref: DST/TMD/MES/2k18/225), for financial support and Amrita Vishwa Vidyapeetham for infrastructural support. We also thank the STIC-CUSAT, Kerala, for support in XRD and ICP analyses.

REFERENCES

1. Ellis, B. L., & Nazar, L. F. (2012). Sodium and sodium-ion energy storage batteries. *Current Opinion in Solid State and Materials Science*, 16(4), 168–177.
2. Meszaros, F., Shatanawi, M., & Ogunkunbi, G. A. (2021). Challenges of the electric vehicle markets in emerging economies. *Periodica Polytechnica Transportation Engineering*, 49(1), 93–101.
3. Garole, D. J., Hossain, R., Garole, V. J., Sahajwalla, V., Nerkar, J., & Dubal, D. P. (2020). Recycle, recover and repurpose strategy of spent Li-ion batteries and catalysts: Current status and future opportunities. *ChemSusChem*, 13(12), 3079–3100.
4. Mohanty, A., Sahu, S., Sukla, L. B., & Devi, N. (2021). Application of various processes to recycle lithium-ion batteries (LIBs): A brief review. *Materials Today: Proceedings*, 47(5), 1203–1212.
5. Harper, G., Sommerville, R., Kendrick, E., Driscoll, L., Slater, P., Stolkin, R., ... & Anderson, P. (2019). Recycling lithium-ion batteries from electric vehicles. *Nature*, 575(7781), 75–86.
6. Hossain, E., Murtaugh, D., Mody, J., Faruque, H. M. R., Sunny, M. S. H., & Mohammad, N. (2019). A comprehensive review on second-life batteries: Current state, manufacturing considerations, applications, impacts, barriers & potential solutions, business strategies, and policies. *IEEE Access*, 7, 73215–73252.
7. Zhao, Y., Pohl, O., Bhatt, A. I., Collis, G. E., Mahon, P. J., Rüther, T., & Hollenkamp, A. F. . (2021). A review on battery market trends, second-life reuse, and recycling. *Sustainable Chemistry*, 2(1), 167–205.
8. Sun, P., Bisschop, R., Niu, H., & Huang, X. (2020). A review of battery fires in electric vehicles. *Fire Technology*, 1–50.
9. Dhakal, T., & Min, K. S. (2021). Macro study of global electric vehicle expansion. *Foresight STI Gov.*, 15, 67–73.
10. Hwang, J. Y., Myung, S. T., & Sun, Y. K. (2017). Sodium-ion batteries: Present and future. *Chemical Society Reviews*, 46(12), 3529–3614.
11. Weil, M., Ziemann, S., & Peters, J. (2018). The issue of metal resources in Li-ion batteries for electric vehicles. In Gianfranco Pistoia and Boryann Liaw (eds), *Behaviour of Lithium-Ion Batteries in Electric Vehicles* (pp. 59–74). Springer, Cham.
12. Chayambuka, K., Mulder, G., Danilov, D. L., & Notten, P. H. (2020). From Li-ion batteries toward Na-ion chemistries: Challenges and opportunities. *Advanced Energy Materials*, 10(38), 2001310.
13. Zheng, X., Zhu, Z., Lin, X., Zhang, Y., He, Y., Cao, H., & Sun, Z. (2018). A mini-review on metal recycling from spent lithium ion batteries. *Engineering*, 4(3), 361–370.
14. Steward, D., Mayyas, A., & Mann, M. (2019). Economics and challenges of Li-ion battery recycling from end-of-life vehicles. *Procedia Manufacturing*, 33, 272–279.
15. Kim, H. J., Krishna, T. N. V., Zeb, K., Rajangam, V., Gopi, C. V., Sambasivam, S., ... & Obaidat, I. M. (2020). A comprehensive review of Li-ion battery materials and their recycling techniques. *Electronics*, 9(7), 1161.
16. Sloop, S., Crandon, L., Allen, M., Koetje, K., Reed, L., Gaines, L., ... & Lerner, M. (2020). A direct recycling case study from a lithium-ion battery recall. *Sustainable Materials and Technologies*, 25, e00152.

17. Beaudet, A., Larouche, F., Amouzegar, K., Bouchard, P., & Zaghib, K. (2020). Key challenges and opportunities for recycling electric vehicle battery materials. *Sustainability*, 12(14), 5837.
18. Zeng, X., Li, M., Abd El-Hady, D., Alshitari, W., Al-Bogami, A. S., Lu, J., & Amine, K. (2019). Commercialization of lithium battery technologies for electric vehicles. *Advanced Energy Materials*, 9(27), 1900161
19. Ellingsen, L. A. W., Hung, C. R., & Strømman, A. H. (2017). Identifying key assumptions and differences in life cycle assessment studies of lithium-ion traction batteries with focus on greenhouse gas emissions. *Transportation Research Part D: Transport and Environment*, 55, 82–90.
20. Natarajan, S., & Aravindan, V. (2018). Burgeoning prospects of spent lithium-ion batteries in multifarious applications. *Advanced Energy Materials*, 8(33), 1802303
21. Shen, X., Zhang, X. Q., Ding, F., Huang, J. Q., Xu, R., Chen, X., … & Zhang, Q. (2021). Advanced electrode materials in lithium batteries: Retrospect and prospect. *Energy Material Advances*, 2021, 1–15.
22. Duffner, F., Wentker, M., Greenwood, M., & Leker, J. (2020). Battery cost modeling: A review and directions for future research. *Renewable and Sustainable Energy Reviews*, 127, 109872.
23. Ramoni, M. O., & Zhang, H. C. (2013). End-of-life (EOL) issues and options for electric vehicle batteries. *Clean Technologies and Environmental Policy*, 15(6), 881–891.
24. Ahmadi, L., Yip, A., Fowler, M., Young, S. B., & Fraser, R. A. (2014). Environmental feasibility of re-use of electric vehicle batteries. *Sustainable Energy Technologies and Assessments*, 6, 64–74.
25. Reinhardt, R., Christodoulou, I., Gassó-Domingo, S., & García, B. A. (2019). Towards sustainable business models for electric vehicle battery second use: A critical review. *Journal of Environmental Management*, 245, 432–446.
26. Ahmadian, A., Sedghi, M., Elkamel, A., Fowler, M., & Golkar, M. A. (2018). Plug-in electric vehicle batteries degradation modeling for smart grid studies: Review, assessment and conceptual framework. *Renewable and Sustainable Energy Reviews*, 81, 2609–2624.
27. Rehme, M., Richter, S., Temmler, A., & Götze, U. (2016). CoFAT 2016-Second-Life Battery Applications-Market potentials and contribution to the cost effectiveness of electric vehicles. In *Conference on Future Automotive Technology*.
28. Deng, Y., Zhang, Y., Luo, F., & Mu, Y. (2020). Operational planning of centralized charging stations utilizing second-life battery energy storage systems. *IEEE Transactions on Sustainable Energy*, 12(1), 387–399.
29. Fan, E., Li, L., Wang, Z., Lin, J., Huang, Y., Yao, Y., … & Wu, F. (2020). Sustainable recycling technology for Li-ion batteries and beyond: Challenges and future prospects. *Chemical Reviews*, 120(14), 7020–7063.
30. Mossali, E., Picone, N., Gentilini, L., Rodrìguez, O., Pérez, J. M., & Colledani, M. (2020). Lithium-ion batteries towards circular economy: A literature review of opportunities and issues of recycling treatments. *Journal of Environmental Management*, 264, 110500.
31. Gaines, L., Richa, K., & Spangenberger, J. (2018). Key issues for Li-ion battery recycling. *MRS Energy & Sustainability*, 5, E14.
32. Oh, S. M., Myung, S. T., Hassoun, J., Scrosati, B., & Sun, Y. K. (2012). Reversible $NaFePO_4$ electrode for sodium secondary batteries. *Electrochemistry Communications*, 22, 149–152.
33. Heubner, C., Heiden, S., Matthey, B., Schneider, M., & Michaelis, A. (2016). Sodiation vs. Lithiation of $FePO_4$: A comparative kinetic study. *Electrochimica Acta*, 216, 412–419.
34. Hasa, I., Hassoun, J., Sun, Y. K., & Scrosati, B. (2014). Sodium-ion battery based on an electrochemically converted $NaFePO_4$ cathode and nanostructured tin–carbon anode. *ChemPhysChem*, 15(10), 2152–2155

35. Rahmawati, F., Faiz, Z., Romadhona, D. A. N., Saraswati, T. E., & Lestari, W. W. (2020, July). The performance of sodium ion battery with $NaFePO_4$ cathode prepared from local iron sand. In *IOP Conference Series: Materials Science and Engineering* (Vol. 902, No. 1, p. 012008). IOP Publishing.
36. Ali, G., Lee, J. H., Susanto, D., Choi, S. W., Cho, B. W., Nam, K. W., & Chung, K. Y. (2016). Polythiophene-wrapped olivine $NaFePO_4$ as a cathode for Na-ion batteries. *ACS Applied Materials & Interfaces*, 8(24), 15422–15429.
37. Karthik, M., Sathishkumar, S., BoopathiRaja, R., Meganathan, K. L., & Sumathi, T. (2020). Design and fabrication of $NaFePO_4$/MWCNTs hybrid electrode material for sodium-ion battery. *Journal of Materials Science: Materials in Electronics*, 31(23), 21792–21801.
38. Gangaja, B., Nair, S., & Santhanagopalan, D. (2021). Reuse, recycle, and regeneration of $LiFePO_4$ cathode from spent lithium-ion batteries for rechargeable lithium-and sodium-ion batteries. *ACS Sustainable Chemistry & Engineering*, 9(13), 4711–4721.

20 Polymeric Wastes for Metal-Ion Batteries

Ranjusha Rajagopalan and Haiyan Wang
Central South University

CONTENTS

20.1 INTRODUCTION

Polymeric materials contribute significantly to meeting the routine requirements in our daily lives. As a result of this heavy dependency on polymers for different applications, the global production and disposal of polymers have increased enormously over the past decades. Nevertheless, the polymeric wastes are bulkier than the organic remains and a large part of these wastes do not decompose. It is reported that biodegradation begins only when molecular weight value reaches to a few tens of thousands, which can be considered very low for many commercial polymers.[1,2] This has resulted in a continuous demand for disposing of the non-biodegradable polymeric wastes in the landfill, which consumes massive space and can result in environmental and health hazards.[3–8] The increasing demand for polymers has also resulted in the diminution of the petroleum products, which are a part of the non-renewable fossil fuels.[9,10] Polymeric waste materials are usually mixtures of different types, such as high-density polyethylene (HDPE), low-density polyethylene (LDPE), poly(3,4-ethylenedioxythiophene)–poly(styrenesulfonate) (PEDOT:PSS), polyethylene terephthalate (PET), polypropylene (PP), polystyrene (PS), polyamide (PA), polyvinyl chloride (PVC), polycarbonate (PC), polyacrylate (PAC), polyurethanes (PUs), nylon 6, and cured thermosets such as epoxy and rubber waste.[11–14] In order to manage polymeric wastes, different approaches have been employed, such

DOI: 10.1201/9781003178354-24

as incineration, recycling, and energy recovery systems.[15–20] The backbone chains of addition polymers such as polyolefins, PE, PP, PS, and PVC are built of carbon atoms that are reasonably resistant to decomposition or hydrolytic cleavage. As a result, these polymers become non-biodegradable by nature. The condensation polymers such as nylons, PET, PU, and PC, on the other hand, contain heteroatoms such as O and N on their backbones. These polymers are more susceptible to decomposition by the hydrolytic cleavage due to hydrophilic amide or ester linkage between monomers. However, these condensation polymer wastes still persist for decades and probably for centuries when they get disposed of to landfills. This is mainly attributed to the fact that commercial products made from these condensation polymers resist oxidation and biodegradation due to the additives such as anti-oxidants and stabilizers present in their composition. Crystallinity is another parameter that adversely affects biodegradation.[21–23] Further, when polymers are synthesized by addition polymerization, degradation by natural means is even more difficult because of their chemically inert character. It is well known that incineration of synthetic polymers can lead to unacceptable emissions of hazardous products. The recycling process of them has been associated with high cost due to the separation process and can lead to water contamination,[24] which can reduce the process sustainability.

Henceforth, the utilization of polymeric wastes in carbonized form for energy storage applications stands as a healthier option to solve the overwhelming environmental problem, which can as well as compensate for the prevalent high energy demand.[25–28] Extensive research and development of the technology to convert these non-biodegradable polymeric wastes to energy materials holds great potential. The carbonization process of non-charring polymers generally involves two steps: the decomposition of the polymers and the subsequent carbonization of the decomposed products to form carbon materials of varying sizes and morphologies. Synching the rate of the two steps is important. For instance, if the decomposition reaction in the first step is gradual, the amount of intermediate decomposition products will be insufficient to obtain a large quantity of carbon products. If the first decomposition reaction rate is faster and the second step is slower, many decomposed products will not be converted into carbon materials. Thus, the core scientific issue for the efficient conversion of non-charring polymers into carbon materials with defined sizes and morphologies is to modulate the decomposition behavior of polymers and the vapor deposition of decomposed products. Research on converting polymer wastes to carbon materials is beginning to gain traction in the energy storage industry. Such projects use carbon's structure to store charges, thereby creating electrochemical devices such as batteries and supercapacitors. This approach is acknowledged for its economic and environmental benefits, yet it remains in its developmental stage. The controlled carbonization of polymers has been proven to be an efficient method for synthesizing nanocarbons with well-defined morphologies and structures. It also provides a new strategy to fabricate polymer materials with improved flame-retardant nature and supplies a new approach to transforming waste polymers into high value-added carbon materials. Several carbon materials obtained from polymers or their wastes have been studied as electrode materials for metal-ion batteries.

This chapter conducts analyses on the available strategies that can be selected for making carbons from waste polymers and their applications in different MIBs. Although this progress has been achieved to a certain degree, there are still some challenges to address the major social needs in large-scale industrial applications.

20.2 AN OVERVIEW OF POLYMER WASTES

Different kinds of polymers are fabricated for different applications. The disposal of these used materials is a major concern in many countries. In contrast to naturally occurring polymers, these man-made polymers cannot be easily degraded by microorganisms.[29] Polymeric material wastes such as plastic and rubber cause enormous pollution across the globe. The majority of these waste materials are coming from industry and individual households. Recently, the recycling of these waste materials has already started in many developed countries; however, it is still a huge hurdle for many other countries. Further, the slow decomposition and accumulation of enormous amounts of these materials in landfills and oceans lead to major environmental concerns.[30] To avoid this increasing polymer waste pollution, effective management of polymer wastes is necessary. Apart from the recycling, biodegradation and incineration are also very effective strategies.

20.3 ENVIRONMENTAL, ECOSYSTEMIC, AND ECONOMIC ADVANTAGES

The use of polymers, especially plastic and rubber, are growing tremendously. Hence, it is necessary to consider the use of such polymer wastes for multiple applications. The demand for an energy storage system is increasing day by day. Thus, it is very reasonable to use those waste products to build various rechargeable batteries. During the production of plastics from the raw material, enormous amounts of greenhouse gases are released, which are hazardous to our environment and ecosystem, while the use of such waste materials for different applications instead of manufacturing them from scratch could reduce the production of greenhouse gases. Further, the effective use of waste materials for different applications helps to reduce the energy utilization (hence more economical) and consumption of raw materials. More importantly, it reduces the amount of such wastes getting accumulated in landfills and in water bodies and thus decreases the water and air pollution, resulting in the protection of many land and marine life.

20.4 DRAWBACKS OF USING POLYMER WASTE AND WAYS TO OVERCOME

Even though the waste products are very economical, the recycling process to make the desirable product such as anode materials for MIBs requires some high-end equipment for material production, by-product management, etc. These recycling processes may also produce some health hazards for people who are handling these materials. In addition to the people, it can pollute the environment due to the release of volatile organic compounds and other harmful chemicals during the recycling process. Another notable drawback is that the quality and reproducibility of the product may not be the same always, which could affect the battery performance. Further, the transportation, sorting, and cleaning of these waste materials could require tremendous energy and time. Even by considering all these disadvantages, we believe that the utilization of such waste materials in different MIBs would be an efficient strategy to meet the ever-growing demands for energy storage systems. All the aforementioned risks can be minimized/avoided by taking a proper protocol while handling and processing the waste materials. Further, the utilization of automation could also be a considerable approach to minimizing human contact and errors. As mentioned above, the quality and reproducibility of the products are also big concerns when it comes to battery material production. This can only be addressed through thorough quality checking as well as by maintaining an effective processing technique.

20.5 DIFFERENT TREATMENT STRATEGIES OF POLYMERIC WASTES

Researchers have tried out multiple ways to treat polymeric material wastes for battery applications. There are various techniques to recycle polymeric waste for energy storage applications. Among different approaches, chemical recycling is considered one of the most promising methods.[31] Pyrolysis of polymeric material wastes at high temperatures was observed to be a common technique to synthesize carbon-based materials for battery applications.[31–33] Hydrothermal carbonization is also a well-studied method to fabricate carbon forms from polymers.[33,34] This particular technique resembles the natural coal formation due to the self-buildup of pressure at low temperature.[33] The dry autoclaving technique is also used to generate carbonaceous materials from polymers.[35] For instance, studies have utilized dry autoclaving technique to synthesize solid, dense, low-surface-area functional spherical carbon particles from polyethylene (PE) for lithium-ion and sodium-ion batteries.[35,36] The dry autoclaving is also proven to generate CNTs from PE.[37] Pol and coworkers demonstrated that it was possible to completely recycle the plastic waste using a solvothermal upcycling process to produce high-yield and functional carbonaceous materials for various applications including battery electrodes.[33] Recently, a study has revealed that the electrospinning technique could be utilized to fabricate ultrathin long PET fibers having a uniform diameter.[38]

20.6 APPLICATIONS OF TREATED POLYMERIC WASTE FOR METAL-ION BATTERIES

20.6.1 Lithium-Ion Batteries (LIBs)

LIBs are the most promising battery systems for application in various electronic gadgets, portable devices, and electric vehicles. The electrochemical performances of these LIBs mainly depend on the properties of electrode and electrolyte materials. In general, commercially available LIBs contain metal oxide cathodes, carbon-based anodes, and $LiPF_6$ electrolytes. Among different carbonaceous compounds, graphite is the widely used anode for LIBs. This material possesses high electronic and ionic conductivity. Another promising property of graphite is its ability to intercalate and deintercalate Li-ions in the interlayer space, leading to a reversible Li storage capacity of 372 mAh/g. Other widely studied carbon-based anode materials for LIBs are carbon nanotubes (CNTs), graphene, and carbon nanofibers. Nevertheless, contrary to expectations, most of these systems showed lower specific capacity as compared to graphite. Thus, it is logical to design and investigate more carbonaceous materials to meet the ever-growing energy demands. Pol et al.[33] prepared carbon chips from the waste PE-based plastics bags using a novel solvothermal technique (Figure 20.1). The obtained carbonaceous material was employed as an anode in LIBs. In general, the melting temperature of the PE was between 115 °C and 135 °C. Above this temperature range, the material decomposed into hydrocarbons and released toxic gases such as CO, CO_2. This heating process alone would not provide any useful carbonaceous material for battery applications. However, in this study, they performed oxidative dehydrogenation along with SO_3H functionalization during the solvothermal process. Further, the subsequent elevated temperature annealing of the solvothermal processed product helped to remove the sulfonic acid, $-H_2$, $-CO$, $-CO_2$, $-H_2O$, and SO_2, thus generating pure carbon.[33] In this study, they have compared the electrochemical performances of carbon derived from low-density PE (LDPE-C) and high-density PE (HDPE-C). The obtained specific capacities of both the derived carbons (LDPE-C: ~400 mAh/g (at C/10 and 50 °C); HDPE-C: ~700 mAh/g (at C/10 and at 50 °C)) are higher than the theoretical specific capacity of graphite (372 mAh/g). Note that HDPE-C shows a much higher capacity than LDPE-C due to the presence of additional inorganic impurities that participate in the redox or alloying processes.[33,39] Both samples could retain over 200 mAh/g capacity even at 1C due to the smaller

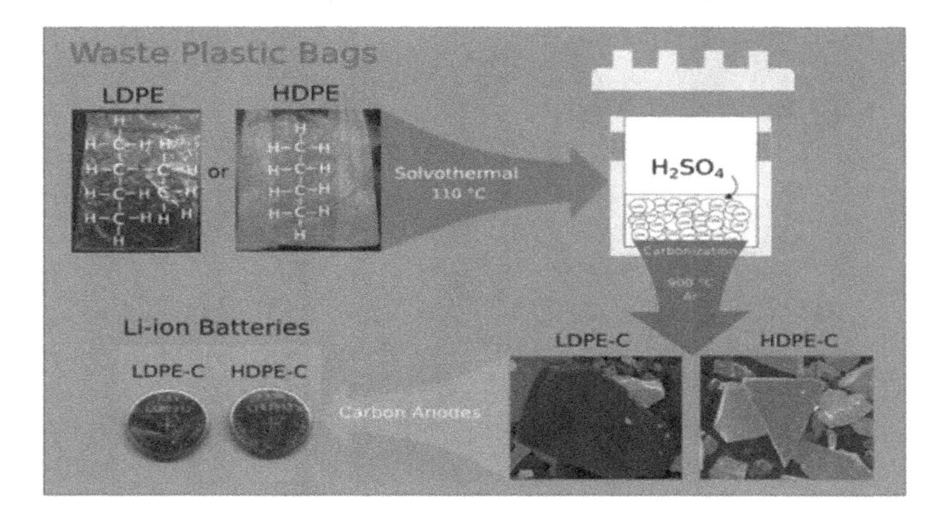

FIGURE 20.1 Schematic illustration of the solvothermal process of carbon chips obtained from used low- and high-density polyethylene (PE) plastic bags. (Reprinted with permission from *ACS Omega* 2018, 3, 12, 17520–17527, Copyright (2018) American Chemical Society.)

diffusion distance in these carbons as compared to a typical graphite sample.[33] A research group prepared carbon spheres (yield 50%) and nanotubes (yield 40%) using autogenic reactions from PE-based waste plastics and evaluated their electrochemical performance by employing them as anode materials in LIBs.[35] These materials could demonstrate a constant capacity of ~240 mAh/g for about 200 cycles at 1C in the voltage window of 1.5–0.005 V. Further, a capacity of 372 mAh/g was achieved for nanotube morphology when the upper cutoff voltage was raised from 1.5 to 3 V. To enhance the graphitic order and reduce the irreversible capacity loss in the first cycle, the spherical carbon particles were treated at an elevated temperature of 2,400 °C at inert conditions.[35] A scientific group recycled the plastic waste to solid-state electrolyte for lithium batteries.[31] They prepared bis(3-hydroxypropyl) carbonate and bis(5-hydroxypentyl)carbonate from plastic waste. In this study, they revealed that the plastic waste could be used as both a carbon feedstock via high-temperature decomposition and a potential precursor for the preparation of functional monomers.[31] Paranthaman and coworkers recovered carbon particles from waste tires and employed them as an anode material for LIBs (Figure 20.2).[40] They prepared a carbon using two different approaches: (i) chemical pre-treatment using hot oleum bath before pyrolysis (sulfonated sample) and (ii) controlled pyrolysis, both of which showed higher capacities than graphite. The sulfonated rubber-tire-derived carbon sample showed a better initial coulombic efficiency (71%) than the control rubber-tire-derived carbon (45%), but less than commercial graphite (90%). Further,

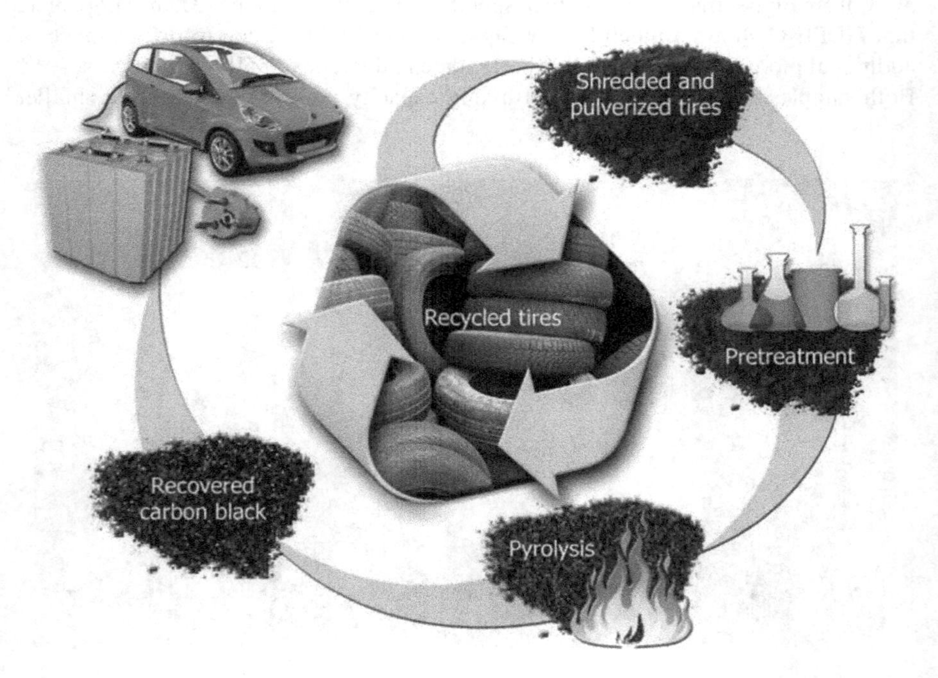

FIGURE 20.2 Schematic illustration of the recycling of sulfonated carbon black from waste rubber-tires. (Reprinted with permission from *RSC Advances*, 2014, 4, 38213–38221, Copyright (2014) Royal Society of Chemistry.)

the sulfonated sample showed a better yield and reversible capacity (390 mAh/g after 100 cycles) than the control sample.[40]

The human population is increasing rapidly; thus, the demand for food has also increased sharply. Hence, it is logical to use waste biomass to synthesize carbonaceous materials for battery applications. Studies have used seafood waste such as chitin and chitosan and other biomass to prepare carbonaceous anode materials for LIBs.[41–47] However, most of the carbons derived from biomass showed poor electrochemical performance when used as an anode material in metal-ion batteries. Nevertheless, a recent study produced carbon compounds with good performance from waste avocado seeds using a simple pyrolysis technique.[48] This biomass-derived carbon showed a good cycling stability of over 100 cycles with a reasonably good capacity of 315 mAh/g at 100 mA/g, which is comparable to the commercial graphite anode. Even though the carbon derived from waste materials shows very promising electrochemical performance in half-cell format, full-cell studies are missing (in all these studies), which is the important aspect when we consider these materials for any practical or commercial application. Hence, we believe that more and more studies should be encouraged in this area to transform these waste plastics and rubbers into electrode materials for highly efficient LIBs.

20.6.2 Sodium-Ion Batteries (SIBs)

The ever-increasing demands for energy storage systems and the unsustainable nature of lithium resources lead scientists across the world to think of an alternate metal-ion battery to LIBs such as SIBs. Further, over the last few decades, researchers have enormously been investigating SIBs, particularly in the areas of design, synthesis, and application of electrodes and electrolytes. In recent years, owing to the similar chemistry of SIBs to LIBs, the investigation into SIBs has rapidly increased. Moreover, the application of hard carbon as anode also sheds light on the research field of SIBs. The widely studied cathode materials for SIBs are polyanionic materials and layered metal oxides. However, so far studies on anode materials have mostly been limited to hard carbon. Even though electrochemical energy storage mechanisms of SIBs are similar to that of LIBs, the lower electrochemical performance (particularly energy density) of SIBs is always a big challenge. Hence, it is reasonable to explore more and more materials as electrodes for SIBs.

Scientists have explored various carbonaceous materials using waste plastics. For instance, Johnson and coworkers prepared carbon spheres using an autogenic technique from used low-density polyethylene bags.[36] This waste plastic-derived carbon showed reversible Na intercalation into graphene regions (this carbon did not possess any nanocavity or nanopore filling of sodium). The aforementioned characteristics and the spherical morphology led to the low reactivity nature with the electrolyte, which contributed to the highly reversible (de)sodiation in the material.[36] The density of the formed particle was 2.3 g/cc, which makes this waste particle-derived carbon suitable for SIBs as a dense anode. A study conducted by Sahajwalla's group utilized styrene-acrylonitrile (SAN) plastics from an end-of-life printer to prepare activated carbon anodes for SIBs.[49] These carbons could demonstrate a high rate performance. It could exhibit a capacity of 190 mAh/g at a current density of 3 mA/g after 25 cycles.

These carbon forms could also retain ~100% of their second discharge capacity even after 100 cycles at 20 mA/g.[49] Figure 20.3a demonstrates the schematic illustration of the preparation of activated carbon from SAN plastics using fast and low-temperature thermal transformation. This scheme also illustrates the effective assembly of this material as anode for coin-cell format SIBs. The discharge curves depict the capacity variation of the sample when heated at different temperatures (700 °C, 800 °C, and 900 °C).[49] A study conducted by Deng et al. reported amorphous carbon from waste polystyrene cups (Figure 20.3b), which exhibited a capacity of 116 mAh/g at a current density of 20 mA/g.[50] However, the specific capacity of this system seems to be low as compared to commercially available hard carbons. Many studies reported the upcycling of waste rubber-tire to prepare carbonaceous materials for SIBs.[51,52] For instance, Wang and coworkers prepared a highly efficient carbon composite with *in situ* generated ZnS from waste rubber-tire (Figure 20.3c).[52] When employed as an anode material for SIBs, it could deliver a capacity of 267 mAh/g even after 100 cycles at 50 mA/g. It was reported that the good dispersion of the ZnS nanoparticles in the carbon matrix could considerably improve the electrochemical properties of the composite material.[52] Another study conducted by Paranthaman et al. derived hard carbon compounds for SIB anodes from waste rubber-tire. The carbons synthesized at various temperatures such as 1,100 °C, 1,400 °C, and 1,600 °C and after acid treatment could demonstrate varying capacities of 179, 185, and 203 mAh/g, respectively, at 20 mA/g when used as anode in SIBs.[51] Many studies have utilized biomass and seafood waste to prepare carbon compounds for SIBs.[53–57] For instance, a study conducted by Ghimbeu et al. prepared a hard carbon anode material for SIBs from chitin and chitosan.[57] Although the hard carbon obtained from chitin and chitosan showed similar *d*-spacing and crystallite size, the pyrolysis of chitin led to a high-surface-area micro-/mesoporous carbon form, while the latter led to a non-porous carbon.[57] Both these carbon anodes showed a similar initial capacity of 280 mAh/g at 0.1C; however, their capacities started to deviate after long cycles at high C rates.[57] Most of these carbonaceous materials derived from waste polymer showed superior electrochemical performances as compared to the conventional hard carbons for SIBs. However, as mentioned above, we could not simply draw a conclusion by just looking into the half-cell performance of these materials. To evaluate the practical performance of any battery system, a full-cell analysis needs to be conducted. Hence, we hope that more studies would come up in the future with the potential full-cell performance of these materials for practical SIB applications.

20.6.3 Potassium-Ion Batteries (PIBs)

The ever-increasing demand for energy leads researchers across the world to think beyond LIBs. Among different MIBs, potassium-ion batteries (PIBs) have attracted significant consideration because of their distinctive properties and are hence recognized as one of the future energy storage systems of interest in the scientific community. One of the major advantages of PIB over SIB is that the graphite anode (which is a proven technology in LIBs) can be employed as an anode material in PIBs. Hence, unlike SIBs, developing new anode chemistry is not a big hurdle for PIBs. The growing use of private vehicles in recent years has started creating a large

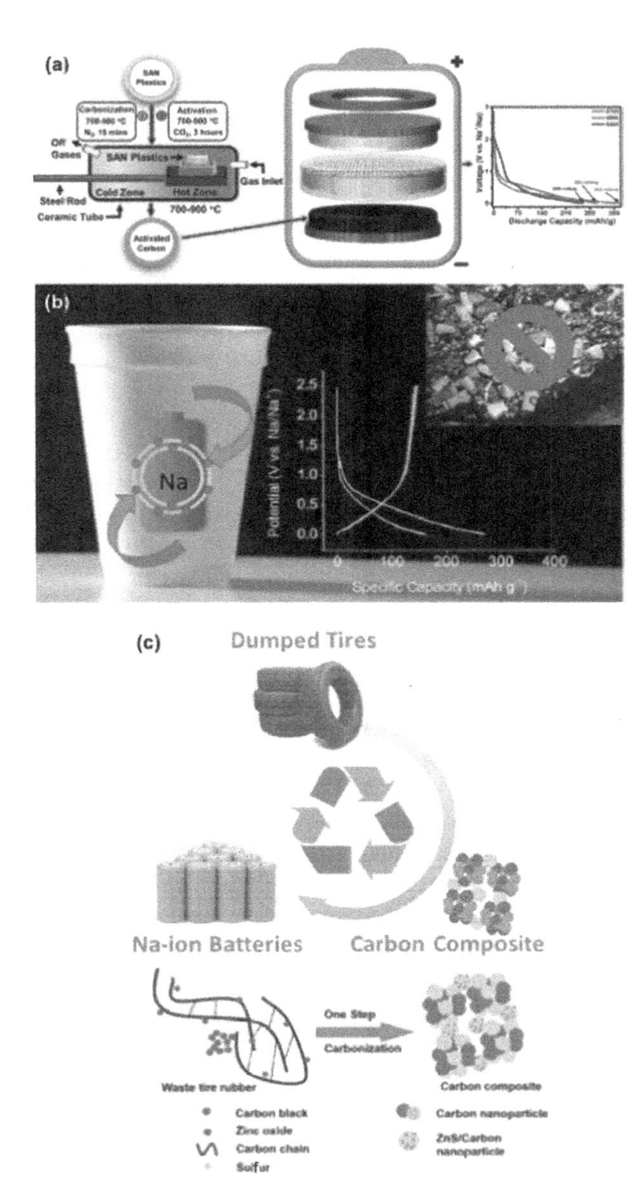

FIGURE 20.3 (a) Schematic illustration of the preparation of activated carbon from SAN plastics, their assembly as an anode for SIBs, and their discharge curves showing the specific capacities of the samples treated at different temperatures. (Reprinted with permission from *ACS Sustainable Chemistry & Engineering*, 2019, 7, 10310–10322, Copyright (2019) American Chemical Society.) (b) Illustration of recycling of plastic cup to fabricate effective SIBs and its charge/discharge characteristics. (Reprinted with permission from *ACS Sustainable Chemistry & Engineering*, 2015, 3, 9, 2153–2159, Copyright (2015) American Chemical Society.) (c) Schematic diagram represents the recycling of tires to make carbon composites for SIBs. (Reprinted with permission from *Dalton Transactions*, 2018, 47, 4885–4892, Copyright (2018) Royal Society of Chemistry.)

quantity of waste tires. It is also reported that about 1.5 billion tires are getting discarded across the globe, which causes severe health and environmental hazards due to their non-biodegradable nature.[58,59] Hence, it is essential to make these waste tire materials into some useful products such as carbonaceous compounds for batteries. For instance, a study conducted by Paranthaman et al. prepared a hard carbon from these waste rubber-tires using pyrolysis.[59] Xie et al. prepared nitrogen-doped carbon microspheres from seafood waste to address the inferior cycling life and rate capability in PIBs.[60] This chitin-derived carbon could deliver a superior rate performance (154 mAh/g at 72 C) and cycling stability (180 mAh/g at 1.8 C for 4,000 cycles without obvious capacity degradation). These excellent electrochemical performances could be attributed to (i) the dominant capacitive surface-driven process of the material due to its hierarchical porous structure and (ii) the improved K absorption and ionic/electronic conductivities because of the nitrogen-doped carbon structure.[60] PIB is relatively a nascent technology as compared to LIBs and SIBs; thus, it is very difficult to draw a conclusion regarding the practical application of waste plastic materials. Further, only limited studies have reported the utilization of waste polymers as electrode materials in PIBs. We believe that more studies would come up in the near future with cutting-edge technologies to make maximum utilization of waste polymer-based materials in PIBs.

20.7 SUMMARY AND OUTLOOK

Polymeric waste management, especially plastic, has become very difficult because the conventional landfilling methods are turning unfavorable because of the adverse environmental impact and the scarcity of landfills. Further, these polymers are prepared to resist biodegradation, resulting in more concerns. However, as we all know, the production and demand for such polymers are increasing day by day, resulting in the piling up of their wastes. This is mainly because of the ease of handling and the economic benefits of these polymers as compared to other products such as metal, wood, and glasses. Thus, it is logical to recycle these polymeric wastes to be used in different applications. In this regard, scientists across the world have revealed the potential applications of these polymeric waste materials as a promising candidate for making electrodes for different electrochemical energy storage systems. Plastic and rubber are the two most commonly used polymeric wastes to produce carbon materials for application as anode in different MIBs. However, these processed carbon materials possess some specific drawbacks; for example, it is difficult to reproduce good electrochemical results and to achieve stable performance. Another major hurdle to consider these materials as a potential source for future commercial application is that there are inadequate full-cell studies to prove the practical application of these materials in various MIBs. However, it is worth investigating in detail the electrochemical performance of these waste polymers in MIB applications to have a better understanding of these materials' battery performance. Prior to this, it is advisable to design controlled/suitable processing techniques to achieve preferable morphology with stable performance. While strategizing the synthesis procedures, it is advisable to consider an industrially viable processing technique. Nowadays, designing optimal processing techniques (not only to prepare anode materials for

MIBs, but also for other various applications) to recycle polymeric waste materials is considered a global challenge. We hope that soon it could come true due to the continuous effort from different industries, academic/research institutions, and government entities.

ACKNOWLEDGMENTS

This work was financially supported by the National Natural Science Foundation of China (No. 21975289 and No. U19A2019) and Hunan Provincial Science and Technology Plan Project, China (No. 2017TP1001, No. 2018RS3009, and No.2020JJ2042). This work was also financially supported by Postdoctoral International Exchange Program Funding of China (No. 115) and China Postdoctoral Science Foundation (2019M652802).

REFERENCES

1. Rowatt, R. J. The plastics waste problem, 1993.
2. Scott, G. *Polymers and the Environment*; Royal Society of Chemistry: Cambridge, 1999.
3. The compelling facts about plastics 2007, In *Plastics Europe*, 2007: Belgium.
4. Plastics – The facts 2012, In *Plastics Europe*, 2012: Belgium.
5. Plastic Pollution, United States Environmental Protection Agency: USA, 2013.
6. Report on the Indian Plastics Industry, Plastic India foundation, India, 2018.
7. Mutha, N. H.; Patel, M.; Premnath, V. Plastics materials flow analysis for India. *Resources, Conservation and Recycling* **2006**, *47*, 222.
8. Anuar Sharuddin, S. D.; Abnisa, F.; Wan Daud, W. M. A.; Aroua, M. K. A review on pyrolysis of plastic wastes. *Energy Conversion and Management* **2016**, *115*, 308.
9. Baheti, P. British Plastics Federation: London, 2021.
10. Waddams, A.; Petrochemicals L. By P. Wiseman, Ellis Horwood Ltd, Chichester, 1985. pp. 182. ISBN 0–85312-741–7. *British Polymer Journal* **1986**, *18*, 277.
11. Pearce, E. M. Principles of polymerization (third edition), by George Odian, Wiley-Interscience, New York, 1991, 768 pp. *Journal of Polymer Science Part A: Polymer Chemistry* **1992**, *30*, 1508.
12. Shastri, V. P. Non-degradable biocompatible polymers in medicine: Past, present and future. *Current Pharmaceutical Biotechnology* **2003**, *4*, 331.
13. Utracki, L. A. *International Abbreviations for Polymers and Polymer Processing*; Kluwer Academic Publishers: Netherlands, 2003.
14. Ojeda, T. *Polymers and the Environment*; IntechOpen, United Kingdom, 2013.
15. Kaminsky, W.; Menzel, J.; Sinn, H. Recycling of plastics. *Conservation & Recycling* **1976**, *1*, 91.
16. Peeters, J. R.; Vanegas, P.; Tange, L.; Van Houwelingen, J.; Duflou, J. R. Closed loop recycling of plastics containing flame retardants. *Resources, Conservation and Recycling* **2014**, *84*, 35.
17. Kaminsky, W.; Rossler, H. Olefins from wastes **1992**.
18. Kaminsky, W.; Schlesselmann, B.; Simon, C. Olefins from polyolefins and mixed plastics by pyrolysis. *Journal of Analytical and Applied Pyrolysis* **1995**, *32*, 19.
19. Akovali, G.; Bernardo, C. A.; Leidner, J.; Utracki, L. A.; Xanthos, M. *Frontiers in the Science and Technology of Polymer Recycling*; Springer, Switzerland, 1998, 351, 1.
20. Thiounn, T.; Smith, R. C. Advances and approaches for chemical recycling of plastic waste. *Journal of Polymer Science* **2020**, *58*, 1347.

21. Tokiwa, Y.; Calabia, B. P.; Ugwu, C. U.; Aiba, S. Biodegradability of plastics. *International Journal of Molecular Sciences* **2009**, *10*, 3722.
22. Chamas, A.; Moon, H.; Zheng, J.; Qiu, Y.; Tabassum, T.; Jang, J. H.; Abu-Omar, M.; Scott, S. L.; Suh, S. Degradation rates of plastics in the environment. *ACS Sustainable Chemistry & Engineering* **2020**, *8*, 3494.
23. Pantani, R.; Sorrentino, A. Influence of crystallinity on the biodegradation rate of injection-moulded poly(lactic acid) samples in controlled composting conditions. *Polymer Degradation and Stability* **2013**, *98*, 1089.
24. Nagy, Á.; Kuti, R. The environmental impact of plastic waste incineration **2016**, *15*, 231.
25. Dwivedi, P.; Mishra, P. K.; Mondal, M. K.; Srivastava, N. Non-biodegradable polymeric waste pyrolysis for energy recovery. *Heliyon* **2019**, *5*, e02198.
26. Williams, E. A.; Williams, P. T. Analysis of products derived from the fast pyrolysis of plastic waste. *Journal of Analytical and Applied Pyrolysis* **1997**, *40–41*, 347.
27. Al-Salem, S. M.; Antelava, A.; Constantinou, A.; Manos, G.; Dutta, A. A review on thermal and catalytic pyrolysis of plastic solid waste (PSW). *Journal of Environmental Management* **2017**, *197*, 177.
28. Mastral, F. J.; Esperanza, E.; García, P.; Juste, M. Pyrolysis of high-density polyethylene in a fluidised bed reactor. Influence of the temperature and residence time. *Journal of Analytical and Applied Pyrolysis* **2002**, *63*, 1.
29. Huang, S. J. Polymer waste management–Biodegradation, incineration, and recycling. *Journal of Macromolecular Science, Part A* **1995**, *32*, 593.
30. Horak, C.; 14 Oct 2020 ed.; International Atomic Energy Agency: Austria, 2020; Vol. F23036.
31. Saito, K.; Jehanno, C.; Meabe, L.; Olmedo-Martínez, J. L.; Mecerreyes, D.; Fukushima, K.; Sardon, H. From plastic waste to polymer electrolytes for batteries through chemical upcycling of polycarbonate. *Journal of Materials Chemistry A* **2020**, *8*, 13921.
32. Kim, P. J.; Fontecha, H. D.; Kim, K.; Pol, V. G. Toward high-performance lithium-sulfur batteries: Upcycling of LDPE plastic into sulfonated carbon scaffold via microwave-promoted sulfonation. *ACS Applied Materials & Interfaces* **2018**, *10*, 14827.
33. Villagómez-Salas, S.; Manikandan, P.; Acuña Guzmán, S. F.; Pol, V. G. Amorphous carbon chips Li-ion battery anodes produced through polyethylene waste upcycling. *ACS Omega* **2018**, *3*, 17520.
34. Titirici, M.-M.; Antonietti, M. Chemistry and materials options of sustainable carbon materials made by hydrothermal carbonization. *Chemical Society Reviews* **2010**, *39*, 103.
35. Pol, V. G.; Thackeray, M. M. Spherical carbon particles and carbon nanotubes prepared by autogenic reactions: Evaluation as anodes in lithium electrochemical cells. *Energy & Environmental Science* **2011**, *4*, 1904.
36. Pol, V. G.; Lee, E.; Zhou, D.; Dogan, F.; Calderon-Moreno, J. M.; Johnson, C. S. Spherical carbon as a new high-rate anode for sodium-ion batteries. *Electrochimica Acta* **2014**, *127*, 61.
37. Pol, V. G.; Thiyagarajan, P. Remediating plastic waste into carbon nanotubes. *Journal of Environmental Monitoring* **2010**, *12*, 455.
38. Mirjalili, A.; Dong, B.; Pena, P.; Ozkan, C. S.; Ozkan, M. Upcycling of polyethylene terephthalate plastic waste to microporous carbon structure for energy storage. *Energy Storage* **2020**, *2*, e201.
39. Etacheri, V.; Seisenbaeva, G.; Caruthers, J.; Daniel, G.; Nedelec, J.-M.; Kessler, V.; Pol, V. Ordered network of interconnected SnO_2 nanoparticles for excellent lithium-ion storage. *Advanced Energy Materials* **2014**, *5*, 1401289.
40. Naskar, A. K.; Bi, Z.; Li, Y.; Akato, S. K.; Saha, D.; Chi, M.; Bridges, C. A.; Paranthaman, M. P. Tailored recovery of carbons from waste tires for enhanced performance as anodes in lithium-ion batteries. *RSC Advances* **2014**, *4*, 38213.

41. Hernández-Rentero, C.; Marangon, V.; Olivares-Marín, M.; Gómez-Serrano, V.; Caballero, Á.; Morales, J.; Hassoun, J. Alternative lithium-ion battery using biomass-derived carbons as environmentally sustainable anode. *Journal of Colloid and Interface Science* **2020**, *573*, 396.
42. Liu, J.; Yuan, H.; Tao, X.; Liang, Y.; Yang, S. J.; Huang, J.-Q.; Yuan, T.-Q.; Titirici, M.-M.; Zhang, Q. Recent progress on biomass-derived ecomaterials toward advanced rechargeable lithium batteries. *EcoMat* **2020**, *2*, e12019.
43. Fromm, O.; Heckmann, A.; Rodehorst, U. C.; Frerichs, J.; Becker, D.; Winter, M.; Placke, T. Carbons from biomass precursors as anode materials for lithium ion batteries: New insights into carbonization and graphitization behavior and into their correlation to electrochemical performance. *Carbon* **2018**, *128*, 147.
44. Tao, L.; Huang, Y.; Yang, X.; Zheng, Y.; Liu, C.; Di, M.; Zheng, Z. Flexible anode materials for lithium-ion batteries derived from waste biomass-based carbon nanofibers: I. Effect of carbonization temperature. *RSC Advances* **2018**, *8*, 7102.
45. Wang, B.; Li, S.; Wu, X.; Liu, J.; Chen, J. Biomass chitin-derived honeycomb-like nitrogen-doped carbon/graphene nanosheet networks for applications in efficient oxygen reduction and robust lithium storage. *Journal of Materials Chemistry A* **2016**, *4*, 11789.
46. Li, R.; Zhou, Y.; Li, W.; Zhu, J.; Huang, W. Structure engineering in biomass-derived carbon materials for electrochemical energy storage. *Research* **2020**, *2020*, 8685436.
47. Lotfabad, E. M.; Ding, J.; Cui, K.; Kohandehghan, A.; Kalisvaart, W. P.; Hazelton, M.; Mitlin, D. High-density sodium and lithium ion battery anodes from banana peels. *ACS Nano* **2014**, *8*, 7115.
48. Yokokura, T. J.; Rodriguez, J. R.; Pol, V. G. Waste biomass-derived carbon anode for enhanced lithium storage. *ACS Omega* **2020**, *5*, 19715.
49. Kumar, U.; Goonetilleke, D.; Gaikwad, V.; Pramudita, J. C.; Joshi, R. K.; Sharma, N.; Sahajwalla, V. Activated carbon from e-waste plastics as a promising anode for sodium-ion batteries. *ACS Sustainable Chemistry & Engineering* **2019**, *7*, 10310.
50. Fonseca, W. S.; Meng, X.; Deng, D. Trash to treasure: Transforming waste polystyrene cups into negative electrode materials for sodium ion batteries. *ACS Sustainable Chemistry & Engineering* **2015**, *3*, 2153.
51. Li, Y.; Paranthaman, M. P.; Akato, K.; Naskar, A. K.; Levine, A. M.; Lee, R. J.; Kim, S.-O.; Zhang, J.; Dai, S.; Manthiram, A. Tire-derived carbon composite anodes for sodium-ion batteries. *Journal of Power Sources* **2016**, *316*, 232.
52. Wu, Z.-Y.; Ma, C.; Bai, Y.-L.; Liu, Y.-S.; Wang, S.-F.; Wei, X.; Wang, K.-X.; Chen, J.-S. Rubber-based carbon electrode materials derived from dumped tires for efficient sodium-ion storage. *Dalton Transactions* **2018**, *47*, 4885.
53. Zhu, J.; Roscow, J.; Chandrasekaran, S.; Deng, L.; Zhang, P.; He, T.; Wang, K.; Huang, L. Biomass-derived carbons for sodium-ion batteries and sodium-ion capacitors. *ChemSusChem* **2020**, *13*, 1275.
54. Zhu, Y.; Chen, M.; Li, Q.; Yuan, C.; Wang, C. A porous biomass-derived anode for high-performance sodium-ion batteries. *Carbon* **2018**, *129*, 695.
55. Hong, K.-l.; Qie, L.; Zeng, R.; Yi, Z.-Q.; Zhang, W.; Wang, D.; Yin, W.; Wu, C.; Fan, Q.-J.; Zhang, W.-x. et al. Biomass derived hard carbon used as a high-performance anode material for sodium ion batteries. *Journal of Materials Chemistry A* **2014**, *2*, 12733.
56. Liu, P.; Li, Y.; Hu, Y.-S.; Li, H.; Chen, L.; Huang, X. A waste biomass derived hard carbon as a high-performance anode material for sodium-ion batteries. *Journal of Materials Chemistry A* **2016**, *4*, 13046.
57. Conder, J.; Vaulot, C.; Marino, C.; Villevieille, C.; Ghimbeu, C. M. Chitin and chitosan—Structurally related precursors of dissimilar hard carbons for Na-Ion battery. *ACS Applied Energy Materials* **2019**, *2*, 4841.

58. Danon, B.; van der Gryp, P.; Schwarz, C. E.; Görgens, J. F. A review of dipentene (dl-limonene) production from waste tire pyrolysis. *Journal of Analytical and Applied Pyrolysis* **2015**, *112*, 1.
59. Li, Y.; Adams, R. A.; Arora, A.; Pol, V. G.; Levine, A. M.; Lee, R. J.; Akato, K.; Naskar, A. K.; Paranthaman, M. P. Sustainable potassium-ion battery anodes derived from waste-tire rubber. *Journal of The Electrochemical Society* **2017**, *164*, A1234.
60. Chen, C.; Wang, Z.; Zhang, B.; Miao, L.; Cai, J.; Peng, L.; Huang, Y.; Jiang, J.; Huang, Y.; Zhang, L. et al. Nitrogen-rich hard carbon as a highly durable anode for high-power potassium-ion batteries. *Energy Storage Materials* **2017**, *8*, 161.

21 Biowaste-Derived Components for Zn–Air Battery

Yiyang Liu
University College London (UCL)

Tasnim Munshi
University of Lincoln

Jennifer Hack
University College London (UCL)

Ian Scowen
University of Lincoln

Paul R. Shearing
University College London (UCL)
The Faraday Institution

Guanjie He
University College London (UCL)
University of Lincoln

Dan J. L. Brett
University College London (UCL)
The Faraday Institution

CONTENTS

DOI: 10.1201/9781003178354-25

21.1 INTRODUCTION

Climate change has become the focus of international initiatives such as the Kyoto Protocol (1997) and the Paris Agreement (2015), which aim to reduce greenhouse gases (GHGs) and keep global warming below 2 °C through the carbon offsets. Generally, this is achieved by increasing the utilization and efficiency of renewable energy and fuel conversion, which have stimulated the development of novel energy storage and conversion devices (e.g., batteries, supercapacitors, and fuel cells) [1]. Rechargeable batteries, especially lithium-ion batteries (LIBs) based on the intercalation mechanism, account for more than 95% of the market share due to their advantages such as high energy density (120–265 Wh/kg) and long service life. Although LIBs have become the powerful driving force for electrification and smart economy, commercial LIBs are approaching their theoretical limits, making them difficult to meet the increasing demand for higher energy density devices. In addition, other inherent shortcomings of LIBs, such as high cost (100–150 USD/kg), potential safety risks, and difficulties in the recycling of raw materials, limit their further commercial deployment [2].

As promising candidates for energy storage in the post-lithium battery era, metal–air batteries have received great attention due to their wide range of applications. Their suitable energy storage scenarios include portable electronic products (100–20,000 mAh), electric vehicles (1–300 kWh), aerospace equipment (300 kWh–several MWh), and grid-scale storage (>MWh level). At present, a variety of metal–air batteries have been proposed, based on various metals including Li, Zn, Mg, and Al. Among these metal–air batteries, Zn–air batteries (ZABs) have received particular attention due to their many advantages such as: (i) high theoretical energy density (1,353 Wh/kg); (ii) high density (7.133 g/cm^3) and high theoretical specific capacity of Zn metals (820 mAh/g, 5,855 mAh/cm^2); (iii) suitable metal activity that enables to operate in aqueous electrolytes; (iv) good reserves in the earth's crust, low cost (~3,000 USD per ton), and convenience of recycling, which are conducive to mass production; and (v) low operational risk and high environment-friendliness [3]. Although this technology has been proposed for nearly two centuries, the first commercialization of rechargeable zinc–air batteries was not realized until 2012. This is mainly limited by the following key factors: (i) the sluggish kinetics and the high overpotential of both the oxygen reduction reaction (ORR) and oxygen evolution reaction (OER); (ii) the lack of high-performance and low-cost ORR/OER bifunctional electrocatalysts; and (iii) dendrite growth, hydrogen evolution, and passivation on Zn anodes that reduce long cycling durability and coulombic efficiency [4].

In addition to electrochemical performance, another factor that affects the future commercial application of batteries is sustainability, including the sum of chemical composition, synthesis process, implementation in the system, and recycling.

At present, most commercial batteries represented by LIBs usually fail to meet the sustainability standards. They typically contain powdery inorganic active materials consisting of multiple different heavy metals in the electrodes, very thin separators (separating the cathode from the anode), and highly flammable organic electrolytes; these materials are often difficult to recycle and may have negative impacts on the environment. Biowaste-derived materials have attractive advantages in terms of sustainability, environmental benignity, and more importantly, the versatility of structures/compositions offering unique physical and chemical properties, which have aroused great interest as electrode materials for high-performance rechargeable batteries. It is worth noting that unprocessed biomass usually has an inherent specific porous structure and a certain content of non-carbon elements, such as N, P, and B. Therefore, they can serve as natural precursors for heteroatom-doped porous carbon materials, which are potentially suitable bifunctional electrocatalysts for air cathodes in ZABs. Meanwhile, biowaste-based aqueous binders can replace toxic halogenated commercial binders, thereby realizing the future of sustainable energy storage devices. In addition, electrolytes and separators can be also synthesized from biowaste [5].

To provide a necessary and comprehensive understanding for new researchers in the fields of biowaste-derived components for ZABs, this chapter provides the introduction and discussion covering: (i) the working principle of ZABs in the aspects of Zn anodes, electrolytes, and air cathodes (Section 21.2); (ii) biowaste-derived bifunctional electrocatalysts for ZABs (Section 21.3); (iii) other biowaste-derived components for ZABs (Section 21.4); and (iv) conclusions and perspectives (Section 21.5).

21.2 WORKING PRINCIPLES OF Zn–AIR BATTERIES

For typically closed systems, such as insertion-type batteries (e.g., LIBs and aqueous Zn-ion batteries) and conversion-type batteries (*e.g.*, nickel–metal hydride batteries and primary alkaline Zn–MnO_2 batteries), their capacity is usually dependent on the solid-state electrode materials. Generally, the conventional solid-state electrodes are confronted with a limited specific capacity, owing to the insufficient depth of discharge (DoD) and rate of utilization. In contrast, traditional open systems represented by the redox flow batteries (RFBs) store charge and energy *via* the redox pairs in the flowing electrolytes, which completely decouples the energy and power. In such a system, the device capacity is only determined by the amount of electrolytes. Therefore, for typical RFBs, large electrolyte storage tanks are commonly used, thereby limiting their applications in scenarios with volume and weight restrictions. In addition, RFBs are also limited by several intrinsic issues: (i) limited solubility of active species in the electrolytes (*e.g.*, 0.8 M for Ce^{3+} in Zn–Ce flow batteries); (ii) the use of ion exchange membranes with high expenditure and poor durability; and (iii) high device cost (*e.g.*, > 800 USD per kWh for all-vanadium RFBs).

The ZAB is a special semi-open system consisting of four key components: Zn anodes, separator, air cathodes (containing bifunctional electrocatalysts), and electrolyte (*e.g.*, 6 M KOH). In such a system, the charge storage depends on the reversible electrochemical deposition and dissolution of the Zn, and the O_2-related conversion reactions catalyzed by the air cathode [6]. Different from the battery with a closed system, the use of air electrodes makes the capacity of ZAB independent of the cathode, which enables the Zn anode to achieve an extremely high DoD (>95%). Also, Zn metals possess a high theoretical volumetric capacity (5,854 mAh/L) compared to that of Li metals (2,277.4 mAh/L), owing to their high density and capability of two-electron transfer process, which enable ZABs to achieve a unique quasi-decoupling state of energy and power. In addition, it is worth noting that the use of O_2 in the air as one of the reactants significantly reduces the mass required by the cathode, thereby greatly increasing the energy density of the device.

Theoretically, the ZAB could achieve a very high energy density (1,353 Wh/kg) compared to that of LIBs (100–265 Wh/kg). In practice, since its first invention in 1878, commercial primary ZABs could usually achieve a high energy density range of 200–500 Wh/kg. Nevertheless, the state-of-the-art secondary ZABs have demonstrated only limited energy densities (<40 Wh/kg) and cycling durabilities (<500 cycles), owing to several substantial challenges: (i) the thermodynamic instability and dendrite formation on Zn anodes in aqueous electrolytes and (ii) the insufficient catalytic activity of bifunctional air cathodes, which leads to a high overpotential for both ORR and OER. It is worth noting that the reactions of ZABs during the cycling are shown in Equations 21.1–21.3. In theory, the discharge voltage of ZABs is 1.69 V; however, their working potentials are usually restricted to <1.3 V, owing to the high ORR and OER overpotential.

$$\text{Cathode: } Zn(s) + 4OH^-(aq) \overset{\text{discharge}}{\Rightarrow} Zn(OH)_4^{2-}(aq) + 2e^- \tag{21.1}$$

$$\text{Anode: } O_2(g) + 2H_2O(l) + 4e^- \overset{\text{discharge}}{\Rightarrow} 4OH^-(aq) \tag{21.2}$$

$$\text{Overall: } 2Zn(s) + O_2(g) + 2H_2O(l) + 4OH^-(aq) \overset{\text{discharge}}{\Rightarrow} 2Zn(OH)_4^{2-}(aq) \tag{21.3}$$

21.3 ENERGY STORAGE MECHANISMS FOR AIR CATHODES

The charge storage on the cathodic side of secondary ZABs is achieved by bifunctional electrocatalysts that involve oxygen-related conversion reactions, including the ORR and OER. Both ORR and OER are multi-step and multi-electron transfer processes that can occur in acidic, neutral, and alkaline electrolytes. Since OER is an intrinsically reverse reaction of ORR, this section will mainly focus on the introduction and discussion of ORR mechanisms. Figure 21.1 demonstrates the ORR pathways in alkaline electrolytes.

Nørskov and colleagues suggested ORR in alkaline electrolytes could be classified into two based on the following mechanisms: (i) associative mechanisms (Equations 21.4 and 21.5) that relate to the formation of $HO_{2,ad}$ or $O^-_{2,\,ad}$, which is more thermodynamically favorable, and (ii) dissociative mechanisms (Equation 21.6) that relate to the dissociation of O_2 and the formation of O^-_{ad} [7].

$$O_2 + e^- \leftrightarrow O^-_{2,\,ad} \tag{21.4}$$

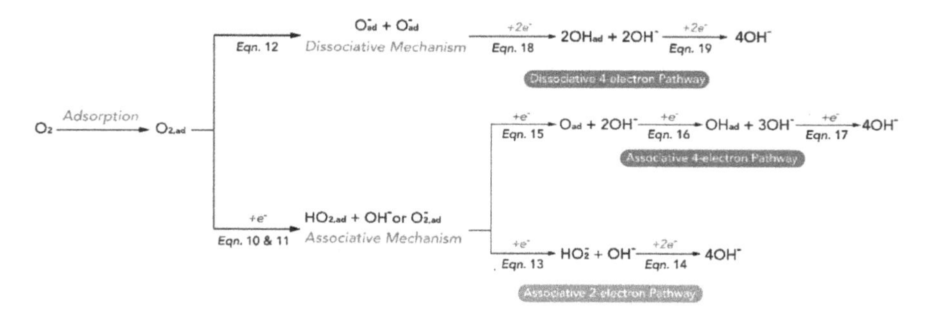

FIGURE 21.1 ORR pathways in alkaline electrolytes.

$$O_2 + 2e^- \leftrightarrow O_{ad}^- + O_{ad}^- \tag{21.5}$$

$$O_2 + H_2O + e^- \leftrightarrow OH^- + HO_{2,ad} \tag{21.6}$$

Subsequently, it will proceed through either a four-electron pathway (Equations 21.7–21.11), or a two-electron pathway (Equations 21.12 and 21.13). Generally, the direct four-electron pathway is more favorable for ZABs, owing to its faster kinetics than the indirect two-electron pathway. For the four-electron process, it can be classified as dissociative and associative mechanisms:

Dissociative four-electron pathway:

$$O_{ad} + O_{ad} + 2H_2O + 2e^- \leftrightarrow 2OH_{ad} + 2OH^- \tag{21.7}$$

$$2OH_{ad} + 2OH^- + 2e^- \leftrightarrow 4OH^- \tag{21.8}$$

Associative four-electron pathway:

$$HO_{2,ad} + OH^- + e^- \leftrightarrow O_{ad} + 2OH^- \tag{21.9}$$

$$O_{ad} + 2OH^- + H_2O + e^- \leftrightarrow OH_{ad} + 3OH^- \tag{21.10}$$

$$OH_{ad} + 3OH^- + e^- \leftrightarrow 4OH^- \tag{21.11}$$

For the indirect two-electron process, it will first form the HO_2^-, which will subsequently be reduced to OH^- or the chemical disproportionation to O_2 and H_2O. Owing to the instability of HO_2^- in alkaline electrolytes, the chemical disproportionation has sluggish kinetics and poor efficiency; therefore, most of ZABs rely on the four-electron pathway.

Associative two-electron pathway:

$$HO_{2,ad} + e^- \leftrightarrow HO_2^- \tag{21.12}$$

$$2OH_{ad} + 2OH^- + 2e^- \leftrightarrow 4OH^- \quad (21.13)$$

$$4OH_{ad} \rightarrow 2H_2O + O_2 \quad (21.14)$$

Recently, Wang and colleagues have reported a Zn-O_2/ZnO_2 chemistry showing higher reversibility and energy density, which provides the possibility for a two-electron pathway device design [6]. Coupling material characterization and theoretical simulation, they proposed that the unique water-poor and (Zn^{2+})-rich inner Helmholtz plane (IHP) on air cathodes will alleviate the four-electron pathway and enable the stable existence of the ZnO_2 on the surface of cathodes.

One major challenge for the air cathode is the high overpotential of both two-electron and four-electron pathways, which leads to several issues, including: (i) sluggish reaction kinetics; (ii) low device power density (50–100 W/kg); and (iii) low round-trip efficiency (<65%) compared to that of commercial LIBs (typically 80%–90%). To address these issues, numerous efforts have focused on the development of highly efficient bifunctional electrocatalysts. However, compared to single functional electrocatalysts, the design of ORR/OER bifunctional electrocatalysts is much more difficult, since, to achieve the optimized activity, the desired adsorption energies of the oxygen-containing species are different. Current efforts mainly focus on the non-precious metal-based and carbon-based materials, which aim to improve the materials in the following aspects: (i) enhancing the ORR/OER bifunctional catalytic activity simultaneously; (ii) increasing the electrical conductivity to promote electron transfer; (iii) increasing the porosity to facilitate mass transport; (iv) improving the electrochemical and mechanical stability to broaden operation ranges (*e.g.*, temperature and pH value).

21.4 BIOWASTE-DERIVED BIFUNCTIONAL ELECTROCATALYSTS

The possibility of widespread commercialization of ZABs is closely related to the cost of electrodes, especially air cathodes. Currently, the conversion of biowaste into active carbon materials is regarded as an attractive method to obtain bifunctional electrocatalysts from more economical, abundant, and renewable resources. As shown in Figure 21.4, various types of biowaste have been used in the synthesis of bifunctional electrocatalysts. The common resources could be categorized as plants (lignocellulose) or animals. Plant biowaste is synthesized through biological photosynthesis, which includes trees, grass, and leaves, whereas animal biowaste mainly contains proteins and minerals, which exist in a variety of forms, such as shells, blood, human wastes, and feathers [5]. This section will provide an overview, introduction, and discussion of typical biowaste-derived material synthesis routes and representative biowaste-derived bifunctional electrocatalysts.

21.4.1 Treatment of Biowaste-Derived Bifunctional Electrocatalysts

In order to convert the biowaste into a usable form for electrocatalysts, pre-treatment of the biowaste precursor is required. Various methodologies have been proposed for the treatment of biowaste precursors, including hydrothermal carbonization (HTC), direct pyrolysis, chemical/physical/catalytic activation, ammonia gasification, hard/soft templating, carbonization of organic aerogels, and ionothermal carbonization (ITC) [5]. HTC (also referred to as "aqueous carbonization at elevated temperature and pressure") converts biomass into carbon through four steps: dehydration, condensation, polymerization, and aromatization. Generally, this process is conducted in aqueous solutions, under a mild temperature (<200 °C) and self-generated pressure. Although the HTC process is easy to control and conducive to mass production, biowaste-derived carbon usually demonstrates a low surface area (<10 m^2/g) and poor electron conductivity. Therefore, further activation and graphitization steps are necessary to convert the carbon to a suitable form.

The activation processes usually mix different types of activators with biowaste precursors that subsequently undergo a series of reactions to alter the arrangement of carbon atoms. During the activation process, partially reacted carbon atoms will vaporize or fall off, thereby increasing the specific surface area and porosity of biowaste-derived materials. The physical activation refers to the use of carbon dioxide, water vapor, air (oxygen), and other oxidizing gases to react with carbon atoms inside the biowaste precursor. Under high temperatures (typically 800 °C–1,000 °C), hydrogen and oxygen elements in biowastes are gradually decomposed and carbon atoms are continuously cyclized and aromatized [8]. Although the physical activation method has many advantages, such as low pollution and simple process, the as-prepared carbon materials are usually relatively low in specific surface area (1,000–1,500 m^2/g) and pore volume (<1.0 cm^3/g). This is mainly because the gas activator adopted in the physical activation possesses a relatively weak reactivity, and it is difficult to form a highly porous structure inside biowaste precursors.

Different from physical activation, chemical activation processes mix a chemical activating agent and biowaste precursors uniformly in a certain proportion, and then carbonize and activate them in an inert atmosphere to prepare porous carbon materials. Essentially, in these processes, chemical activation agents penetrate biowaste precursors and conduct reactions at high temperatures to form abundant pores. The chemical activation has the merits of a low reaction temperature, good materials uniformity, and high specific surface area. However, a certain amount of the chemical activating agents may remain in the as-prepared carbon material, which could block the pores; therefore, the chemical activation must be accompanied by a washing process to remove chemical activating agents. In addition, the by-products produced in the chemical activation process are likely to corrode equipment and cause environmental pollution, which limits the practical application of this method.

The catalytic activation is another effective method for preparing mesoporous biowaste-derived carbon materials. This method usually mixes precursors and metal compounds (*e.g.*, iron, cobalt, nickel, and other precious metal compounds) by two types of processes: (i) directly adding metal compounds to biowaste precursors and then conducting carbonization and activation, and (ii) immersing biowaste precursors into metal inorganic salt solutions, then subsequently drying to remove the solvent, and activating at a high temperature. During the activation, the presence of metal ions will selectively vaporize carbon atoms with higher crystallinity to increase the active sites inside the micropores of carbon materials, thereby expanding micropores to mesopores. Meanwhile, the pores formed when the gasification product escapes to the surface of the materials will also remain in the final prepared carbon materials. Although catalytic activation is a promising strategy for converting biowaste into carbon materials, it also faces disadvantages. For instance, metal compounds will inevitably remain inside biowaste-derived carbon as impurities, which may affect the catalytic activity and significantly increase the complexity of mechanism investigations.

However, there are some inherent defects in conventional activation methods that are difficult to overcome, including poor pore structure orderliness and difficult-to-regulate pore structure. The templating methods can regulate the synthesis of biowaste-derived carbon materials at the nanoscale, which is widely regarded as the most promising method for regulating the pore structures of carbon materials. Besides, biowaste-derived porous carbon materials prepared by the templating method possess advantages such as narrow distribution of pore size, large specific surface area, and interconnected pore network [9]. According to the different structures and properties of the templates used, the templating method can be divided into hard templating and soft templating.

Hard templating, also known as "nanocasting", involves a series of steps, including coating or filling of a rigid template with a biowaste precursor material, the treatment of the precursor to form the desired material, and the removal of the template. The morphology and structure of the porous carbon material obtained by the hard template method are determined by the morphology and structure of the template, such as nanorods, nanofibers, nanospheres, and nanoparticles. Generally, the selection of a hard template requires a certain rigidity of the template material to prevent pore structures from collapsing in subsequent reactions. Meanwhile, the internal connectivity of the template is another critical factor to determine if the reverse-replicated material can be perfectly presented. At present, zeolites, SiO_2, and Al_2O_3 have widely been used as hard templates for microporous, mesoporous, and macroporous biowaste-derived carbon materials, respectively. Although the biowaste-derived carbon materials prepared by hard templating can be well controlled for the morphology and porosity, it still faces a series of challenges. First, the hard template used needs additional preparation, which increases the complexity of the production. Second, the hard template will be sacrificed during the preparation process, thereby increasing the production cost and the difficulty in industrial production. Finally, the removal of typical hard templates (*e.g.*, SiO_2) requires the use of highly hazardous agents (*e.g.*, HF and NH_4HF_2), which increases operation risks and causes environmental pollution [10].

The soft templating uses amphiphilic molecules or block copolymers as soft templates and conducts organic–organic self-assembly with the biowaste precursor in an organic solvent or water phase to form nanomicelles. Then, the precursor is cured to form a three-dimensional cross-linked rigid structure and then carbonized to obtain an ordered porous carbon material. Compared with hard templates, soft templates can control the spatial structure of the template through hydrogen bonding, hydrophilic and hydrophobic interactions, and ion coordination and thus can realize materials design at the molecular level [11]. The synthesis of biowaste-derived porous carbon materials by soft templating can be roughly divided into two routes, including evaporation-induced self-assembly (EISA) and aqueous-phase synthesis (also known as hydrothermal method). The EISA method uses a volatile substance (*e.g.*, ethanol and tetrahydrofuran) as a solvent and slowly volatilizes it at room temperature. With the gradual increase in the surfactant and precursor concentration in the system, the precursor self-assembles into the desired structure, which is guided by the templating agent; subsequently, a composite liquid crystal phase will be formed under high-temperature treatment. After removing the soft template, porous carbon material can be obtained [12]. The aqueous-phase synthesis method is usually realized through a sol–gel process. First, a certain concentration of template agents and a biowaste precursor are mixed to prepare the solution, which is then hydrolyzed and condensed at a lower temperature to form a precipitate. Next, the precursor is further cross-linked and polymerized by hydrothermal treatment at a higher temperature. The biowaste-derived porous carbon material can be prepared through a series of processes including filtration, washing, drying, and roasting.

Soft templating does not require a template preparation process, thereby simplifying the synthesis steps of the biowaste-derived porous carbon materials and reducing production costs. In addition, the as-prepared material has the spatial symmetry of a variety of pore structures, which greatly expands the application range of porous carbon materials. However, compared with hard templating, in soft templating it is more difficult to achieve a precise control of the pore structure [10].

21.4.2 Representative Biowaste-Derived Bifunctional Electrocatalysts

Once the biowaste-derived carbon has been prepared, there is a wide range of opportunities for creating biowaste-based bifunctional electrocatalysts, as will be discussed here. Generally, heteroatom doping can greatly improve the catalytic activity of carbon-based metal-free catalysts for ORR. N-doped carbon materials are promising candidates as bifunctional electrocatalysts for ZABs. It should be noted that the catalytic performance of N-doped metal-free carbon is closely relevant to the nitrogen-containing precursor and synthetic conditions. Compared with other chemicals, nitrogen-rich biowastes are extremely abundant, renewable, and inexpensive. Therefore, the conversion of nitrogen-rich biowastes into nitrogen-doped carbon has attracted widespread attention. Qiao and coworkers prepared a porous N-doped carbon-based bifunctional electrocatalyst *via* the molten salt (KCl) pyrolysis of chitosan precursors at 800°C [13]. After pyrolysis, the content of N species, defects, and specific surface area of catalyst are raised, thus resulting in the high active site density. Compared to 20 wt.% Pt/C, the as-prepared material demonstrates a superior stability, methanol tolerance, and a higher ORR catalytic activity (half-wave potential of 0.86 *vs.* RHE) in alkaline electrolytes. Meanwhile, the ZAB with the prepared electrocatalyst exhibits a high peak power density (178 mW/cm^2), specific capacity (780 mAh/g Zn), and good cycling durability.

The rational design of the structure and active sites of the material will significantly improve the performance of metal-free bifunctional electrocatalysts. Through the pyrolysis and hydrothermal treatment of a traditional Chinese medicine Acori

Tatarinowii Rhizoma (ATR), Li et al. developed a 3D hierarchical porous non-metal carbon electrocatalyst [14]. The as-prepared catalyst exhibits an excellent electrocatalytic performance toward both ORR and OER ($E_{1/2}=0.85$ V, $E_i=10=1.68$ V *vs.* RHE) in alkaline electrolyte, which is superior to that of commercial Pt/C-RuO_2. They attribute these to several favorable characteristics, including 3D hierarchical structures and rich active sites of the prepared catalysts (N functional groups, oxygen vacancies, and carbon defects).

To improve the catalytic activity of the biowaste-derived metal-free carbon materials and further expand their application for ZABs, it is an effective methodology to introduce transition metals (*e.g.*, Fe, Mn, Co, and Ni) and their derivatives into the porous carbon catalyst. As shown in Figure 21.2, Yang and colleagues successfully synthesized a FeNi alloy and N-co-doped porous carbon (FeNi-NC) using peanut shells as precursors [15]. The as-prepared FeNi-NC electrocatalyst demonstrated a high ORR catalytic activity with almost the same onset (0.98 V, *vs.* RHE) and half-wave potentials (0.83 V, *vs.* RHE) as that of commercial 20% Pt/C, but also a remarkable OER performance with the similar activity compared to IrO_2 in alkaline electrolyte. In addition, the FeNi-NC catalyst possesses an overpotential ($E_{OER}-E_{ORR}$) of 0.81 V at 10 mA/cm^2, which is smaller than the 20% Pt/C and IrO_2 systems. Moreover, the assembled ZAB using FeNi-NC as the air cathode demonstrated a superior charging and discharging polarization curve and a larger peak power density than 20% Pt/C and IrO_2 systems, as well as a better long-term durability.

In addition to heteroatom doping and composite fabrication, defects engineering is another promising methodology to improve the catalytic activity of biowaste-derived bifunctional electrocatalysts for ZABs. Recently, Liu and coworkers synthesized graphene-like and defect-rich carbon sheets with N-doping (GPNCS), using the fruits of glossy privet as a precursor (Figure 21.3) [16]. Owing to the graphene-like and porous structure, GPNCS exhibits a high electrical conductivity and specific surface area (1,559 m^2/g), which are beneficial for electron transfer and mass transport. In addition, the synergistic effect between N-doping atoms and topological defects endows the GPNCS with a high ORR electrocatalytic activity, an onset potential of 0.92 V, and a small 7 mV negative shift in the half-wave potential over 10,000 cycles. In addition, the as-assembled ZABs demonstrated a low charge–discharge voltage gap that remains nearly unchanged after 1,340 cycles at 10 mA/cm^2.

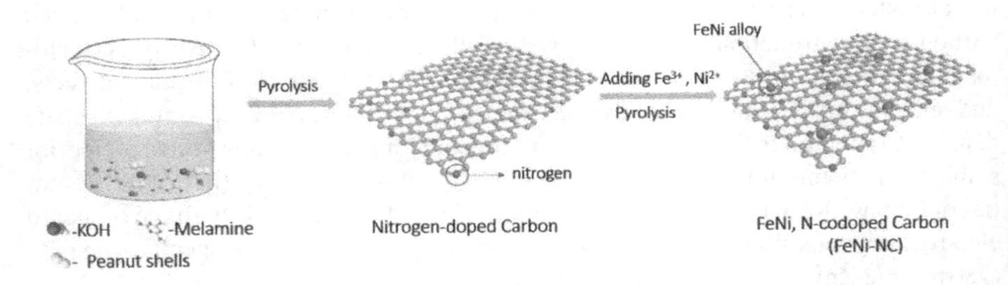

FIGURE 21.2 Schematic illustration of the synthesis of FeNi and nitrogen-co-doped porous carbons from peanut shells. (Adapted with permission from Ref. [15]. Copyright (2018) Elsevier.)

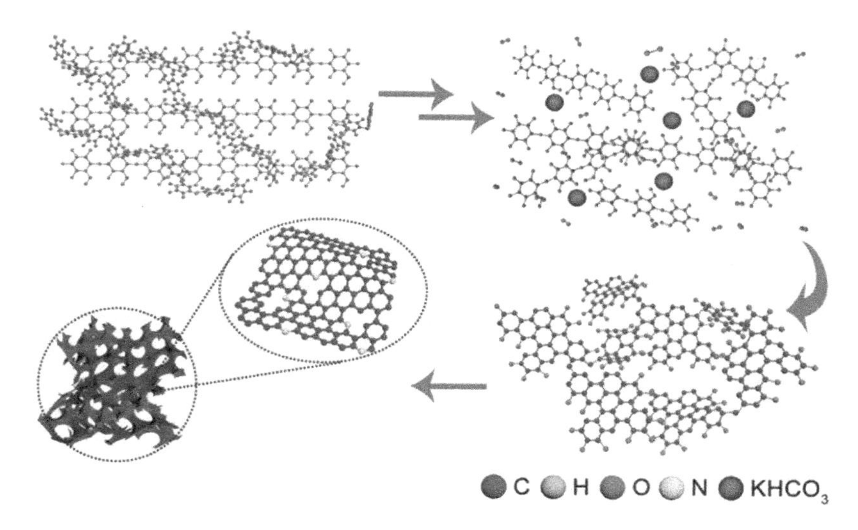

FIGURE 21.3 Simulation of the formation mechanisms of graphene-like and defect-rich carbon sheets with N-doping (GPNCS). (Adapted with permission from Ref. [16]. Copyright (2020) American Chemical Society.)

21.5 OTHER BIOWASTE-DERIVED MATERIALS FOR ZABs

In addition to their promise for use in bifunctional electrocatalysts, biowaste-derived materials hold promise for use in other components of ZABs. Their use would help both to further drive down costs and to cement ZABs as a less toxic battery alternative. A range of applications of biowaste-derived materials in ZABs will be discussed here.

21.5.1 Aqueous Binder

As a key component in electrodes, the main responsibility of the polymer binder is to maintain the stability and integrity of the structure and build an effective conductive skeleton. Currently, fluorinated vinyl polymers (*e.g.*, polyvinylidene fluoride (PVDF)) and organic solvents (*e.g.*, N-methylpyrrolidone (NMP)) are widely used as binders and dispersants for the electrode fabrication. However, they still have limitations. First, the non-polar structure of PVDF can only form weak intermolecular interactions with active materials and current collectors. Therefore, during cycling, the homogeneous composite structure of the pristine electrode will be destroyed, resulting in mechanical failure and capacity degradation. Second, PVDF is an electrically insulating material, so it requires carbon additives to improve the electronic conductivity of the electrode. However, because the PVDF/C mixture does not possess catalytic activity, it will reduce the overall electrochemical performance of ZABs. Finally, the volatile and toxic NMP solvent will result in health and environmental problems due to the poisonous nature of the organic solvent and the elevated safety risks from their flammability at elevated temperatures [17].

Generally, an ideal binder for ZABs should have the following favorable characteristics: (i) strong interaction with active materials to maintain adhesion during the cycling; (ii) strong adhesion to the current collectors to prevent electrode delamination; (iii) a certain electrical conductivity to minimize the amount of conductive additives; (iv) the capability to form a continuous conductive network in the electrode; (v) high chemical/thermodynamic stability to adapt to a wide range of operating environments; and (vi) low-cost and simple manufacturing process to achieve large-scale commercialization.

Recently, natural biopolymers have attracted numerous scientific attention, owing to their diversity in structures and functions. In particular, abundant hydrophilic polar functional groups (e.g., –OH, –COOH, and $–NH_2$) in the skeleton enhance the affinity with water molecules and endow natural polymers with good aqueous electrolyte uptake, thus improving the compatibility with aqueous electrolytes [18]. Meanwhile, the formed intermolecular cross-linked framework can contribute to structural or interfacial stability and retention of electrolyte solution in the electrodes. In addition, the implementation of natural polymer-based aqueous binders can reduce or even avoid the instability or the collapse of the electrode structure caused by the dissolution and swelling of conventional binders in organic electrolytes. So far, a variety of natural polymers have been used as promising aqueous binders (Figure 21.4), such as sodium carboxymethyl chitosan (SCC), sodium alginate (SA), styrene-butadiene rubber (SBR), sodium carboxymethyl cellulose (CMC), guar gum (GG), and xanthan gum (XG) [19].

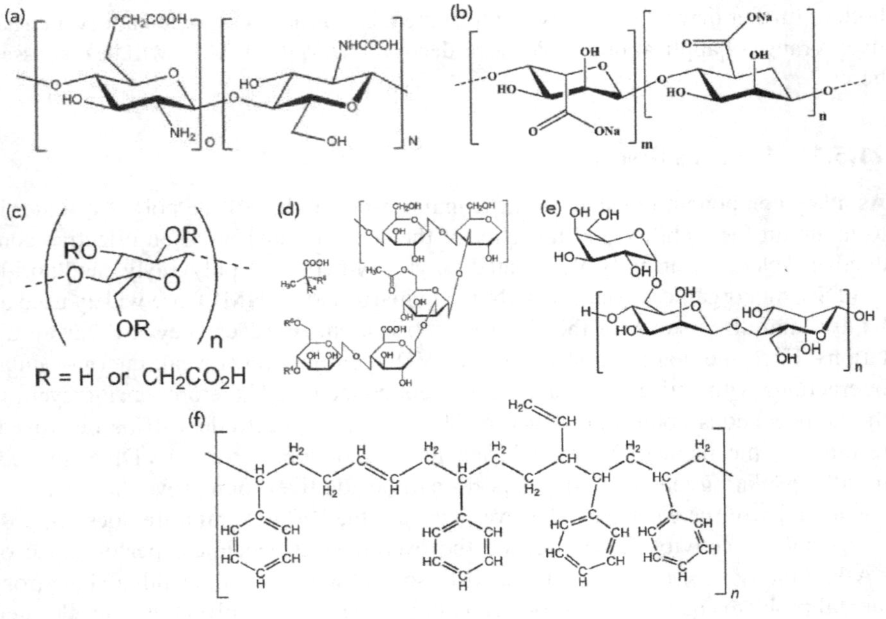

(a) (b) (c) R = H or CH_2CO_2H (d) (e) (f)

FIGURE 21.4 Treatment of biowaste-derived electrocatalysts.

For instance, polysaccharides are widely studied biopolymer-based aqueous binders with several attractive advantages: (i) It is easy to form homogeneous films and layers on different materials; (ii) the source of raw materials is wide and cheap, which could be derived from biowaste; and (iii) many polysaccharides can be processed *via* aqueous solutions, which avoids the use of organic solvents in the electrode assembly process, thus reducing the manufacturing difficulty and cost [20]. However, although they are environmentally friendly, a potential disadvantage of polysaccharide binders is that they can cause battery failure during the drying process. This is due to the shrinkage of the material during drying, which may cause the binder to break or peel off, thus reducing the interface contact area.

21.5.2 Gel Polymer Electrolyte and Separator

The separator is an important component in ZABs, and it usually has the following characteristics: a high mechanical stability to reliably isolate the anode and cathode; a high ionic conductivity to enable ions to flow through the separator and achieve charge compensation, and a thin layer thickness to minimize ohmic loss. Porous petrochemical polymer materials are currently commercialized and widely used separators, such as polypropylene and polyethylene [21]. In addition to conventional petrochemical polymers and designer polymers, recently, biopolymers have also been used as separators or functional units in solid polymer electrolytes.

Cellulose-based materials have commercially been used in aqueous alkaline batteries, but due to their high moisture content, their commercial applications in organic electrolyte-based batteries (*e.g.*, LIBs) have been hindered. However, for ZABs using aqueous electrolytes, a large amount of moisture in the separator is beneficial for achieving optimal ion transport. Cellulose can be used as a substrate for the separator and can be combined with other functional materials to form a composite separator, thereby optimizing the performance of the separator. For instance, mixing cellulose pulp with sodium alginate (SA) and a flame retardant (FR) can improve the fire resistance of the separator [22].

In addition, chitosan can be used as a component of the separator, which can reduce sharp interfaces and maximize the ionic conductivity. Similar to cellulose-based materials, it can usually be blended or compounded with other materials to improve the performance of the separators. The combination with inorganic fillers or thermally stable polymers is particularly attractive, because such combination can be inherently stable and will not shrink under heating, thereby improving battery safety by preventing short-circuit failure. However, this type of material is a relatively complex system and requires further investigation. Other biopolymers, such as alginate, gelatin, and lignin, also showed attractive prospects.

21.6 CONCLUSIONS AND PERSPECTIVES

This chapter provides an introduction and discussion of the working principles and representative biowaste-derived ecological components for ZABs. Although many enlightening works have proved the promising prospects of biowaste-derived materials for ZABs, more effort is required to bridge the significant gap between laboratory-based research and commercial applications. The overview and prospects can be summarized as follows:

a. Improvement in basic understanding of biowastes. Since the properties of the biowaste precursor have a great influence on the derived materials, a thorough understanding of key parameters (*e.g.*, molecular structure, degree of polymerization, and composition) is crucial for obtaining suitable renewable biowaste-derived ecological materials. In addition, for components using pristine biowastes, such as aqueous binders, their properties will directly affect the electrochemical characteristics of the as-assembled electrodes.
b. Development of a simple preparation process. Generally, the conversion process of biowaste into desired functional materials is not straightforward and it may require multiple processes such as cleaning, decomposition, pyrolysis, and fermentation. This will reduce the utilization efficiency and yield of biowaste resources and increase manufacturing costs. Therefore, future research can be focused on simplifying conversion processes.
c. Conduction of advanced characterizations. At present, there is a lack of effective characterization techniques to track the structural evolution and degradation mechanisms of biowaste-derived materials in ZABs, owing to the difficulties caused by different structures and complex characteristics of biowaste-derived materials. The application of advanced characterization techniques will provide new insights into the highly oriented design of biowaste-derived biological materials.
d. Improvement in the rational design of functional materials. A reasonable design of biowaste-derived functional materials is crucial for improving the performance of ZABs, in terms of the key properties such as porosity, adhesive force, triple-phase boundary, and electrochemically active surface area. Owing to the intrinsic diversity of biowastes, relying solely on trial and error is a time-consuming strategy. The introduction of machine learning and artificial neural networks into this field will help to achieve effective screening of structure-oriented, morphology-oriented, surface/interface chemistry-oriented, and function-oriented biowaste precursors.

ACKNOWLEDGEMENTS

The authors would like to thank the Engineering and Physical Sciences Research Council (EPSRC, EP/V027433/1; EP/533581/1), the Royal Society (RGS\R1\211080; IEC\NSFC\201261), and Faraday Institution Degradation Project (EP/S003053/1; FIRG001) for financial support. Jennifer Hack acknowledges the EPSRC for her Doctoral Prize Fellowship (EP/T517793/1).

REFERENCES

1. Y. Lan, Y. Liu, J. Li, D. Chen, G. He, and I. P. Parkin, "Natural clay-based materials for energy storage and conversion applications," *Adv. Sci.*, p. 2004036, Mar. 2021, doi: 10.1002/advs.202004036.

2. Y. Liu, G. He, H. Jiang, I. P. Parkin, P. R. Shearing, and D. J. L. Brett, "Cathode design for aqueous rechargeable multivalent ion batteries: Challenges and opportunities," *Adv. Funct. Mater.*, p. 2010445, Jan. 2021, doi: 10.1002/adfm.202010445.
3. J. Li et al., "High-performance zinc-air batteries with scalable metal-organic frameworks and platinum carbon black bifunctional catalysts," *ACS Appl. Mater. Interfaces*, vol. 12, no. 38, pp. 42696–42703, Sep. 2020, doi: 10.1021/acsami.0c10151.
4. X. Guo et al., "Alleviation of dendrite formation on zinc anodes via electrolyte additives," *ACS Energy Lett.*, pp. 395–403, Jan. 2021, doi: 10.1021/acsenergylett.0c02371.
5. M. Borghei, J. Lehtonen, L. Liu, and O. J. Rojas, "Advanced biomass-derived electrocatalysts for the oxygen reduction reaction," *Adv. Mater.*, vol. 30, no. 24, Wiley-VCH Verlag, p. 1703691, Jun. 13, 2018, doi: 10.1002/adma.201703691.
6. W. Sun et al., "A rechargeable zinc-air battery based on zinc peroxide chemistry," *Science*, vol. 371, no. 6524, pp. 46–51, Jan. 2021, doi: 10.1126/science.abb9554.
7. J. K. Nørskov et al., "Origin of the overpotential for oxygen reduction at a fuel-cell cathode," *J. Phys. Chem. B*, vol. 108, no. 46, pp. 17886–17892, Nov. 2004, doi: 10.1021/jp047349j.
8. J. M. Dias, M. C. M. Alvim-Ferraz, M. F. Almeida, J. Rivera-Utrilla, and M. Sánchez-Polo, "Waste materials for activated carbon preparation and its use in aqueous-phase treatment: A review," *J. Environ. Manag.*, vol. 85, no. 4, Academic Press, pp. 833–846, Dec. 01, 2007, doi: 10.1016/j.jenvman.2007.07.031.
9. B. Sakintuna and Y. Yürüm, "Templated porous carbons: A review article," *Ind. Eng. Chem. Res.*, vol. 44, no. 9, American Chemical Society, pp. 2893–2902, Apr. 27, 2005, doi: 10.1021/ie049080w.
10. Z. Yang, Y. Zhang, and Z. Schnepp, "Soft and hard templating of graphitic carbon nitride," *J. Mater. Chem. A*, vol. 3, no. 27, Royal Society of Chemistry, pp. 14081–14092, July 21, 2015, doi: 10.1039/c5ta02156a.
11. L. Chuenchom, R. Kraehnert, and B. M. Smarsly, "Recent progress in soft-templating of porous carbon materials," *Soft Matter*, vol. 8, no. 42, Royal Society of Chemistry, pp. 10801–10812, Nov. 14, 2012, doi: 10.1039/c2sm07448f.
12. W. Xin and Y. Song, "Mesoporous carbons: Recent advances in synthesis and typical applications," *RSC Adv.*, vol. 5, no. 101, Royal Society of Chemistry, pp. 83239–83285, Sep. 11, 2015, doi: 10.1039/c5ra16864c.
13. Y. Qiao, F. Kong, C. Zhang, R. Li, A. Kong, and Y. Shan, "Highly efficient oxygen electrode catalyst derived from chitosan biomass by molten salt pyrolysis for zinc-air battery," *Electrochim. Acta*, vol. 339, p. 135923, Apr. 2020, doi: 10.1016/j.electacta.2020.135923.
14. Q. Li et al., "Biomass waste-derived 3D metal-free porous carbon as a bifunctional electrocatalyst for rechargeable zinc-air batteries," *ACS Sustain. Chem. Eng.*, vol. 7, no. 20, pp. 17039–17046, Oct. 2019, doi: 10.1021/acssuschemeng.9b02964.
15. L. Yang, X. Zeng, D. Wang, and D. Cao, "Biomass-derived FeNi alloy and nitrogen-codoped porous carbons as highly efficient oxygen reduction and evolution bifunctional electrocatalysts for rechargeable Zn-air battery," *Energy Storage Mater.*, vol. 12, pp. 277–283, May 2018, doi: 10.1016/j.ensm.2018.02.011.
16. Y. Liu, K. Sun, X. Cui, B. Li, and J. Jiang, "Defect-rich, graphene-like carbon sheets derived from biomass as efficient electrocatalysts for rechargeable zinc-air batteries," *ACS Sustain. Chem. Eng.*, vol. 8, no. 7, pp. 2981–2989, Feb. 2020, doi: 10.1021/acssuschemeng.9b07621.
17. D. Bresser, D. Buchholz, A. Moretti, A. Varzi, and S. Passerini, "Alternative binders for sustainable electrochemical energy storage-the transition to aqueous electrode processing and bio-derived polymers," *Energy Environ. Sci.*, vol. 11, no. 11, pp. 3096–3127, Nov. 2018, doi: 10.1039/c8ee00640g.

18. J. Liu et al., "Recent progress on biomass-derived ecomaterials toward advanced rechargeable lithium batteries," *EcoMat*, vol. 2, no. 1, p. e12019, Mar. 2020, doi: 10.1002/eom2.12019.
19. V. A. Nguyen and C. Kuss, "Review—Conducting polymer-based binders for lithium-ion batteries and beyond," *J. Electrochem. Soc.*, vol. 167, no. 6, p. 065501, Apr. 2020, doi: 10.1149/1945-7111/ab856b.
20. M. N. Masri, M. F. M. Nazeri, C. Y. Ng, and A. A. Mohamad, "Tapioca binder for porous zinc anodes electrode in zinc–air batteries," *J. King Saud Univ. - Eng. Sci.*, vol. 27, no. 2, pp. 217–224, July 2015, doi: 10.1016/j.jksues.2013.06.001.
21. C. Liedel, "Sustainable battery materials from biomass," *ChemSusChem*, vol. 13, no. 9, Wiley-VCH Verlag, pp. 2110–2141, May 08, 2020, doi: 10.1002/cssc.201903577.
22. J. Zhang et al., "Sustainable, heat-resistant and flame-retardant cellulose-based composite separator for high-performance lithium ion battery," *Sci. Rep.*, vol. 4, no. 1, pp. 1–8, Feb. 2014, doi: 10.1038/srep03935.

22 Recycling of Wastes Generated in Automobile Metal–Air Batteries

Weng Cheong Tan, Lip Huat Saw, Ming Chian Yew, and Ming Kun Yew
Universiti Tunku Abdul Rahman (UTAR)

CONTENTS

DOI: 10.1201/9781003178354-26

22.1 INTRODUCTION

Electric vehicles (EVs) mainly use lithium-ion batteries as the electric power source. Lithium-ion batteries have a high energy density, life span, and affordability. However, the development of lithium-ion batteries had reached its peak performance due to technology limitations. A study suggested that the energy density of lithium-ion batteries can be improved by an additional 30% only in the future [1]. In the EV, lithium-ion battery is a popular choice in the market due to its high power density. It is expected to reach a power density of 235 Wh/kg by the year 2020. However, the travel distance of the EV that uses lithium-ion batteries is limited to about 500 km due to the limitation in the power density of lithium-ion batteries. Besides that, the disposal of lithium-ion batteries remains one of the important unsolved issues. In view of this, the metal–air battery is projected as an alternative solution to energy storage post-lithium-ion battery era for the EV as the battery swapping technology becomes more mature and eminent.

A metal–air battery is a new type of battery that uses a metal as the anode and air as the cathode for generating electricity. Metals such as aluminum (Al), magnesium (Mg), zinc (Zn), lithium (Li), sodium (Na), and potassium (K) can be used as the anode for the electrochemical reaction. Metal–air batteries have a high theoretical power density and the potential to replace conventional lithium-ion batteries. The redox reaction generates electricity and by-products. Most of the by-products generated involve the corresponding metal oxide depending on the type of metal–air battery. On the other hand, electrolytes used in metal–air batteries can be classified as aqueous electrolyte and non-aqueous electrolyte. Aqueous electrolytes are liquid-based electrolytes that include alkaline electrolytes (potassium hydroxide and sodium hydroxide); acidic electrolytes including inorganic acids (sulfuric acid and hydrochloric acid) and organic acids (methanesulfonic acid, polyvinyl sulfonic acid, and polyvinyl sulfuric acid); and neutral salt solutions (sodium chloride, neutral salt, and seawater). Non-aqueous electrolytes involve solid polymer electrolytes and room-temperature ionic liquids. However, due to the complexity of the solvent-based dissolved salt in the non-aqueous electrolyte, it can cause problems to the environment and difficulty in recycling. In addition, electrocatalysts present in the carbon cathode increase the impurity of the air cathode and increase the complexity of the recycling process. Some of the common types of air cathode used in metal–air batteries are carbon-based materials, noble metals, and transitional metal oxides.

The waste management of metal–air batteries is important when metal–air batteries are commercialized and applied in EV to reduce the impact on the environment. Different types of metal–air batteries require different treatment methods. This chapter will introduce different types of metal–air batteries and their waste products for both aqueous and non-aqueous electrolytes. Nevertheless, the environmental impacts of the waste products will be discussed.

22.2 ARCHITECTURE OF METAL–AIR BATTERY

A metal–air battery is a new type of battery in which the metal anode is consumed through the electrochemical reaction with the oxygen around the atmosphere while generating electricity. Metal–air batteries have a high specific energy density, making them a viable choice to replace the lithium-ion battery in the future. The structure of a metal–air battery is simple. It consists of only three main components, which are metal anode, air cathode, and electrolyte splitting the anode and cathode. The electrolyte provides a medium for the movement of ions during the electrochemical process. During the electrochemical reaction, the anode undergoes oxidation by releasing electrons, while the cathode receives the electrons and thus generates electricity due to the electron movement. An example of half-cell chemical equations for anode and cathode is shown in Equations 22.1 and 22.2. A schematic diagram of the different types of metal–air batteries is shown in Figure 22.1.

Anode:

$$M \rightarrow M^{n+} + ne^- \quad (22.1)$$

Cathode:

$$O_2 + 2H_2O + ne^- \rightarrow 4OH^- \quad (22.2)$$

In the half-cell equations, M represents the metals that are commonly used in metal–air batteries, such as zinc, magnesium, lithium, sodium, and aluminum, and n represents the oxidation number. The operating voltage and energy density depend on the type of metal used in the metal–air battery and are tabulated in Table 22.1 [3].

22.3 ALUMINUM–AIR BATTERY

Aluminum–air batteries have gained attention due to their high energy density as compared to lithium-ion batteries. Besides, aluminum is commonly available in the market at a cheaper price and available abundantly in the earth's crust. Moreover, aluminum can be transported easily and recycled through the industrial pre-treatment process, further adding to the attractiveness of using aluminum as the anode in metal–air batteries.

A state-of-the-art structure of an aluminum–air battery consists of an aluminum anode, separator, air cathode, and electrolyte. The aluminum anode will be consumed through the electrochemical reaction and needs to be replaced when it is eventually used up or covered with aluminum hydroxide and needs to be mechanically recharged by replacing a new aluminum anode. The electrolyte provides a medium for the movement of ions, while the air cathode supplies oxygen for the electrochemical process. Separators isolate the anode and the cathode to prevent short circuit while allowing for the transfer of ions. In general, the electrochemical processes of an aluminum–air battery are shown below [4].

At the anode:

$$Al + 3OH^- \rightarrow Al(OH)_3 + 3e^- \quad (22.3)$$

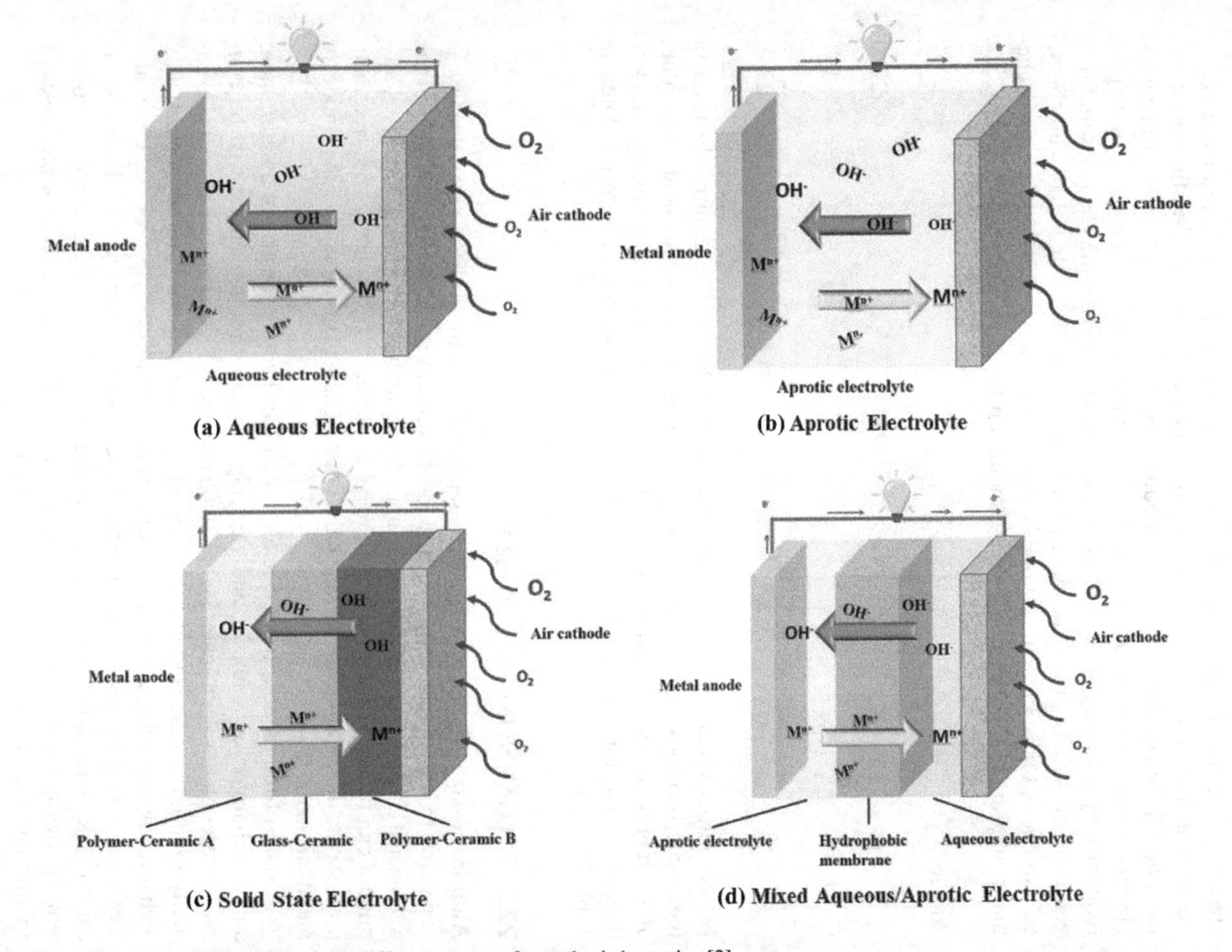

FIGURE 22.1 Schematic diagram of the different types of metal–air batteries [2].

TABLE 22.1
Electrical Properties of Various Types of Metal–Air Batteries

Type of Battery	Practical Operating Voltage, V	Theoretical Voltage, V	Theoretical Energy Density, Wh/kg	Theoretical Specific Capacity, Ah/kg
Li–air	2.4	3.4	13,000	1,170
Na–air	2.3	2.3	1,600	687
Al–air	1.2–1.6	2.7	8,100	1,030
Mg–air	1.2–1.4	3.1	6,800	920
Zn–air	1.0–1.2	1.6	1,300	658

At the cathode:

$$O_2 + 2H_2O + 4e^- \rightarrow 4OH^- \tag{22.4}$$

Overall reaction:

$$4Al + 3O_2 + 6H_2O \rightarrow 4Al(OH)_3 \tag{22.5}$$

Side reaction:

$$2Al + 6H_2O \rightarrow 2Al(OH)_3 + 3H_2 \tag{22.6}$$

It can be seen that the main product after the electrochemical reaction is aluminum hydroxide ($Al(OH)_3$). $Al(OH)_3$ is formed and covers the surface of the aluminum anode and hinders the performance of the aluminum–air battery as time progresses. Hence, a proper recycling method is needed to treat $Al(OH)_3$ in the aluminum–air battery. In the recycling of the aluminum–air battery, KOH electrolyte solution is not recommended for the retreatment of alumina via the industrial Hall–Heroult process. This is because the potassium ions have a harmful effect on the cathode of the electrolysis cell.

22.3.1 Recycling of Aluminum Hydroxide

An aluminum plate is covered with aluminum hydroxide during the electrochemical reaction. Calcination is needed to treat the aluminum hydroxide and obtain aluminum oxide before the pure aluminum can be obtained again. The aluminum hydroxide is heated up to 980 °C in the absence of oxygen or in the presence of limited oxygen to form aluminum oxide in the Hall–Heroult process [5]. Figure 22.2 shows the recycling process of aluminum.

$$2Al(OH)_3 \rightarrow Al_2O_3 + 3H_2O \tag{22.7}$$

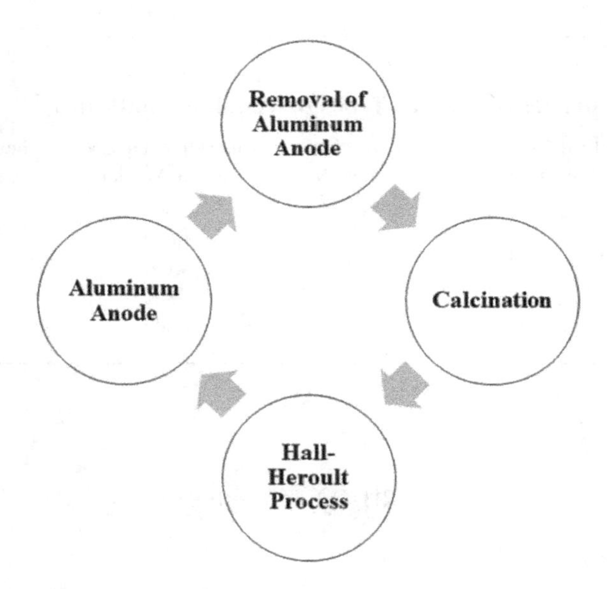

FIGURE 22.2 Aluminum recycling process.

22.3.2 Hall–Heroult Process

The Hall–Heroult process is a widely used industry smelting method to extract aluminum from aluminum oxide [5]. It is an electrolysis technique to convert the aluminum oxide to aluminum and oxygen. This process can produce up to 99.9% of pure aluminum at 940 °C–980 °C. The aluminum oxide is first dissolved at 950 °C in molten cryolite (Na_3AlF_6) to lower the melting point as indicated below.

$$Al_2O_3 + 4AlF_6^{3-} \rightarrow 3Al_2OF_6^{2-} + 6F^- \quad (22.8)$$

The molten solution is stored in reduction cells and placed at the bottom of a carbon cathode. The pots are covered with an anode that is usually made of coke. During the electrolysis process, current is supplied to the molten solution and aluminum is formed at the cathode.

$$AlF_6^{3-} + 3e^- \rightarrow Al + 6F^- \quad (22.9)$$

On the other hand, at the anode, carbon monoxide and carbon dioxide are produced.

$$C + 2Al_2OF_6^{2-} + 12F^- \rightarrow CO_2 + 4AlF_6^{3-} + 4e^- \quad (22.10)$$

The whole process involves redox reaction. The whole process can be simplified to the equation below.

$$2Al_2O_3 + 3C \rightarrow 3CO_2 + 4Al \quad (22.11)$$

The molten pure aluminum (99.9%) is then readily available at the bottom of the pot and can be used again. Aluminum hydroxide can be fully recycled through the Hall–Heroult process to pure aluminum, making aluminum–air battery a green battery with minimal waste.

22.3.3 Energy Saving and Carbon Footprint of Aluminum Recycling

Aluminum is one of the metals that have the highest recyclable value. It is 100% recyclable among all the materials. About 90% of the energy cost in aluminum production can be saved when aluminum is recycled [6]. It is a sustainable metal and can be recycled with a little loss of material due to oxidation (1%–2%). The global recycling rate has recently been increased to about 60%, which shows that the recycling of aluminum is gaining popularity nowadays. The study has revealed that the commercial production of aluminum will result in 45 kWh of energy emission and will release 12 kg of CO_2 for 1 kg of aluminum produced using the aluminum smelting process [7]. However, it requires only about 2.8 kWh of energy and emits roughly 0.6 kg of CO_2 for recycling 1 kg of aluminum. This huge reduction reduces the greenhouse effect by reducing the generation of carbon dioxide. Besides that, it also helps to reduce landfill waste.

22.3.4 Waste Generated in Electrolyte and Air Cathode

The electrolytes of an aluminum–air battery can be an alkaline, acid, and saline solution. The alkaline electrolytes such as potassium hydroxide (KOH) and sodium hydroxide (NaOH) provide a better performance than the acid electrolytes and saline electrolytes, and they are a popular choice. Acid electrolytes such as hydrochloric acid (HCl) and sulfuric acid (H_2SO_4) and saline electrolytes consisting mainly of sodium chloride (NaCl) are rarely used due to their weak performance. Gel electrolyte is introduced to prevent leakage. It is not soluble in water and requires proper treatment in the disposal. Due to the mixing of different chemicals in producing gel electrolytes or solid electrolytes, it is hard to have a general disposal solution. A gel electrolyte can be agarose, polyacrylic acid, polyacrylamide, and polyvinyl. In terms of air cathode, the most common materials are activated carbon and carbon black. This is because carbon materials show high stability and can be used as a conductive agent. Additives such as Pt, $La_{0.8}Sr_{0.2}MnO_3$, Fe_2O_3, NiO, Fe_3O_4, Co_3O_4, CuO, $CoFe_2O_4$, and MnO_2 can act as a catalyst for better electrochemical reaction.

22.4 ZINC–AIR BATTERY

Zinc–air batteries offer the advantages of lower cost and better safety compared to aluminum–air batteries, magnesium–air batteries, and lithium–air batteries. However, they suffer from the issues of dendrite growth due to electrodeposition and change in morphology in the zinc electrode, which can lower the performance gradually and cause short life cycle, preventing them from being widely used in the field of transportation. They require further research to improve the cycle life of the battery for EV application. Zinc–air batteries consist of a zinc electrode, electrolyte, separator, and air cathode that is exposed to air for collecting oxygen. The electrochemical process of the zinc–air battery is shown below [8].

Anode:

$$Zn \rightarrow Zn^{2+} + 2e^- \quad (22.12)$$

$$Zn^{2+} + 4OH^- \rightarrow Zn(OH)_4^{2-} \quad (22.13)$$

$$Zn(OH)_4^{2-} \rightarrow ZnO + H_2O + 2OH^- \quad (22.14)$$

Cathode:

$$O_2 + H_2O + 4e^- \rightarrow 4OH^- \quad (22.15)$$

Overall:

$$2Zn + O_2 \rightarrow 2ZnO \quad (22.16)$$

During the electrochemical reaction, zinc metal is consumed to form zinc oxide as a by-product. The recycling of zinc can be conducted in two ways: hydrometallurgical process and pyrometallurgical process [9,10]. However, hydrometallurgical processes are more popular than pyrometallurgical processes and contribute to about 90% of zinc produced globally. Zinc–air batteries are widely used in miniature hearing aids. This is because they possess a high specific density with low cost. Panasonic, Power One, and Duracell are some of the companies that offer zinc–air button batteries in the market.

22.4.1 Hydrometallurgical Process

The hydrometallurgical process uses an aqueous solution to separate metals from residual materials. In this case, it is used to separate zinc from zinc oxide [9]. First, leeching is conducted to dissolve the zinc oxide in dilute sulfuric acid.

$$ZnO + H_2SO_4 \rightarrow ZnSO_4 + H_2O \quad (22.17)$$

Next, molten zinc sulfate undergoes purification to purify the zinc. Zinc dust and steam are used to remove impurities that will affect the electrolysis process that will be conducted later. Impurities present in the zinc sulfate solution will result in the formation of hydrogen gas rather than zinc metal in the electrolysis process. An agitating tank is used for the purification process with the operating temperature of 40°C–85°C at a pressure of about 240 kPa. The purified zinc sulfate then undergoes electrolysis to separate pure zinc from zinc sulfate through electrowinning. An aluminum sheet is used as the cathode in the process. During the electrolysis process, the electrolyte is circulated through a cooling tower to lower the temperature while maintaining high concentration through evaporation of water in the cooling tower.

$$2ZnSO_4 + 2H_2O \rightarrow 2Zn + 2H_2SO_4 + O_2 \quad (22.18)$$

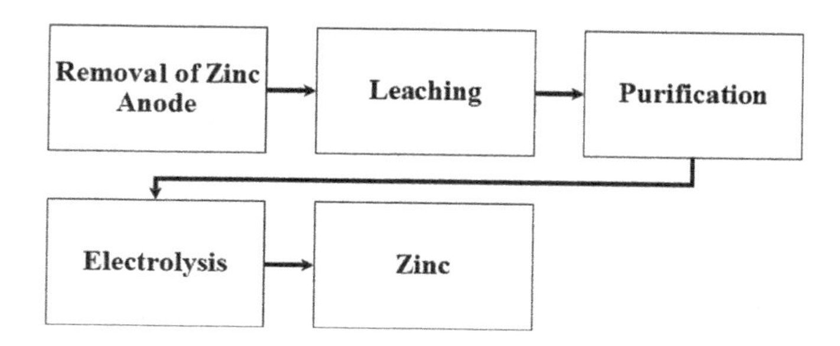

FIGURE 22.3 Flowchart of the hydrometallurgical process.

At the end of the process, sulfuric acid is removed and can be reused again as a solvent to dissolve zinc oxide. Meanwhile, zinc is deposited at the aluminum sheet and is mechanically removed from the aluminum sheet. The zinc produced by this method is of at least 99.96% purity. The fresh zinc is then readily available for further processing and can be used in zinc–air batteries again. A flowchart of the hydrometallurgical process is shown in Figure 22.3.

22.4.2 Pyrometallurgical Process

In the pyrometallurgical process, crushed coke (carbon) is added to the zinc oxide and mixed. Next, it is heated at 1,400°C to reduce the zinc oxide into zinc. The carbon serves as a reducing agent to reduce the zinc oxide [10].

$$2ZnO + C \rightarrow 2Zn + CO_2 \tag{22.19}$$

During the heating process, the zinc produced is in vapor form. Both the zinc vapor and carbon dioxide are then channeled to a vacuum condenser for distillation, in which the zinc is recovered by bubbling through a molten zinc bath and 95% of zinc vapor is converted to liquid zinc. On the other hand, the carbon dioxide is reduced back to carbon and can be reused as the reducing agent during the heating process. The purity of the zinc produced is only about 98%, which is lower than that of the hydrometallurgical process. This process also requires high energy and emission costs, making it less attractive. A flowchart of the pyrometallurgical process is shown in Figure 22.4.

22.4.3 Energy Saving and Carbon Footprint of Zinc Recycling

Zinc can be recycled for an indefinite amount of time without degradation in quality. Currently, about 30% of zinc available in the market comes from recycled zinc [11].

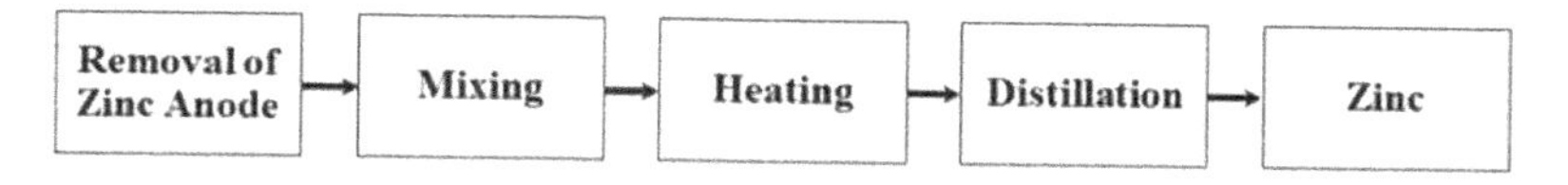

FIGURE 22.4 Flowchart of the pyrometallurgical process.

Recycling zinc requires less energy for secondary production. The primary production of zinc requires 24 MJ per kg zinc, while only 18 MJ per kg zinc is required for secondary production [12]. There is a 25% energy saving. In terms of carbon footprint, there is a drop in carbon dioxide from 2.36 tCO_2 per total zinc during primary production to 1.4 tCO_2 per total zinc during secondary production. Hence, zinc recycling saves energy and reduces the emission of carbon dioxide.

22.4.4 Waste Generated in Electrolyte and Air Cathode

Zinc–air batteries require alkaline solutions such as KOH and NaOH as electrolytes. Besides that, gel electrolyte is also introduced to solve the leakage issues. Some examples of the gel electrolyte are KOH-based hydroponics gel and polymer electrolyte.

The carbon-based materials are a popular choice for air cathode in zinc–air batteries. Carbon composites such as silver-coated carbon and manganese oxide-coated carbon are used to improve the performance. The recycling of carbon cathode is difficult due to the complexity of chemicals applied to the air cathode.

22.5 MAGNESIUM–AIR BATTERY

Magnesium–air batteries show a relatively high theoretical voltage of 3.1 V as compared to other types of metal–air batteries with recharging capability. The electrochemical process involved in magnesium–air batteries is shown below [13].

Anode:

$$2Mg + 4OH^- \rightarrow 2Mg(OH)_2 + 4e^- \quad (22.20)$$

Cathode:

$$O_2 + 2H_2O + 4e^- \rightarrow 4OH^- \quad (22.21)$$

Overall:

$$2Mg + O_2 + 2H_2O \rightarrow 2Mg(OH)_2 \quad (22.22)$$

The performance of magnesium–air batteries is limited due to some scientific difficulties such as poor thermodynamics and kinetics properties of magnesium oxide (MgO) and magnesium dioxide (MgO_2). To date, the application of magnesium–air batteries has been limited. They are used for power supply in the subsea instruments such as subsea monitoring equipment and floats with seawater as electrolytes.

22.5.1 Recycling of Magnesium Hydroxide

The by-product of magnesium–air battery is magnesium hydroxide. A pre-treatment is needed to produce magnesium oxide from magnesium hydroxide before magnesium can be produced. The simplest way is through calcination [14]. In this process, a high temperature is used to burn the magnesium hydroxide with a limited oxygen supply so that magnesium oxide is formed. To further process magnesium oxide to magnesium, two approaches can be considered, which are known as thermal reduction process and electrolytic process.

22.5.2 Thermal Reduction Process

About 85% of magnesium production is through the thermal reduction process [15]. The Pidgeon process is one of the commonly used thermal reduction processes. Magnesium oxide is reduced to magnesium in the presence of ferrosilicon during the reduction process [16]. The reduction reaction is shown in the equation below:

$$2CaO + 2MgO + Si \rightarrow 2Mg + Ca_sSiO_4 \tag{22.23}$$

The reducing process is an endothermic reaction and requires a large amount of heat to keep the operating temperature at about 1,160 °C at a very low pressure of 0.05 atm (near-vacuum) and requires about 35–40 kWh/kg of energy. Magnesium formed during the reduction process is in vapor form.

The Mintek process is introduced to solve the limitation of the Pidgeon process [16]. This process produces a higher yield of magnesium than the Pidgeon process at the expense of higher operating temperature and higher impurity of the product. Instead of relying upon only ferrosilicon as the reducing agent, aluminum and slag are added in the reducing process in an arc furnace at a temperature of around 1,700 °C–1,750 °C at 1 atm. Due to the higher impurities in the product, a refining stage is needed, which will increase the operating and capital costs.

22.5.3 Electrolytic Process

Solid oxide membrane process is an electrolytic process in which electrolysis is used in the production of magnesium from magnesium oxide [16]. A fluoride-based electrolyte, MgF_2–CaF_2–MgO, is used to dissolve the magnesium oxide in the electrolysis cell. An yttria-stabilized zirconia (YSZ) membrane is used to separate the anode from the electrolyte. At the operating temperature of 1,150 °C–1,300 °C, current is supplied to the cell to separate the magnesium from magnesium oxide. The magnesium vapor is collected at the cathode and channeled to a separate chamber for condensation. As compared to the Pidgeon process, the solid oxide membrane process requires only one-third of the energy, which is about 10 kWh/kg.

Anode:

$$H_2 + O^{2-} \rightarrow H_2O + 2e^- \tag{22.24}$$

Cathode:

$$Mg^{2+} + 2e^- \rightarrow Mg \tag{22.25}$$

22.5.4 Alternative Routes

Carbothermal reduction provides a cheaper alternative to the thermal reduction process and electrolysis process [16]. The reducing agent used is carbon.

$$MgO + C \leftrightharpoons Mg + CO \tag{22.26}$$

However, the reaction is reversible. A mixture of magnesium oxide with magnesium powder is observed at the end product. Besides, the low yield also limits the wide application of this method in the manufacturing of magnesium. The Heggie–Iolaire process using aluminum as a reducing agent provides an alternative route to magnesium production [16]. Aluminum is a stronger reducing agent as compared to carbon and ferrosilicon. However, the price of aluminum is high, making it less economically viable.

$$4MgO + 2Al \leftrightharpoons 3Mg + MgAl_2O_4 \tag{22.27}$$

There are no issues with the contamination of magnesium oxide in magnesium as the end product. This is because the magnesium produced is in vapor form and the magnesium oxide is in solid state.

22.5.5 Energy Saving and Carbon Footprint of Magnesium Recycling

For the primary production of magnesium, a total energy of 35 kWh/kg is required. However, recycling of magnesium reduces the energy to about 3 kWh/kg, which is about 85% reduction [17]. There are many methods used in the production of primary magnesium; hence, it is difficult to estimate the carbon footprint as different methods produce different carbon dioxide emissions. In general, the Pidgeon process contributes to about 21.8 kg of CO_2-eq for every kilogram of magnesium produced. On the other hand, the production of magnesium using the electrolysis process produces 5.3 kg of CO_2-eq for every kilogram of magnesium produced. Recycling magnesium can reduce the carbon dioxide emission to about 0.45 kg CO_2-eq [18].

22.5.6 Waste Generated in Electrolyte and Air Cathode

The aqueous electrolyte used in magnesium–air batteries is usually a salt or weak acid. Some of the common salt electrolytes used are NaCl, Na_2SO_4, and $NaNO_3$, while the examples of acid electrolytes are $MgCl_2$, HNO_3, and $Mg(Cl_4)_2$. On the other hand, the non-aqueous electrolyte used in the magnesium–air battery includes phosphonium chloride ionic liquid and chitosan–choline nitrate [16]. The chemistry behind the electrolyte is complicated and requires extensive care in the disposal procedure. The application of carbon-based air cathode is feasible in the magnesium–air battery. Catalysts such as cobalt and iron tetramethoxyphenylporphyrin (CoTMPP and FeTMPP) are added on the carbon cathode to improve the performance. Besides that, transitional metal oxides such as MnO_3 and $CoMn_2O_4$ are also used as an air cathode in the magnesium–air batteries. Normally, the air cathode will be treated as waste since there is no proper recycling procedure available at the current stage.

22.6 LITHIUM–AIR BATTERY

Lithium–air batteries are very competitive as they have a higher energy density than the normal lithium-ion batteries. The electrochemical process of the lithium–air battery is tabulated in Table 22.2 [19].

22.6.1 RECYCLING OF LITHIUM HYDROXIDE

Lithium hydroxide is the waste product produced in lithium–air batteries. Due to the poor recycling capability of the lithium-based battery, lithium hydroxide is rarely recycled to the lithium metal for secondary use. First, lithium hydroxide is converted to lithium carbonate (Li_2CO_3) through leaching [20]. The leaching process continues for 6–10 h with the pH value of 8–9. In this process, lithium hydroxide reacts with carbon dioxide to produce lithium carbonate.

$$LiOH + CO_2 \rightarrow 2Li_2CO_3 + H_2O \tag{22.28}$$

Next, the lithium chloride is produced from Li_2CO_3 through the reaction with hydrochloric acid [21].

$$Li_2CO_3 + 2HCl \rightarrow 2LiCl + CO_2 + H_2O \tag{22.29}$$

The lithium chloride then undergoes further reaction to produce lithium metal. The molten aluminum chloride and molten potassium chloride are used in the electrolysis process. The lithium chloride will be decomposed and form lithium at the cathode. A total of 35 kWh energy is required to produce 1 kg of lithium.

Besides, vacuum aluminothermic reduction can also be used to produce lithium metal from lithium carbonate [22]. In this process, lithium carbonate, calcium oxide, and aluminum oxide are compressed to form a pellet. It is then roasted for 2 h at 800°C to produce $LiAlO_2$. Next, aluminum powder is added and compressed with

TABLE 22.2
Electrochemical Process of the Lithium–Air Battery [19]

Type of Electrolyte	Cell Reaction	Electrolytes	Waste Product
Aqueous (alkaline)	$4Li + O_2 + H_2O \rightarrow 4LiOH$	Lithium salt in water	LiOH
Aqueous (acidic)	$4Li + O_2 + 4H^+ \rightarrow 4LiOH + H_2O$	Lithium salt in water	LiOH
Aprotic	$2Li + O_2 \rightarrow Li_2O_2$ $4Li + O_2 \rightarrow 2Li_2O$	Lithium salt	Li_2O_2 or Li_2O
Solid-state	$O_2 + 2e^- + 2Li^+ \rightarrow Li_2O_2$	Polymer ceramic or glass	Li_2O_2

$LiAlO_2$ followed by a reduction stage for 3 h at a temperature of 1,150 °C. Lithium is produced as the end product in the reduction process. The chemical process involved in vacuum aluminothermic reduction is shown in the equations below.

$$Li_2CO_3 + Al_2O_3 \rightarrow 2LiAlO_2 + CO_2 \quad (22.30)$$

$$3LiAlO_2 + 2CaO + Al \rightarrow 3Li + 2(CaO.Al_2O_3) \quad (22.31)$$

22.6.2 Recycling of Lithium Oxide and Lithium Peroxide

The production of lithium is possible only if the stock feed is lithium oxide. Hence, further treatment is needed to reduce the lithium peroxide to lithium oxide. Thermal decomposition is performed to reduce the lithium peroxide to lithium oxide. A temperature of about 300 °C–400 °C is used in the process [21].

$$2Li_2O_2 \rightarrow 2Li_2O + O_2 \quad (22.32)$$

To produce lithium metal, roasting of lithium oxide is carried out [23]. The lithium oxide in the presence of calcium oxide and ferrosilicon (FeSi) undergoes roasting at a temperature of about 970 °C–1,025 °C. This process produced lithium metal in vapor form. The vapor is gathered and condensed to obtain solid lithium.

$$2Li_2O + 2CaO + Si \rightarrow 4Li + 2CaO.SiO_2 \quad (22.33)$$

22.6.3 Recent Developments in the Recycling of Lithium-Based Battery

Industry-wide recycling of lithium-based batteries began in 2015 due to the consideration of limited resource supply in the future and the new law enforcement in Europe. The LithoRec project [24] is a German-based project that targets to develop a new recycling process for the used lithium battery in EVs. The recycling process consists of four stages, which are as follows: (i) battery and module disassembly, (ii) cell disassembly, (iii) cathode separation, and (iv) hydrometallurgical treatment. First, the battery undergoes discharging and short circuit to empty its remaining energy. Next, the battery is crushed under an inert atmosphere to avoid burning during the crushing process. The remaining components in the electrolyte are then removed through solid–liquid extraction via dimethyl carbonate, which can regain the solvent and conducting salt in the electrolyte. The conducting salt is diluted in a rotary kiln and dried afterward so that dimethyl carbonate can be recycled. Meanwhile, the electrolyte recovery is conducted through evaporation to obtain a concentrated solution.

The second approach is to use thermal drying. The electrolyte is vaporized in the temperature range of 80 °C–140 °C and decreased pressure. It then undergoes recovery through condensation or burning in a thermal post-combustion process. The electrolyte can also be recovered using supercritical carbon dioxide under the pressure of 120 bars.

Mechanical separation is used to separate the electrodes in the battery. It involves air sifting, crushing, and sieving processes. In the air sifting process, dried battery fragments are moved to a magnetic and a simple air separator to isolate the heavy parts such as cell module, housing, and iron. Next, a cutting mill is used to homogenize the battery fragment while increasing the mechanical stress of the fragments to obtain a higher yield of coating materials. Zigzag sifting, a type of air classification, is used to separate the fragments into current collector foils and separator/coating materials. On the other hand, optical sorting is used to separate current collector foils, copper, and aluminum from the fragments. All the remaining valuable coating materials are then separated using a sieving machine.

Lastly, hydrometallurgical treatment is performed depending on the type of impurities. The remaining lithium present in the battery will be cleaned to form lithium hydroxide through crystallization. Lithium hydroxide is then ready to be used for other purposes such as new battery active material. The LithoRec process has a recycling rate of about 75%–80%. Separator and graphite are the components that cannot be recycled in the process.

22.6.4 Climate Impact of Lithium–Air Battery

The climate impact of the production and recycling of lithium–air batteries is discussed in terms of carbon dioxide footprint per kilometer. Based on a study, the carbon footprint to produce a lithium–air battery is only 3 g CO_2-eq/km and it is lower than that of the lithium-ion battery, which is recorded at 27 g CO_2-eq/km [25]. In the long term, producing lithium–air batteries can help to reduce the climate impact by four to nine times compared to the carbon footprint produced by lithium-ion batteries. In terms of recycling, lithium–air batteries show only 10% reduction in carbon footprint. This reduction is due to the production of lithium metal from raw materials.

22.6.5 Waste Generated in Electrolyte and Air Cathode

Lithium–air batteries provide four different types of electrolytes in their structure. In general, they can be classified into aqueous, non-aqueous (aprotic), mixed (aprotic–aqueous), and solid-state electrolytes. In aqueous electrolytes, a lithium–air battery involves dissolving a lithium salt in water. In non-aqueous electrolytes, the lithium salt is dissolved in different solvents. Some common examples of lithium salt used in lithium–air batteries are $LiPF_6$, $LiSO_3CF_3$, and $LiAsF_6$, while some of the common solvents are carbonate, esters, and ethers. At last, solid-state electrolytes involve polymer ceramic or glass. In the lithium–air battery, carbon is the most common air cathode and different catalysts are added to the carbon for better electrochemical reaction. These catalysts include MnO_2, Fe_2O_3, and CuO and improve the performance of the lithium–air battery. Similar to other types of metal–air batteries, the air cathode will be treated as waste since there is no proper recycling procedure available at the current stage.

22.7 NEW APPROACH TO RECYCLING THE AIR CATHODES

The recycling of air cathodes can be done through roasting [26] and the vacuum distillation process. In the roasting process, the carbon cathode is separated from the battery and crushed. Next, the crushed carbon cathode is ground using a ball mill to reduce the size. The process is continued by roasting the carbon cathode in a furnace. The roasted carbon tends to contain higher carbon contents as the roasting temperature increases, and the roasting temperature should be above 1,200 °C and the roasting time should be 1 h. The second method is to treat the spent carbon cathode through distillation in vacuum conditions [27]. The spent carbon cathode is crushed to the size of less than 2 cm before it is distilled in a vacuum. Similar to roasting, a high distillation temperature is more favorable as it can provide a higher amount of carbon content as residue in the end product. The distillation temperature should be above 1,200 °C. The advantage of the vacuum distillation process is it does not require a very fine carbon feedstock.

22.8 CONCLUSIONS

The metal–air battery is considered a very promising energy storage solution for the post-lithium-ion battery era. However, there is still a long way to go before commercializing metal–air batteries for electric vehicles. Currently, only aluminum–air battery technology is being tested on electric vehicles, but it is still not good enough to replace lithium-ion batteries as the primary energy source. On the other hand, other types of metal–air batteries still require extensive research before they can be used in electric vehicles. In battery research, researchers should also focus on the waste generated from the battery and its impact on the environment. The product used to fabricate the battery should be environmentally friendly. A proper recycling process or technique should be designed and implemented before the commercialization of metal–air batteries in the market. Compared to lithium-ion batteries, the recycling of metal–air batteries is less complex and does not involve high costs and advanced technology. The existing technologies can recover most of the material from the used metal–air battery. In marching toward the post-lithium-ion battery era, the mistake of ignoring the recycling process during the lithium-ion battery era should not be repeated and the circular economy concept should be the main priority in any energy storage system design.

REFERENCES

1. Van Noorden R (2014) The rechargeable revolution: A better battery. *Nature* 507:26.
2. Girishkumar G, McCloskey B, Luntz AC, Swanson S, Wilcke W (2010) Lithium–air battery: Promise and challenges. *J Phys Chem Lett* 1:2193–2203.
3. Liu Y, Sun Q, Li W, Adair KR, Li J, Sun X (2017) A comprehensive review on recent progress in aluminum–air batteries. *Green Energy Environ* 2:246–277.
4. Goel P, Dobhal D, Sharma RC (2020) Aluminum–air batteries: A viability review. *J Energy Storage* 28:101287.
5. Husband T (2012) Recycling aluminum, A way of life or a lifestyle? *Chemmatters*:15–17.
6. Haraldsson J, Johansson MT (2018) Review of measures for improved energy efficiency in production-related processes in the aluminium industry-from electrolysis to recycling. *Renew Sustain Energy Rev* 93:525–548.

7. Wagiman A, Mustapa MS, Asmawi R, Shamsudin S, Lajis MA, Mutoh Y (2020) A review on direct hot extrusion technique in recycling of aluminium chips. *Int J Adv Manuf Technol* 106:641–653.
8. Zhang W, Liu Y, Zhang L, Chen J (2019) Recent advances in isolated single-atom catalysts for zinc air batteries: A focus review. *Nanomaterials* 9:1402.
9. Jha MK, Kumar V, Singh RJ (2001) Review of hydrometallurgical recovery of zinc from industrial wastes. *Resour Conserv Recy* 33:1–22.
10. Stewart DJ, Barron AR (2020) Pyrometallurgical removal of zinc from basic oxygen steelmaking dust–A review of best available technology. *Resour Conserv Recy* 157:104746.
11. Nilsson AE, Aragonés MM, Torralvo FA, Dunon V, Angel H, Komnitsas K, Willquist K (2017) A review of the carbon footprint of Cu and Zn production from primary and secondary sources. *Minerals* 7:168.
12. Grimes S, Donaldson J, Gomez GC (2008) *Report on the Environmental Benefits of Recycling.* Bureau of International Recycling, Belgium.
13. Guo Z, Zhao S, Li T, Su D, Guo S, Wang G (2020) Recent advances in rechargeable magnesium-based batteries for high-efficiency energy storage. *Adv Energy Mater* 10:1903591.
14. Froehlich P, Lorenz T, Martin G, Brett B, Bertau M (2017) Valuable metals—Recovery processes, current trends, and recycling strategies. *Angew* 56:2544–2580.
15. Brown RE (2011) Environmental challenges for the magnesium industry. In: *Magnesium Technology 2011.* Sillekens WH, Agnew SR, Neelameggham NR, Mathaudhu SN (eds). Springer, Cham, Switzerland.
16. Wulandari W, Brooks G, Rhamdhani M, Monaghan B (2010) Magnesium: Current and alternative production routes. *Chemeca* 2010: 347–357.
17. Antrekowitsch H, Hanko G, Ebner, P (2002) Recycling of different types of magnesium scrap. TMS Annual Meeting, 43–48.
18. Simone E (2020) *Update of Life Cycle Assessment of Magnesium Components in Vehicle Construction.* German Aerospace Centre, Germany.
19. Rahman MA, Wang X, Wen C (2013) High energy density metal-air batteries: A review. *J Electrochem Soc* 160:A1759–A1771.
20. Nazarov VI, Gonopolsky AM, Makarenkov DA, Klyushenkova MI, Popov AP (2020) Production of lithium hydroxide and lithium carbonate from spent lithium batteries. *Coke Chem* 63:97–103.
21. Kamienski CW, McDonald DP, Stark MW, Papcun JR (2004) Lithium and lithium compounds. In: *Kirk-Othmer Encyclopedia of Chemical Technology.* Kirk-Othmer (ed). John Wiley & Sons, Inc. New York, United States.
22. Di YZ, Wang ZH, Tao SH, Feng NX (2013) A novel vacuum aluminothermic reduction lithium process. *4th International Symposium on High Temperature Metallurgical Processing.* John Wiley & Sons.
23. Morris W, Pidgeon LM (1958) The vapor pressure of lithium in the reduction of lithium oxide by silicon. *Can J Chem* 36:910–914.
24. Diekmann J, Rothermel S, Nowak S, Kwade A (2018) The LithoRec process. In: *Recycling of Lithium-Ion Batteries.* Kwade A, Diekmann J (ed). Springer, Cham, Switzerland.
25. Zackrisson M, Fransson K, Hildenbrand J, Lampic G, O'Dwyer C (2016) Life cycle assessment of lithium-air battery cells. *J Clean Prod* 135:299–311.
26. Yang K, Gong P, Tian Z, Lai Y, Li J (2020) Recycling spent carbon cathode by a roasting method and its application in Li-ion batteries anodes. *J Clean Prod* 261:121090.
27. Wang Y, Peng J, Di Y (2018) Separation and recycling of spent carbon cathode blocks in the aluminum industry by the vacuum distillation process. *JOM* 70:1877–1882.

23 Biowastes for Metal–Sulfur Batteries

Chaofeng Zhang, Quanwei Ma, Longhai Zhang, Rui Wang, Hao Li, and Tengfei Zhou
Anhui University

Changzhou Yuan
University of Jinan

CONTENTS

23.1 INTRODUCTION

With the growing demand for energy from electric vehicles and smart grids, it becomes important to explore sustainable energy storage systems with high energy density and long service life.[1] Li–S batteries possess a potential commercial value for energy storage equipment, because they have a high theoretical capacity of 1,675 mAh/g and theoretical specific energy density of 2,600 Wh/kg, which are far better than many current lithium-ion batteries (LIBs).[2,3] In addition, sulfur element has many advantages, such as abundant resources, low cost, and environmentally friendly properties, which make the metal–S battery become a research focus for energy storage. However, Li–S batteries can't be easily commercialized due to the following drawbacks, including low electrical conductivity, large volume expansion during cycling process, dissolution, and shuttle effect of

DOI: 10.1201/9781003178354-27

polysulfide intermediates (the dissolved polysulfide side products (Li_2S_n) from the cathode to the anode).[4,5]

To address the current challenges of metal–S batteries, many strategies have been employed to suppress the shuttle effect of polysulfides and enhance the cycle life. The main methods include designing the sulfur host, facilitating the conversion of polysulfides. These sulfur hosts include graphene, MXenes, carbon nanotubes, functional polymers, and porous carbon. Another strategy is using separator in the cell to enhance the cycling performance and limit the shuttle effect in metal–S batteries. Numerous carbon materials have been tried as the interlayer of metal–S batteries, such as carbon sphere, carbon nanofiber, and graphene.[6,7] However, the complex synthesis methods and the utilization of toxic substances or expensive chemicals lead to an increase in production costs and dramatically limit the commercialization process of metal–S batteries.[8]

The application of biowaste materials in metal–S batteries is developing promptly due to their low cost and special physical/chemical properties.[8] Meanwhile, biowaste materials have abundant resources in nature,[7] for example corncob, oak tree fruit shells, grapefruit peel, cherry pit, and wheat straw. Due to unique nanostructures, good chemical stability, and outstanding adsorption ability, biowaste-derived carbon can be employed as the S host and interlayer of metal–S batteries. The carbon with porous structure, high specific surface area, and abundant surface functional groups can be fabricated by simple carbonization and activation. The biowaste-derived carbon with a highly porous structure can obviously block the dissolution of polysulfides, enhance the ion/electronic conductivity, and mitigate the volume change of S cathode.[3,9] Biopolymers derived from biowaste materials can be employed as the binder of S cathode, which can obviously boost the electrochemical performance of Li–S batteries.[10] Moreover, the biowaste-derived materials also can be adopted as the S host for Na–S batteries, which remarkably enhance the electrochemical performance of Na–S batteries. Therefore, the special properties of the biowaste materials make them good from the application perspective in metal–sulfur batteries.

Herein, we summarized the applications of biowaste materials in metal–S batteries in recent years, including sulfur hosts, separator materials, and biopolymer binders of S cathode. We also discuss the effects of heteroatom doping, nanostructure, and recombination with other materials in detail when the biowaste carbon is used as the S hosts or separator for metal–S batteries. Finally, the perspectives and suggestions on this field for the further development are concluded.

23.2 BIOWASTE CARBON ACTS AS SULFUR HOST IN LI–S BATTERIES

The biowaste-derived carbon is prepared from carbonization of biowaste materials, for example grass, almond shell, and wheat straw.[6] The biowaste-derived carbon demonstrates a large amount of nanostructure with various morphology and high specific surface area.[6,11] Due to its unique structure, the carbon is able to accommodate sulfur, thus inhibiting the negative effects of S for metal–S batteries.

23.2.1 Structural Design

23.2.1.1 Biowaste-Derived Porous Carbon

The pore diameter of microporous carbon is less than 2 nm. Microporous carbon/sulfur composites can effectively immobilize sulfur and avoid the dissolution of polysulfide on account of spatial constraints of microporous structures.[12] Yang et al.[13] prepared a unique activated carbon (AAC) by a simple and cost-effective method from natural waste apricot shell. First, the apricot shell was heated at 300°C and carbonized at 750 °C under a flowing Ar atmosphere, and then it was activated with KOH at 750 °C. The pore size distribution of AAC ranged from 0.6 to 2.0 nm, which is favorable for the penetration of sulfur molecules by capillary action in a highly dispersed state. Therefore, these physical features are beneficial for immobilizing sulfur and restraining the shuttle effect of polysulfides during the charge/discharge processes. When tested as cathode for Li–S batteries, the optimized AAC/S composite presented a high initial capacity of 1,277 mAh/g at 0.1 C and reversible capacity of 613 mAh/g at 1 C after 200 cycles.

The microporous structure guarantees the strong confinement of small sulfur molecules (S_{2-4}), improving the utilization and stability of the active materials. Also, the microporous carbon can effectively limit the side reactions of S_{2-4} with carbonate solvents.[14,15] For example, Zhou et al. reported S_{2-4} can be confined in microporous graphitic carbon (MGC) that comes from peanut shell char (Figure 23.1a).[14] The obtained MGC possessed abundant interconnected ultra-micropores with a diameter less than 0.4 nm. The S/MGC composite showed a good cycling stability and rate capability in Li–S batteries. The S_{2-4} was well confined in the interconnected ultra-microporous and graphitic carbon walls. At the same time, the unoccupied pores of S/MGC composite can facilitate fast Li^+ transport and kinetics of electrochemical reactions.

The pore diameter of mesoporous carbon is between 2 and 50 nm. The structure can allow a high S-loading, promote the ionic and electronic transportation, and enhance the infiltration of electrolyte.[16] In 2009, Nazar et al. reported a pioneering work on the use of mesoporous carbon in Li–S batteries.[17] Then, many mesoporous carbons have been prepared by pyrolyzing natural biomass materials. Lee et al. successfully prepared garlic-peel-derived porous carbons by direct precarbonization (GPC) and hydrothermal treatment (GPC/HT) from biowaste garlic-peels, respectively.[16] In the preparation process of GPC/HT, the hydrothermal process played an important role in the activation of the porous carbon. The GPC/HT with a high surface area offered an interconnected porous structure, accommodating a S-loading of 87.6 wt.%. When tested in Li–S batteries, it exhibited an initial capacity of 1,087 mAh/g and long-term cycle performance.

Biowaste-derived hierarchical porous carbon possessed different sizes of pores. To promote the electrochemical performance of Li–S batteries, carbon materials with a hierarchical porous structure were proposed. In the hierarchical structure, meso- and macropores can facilitate the contact between the active materials and electrolyte, buffer the volume change during charging/discharging, and promote the ion transportation.[18] Compared with individual macro-, meso-, or microporous carbon, the hierarchical porous carbon as a sulfur host demonstrate synergistic effect.[15]

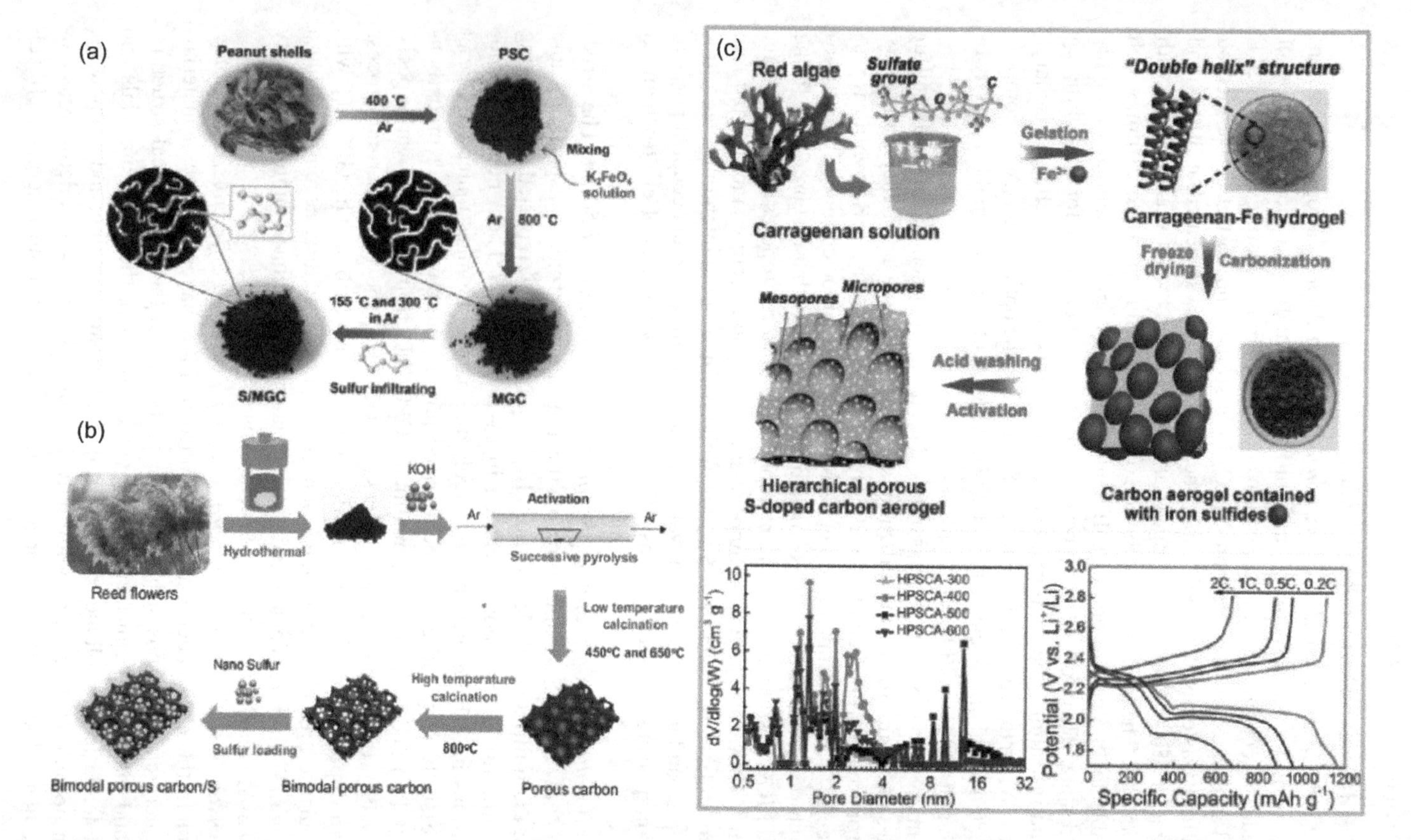

FIGURE 23.1 (a) The fabrication method of the S/MGC composite.[14] (Copyright 2018, Elsevier Ltd.) (b) Preparation process of BPC.[19] (Copyright 2020, Elsevier Inc.) (c) The synthesis process of HPSCA. Pore size distribution of HPSCA-T and the voltage profiles of S/HPSCA-400 electrode at different current densities.[20] (Copyright 2018, Elsevier B.V.)

Wang et al. successfully synthesized a bimodal porous carbon (BPC) by a simple and low-cost process from reed flowers (Figure 23.1b).[19] First, the reed flowers were pretreated by hydrothermal reaction and then were heated to form the first-class pores (macropores) and second-class pores (mesopores). The BPC/S electrode presented an outstanding rate capability and cycling stability.

The activated carbon with abundant meso-/macro-/micropores and three-dimensional (3D) framework possesses a high S-loading and effective suppression of polysulfides.[20] Li et al. used a sustainable biomass conversion method to produce a 3D hierarchical macro-/meso-/microporous carbon from red algae precursor (Figure 23.1c).[20] In detail, the carrageenan-Fe hydrogel can be formed by mixing the ι-carrageenan solution and $FeCl_3{\cdot}6H_2O$ aqueous solution. After freeze-drying, pyrolysis, iron removal with hydrochloric acid, and activation with KOH, the product was obtained. In this process, carrageenan could form a conductive carbon framework and porous structure. The as-prepared 3D hierarchical porous sulfur-doped carbon aerogel (HPSCA) presented an ultra-high surface area of 4,037.0 m^2/g and interconnected conductive network. Moreover, the S/HPSCA exhibited a high capacity of 1,110 mAh/g at 0.2 C and cycling stability of 400 cycles. The hierarchical porous carbon structure shows a large specific surface area and large pore volume. Meanwhile, porous carbon can provide a large number of channels to improve the electronic and ionic transfer, thus enhancing the rate capability.[21] Furthermore, the micro-/mesopores in hierarchical porous carbon can effectively adsorb the lithium polysulfides, thus weakening the shuttle effect.[22]

23.2.1.2 Biowaste Carbon with Regular Morphology Structure

The carbon materials with anomalous morphology have more grain boundaries, leading to a longer charge transfer path and higher charge transfer resistance.[7] Generally, biowaste can be used to prepare the carbon materials with regular structure, such as hole array, nanosheets, nanofibers, and hollow spheres.

Two-dimensional (2D) materials have numerous unique advantages, including ultrathin thickness, large lateral size, and unusual physical and chemical properties. Therefore, porous carbon nanosheets (PCNs) combining the properties of 2D nanostructures will present the synergetic properties different from those of either common porous carbons or ultrathin nanosheets.

Guo et al.[11] prepared the PCNs by a simple, low-cost, and high-yield strategy from waste corncob. The PCNs with large specific surface area was compounded with S (S/PCNs) as the cathode for Li–S batteries. The S/PCNs exhibited a high initial discharge capacity of about 1,600 mAh/g. The PCNs enabled a stable and continuous pathway for rapid electronic/ion transportation and can restrain the shuttle effect. The sheet-like activated carbon material derived from corncob waste material is a promising S-loading matrix for Li–S batteries.

Also, the nanofibrous carbon can effectively restrain the shuttle effect of polysulfides. Luo et al. fabricated the mesoporous carbon fiber from natural wood fiber for Li–S batteries.[23] Bleached softwood pulp was pretreated by slushing and stirring in DI water and then filtered by a vacuum filtration process. The as-obtained sample was stabilized and carbonized at 240 °C and 1,000 °C, respectively. The as-prepared carbon fibers/S composite cathode presented a high initial capacity and retained a reversible capacity of 859 mAh/g for 450 cycles, corresponding to a capacity decay rate of 0.046% per cycle. Importantly, the free-standing carbon fibers/S composite didn't contain inactive component, such as current collector, binder, and conductive additive. The large specific surface is beneficial to improving the interface contact between active material and electrolyte, providing more active sites. Additionally, the mesoporous carbon is conducive to electrolyte diffusion and Li^+ transportation and can relieve the volume expansion during cycling. The microporous structure is of help for immobilizing sulfur and avoiding the dissolution of polysulfide.

23.2.2 Heteroatom Doping

Introducing heteroatoms such as N, O, P, or B into porous carbon can effectively enhance the chemical adsorption effect in Li–S batteries, which is one of the powerful means to suppress the shuttle effect. The porous carbon containing heteroatom can be easily obtained by suitable materials and synthesis methods.[2]

Liu et al.[5] fabricated N-doped hollow porous carbon spheres (N-HPCS) by a simple spray-drying method from biomass lignin and cyanuric acid as N-dopant. The N-HPCS presented a high specific surface area (446.2 m^2/g) and high-level pyrrolic-N-doping. In electrochemical tests, the N-HPCS/S composite showed a high initial discharge capacity. After 1,000 cycles at 1 C, the cathode manifested a low capacity decay rate of 0.041% per cycle. The N-HPCS material with N-doping, especially pyrrolic-N, made a strong Li_2S_x-N chemical adhesion, which effectively suppressed the dissolution and shuttle effect of polysulfides in Li–S batteries. Ji et al.[24] prepared a coralline-like N-doped hierarchically porous carbon (CNHPC) via hydrothermal and carbonization processes. The CNHPC/S showed a high initial capacity and low capacity decay rate. The outstanding performance can be attributed to the coralline-like hierarchically porous structure and inherently abundant N-doping.

Wu et al.[1] fabricated a N, O-co-doped carbon from biomass bagasse. The N, O-co-doped porous carbon obviously enhanced the adsorption and conversion of polysulfides via chemical interaction, resulting in a stable cycling of Li–S batteries. The electrode presented a high initial capacity of 1,123 mAh/g and low decay rate over 800 cycles.

With the development of heteroatom doping methods, more kinds of heteroatom are selected to be doped into carbon materials. Ren et al.[25] prepared a cauliflower-like N, O, P-tri-doped porous carbon (NOPC) by a simple activation and carbonization process from goat hair (Figure 23.2). The P element was introduced by activation with H_3PO_4. The synergistic effect of structure and heteroatom doping improved the conductivity and utilization of sulfur and restrained the shuttle effect of polysulfides in Li–S batteries. Therefore, the S/NOPC electrode exhibited an outstanding electrochemical performance. Zhong et al.[26] synthesized an alveolation-like N, O, S-tri-doped hierarchical porous carbon (a-NOSPC) by a facile pyrolysis of tobacco stem. The a-NOSPC provided an abundant electrocatalytic active sites and can accelerate the redox kinetics of chemical binding of polysulfides. In electrochemical tests, a-NOSPC/S cathode exhibited a high initial capacity and long-term cycling capacity

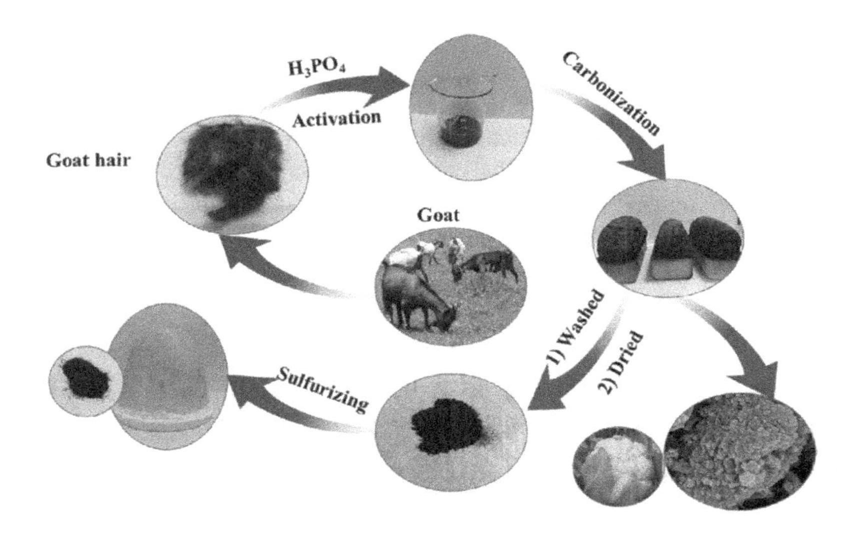

FIGURE 23.2 The fabrication process of S/NOPC composite materials.[25] (Copyright 2018, Science Press and Dalian Institute of Chemical Physics, Chinese Academy of Science.)

of 754 mAh/g after 400 cycles at 0.5 C. The hierarchical porous carbon containing heteroatom can also be prepared from pomegranate residues.[27] A large number of studies demonstrated that the introduction of heteroatom in porous carbon can be a good method to suppress the dissolution and shuttle effect in Li_2S_x in batteries, because heteroatoms can form a strong chemical bond with lithium polysulfide, and the heteroatom-doped porous structure can improve the chemical adsorption of polysulfide. Therefore, the heteroatom-doped porous carbon can enhance the cycling stability and rate capability of Li–S batteries.

23.2.3 Composites as Sulfur Host

Wang et al.[28] developed a triple protection strategy through graphene, N, P-co-doped biological carbon (NPBCS), and organic component PEDOT to encapsulate S species. The unique hierarchical structure not only remarkably encapsulated sulfur, but also improved the electrical conductivity and ensured efficient lithium-ion transport. The PEDOT and graphene effectively mitigated the volume change in the cycling process. In addition, the inherent N and P in carbon materials from natural bacteria possessed a strong chemical absorption to lithium polysulfides, which could effectively restrain the dissolution and shuttle effect of polysulfides.

Except for organic polymers, metal oxides, sulfides, or other inorganic materials can be mixed with biomass-derived carbon, such as MnO_2,[29] Co,[30] and NiS_2.[31] Muhammad et al.[32] fabricated a Co nanocrystal encapsulated in 3D N-doped porous carbon (Co@NC) by a simple pressure-cooking strategy from chickpea. The 3D N-doped porous carbon could withstand a volume change of S during cycling process and enhance the ionic/electronic transportation. In the meantime, the homogeneously dispersed Co nanocrystal promoted the fast conversion of polysulfides, which was beneficial to enhancing the electrochemical performance of Li–S batteries. The S/Co@NC with a high S-loading showed a high capacity and cycle stability. Introducing nanocatalytic materials and designing structures for carbon materials are beneficial to enhancing the electrochemical performance of Li–S batteries.

The recent studies have shown that combining different functional materials with heteroatom, inorganic, and organic compounds can effectively suppress the dissolution and shuttling of polysulfides, as well as increase the utilization of S.

23.3 BIOWASTE-DERIVED MATERIALS USED AS SEPARATORS FOR LI–S BATTERIES

It is well known that the biowaste-derived carbon as the S host in Li–S batteries can effectively limit the dissolution of polysulfides in electrolyte and their migration between the anode and cathode (i.e., shuttle effect).[8] Besides, the rational design of the biomass-derived separators is also a promising method. Currently, the reported designs of biowaste-derived separators have mainly been divided into two categories, including commercial separator coated with porous carbon and free-standing carbon interlayer.[33]

23.3.1 Biowaste-Derived Carbon Film Coated on Separator

In general, the commercial separator coated with porous carbon is a widely used simple and effective method to suppress the diffusion of polysulfide. Among them, the biomass-derived carbon has attracted great attention owing to its diversity.

Li et al.[33] synthesized a N, O-co-doped chlorella-based biowaste carbon (CBBC) with a micro-/mesoporous structure by a facile chemical activation (Figure 23.3a). Afterward, the as-prepared CBBC, acetylene black (AB), and PVDF binder were mixed and homogenously casted on one side of the PP membrane with a dosage of 60%. Such a CBBC interlayer can not only enhance the wettability of electrolyte and facilitate the diffusion of Li^+, but also suppress the shuttle effect due to the strong physical barriers and chemical adsorption for polysulfides, thereby leading to an improved electrochemical performance of Li–S batteries. The assembled Li–S batteries with a CBBC-modified separator exhibited a high reversible capacity, good cycling stability (only 0.067% capacity decay per cycle at 0.5 C), and remarkable rate performance.

Liu et al.[34] successfully prepared a novel biowaste carbon fiber@SiO_2 (BCF@SiO_2) composite as the interlayer of Li–S batteries for boosting the cycling performance *via* using the absorbent cotton as template (Figure 23.3b). The SiO_2 nanoparticles were uniformly distributed on BCF and acted as a polar functional group, which apparently improved the charge conductivity and strengthened the chemical adsorption for polysulfides. In addition, the combination of BCF and polar SiO_2 could also enhance the sulfur utilization and suppress the shuttle effect. Thus, the batteries using the separator coated with BCF@SiO_2 exhibited a high reversible capacity and promising cycling performance.

Besides, the following carbon materials derived from biowaste had been investigated in Li–S batteries, including the pomelo peel-derived N, B-dual-doped carbon

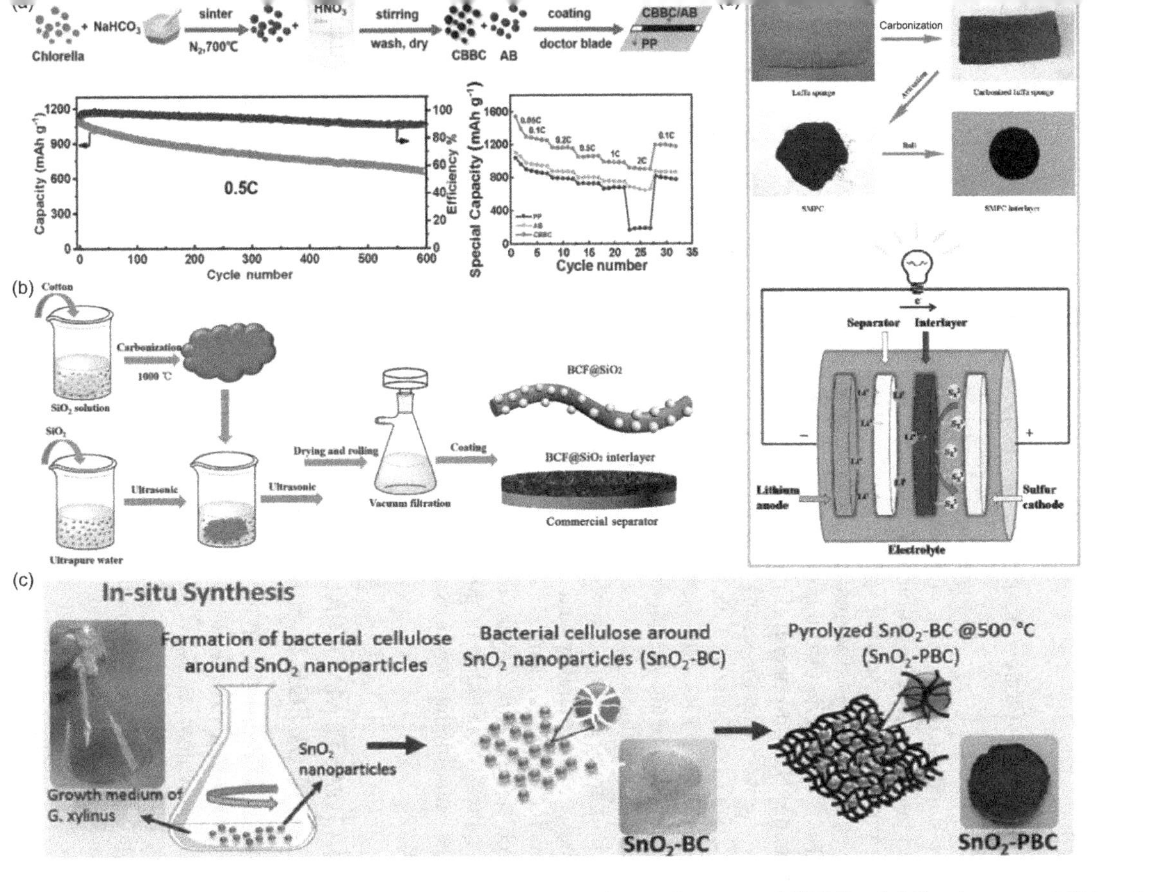

FIGURE 23.3 (a) The preparation process of CBBC-coated separator. The cycling performance of CBBC at 0.5 C and rate capabilities of three types of batteries.[33] (Copyright 2020, Elsevier Inc.) (b) Preparation process of BCF@SiO_2 interlayer.[34] (Copyright 2018, Elsevier Ltd.) (c) Preparation process of SnO_2-PBC by in situ synthesis.[37] (Copyright 2018, Elsevier Inc.) (d) The formation process of the SMPC interlayer. Schematic configuration of Li–S battery with SMPC interlayer.[38] (Copyright 2016, Royal Society Chemistry.)

aerogel (NB-PPCA) with abundant specific surface area and pores and polar MnO with enhanced chemical adsorption for polysulfides.[9,35] Remarkably, the above carbon-modified separators can effectively accommodate the volume change during charge/discharge processes, increase the utilization of active materials, reduce the internal resistance, and inhibit the shuttle effect, thereby offering a good specific capacity, cyclability, and rate performance.

23.3.2 Biowaste-Derived Carbon as Free-Standing Interlayer

As for the free-standing biowaste-derived carbon interlayer in Li–S batteries, polar free-standing interlayer could not only act as an efficient physical barrier to strengthen polysulfides adsorption, but also work as a second current collector, significantly improving the specific capacity and cycle stability.[36]

Rezan et al.[37] successfully obtained a SnO_2/carbon derived from the bacterial cellulose (PBC) via a facile in situ growth of PBC on SnO_2 surface following pyrolysis (500 °C under N_2 atmosphere) (Figure 23.3c). Benefiting from the PBC-derived 3D conductive network and the strong polysulfides adsorption ability of SnO_2, the introduction of SnO_2/PBC-derived carbon free-standing interlayer effectively improved the cycle stability of Li–S batteries.

Porous carbon can also be employed as free-standing interlayers by using the rolling method. For example, Yang et al.[38] rolled the sulfur-doped microporous carbon (SMPC) into free-standing interlayers. And the batteries using this film show an outstanding cycling stability and ultra-high rate performance (Figure 23.3d). Due to the special physical/chemical adsorption properties and high conductivity, the SMPC free-standing interlayer could effectively accelerate electron and Li^+ transfer and suppress dissolution of polysulfides. As a result, the Li–S batteries with a SMPC free-standing interlayer exhibited a high initial capacity and outstanding rate performance.

23.4 BIOWASTE MATERIALS AS BINDER OF SULFUR CATHODE FOR LI–S BATTERIES

Polyvinylidene fluoride (PVDF) as a primary binder material has widely been applied in LIBs, which can effectively keep the integrality of electrode during repeated charge/discharge process. However, the organic solvent (N-methyl-2-pyrrolidone) for dissolving PVDF is toxic. Thus, exploring a promising alternative binder for high-performance Li–S batteries is required. Recently, several biopolymers accompanied with water solvent have attracted more attention as binders owing to their high stability, low cost, nontoxicity, and sustainability. For example, Li et al.[39] employed a natural gum arabic (GA) as a binder for S cathode in Li–S batteries, which mainly contains polysaccharides and glycoproteins with rich functional groups of hydroxyl, ether, and carboxyl groups (Figure 23.4a). Owing to its excellent mechanical property, high binding strength, and suitable flexibility, the GA binder can significantly suppress the dissolution of polysulfides and accommodate volume change, thereby confirming the integrality of cathode and cycling stability of Li–S batteries (841 mAh/g at 0.2 C after 500 cycles).

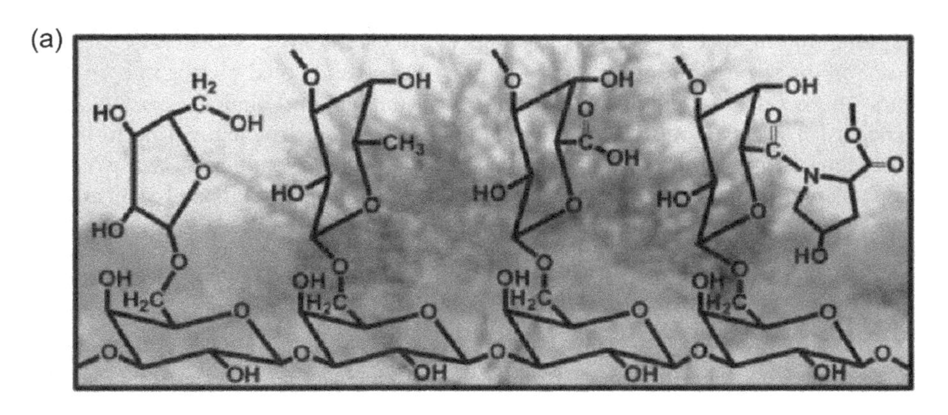

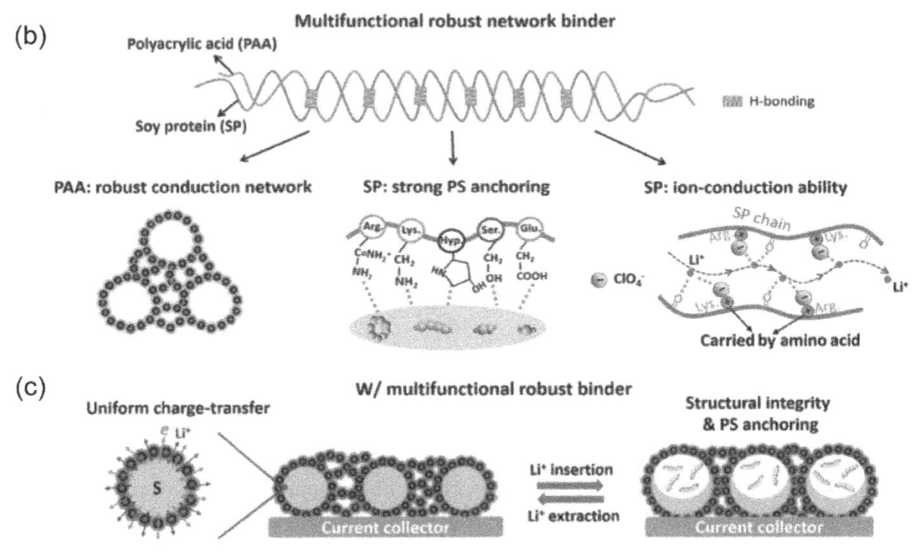

FIGURE 23.4 (a) Chemical structure of GA.[39] (Copyright 2015, Wiley-VCH Verlag GmbH & Co. KGaA, Weinheim.) (b) A diagram of the design strategy for preparing the multi-functional robust binder by using the advantages of PAA and SP.[10] (Copyright 2018, Royal Society Chemistry.) (c) A schematic diagram of the structural changes in S cathodes with a multi-functional robust binder during the Li^+ insertion/extraction process. (Note: PS refers to polysulfide.)[10] (Copyright 2018, Royal Society Chemistry.)

Additionally, the soy protein with abundant polar groups (amines and carboxyl groups) exhibited a stronger chemical adsorption for polysulfides and has been considered as another promising binder candidate for high-performance Li–S batteries. For example, Fu et al.[10] adopted a multi-functional binder, i.e., soy protein (SP) and poly(acrylic acid) (PAA) hybrid binder (Figure 23.4b). Benefiting from the strong polysulfide binding/good ion conductivity of SP and the good mechanical strength of PAA, the SP–PAA binder can effectively retain the structural stability of the S cathode (Figure 23.4c) and promote the electrochemical reaction kinetics, finally resulting in improved electrochemical performance for Li–S batteries. Besides, based on the strong intermolecular binding effect of functional groups in both polymers, another dual biopolymer (guar gum and xanthan gum) binder also represented a robust network structure and exhibited an excellent electrochemical performance with high sulfur loading.[40] Thus, this type of biowaste-derived binders with nontoxic property and sustainability holds great promise for high-performance Li–S batteries because of their outstanding physical/chemical adsorption properties, high stability, and ion conductivity.

23.5 BIOWASTE-DERIVED CARBON FOR Na–S BATTERIES

As for the Na–S batteries, apart from their low electrical conductivity and soluble polysulfide, the larger size of Na^+ than Li^+ always makes them suffer from more severe volume variation and sluggish reaction kinetics, thereby leading to poor electrochemical performance. To address these issues, many novel biowaste carbon materials are designed as suitable S host for high-performance Na–S batteries. For example, *via* a facile chemical activation of waste bamboo char, Liu et al.[41] synthesized a dual-porosity carbon (DPC) with abundant oxygen for Na–S batteries. Such unique dual-porosity carbon can effectively physically confine the S species, tolerate the shuttle effect, and improve the contact between cathode and electrolyte, ensuring the electrochemical stability and rapid reaction kinetics. In addition, *via* strong chemical adsorption, the oxygen doping can further restrain the dissolution of polysulfides. As a result, DPC/S composite delivered a high initial capacity and excellent cyclability.

23.6 CONCLUSIONS AND OUTLOOK

In summary, the biowaste-derived materials with rich source, low cost, environment-friendly nature, and adjustable physical/chemical properties have a potential application value for metal–S batteries. The reuse of biowastes materials is favorable for reducing the environmental pollution and waste of natural resources. In this review, we summarize the researches of biowaste materials for Li–S batteries, for example the biowaste-derived carbon as the S host and separator membranes, and the biowaste material serving as binders in Li–S batteries.

The biowaste-derived materials possess the following advantages: (i) The porous structure of carbon can accommodate the volume change of S cathode in the cycling process, and the meso- and microporous carbon can confine S. (ii) The hierarchical porous carbon with a high specific surface area can expose more active sites and facilitate electrolyte diffusion and Li^+ transport. (iii) The conductive network can obviously improve the charge transfer to solve poor conductivity of sulfur. (iv) The heteroatom-doped (N, O, P, S, etc.) carbon can offer chemical adsorption of polysulfides due to its polar component, which can further restrain the dissolution of polysulfides. (v) The porous carbon composite with metal oxides or metal can effectively accelerate the reaction kinetics. (vi) The biowaste-derived carbon-coated separator can prevent the free diffusion of polysulfides between the cathode and anode and can be used as a second current collector to improve the utilization of active material. (vii) The biowaste materials can not only be applied as a binder to limit the volume change of S cathode in cycling process, but can also suppress the dissolution of polysulfides due to their polar group.

Although the biowaste-derived materials have improved the electrochemical performance of metal–S batteries, there are some challenges that need to be addressed for the commercialization these battery systems: (i) The high-temperature (over 800°C) carbonization is a waste of energy, and the expensive activator also increases the cost. Therefore, some cheaper and suitable carbonization methods should be studied. (ii) The extraction of some biopolymers from biomass is very difficult, and it will increase the production cost. The application of biowaste-derived materials has widely been studied in metal–S batteries, which obviously improves the electrochemical performance of Li–S batteries, such as high capacity, excellent long-term cycling life, and outstanding rate performance. We hope that this review can offer ideas for developing low-cost and sustainable biowaste-derived materials to promote the commercialization of metal–sulfur batteries.

REFERENCES

1. Wu D; Liu J; Chen J; Li H; Cao R; Zhang W; Gao Z; Jiang K (2021) Promoting sulphur conversion chemistry with tri-modal porous N, O-codoped carbon for stable Li–S batteries. *J. Mater. Chem. A* 9: 5497–5506.
2. Wen X; Lu X; Xiang K; Xiao L; Liao H; Chen W; Zhou W; Chen H (2019) Nitrogen/sulfur co-doped ordered carbon nanoarrays for superior sulfur hosts in lithium-sulfur batteries. *J. Colloid Interface Sci.* 554: 711–721.
3. Xiao Q; Li G; Li M; Liu R; Li H; Ren P; Dong Y; Feng M; Chen Z (2020) Biomass-derived nitrogen-doped hierarchical porous carbon as efficient sulfur host for lithium–sulfur batteries. *J. Energy Chem.* 44: 61–67.
4. Jiang S; Chen M; Wang X; Zhang Y; Huang C; Zhang Y; Wang Y (2019) Honeycomb-like nitrogen and sulfur dual-doped hierarchical porous biomass carbon bifunctional interlayer for advanced lithium-sulfur batteries. *Chem. Eng. J.* 355: 478–486.
5. Liu Y; Guo H; Zhang B; Wen G; Vajtai R; Wu L; Ajayan PM; Wang L (2020) Sustainable synthesis of N-doped hollow porous carbon spheres via a spray-drying method for lithium-sulfur storage with ultralong cycle life. *Batteries & Supercaps* 3: 1201–1208.
6. Liu P; Wang Y; Liu J (2019) Biomass-derived porous carbon materials for advanced lithium sulfur batteries. *J. Energy Chem.* 34: 171–185.
7. Zhang H; Lin Y; Chen L; Wang D; Hu H; Shen C (2020) Synthesis and electrochemical characterization of lithium carboxylate 2D compounds as high-performance anodes for Li–ion batteries. *ChemElectroChem* 7: 306–313.
8. Yuan H; Liu T; Liu Y; Nai J; Wang Y; Zhang W; Tao X (2019) A review of biomass materials for advanced lithium-sulfur batteries. *Chem. Sci.* 10: 7484–7495.
9. Zhu L; Jiang H; Ran W; You L; Yao S; Shen X; Tu F (2019) Turning biomass waste to a valuable nitrogen and boron dual-doped carbon aerogel for high performance lithium-sulfur batteries. *Appl. Surf. Sci.* 489: 154–164.
10. Fu X; Scudiero L; Zhong W-H (2019) A robust and ion-conductive protein-based binder enabling strong polysulfide anchoring for high-energy lithium–sulfur batteries. *J. Mater. Chem. A* 7: 1835–1848.
11. Guo J; Zhang J; Jiang F; Zhao S; Su Q; Du G (2015) Microporous carbon nanosheets derived from corncobs for lithium–sulfur batteries. *Electrochim. Acta* 176: 853–860.

12. Benitez A; Gonzalez-Tejero M; Caballero A; Morales J (2018) Almond shell as a microporous carbon source for sustainable cathodes in lithium-sulfur batteries. *Materials* 11: 1428.
13. Yang K; Gao Q; Tan Y; Tian W; Zhu L; Yang C (2015) Microporous carbon derived from Apricot shell as cathode material for lithium–sulfur battery. *Microporous Mesoporous Mater.* 204: 235–241.
14. Zhou J; Guo Y; Liang C; Yang J; Wang J; Nuli Y (2018) Confining small sulfur molecules in peanut shell-derived microporous graphitic carbon for advanced lithium sulfur battery. *Electrochim. Acta* 273: 127–135.
15. Yang K; Gao Q; Tan Y; Tian W; Qian W; Zhu L; Yang C (2016) Biomass-derived porous carbon with micropores and small mesopores for high-performance lithium-sulfur batteries. *Chem. Eur. J.* 22: 3239–3244.
16. Lee S-Y; Choi Y; Kim J-K; Lee S-J; Bae JS; Jeong ED (2021) Biomass-garlic-peel-derived porous carbon framework as a sulfur host for lithium-sulfur batteries. *J. Ind. Eng. Chem.* 94: 272–281.
17. Ji XL; Lee KT; Nazar LF (2009) A highly ordered nanostructured carbon-sulphur cathode for lithium-sulphur batteries. *Nat. Mater.* 8: 500–506.
18. Zhao Y; Ren J; Tan T; Babaa MR; Bakenov Z; Liu N; Zhang Y (2017) Biomass waste inspired highly porous carbon for high performance lithium/sulfur batteries. *Nanomaterials* 7: 260.
19. Wang Z; Zhang X; Liu X; Zhang Y; Zhao W; Li Y; Qin C; Bakenov Z (2020) High specific surface area bimodal porous carbon derived from biomass reed flowers for high performance lithium-sulfur batteries. *J. Colloid Interface Sci.* 569: 22–33.
20. Li D; Chang G; Zong L; Xue P; Wang Y; Xia Y; Lai C; Yang D (2019) From double-helix structured seaweed to S-doped carbon aerogel with ultra-high surface area for energy storage. *Energy Storage Mater.* 17: 22–30.
21. Liang C; Zhang X; Zhao Y; Tan T; Zhang Y; Chen Z (2018) Preparation of hierarchical porous carbon from waterweed and its application in lithium/sulfur batteries. *Energies* 11: 1535.
22. Zhou J; Liu Y; Zhang S; Zhou T; Guo Z (2020) Metal chalcogenides for potassium storage. *InfoMat* 2: 437–465.
23. Luo C; Zhu H; Luo W; Shen F; Fan X; Dai J; Liang Y; Wang C; Hu L (2017) Atomic-layer-deposition functionalized carbonized mesoporous wood fiber for high sulfur loading lithium sulfur batteries. *ACS Appl. Mater. Interfaces* 9: 14801–14807.
24. Ji S; Imtiaz S; Sun D; Xin Y; Li Q; Huang T; Zhang Z; Huang Y (2017) Coralline-like N-doped hierarchically porous carbon derived from enteromorpha as a host matrix for lithium-sulfur battery. *Chem. Eur. J.* 23: 18208–18215.
25. Ren J; Zhou Y; Wu H; Xie F; Xu C; Lin D (2019) Sulfur-encapsulated in heteroatom-doped hierarchical porous carbon derived from goat hair for high performance lithium–sulfur batteries. *J. Energy Chem.* 30: 121–131.
26. Zhong M-e; Guan J; Sun J; Guo H; Xiao Z; Zhou N; Gui Q; Gong D (2019) Carbon nanodot-decorated alveolate N, O, S tridoped hierarchical porous carbon as efficient electrocatalysis of polysulfide conversion for lithium-sulfur batteries. *Electrochim. Acta* 299: 600–609.
27. Chen X; Du G; Zhang M; Kalam A; Su Q; Ding S; Xu B (2019) Nitrogen-doped hierarchical porous carbon derived from low-cost biomass pomegranate residues for high performance lithium-sulfur batteries. *J. Electroanal. Chem.* 848: 113316.
28. Wang T; Zhu J; Wei Z; Yang H; Ma Z; Ma R; Zhou J; Yang Y; Peng L; Fei H; Lu B; Duan X (2019) Bacteria-derived biological carbon building robust Li-S batteries. *Nano Lett.* 19: 4384–4390.
29. Raghunandanan A; Mani U; Pitchai R (2018) Low cost bio-derived carbon-sprinkled manganese dioxide as an efficient sulfur host for lithium–sulfur batteries. *RSC Adv.* 8: 24261–24267.

30. Zhuang Y; Ma J; Chen J; Shi Z; Leng F; Feng W (2021) Fabrication and electrochemical applications of the Co-embedded N&P-codoped hierarchical porous carbon host from yeast for Li-S batteries. *Appl. Surf. Sci.* 545: 148936.
31. Liu J; Xiao SH; Zhang Z; Chen Y; Xiang Y; Liu X; Chen JS; Chen P (2020) Naturally derived honeycomb-like N, S-codoped hierarchical porous carbon with MS_2 (M = Co, Ni) decoration for high-performance Li-S battery. *Nanoscale* 12: 5114–5124.
32. Faheem M; Li W; Ahmad N; Yang L; Tufail MK; Zhou Y; Zhou L; Chen R; Yang W (2021) Chickpea derived Co nanocrystal encapsulated in 3D nitrogen-doped mesoporous carbon: Pressure cooking synthetic strategy and its application in lithium-sulfur batteries. *J. Colloid Interface Sci.* 585: 328–336.
33. Li Q; Liu Y; Yang L; Wang Y; Liu Y; Chen Y; Guo X; Wu Z; Zhong B (2021) N, O co-doped chlorella-based biomass carbon modified separator for lithium-sulfur battery with high capacity and long cycle performance. *J. Colloid Interface Sci.* 585: 43–50.
34. Liu T; Sun X; Sun S; Niu Q; Liu H; Song W; Cao F; Li X; Ohsaka T; Wu J (2019) A robust and low-cost biomass carbon fiber@SiO_2 interlayer for reliable lithium-sulfur batteries. *Electrochim. Acta* 295: 684–692.
35. Feng G; Liu X; Wu Z; Chen Y; Yang Z; Wu C; Guo X; Zhong B; Xiang W; Li J (2020) Enhancing performance of Li–S batteries by coating separator with MnO @ yeast-derived carbon spheres. *J. Alloys Compd.* 817: 152723.
36. Song Z; Lu X; Hu Q; Ren J; Zhang W; Zheng Q; Lin D (2019) Synergistic confining polysulfides by rational design a N/P co-doped carbon as sulfur host and functional interlayer for high-performance lithium–sulfur batteries. *J. Power Sources* 421: 23–31.
37. Celik KB; Cengiz EC; Sar T; Dursun B; Ozturk O; Akbas MY; Demir-Cakan R (2018) In-situ wrapping of tin oxide nanoparticles by bacterial cellulose derived carbon nanofibers and its application as freestanding interlayer in lithium sulfide based lithium-sulfur batteries. *J. Colloid Interface Sci.* 530: 137–145.
38. Yang J; Chen F; Li C; Bai T; Long B; Zhou X (2016) A free-standing sulfur-doped microporous carbon interlayer derived from luffa sponge for high performance lithium–sulfur batteries. *J. Mater. Chem. A* 4: 14324–14333.
39. Li G; Ling M; Ye Y; Li Z; Guo J; Yao Y; Zhu J; Lin Z; Zhang S (2015) Acacia senegal-inspired bifunctional binder for longevity of lithium-sulfur batteries. *Adv. Energy Mater.* 5: 1500878.
40. Liu J; Galpaya DGD; Yan L; Sun M; Lin Z; Yan C; Liang C; Zhang S (2017) Exploiting a robust biopolymer network binder for an ultrahigh-areal-capacity Li–S battery. *Energy Environ. Sci.* 10: 750–755.
41. Liu Y; Li X; Sun Y; Yang R; Lee Y; Ahn J-H (2020) Dual-porosity carbon derived from waste bamboo char for room-temperature sodium-sulfur batteries using carbonate-based electrolyte. *Ionics* 27: 199–206.

24 High-Performance Supercapacitors Based on Biowastes for Sustainable Future

Kwadwo Mensah-Darkwa and Stefania Akromah
Kwame Nkrumah, University of Science and Technology

Benjamin Agyei-Tuffour and David Dodoo-Arhin
University of Ghana

Anuj Kumar
GLA University

Ram K. Gupta
Pittsburg State University

CONTENTS

24.1 INTRODUCTION

Self-renewing natural resources such as the sun, wind, flowing water, the earth's internal energy, and biomass are known as renewable energy sources. Renewable energy offers many appealing economic and environmental advantages. They represent a clean and abundant replacement to the conventional energy sources (fossil fuels), producing and

DOI: 10.1201/9781003178354-28

emitting little to no environmental wastes and pollutants. Plant biomass is a great way to reabsorb CO_2 emitted into the environment by the same biomass-derived energy, resulting in overall near-zero CO_2 emission. It is reported that the replacement of conventional energy in the USA alone can reduce the total CO_2 emission by 70 million metric tons per annum [1]. Thus, global renewable energy implementation can resolve the issues of climate change and the greenhouse effect, environmental pollution and degradation, and so on. Additionally, when compared to conventional energy systems, renewable energy systems are more cost-effective. Thus, they are sustainable in terms of current and future economic and social needs.

The millennial improvement in quality of life, the high rate of industrialization, and the exponential population growth are the causes of the exorbitant demands on energy supply and the associated environmental issues resulting from the combustion of fossil fuels. Sustainable resource management strategies are focused on high energy production efficiency of renewable energy technologies as well as energy-saving through storage. The implementation of renewable energy has been limited to 15%–20% of the global energy supply due to the intermittent nature of these resources and, sometimes, their inaccessibility and potential environmental and socioeconomic impacts on some communities [2]. Energy storage is rapidly becoming an integral area of interest in energy management to tackle the shortcomings of renewable energy by providing an uninterrupted power supply and saving energy. Storage devices are particularly used in electric power and transportation systems, and portable electronics.

Large-scale energy storage systems (or reservoirs) can be classified into mechanical, chemical, electrical, and electrochemical storage systems. The need for electrochemical energy storage (EES) has especially surged in recent times as modern electronics, hybrid electric vehicles, and micro- and nanoelectromechanical systems are becoming more popular. EESs are generally highly efficient and have a relatively long cycle life. Since the invention of the first fuel cell (FC) in 1839, the EES industry has evolved drastically. Currently, capacitors, supercapacitors (SCs), and batteries have taken the spot as the next-generation energy storage devices (ESDs), whereas FCs are considered energy conversion systems [3]. Because of their fast charge/discharge rate, long life duration, and outstanding power density, supercapacitors have been used in power capture and supply, power quality applications, and backup power applications in the last 10 years. SCs differ from batteries in that they store charges at the surface of the active material rather than in its bulk: This explains the low energy density (5 Wh/kg, due to limited storage space) and fast charge delivery. Also, SCs have a longer cycle life due to the electrostatic storage principle and, thus, are not limited by active material volume.

A typical supercapacitor consists of a pair of highly conductive electrodes with an ionic electrolyte in between them. SCs are divided into two groups: electrochemical double-layer capacitors (EDLCs) and pseudocapacitors, which are based on electrostatic and faradaic storage principles, respectively. Aside from the different storage mechanisms, EDLCs and pseudocapacitors can also be distinguished by the type of active electrode materials used: the former use porous carbon electrodes, whereas the latter primarily employ oxide-based electrode materials. The composition, structure, and properties of the electrode materials showed a large impact on the storage mechanism and electrochemical performance of SCs. Therefore, extensive research has been devoted to the syntheses of electrode materials to maximize their energy efficiencies with preferences for sustainable and renewable materials as the ultimate millennial focus as a tool to restore the environment and preserve its natural resources.

Biowastes are sustainable materials used in a wide range of applications including supercapacitor applications due to their diverse chemical and structural compositions [4–7]. Bio-derived carbons are cost-effective, sustainable, abundant, and clean. The recycling of biodegradable materials, thus, promotes (i) the preservation of our environment and natural resources, (ii) the mitigation of climate change, (iii) the implementation of clean waste management protocols, etc. Additionally, the versatility of these materials provides a great avenue for scientific and technological advancement. Aside from the environmental advantages, the components constituting these materials (e.g., cellulose, hemicelluloses, lignins, and proteins) facilitate the synthesis of carbons with distinct microstructures. Cherry calyces, for example, have yielded in highly porous 3D scaffolding frameworks of hierarchically porous carbon nanosheets (PCNSs) containing oxygen and nitrogen heteroatoms, and a good combination of macro-, meso-, and micropores [8].

In the USA, paper alone accounts for 23% of the total domestic solid waste, with trillion tons being disposed of every year, making it the most abundant; the paper industry also generates thousands of tons of pulp sludge waste from paper pulp mills [9]. The standard management of paper wastes involves incineration, which contributes to the emission of greenhouse gases. Paper, however, consists of cellulose networks with an average fiber diameter of 15–30 μm as well as oxygen functional groups, an ideal composition for carbonization and synthesis of porous carbon materials. Because of this, many research works have been conducted in the field of electrochemical storage using paper-derived electrode materials. Figure 24.1 shows the nitrogen adsorption-desorption isotherm (Figure 24.1a) and the SEM micrograph (Figure 24.1b) of pulp sludge-derived activated carbon (AC) [10]. The adsorption curves show a well-defined plateau in the high-pressure region and broad knee in the low-pressure range, indicating micro- and mesoporosity, respectively.

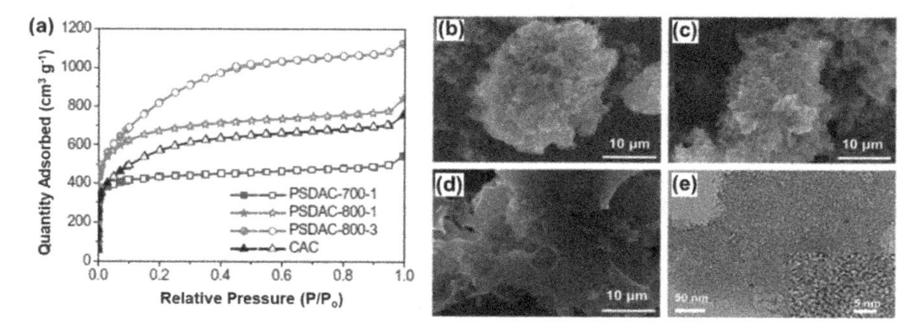

FIGURE 24.1 (a) Nitrogen adsorption-desorption isotherms of the activated carbons. (b–d) SEM and (e) TEM micrographs of paper sludge-derived activated carbon. (Adapted with permission from Ref. [10]. Copyright (2013) Elsevier.)

24.2 CHARGE STORAGE MECHANISM IN BIOWASTE-DERIVED SUPERCAPACITORS

24.2.1 Electrochemical Double-Layer Capacitors

Energy storage in EDLCs occurs by electrostatic charge adsorption at the electrode-electrolyte boundary. The formation of the EDL is governed by two fundamental theories: the Helmholtz and the Gouy–Chapman–Stern theories [3,11]. In the charge mode, an electrostatic force of attraction is generated between the electrode surface charges and the nearby electrolytic ions, resulting in the formation of an oppositely charged double layer known as the Helmholtz double layer. Thus, within the Debye length (1–10 nm), the charge on the electrode surface is balanced by electrolyte polarization. Usually, solvent permittivity, the concentration of electrolyte, and its structure at the electrode surface determine the thickness of the double layer [11]. In discharge mode, the charges constituting the double layer are rearranged, generating a displacement current, the discharge current.

Due to their electrostatic nature, EDLCs enjoy the benefit from the high surface area and the rapid electrostatic response of porous carbonaceous materials. Porous activated carbons (ACs) being the most suitable, ~95% of commercialized EDLCs are based on AC electrodes. ACs offer a high specific capacitance, tunable porosity, high electrical conductivity, and outstanding electrochemical stability, and chemical inertness [11,12]. Recent studies have shown that the capacitance of an ideal EDLC is dependent on not only the surface area of the carbon electrode, but also the pore characteristics of the material: pore structure, size, and distribution. Several strategies have been tried to improve the energy density levels of EDLCs, which limit their usage. Often porous carbons are combined with a pseudocapacitive material to convert an EDLC into a hybrid SC with improved electrochemical characteristics. However, optimization of the carbon pore structure seems to yield the most efficient results. It is believed that nanoporous carbons are more efficient than mesoporous carbons in terms of charge storage in EDLCs [11] and an optimized combination of nano-, meso-, and micropores is expected to yield the best results [13]. To explain this, schematic diagrams are illustrated in Figure 24.2 [3]. According to the Helmholtz storage model (Figure 24.2a), the ideal pore size should be maintained within 3–5 nm (mesoporous) range, twice the size of a solvated electrolyte ion. Consequently, a reduction in pore size is expected to reduce the capacitance of the EDLC.

Recent research has shown, however, that shrinking the pore size to the point where the pore diameter is comparable to the size of the solvated ion boosts the device's capacitance. Because the solvated ions are partially de-solvated and stacked along the pore axis, this occurs. Thus, the Helmholtz theory cannot be applied under these conditions as no diffuse layer (δ) would be formed; instead, the mathematical model (Equation 24.1) proposed by Meunier's team better describes this phenomenon and applies to carbide-derived carbons [3]:

$$\frac{C}{A} = \frac{\varepsilon_o \varepsilon_r}{b \ln\left(b/a_o\right)} \tag{24.1}$$

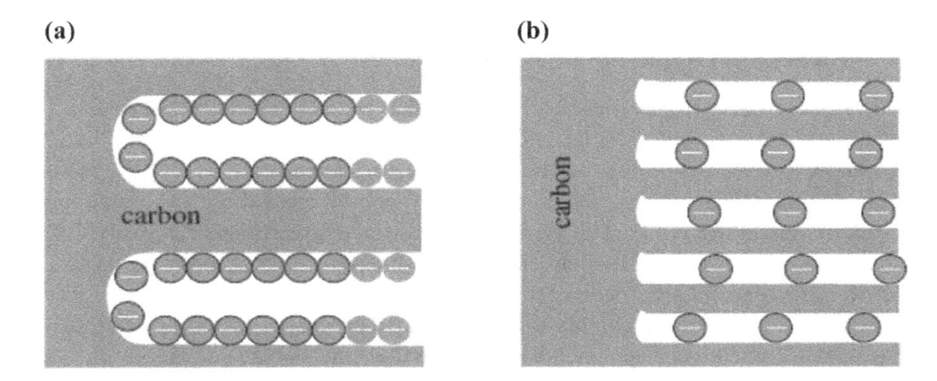

FIGURE 24.2 (a) Helmholtz storage model. (b) Sub-nanometer storage model. (Adapted with permission from Ref. [3]. Copyright (2010) the Royal Society (the UK).)

where ε_r is the electrolyte dielectric constant, ε_o is the dielectric constant of vacuum, b is the pore radius, and a_o is the ion size in the pore.

24.2.2 Pseudocapacitors and Hybrid Supercapacitors

Pseudocapacitive electrodes are based on non-carbonaceous active materials that display an electrochemical storage effect similar (i.e., pseudo) to EDLCs. Pseudocapacitive materials are mostly oxide-based materials (e.g., RuO_2 and MnO_2), which store charges by redox reactions without undergoing phase transformations. Pseudocapacitors exhibit much higher levels of charge storage and energy density than EDLCs due to the extension of pseudocapacitive effects throughout the bulk of the material. Pseudocapacitors can be intrinsic or extrinsic based on the particle size dependence of the capacitance. Power density is inversely proportional to the crystallite size of extrinsic pseudo-materials such as MoO_2 (and other transition metal oxides, sulfides, and nitrides); these materials do not show pseudocapacitive behavior in the bulk state, and the ideal crystallite size for improved electrochemical performance is <20 nm when the diffusion distance is less than $(Dt)^{1/2}$–D being the diffusion coefficient and the time. In this regard, carbons are frequently coupled with pseudocapacitive materials to account for the latter's poor capacitance and energy density characteristics. The resulting hybrid material synergistically combines the physicochemical properties of the two while accounting for their respective weaknesses. Pseudocapacitors (transitional metal oxides and conductive polymers) suffer from capacitive decay, low conductivity, and structural instability. The carbon particles provide support and anchorage for the pseudocapacitive particles and a conductive path for charge transfer across the electrode as well. The most advanced hybrid supercapacitor electrodes are based on $Li_4Ti_5O_{12}$ (LTO), which is best coupled with non-aqueous electrolytes and provides both high power and energy densities at high voltages.

24.3 SUPERCAPACITOR BASED ON BIOWASTE-DERIVED CARBONS

Bio-derived carbons are a renewable, sustainable, and cost-effective alternative to conventional carbons obtained from coal, petroleum, and their derivatives, which additionally involve hazardous and toxic synthesis procedures. Recycling waste carbon to energy materials is an excellent approach to meet the global challenge of properly managing waste. The high specific surface area, tunable porosity, high electrical conductivity, and excellent electrochemical performance make porous ACs the most attractive and pragmatic carbons for EDLCs. ACs can be synthesized from biomasses and biowaste precursors through various processes including hydrothermal processes, chemical vapor deposition, and direct pyrolysis, and in our current era of millennial revolution, lots of green synthesis routes have also been designed. The electrochemical activity and overall performance of SCs are influenced by the surface chemistry and intrinsic pore characteristics of the electrodes, the electrolyte, and their interaction [10,13].

24.3.1 ELECTROCHEMICAL DOUBLE LAYER-BASED SUPERCAPACITORS

An ideal biowaste material for EDLC applications typically consists of a combination of organic and inorganic compounds, which determine the structure (and properties) of the bio-derived carbons and facilitate the process of carbonization, thus enhancing overall energy efficiency and cost-effectiveness. Mesoporous carbons, 3D hierarchically porous carbons, carbon onions, carbon nanotubes, and graphene have demonstrated exceptional electrochemical properties in SCs due to rapid ion transport phenomena. Paper and its associated wastes have successfully been used in the synthesis of carbon nanostructures as a green paper-waste management route that yields highly efficient electrochemical devices. It is worth mentioning that there exist a wide variety of paper types, each with a different composition and structure, and, therefore, the resulting carbon nanostructure is dependent on the physicochemical properties of the paper precursor. Singu et al. [14] synthesized highly conductive graphene paper electrodes with a specific surface area of 63.4 m^2/g and pore volume of 0.24 cm^3/g. The capacitance of the graphene sheets was enhanced by incorporating nitrogen groups through urea doping, thus adding a capacitive effect to the EDLC. The maximum specific capacitance of the as-prepared graphene electrode was 1,122.3 F/g, comparatively higher than the values obtained for the paper pulp sludge-derived ACs previously synthesized by Wang and his colleagues. The latter only yielded a maximum specific capacitance ranging from 160 to 190 F/g. Despite the lower storage capacity, the ACs displayed high energy (up to 51 Wh/kg) and power densities (up to 6,760–7,000 W/kg) depending on the nature of the electrolyte.

The pore size distribution (PSD) of porous carbons is controlled by the activation parameters such as the type and amount of activating agent, activation temperature, and duration of activation. Bhat et al. [12] synthesized an AC from pinecone powder (PP) by chemical activation using $ZnCl_2$ as an activator. The morphological, structural, and porosity analyses showed that a low precursor-to-activating agent weight ratio (i.e., PP:$ZnCl_2$ = 1:1) resulted in partially activated carbons, which appear as a quasi-uniform flat surface of interconnected particles with haphazardly distributed carbon agglomerates on it. Increasing the amount of $ZnCl_2$ to a PP:$ZnCl_2$ ratio of 1:3 produced homogenously distributed spherical particles with a diameter range of 10–30 nm. A ratio of 1:5 produced a 3D network of interconnected large spherical carbon particles with high macropore content (pore size > 50 nm). All samples showed the same amounts of defects and disorders, irrespective of the activation ratio. The optimized PSD of the sample with a ratio of 1:3 having the highest degree of microporosity with some mesoporous character yielded the highest electrochemical characteristics, demonstrating its applicability and efficiency in EDLCs.

Yu et al. [8] analyzed the influence of hierarchical pores on the electrochemical properties of EDLCs using cherry calyces-derived porous carbon nanosheet (PCNS) electrodes. The high specific surface area (1,612 m^2/g), accessible surface area, and hierarchical pore distribution, as well as the hetero-atomic nature of the as-synthesized PCNSs, accounted for the large specific capacitance, energy density, and power density for both liquid and organic electrolytes. The PCNSs formed a 3D network of stacked nanosheets containing a high level of interconnected meso- and macropores for fast ion diffusion and charge storage. In addition, the presence of oxygen and nitrogen groups provided a pseudocapacitive effect, thus enhancing the electrochemical response of the electrode by combined electrostatic attraction and pseudocapacitance.

The electrochemical performance of carbonaceous active materials is largely controlled by the pore structure and distribution. Considering the hierarchical distribution, macropores (>50 nm), mesopores (2–50 nm), and micropores (<2 nm) serve as ion-buffering reservoirs, electrolyte ion transport channels, and charge storage sites, respectively [15]. Thus, a lot of work has been devoted to the quest to balance the distribution of these pores either by the appropriate selection of carbon precursors or synthesis routes. Lignosulfonate, obtained from lignin, is an abundant waste material in the paper industry, and concerns have been raised about its detrimental environmental impact. The unique molecular structure and intrinsic properties of lignosulfonates make them ideal precursors for the synthesis of graphitic carbonaceous materials via hard templating methods. Lignosulfonates are aromatic polymers with a relatively smaller molecular weight, compared to their precursor – lignin, and possess oxygen-containing rings. Coffee residue, the most generated agro-food waste worldwide after tea residue [16], has also been proposed as a precursor for ACs to minimize the waste generated from coffee consumption by synthesizing a cost-effective and sustainable AC.

Bai et al. [15] synthesized a hierarchical ordered porous carbon (HOPC) using KIT-6, an ordered mesoporous silica template, as the hard template and sodium lignosulfonate as the carbon precursor. The scientists employed a five-step process involving (i) the preparation of the KIT-6 hard templates, (ii) the impregnation of the KIT-6 with sodium lignosulfonate, (iii) carbonization of the impregnated KIT-6, (iv) template removal using warm 2.5 M NaOH solution, and (v) chemical activation of OMCs using $ZnCl_2$. Highly porous HOPCs were finally obtained with a specific surface area of 2,602 m^2/g, a micropore volume of 1.03 cm^3/g, and a mesopore volume of 3.49 cm^3/g. Per the obtained microstructural and morphological characteristics, the electrochemical performance of these HOPC was promising with a high specific capacitance of 289 F/g at 0.5 A/g, energy density of 40 Wh/kg, and power density of 900 W/kg. The most common synthesis route involves the two-step carbonization and activation process. Compared to the aforementioned template method, both physical and chemical activations yield porous carbons with randomly distributed pores. For example, Bai et al. [15] prepared a honeycomb-like microstructure, while Adan-mas et al. [16] obtained a carbon with the disordered hierarchically porous structure. The different structures are dependent on both the type of precursors and the synthesis route. Adan-mas and his team members used coffee residue as their precursor and compared the characteristics of chemically and physically activated coffee-derived ACs as an active material for EDLC electrodes. Similarly, their results confirmed that the coexistence of meso- and micropores improves the mass transport of electrolyte and charge storage capacity, respectively. The surface of the chemically activated ACs displayed a higher specific surface area and homogenous porosity than that of the physically activated ACs, whereas the latter displayed a higher degree of graphitization with (002) and (001) preferred crystallographic planes. The physically activated ACs were less porous, with surface pits and cracks. Often, the demerits of heterogeneous ACs can be minimized by combining them with pseudocapacitive materials such as Co_3O_4, thereby forming hybrid electrodes.

Yajun et al. [17] prepared their ACs using agricultural waste soybean dregs (carbon precursor) pre-treated with phosphoric acid, wherein the activator as well as a source of phosphorous improved the hetero-atomic content. The high specific surface area of the porous carbon improved the number of active sites for Co_3O_4 deposition and, thus, the electrochemical behavior of the SCs. The hetero-atomic (N, P, and O) composition of the biowaste further provided an additional pseudocapacitance for a higher pseudocapacitive character as well as conductivity. The energy density of the hybrid electrode reached 42.5 Wh/kg at a power density of 746 W/kg and 40.6 mWh/cm^2 in aqueous and solid-state supercapacitor systems, respectively. Electrochemical properties of physically and chemically activated coffee-derived ACs were compared and studied [16]. The maximum specific capacitance of physically activated coffee-derived ACs was 72 F/g at 1 A/g, while that of chemically activated coffee-derived ACs was 84 F/g at 1 A/g (Figure 24.3). All three ACs displayed a quasi-square shape of the hysteresis loops at low scan rates. At higher scan rates, however, the physically activated carbon showed a shifted voltammogram, due to the high resistance and ohmic drop as a result of the non-uniform structure and defects. On the other hand, the open microstructure of the chemically activated ACs enhanced the electrolyte penetration, shortening the ionic diffusion path and improving the electrolyte–electrode interactions, thus improving the capacitance and electrochemical performance of the electrodes. These characteristics are further enhanced in the lignosulfonate ACs due to the ordered honeycomb structure.

24.3.1.1 Role of Electrolytes

Another important parameter to consider in the design of SCs is the nature of electrolytes as well as the size of electrolyte ions concerning the pore size of electrodes. Despite the promising performance of liquid electrolytes, organic electrolytes are still more desirable because of their higher operating potential windows. Therefore,

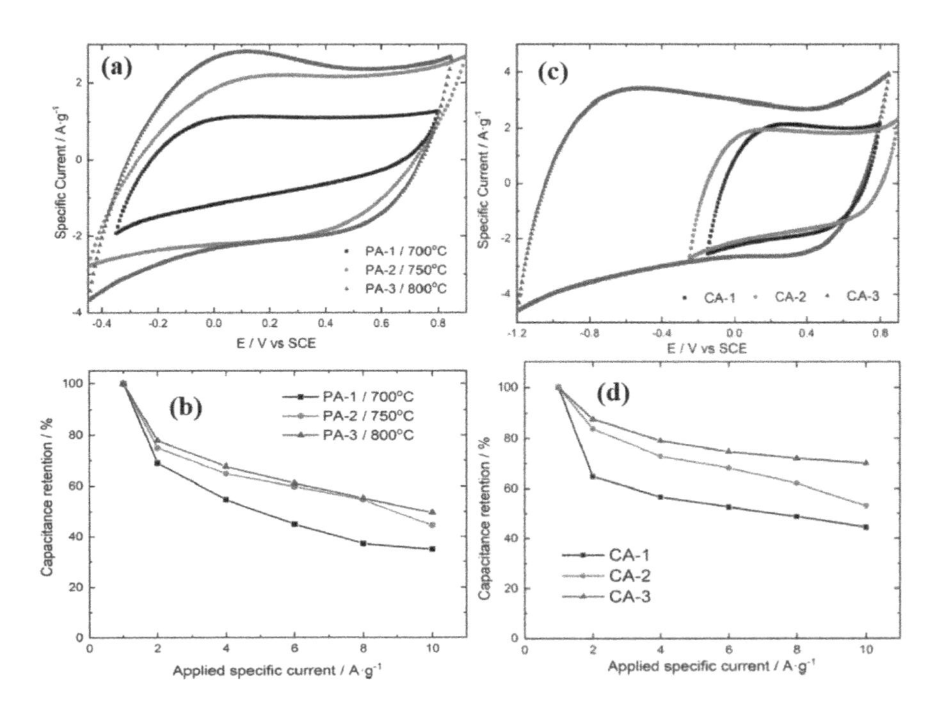

FIGURE 24.3 Electrochemical results in 1 M Na_2SO_4 for physically activated ACs treated at different temperatures: (a) cyclic voltammetry comparison at 50 mV/s and (b) capacitance retention results at different applied specific currents. Electrochemical results in 1 M Na_2SO_4 for chemically activated ACs treated at different temperatures: (c) cyclic voltammetry comparison at 50 mV/s and (d) capacitance retention results at different applied specific currents. (Adapted with permission from Ref. [16]. Copyright (2021) Elsevier.)

there have been lots of attempts to improve the electrochemical storage of SCs through PSD optimization to account for the high viscosity of organic electrolytes. Table 24.1 shows the comparison of some biowastes and the electrochemical performance of bio-derived carbon electrodes based on nanostructure type and electrode/electrolyte interaction. Despite the advancements on traditional SCs with traditional electrolytes, concerns stemming from the chemical, electrochemical, and thermal instability, as well as safety and transportation issues associated with traditional aqueous and organic liquid electrolytes, have resulted in the development of a new emerging class of electrolytes. The gel polymer electrolytes (GPEs) consist of liquid electrolytes entrapped in polymeric networks. GPEs combine the electrochemical and flexible mechanical properties of liquid electrolytes and polymeric materials, respectively, and their flexible nature makes them suitable for use in flexible SCs. GPEs of organic ion plastic crystal (OIPC)/salt mixtures incorporated in succinonitrile (SN) non-ionic plastic crystals, for example, have shown superior ion conductivity and excellent stability due to the unique composition.

Wang and his team tested the performance of their paper pulp sludge-derived AC in 1.5 M tetraethylammonium tetrafluoroborate solution in acetonitrile ($TEABF_4$/AN);

TABLE 24.1
Some Examples of Biowastes and the Electrochemical Performance of the Bio-Derived Carbon Electrodes Based on the Nanostructure Type and Electrode/Electrolyte Interaction

Biowaste Precursor	Bio-Derived Electrode	Electrolyte	Max. Specific Capacitance (F/g)	Max. Energy Density (Wh/kg)	Max. Power Density (W/kg)	References
Waste paper	Graphene	6 M KOH	1,122.3	-	-	[14]
Paper pulp mill sludge	AC	$TEABF_4$/AN	162.2	30	228	[10]
		EMIM TFSI	180–190	51	375	
		BMPY TFSI	180–190	26–31	6,760–7,000	
Cherry calyces	Nanosheets	2/6 M KOH	350	12.2	25	[8]
		Li_2SO_4	259.4	22.8	198.8	
		$EMIMBF_4$	173.1	81.4	446.3	
Soybean dreg	3D honeycomb	6 M KOH	281.4	7.6	-	[19]
Eggshell membrane	AC	1 M Na_2SO_4	478.5	14	150	[20]
Pine cone	AC	GPE (with Li^+)	255	20	55.7	[12]
		GPE (without Li^+)	244	19	39.3	
Cotton stalk	Nanofoam	CMC–Na/Na_2SO_4	282	22.6	-	[18]

1-ethyl-3-methylimidazolium bis(trifluoromethylsulfonyl)imide (EMIM TFSI); and 1-butyl-1-methylpyrrolidinium bis(trifluoromethylsulfonyl)imide (BMPY TFSI) electrolytes, respectively [10]. While the organic electrolyte (i.e., TEABF4/AN)-based EDLC displayed a promising electrochemical performance, the results obtained for EMIM TFSI and BMPY TFSI electrolytes showed superior specific capacitances and power densities. The values of the measured capacitances, energy densities, and power densities were 162.2 F/g, 30 Wh/kg, and 228 W/kg for TEABF4/AN; 180–190 F/g, 51 Wh/kg, and 375 W/kg for EMIM TFSI; and 180–190 F/g, 26–31 Wh/kg, and 6,760–7,000 W/kg for BMPY TFSI. Hence, the SCs with liquid electrolytes showed much-improved characteristics, and these results are consistent in the literature. Yu et al., for example, compared the electrochemical performance of their bio-derived PCNSs in KOH, Li_2SO_4, and $EMIMBF_4$ (1-ethyl-3-methylimidazolium tetrafluoroborate) electrolytes. With respect to the liquid electrolytes, the PCNSs showed higher compatibility with Li_2SO_4 than with KOH. For Li_2SO_4, the maximum energy and power densities were 22.8 Wh/kg and 198.8 W/kg, respectively, whereas KOH yielded 12.2 Wh/kg and 25 W/kg, respectively. The organic electrolyte ($EMIMBF_4$)-based SCs exhibited an ultra-high electrochemical performance with a high specific capacitance of 115.6 F/g even at a high current density of 10 A/g. The maximum energy and power densities were 81.4 Wh/kg and 446.3 W/kg, respectively.

Based on the aforementioned experimental results as well as those in the literature, the following conclusions can be drawn: (i) The ionic size of all electrolytes, in general, ought to be comparable to the pore size of the active electrode material. Li^+ ions in Li_2SO_4 electrolyte possess smaller ionic radii compared to K^+ ions in KOH electrolyte, and they match the pore size of PCNSs facilitating the ionic charge storage in hierarchical pores of porous carbons, and (ii) the hierarchical pores provide an avenue to overcome the high viscosity of organic electrolytes. Organic electrolytes require larger (meso)pores for easy ion diffusion, normally the pore size being larger than 1.2 nm. The sizes of the solvated cations and anions in $TEABF_4$/AN electrolyte, for example, are 1.30 and 1.16 nm, respectively, comparatively larger than the pore size of the derived AC, hence resulting in the relatively lower properties with respect to EMIM TFSI and BMPY TFSI. The scientists, however, believe that the solvated ions underwent partial de-solvation to provide the obtained maximum specific capacitance, energy density, and power density values. On the other hand, the optimum pore size for ionic liquid electrolytes should be above 2 nm to account for ionic diffusional losses and further fast ion diffusion.

As previously mentioned, GPEs have shown superior conductivity compared to conventional liquid and organic electrolytes. For example, carboxymethyl cellulose (CMC)–Na/Na_2SO_4 gel electrolyte can produce a high specific capacitance of 282 F/g with an energy density of 22.6 Wh/kg when coupled with a highly porous carbon nanofoam [18]. Bhat and colleagues prepared a GPE consisting of a mixture of 1-ethyl-1-methylpyrrolidinium bis(trifluoromethylsulfonyl)imide (i.e., EMPTFSI and the OIPC) and non-ionic plastic crystal (SN), immobilized in host polymer PVDF-HFP. In fact, from Table 24.1 it can be observed that, despite the common use of AC electrodes, EDLCs based on GPE electrolytes provided a higher storage capacitance than those based on $TEABF_4$/AN, EMIM TFSI, and BMPY TFSI electrolytes employed with the paper pulp sludge-derived AC electrodes. The comparatively higher energy and power densities can be attributed to the higher mobility of ions and wettability of liquid electrolytes. The further addition of Li salt (lithium bis(trifluoromethylsulfonyl)imide, LiTFSI) to the OIPC/SN boosted the ionic conductivity of the GPE from ~1.53×10^{-3} to ~2.87×10^{-3} S/cm at room temperature as the mobility of Li^+ ions is greater than that of the ionic components of the OIPC. Also, the electrochemical stability window (ESW) was improved from ~3.1 to ~3.8 V against Ag/Ag^+ due to the better stability of Li^+ ions compared to EMP^+ ions in the GPE electrolyte.

24.3.2 Hybrid Supercapacitors

EDLCs are limited by the low power density and poor charge transport properties as a result of the intrinsic grain boundary scattering of the charge carriers. To account for this, carbon nanostructures are often coupled with pseudocapacitive materials (transition metal oxides and conductive polymers) to form hybrid SCs. The desire for enhanced or customizable capacity behavior in electrode materials has prompted research into hybrid systems that combine carbon-based materials with pseudocapacitive materials. The hybrid system enables a charge storage mechanism that is driven by both physical and chemical processes in the electrode material.

Through rational chemical synthesis, unique hybrid structures are generated, combining the electrostatic storage mechanism of carbon nanostructures with the faradaic reactions of pseudocapacitive materials. The resulting hybrids (composites) showed a synergistic improvement in physicochemical and electrochemical properties. Hybrid electrodes have been synthesized through a variety of chemical or electrochemical processes, which are usually complex and expensive. Recent research has been directed toward the design of clean and inexpensive synthesis routes such as the one developed by Goswami et al. In their work, carbon nanoparticle waste (WC) collected from the bottom of ordinary cooking ovens was used as the carbon support, while polyaniline (PANI) was used as the pseudocapacitive material. The resulting PANI–WC electrodes were synthesized via in-situ polymerization by oxidizing aniline in the presence of carbon nanoparticles in a direct step. Besides the attractive physicochemical and electrochemical performance of the electrodes, this approach is highly reproducible, cheaper, and eco-friendly. Due to the prolonged high-temperature treatment, the wastes (from vegetables, meat, etc.) are fully converted to pure carbon nanoparticles (confirmed by XPS and EDS analyses), eliminating all other impurities.

PANI is a low-cost highly conductive polymer with excellent pseudocapacitance, with the additional attractive advantages of easy synthesis and manipulation, interesting doping chemistry, and good environmental stability. Carbon increases the stability and cyclability of PANI-based electrodes by controlling the swelling and shrinkage of the polymer during charge-discharge cycles. Following this, a research team used a sugarcane bagasse-derived carbon (one-step pyrolysis followed by activation), which yielded a 3D interconnected structure of nitrogen-doped PANI [21]. The observed energy density and specific capacitances were 49.4 Wh/kg and 298 F/g (at 1 A/g), respectively. In a similar work [22], carbon derived from sugarcane bagasse was used to form composites with MnO_2 using the hydrothermal method. The metal oxide is expected to enhance the capacitive performance of the electrode materials through improved rapid faradaic redox effects, and they reported a specific capacitance of 359 F/g for the composite electrode (approximately 30% increase in performance compared to a single material electrode).

By using the in situ polymerization method, Chen's research group produced a composite of PANI and activated carbon from sugarcane bagasse [23] (Figure 24.4a and b). They discovered that the mass ratio of constituents affected the electrochemical performance of the electrode and showed a specific capacitance of 447 F/g (0.5 A/g) for the sample with a 40:1 (PANI/activated carbon) mass ratio (Figure 24.4c and d), exhibiting good power and energy densities. Vighnesha's group also prepared activated carbon from coconut shells; the samples were carbonized and treated with KOH and activated at 450°C; the obtained porous carbon had a nitrogen absorption–desorption test surface area of 312 m^2/g [24]. A composite of PANI and the activated carbon with a ratio of 1:1 exhibited a commendable electrochemical performance with a specific capacitance of 99.6 F/g at 0.5 mA/g.

There are numerous possibilities for the formation of hybrid systems resulting from the combination of different capacitive materials with the uniqueness of certain capacitive characteristics. To demonstrate the surface adsorption and intercalation of ions for hybrid supercapacitors, Vijayan's team also prepared a composite of palm kernel shells (carbon content 43%–47%) activated carbon and Mn_2O_3. Figure 24.5a and b illustrate the synthesis procedure used to produce flower-shaped $MnCo_2O_4$ (MC)

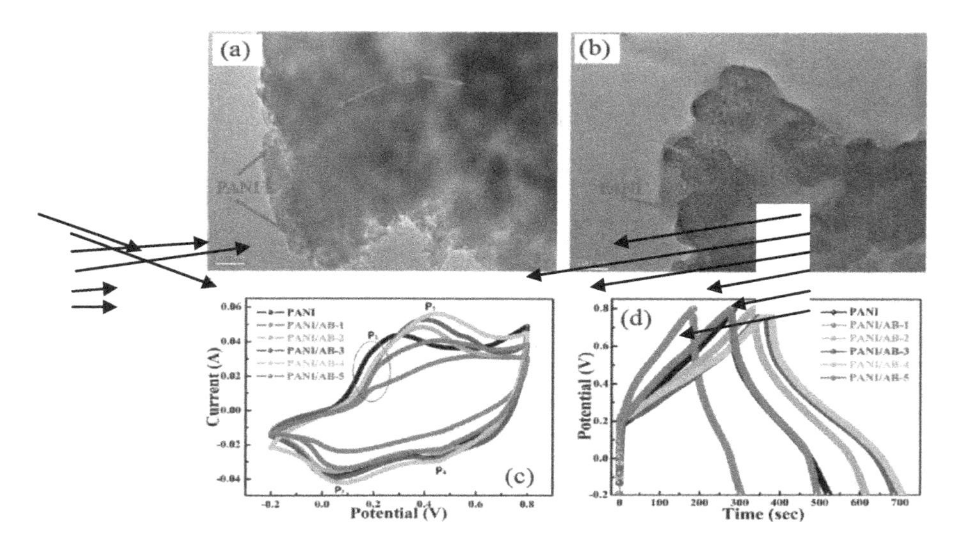

FIGURE 24.4 (a, b) TEM image of PANI/AB-4 composites. (c) CV at a scan rate of 10 mV/s. (d) Charge–discharge characteristics at 1 A/g. (Adapted with permission from Ref. [23]. Copyright (2017) Elsevier.)

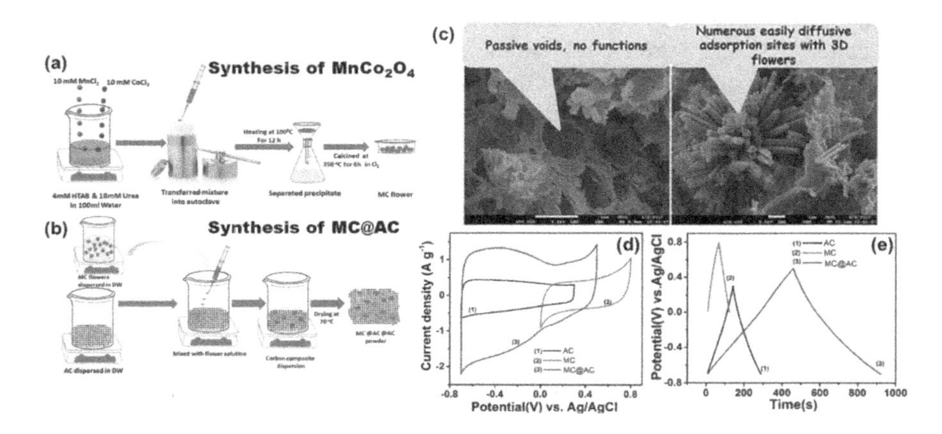

FIGURE 24.5 Schematic representation of the syntheses of (a) MC and (b) MC@AC. (c, left) The microstructure of porous carbon, which contains voids as large as several micrometers in addition to the micro-/mesopores for charge storage. (c, right) The housing of a flower-shaped particle in a void, which provides numerous ion adsorption sites. (d) CV curves of AC, MC, and MC@AC. (e) GCD curves of AC, MC, and MC@AC. (Adapted with permission from Ref. [25]. Copyright (2020) American Chemical Society.)

and MC@AC via the hydrothermal method [25]. The activated carbon sample had passive voids; these voids were filled by the $MnCo_2O_4$ rod-shaped features (Figure 24.5c), which yielded a maximum capacitance of 510 F/g in 1 M Na_2SO_4 at 1 A/g (Figure 24.5d). Similar research works showed how Mn_2O_3–C interconnected

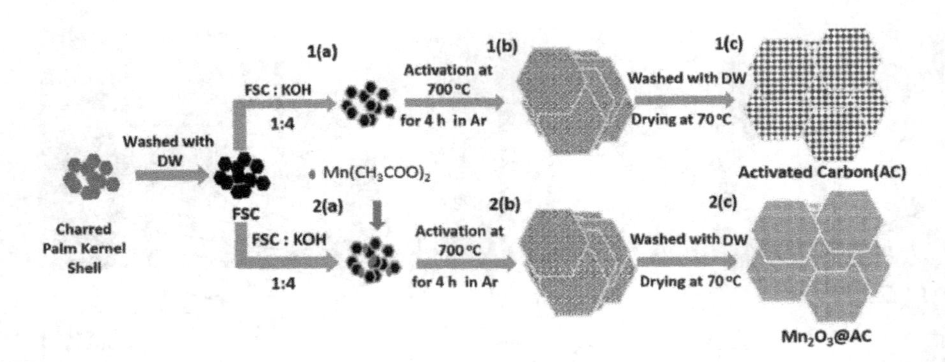

FIGURE 24.6 Schematic representation of the synthesis process. Steps (1a), (1b), and (1c) represent the activation process, and Steps (2a), (2b), and (2c) represent the preparation of composite. The gray boundaries in (1a) and (2a) represents the addition of KOH, which was completely washed out with deionized water (DW) in Steps (1c) and (2c). (Adapted with permission from Ref. [28]. Copyright (2020) Elsevier.)

porous network and MnO_2/ACFs MnO_2 nanosheets grown on carbon fibers yielded 776 F/g in 1 M Na_2SO_4 at 1 A/g and 135 F/g in 1 M Na_2SO_4 at 10 mA/cm^2 [26,27]. In another work by Vijayan and his colleagues, composite samples of activated carbon (derived from palm kernel shells) and $MnCo_2O_4$ were used to produce hybrid electrode materials [28]. Figure 24.6 shows the synthesis process. The electrodes showed a nearly fourfold increase in energy density compared to the pure carbon with the composite; this observation was attributed to the enhancements in the voltage window. The electrochemical properties of some select biowaste residues-derived carbons and pseudo-materials are shown in Table 24.2.

24.4 APPLICATION OF BIO-DERIVED CARBON IN FLEXIBLE DEVICES

Flexible supercapacitors (FSCs) are used to replace the conventional supercapacitors in applications that require high strength coupled with flexibility, lightweight, and a compact size. The main idea behind the development of FSCs is to integrate them into advanced wearable electronic devices, printable electronics, integrated systems, and textile materials. Because of the working conditions, FSCs must be reliable in bending conditions and serial connections. One way to account for this requirement involves the use of conductive fibrous current collectors such as stainless steel (SS) and carbon fibers (CFs). The latter, especially, are flexible, strong, highly conductive, chemically inert, and cost-effective. CNT-coated cotton threads and plastic-coated wires are either unstable in acid- and alkali-based electrolytes, or too heavy and brittle, making them impractical. Despite the advances in this field, the currently reported energy densities, especially for monolithic FSCs, are still very low and great attention is now being paid to hybrid flexible supercapacitors (HFSCs). Asymmetric HFSCs show a synergistic enhancement of energy density compared to the individual components due to the high combined cell voltage, with values being particularly high for HFSCs based on $NiCo_2O_4$, $ZnCo_2O_4$, and $NiMoO_4$.

TABLE 24.2
Electrochemical Performance of Select Biowastes and Pseudo-Materials for Supercapacitors

Biowaste Precursor	Bio-Derived Carbon	Pseudo-Material	Electrolyte	Specific Cap. (F/g)	Energy Density (Wh/kg)	Power Density (W/kg)	References
Banana peels	Nanowires	$NiCo_2O_4$	6 M KOH	1,670.3	40.7	8.4×10^3	[29]
Tar	Nanosheets	NPC@LDH	PVA/KOH	9.6	1.28×10^{-3}	8.04×10^{-3}	[30]
Pencil shavings	Graphite	Zinc foil	$Zn(CF_3SO_3)_2$	413.3	147.0	15.7×10^3	[31]
Fish bones	Fibrous biocarbon foams	N–S–P–O co-doping	1.0 M LiPF6 in EC/DMC mixture	58	79	62×10^3	[32]
Bacteriorhodopsin (bR)	Nanofilms	PANI	1.0 M Na_2SO_4/H_2SO_4 solution	1,146	-	-	[33]
Agave Americana	AC	$LiCo_{1/3}Ni_{1/3}Mn_{1/3}O_2$@C	0.5 M Li_2SO_4	56	20	264	[34]

CNTs are undoubtedly the most suitable carbons for FSCs due to the large aspect ratio, a parameter for flexibility, and long conductive paths. Single-walled CNT (SWCNT)-coated PET electrodes display a high electrical conductivity (40–50 Ω/sq.) and an energy density of 6 Wh/kg, despite the poor active material mass loading capacity of the PET due to the weak bonding with CNTs. To compensate for this challenge and the issue of delamination, CNTs ought to be coated onto more compatible current collectors such as polyvinylidene fluoride (PVDF)-pre-treated paper, which increases the specific mass deposited from 0.03 mg/cm^2 (SWCNT-PET) to 0.3 mg/cm^2 and yield energy and power densities as high as 41 Wh/kg and 164 kW/kg, respectively [35].

Senthilkumar and Selvan [36] manufactured a porous carbon-coated stainless steel FSC by using *Ficus religiosa* leaves as the carbon precursor and PVA–H_3PO_4 as the gel electrolyte. Figure 24.7 shows the images of the prepared samples at various bending conditions (Figure 24.7a–e) and SEM micrographs (Figure 24.7f–g) of the bare and coated SS wires. The electrochemical analysis shows ideal capacitance behavior, characteristic of EDL-FSCs with a linear charge-discharge curve and rectangular CV. The shape of the CV curves was retained even after the various angles of bending (Figure 24.7h). The gravimetric capacitance and energy densities are 3.4 F/g and 311 mWh/kg, respectively, which are considerably low compared to the values obtained for conventional SCs. Regardless, the obtained results were consistent even in bending, thus showing good in-service reliability. The performance of the device improved after joining them in series as shown in Figure 24.7i and j.

As previously mentioned, HFSCs show a considerably improved electrochemical performance. For example, an ultra-high power density of 425 W/kg, an energy

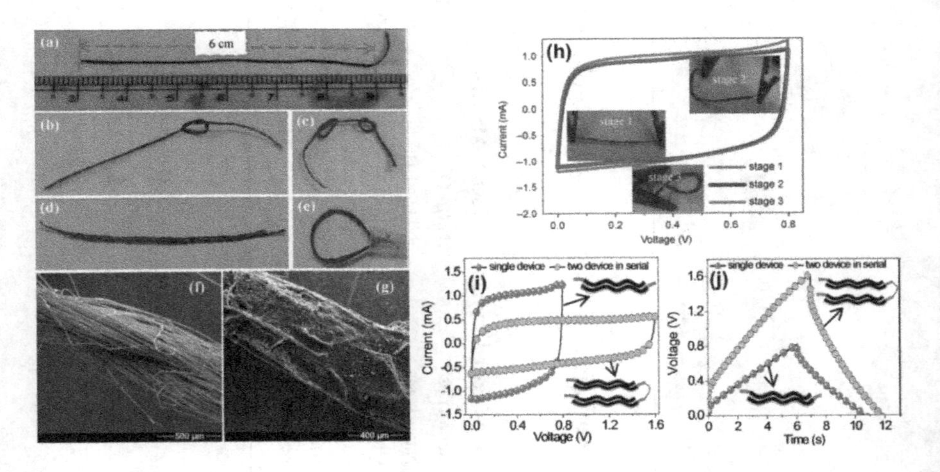

FIGURE 24.7 Porous carbon-coated fiber electrode in various bending conditions (a–e). SEM micrographs of a bare SS thread electrode (f) and porous carbon-coated fiber electrode (g). CV curves of the electrode at various bending stages (h). Series-connected electrodes: (i) CV and (j) GCD. (Adapted with permission from Ref. [36]. Copyright (2015) John Wiley and Sons.)

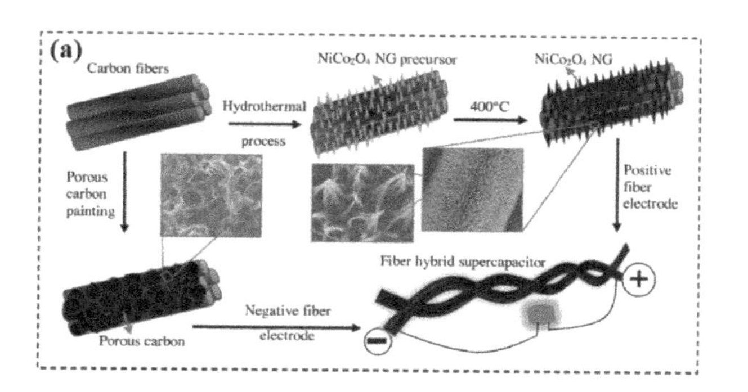

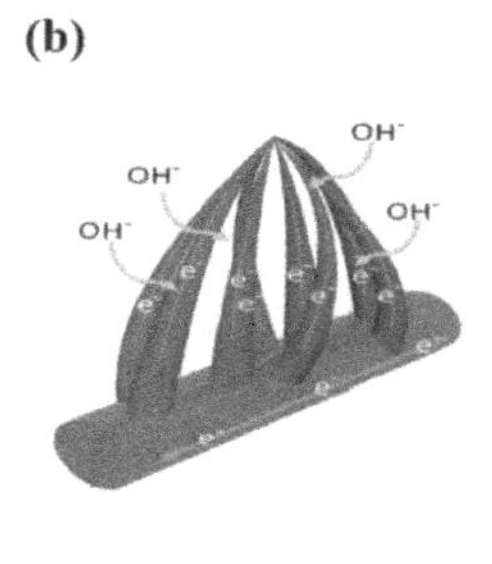

FIGURE 24.8 (a) Schematic illustrations of the preparation of the $NiCo_2O_4$ NG@CF fiber electrode, the porous carbon-coated CF fiber electrode, and the fabricated fiber HSC. (b) Schematic representation of charge storage mechanism of $NiCo_2O_4$ NG@CF electrode. (Adapted with permission from Ref. [37]. Copyright (2016) Elsevier.)

density of 6.61 Wh/kg, and a specific capacitance of 17.5 F/g were obtained for a hybrid consisting of a porous carbon FSC twisted with a $NiCo_2O_4$ nanograss (NG)@ carbon fiber (CF) with polyvinyl alcohol (PVA)/KOH gel electrolyte [37]. The sample preparation process with insets of the SEM micrograph of the microstructures of the two electrodes is illustrated in Figure 24.8a. The NGs on the positive electrode provide a large active surface area and short pathway for electrolytic ion diffusion and charge transfer. Additionally, the network configuration of the NG bundles onto the CF substrates (Figure 24.8b) provides a good electrical connection for fast electron transport as well as active electro-active sites for electrolyte diffusion. Simultaneously, the porous carbon derived from lemon peels also provides a highly accessible and interconnected structure for easy electrolyte infiltration.

24.5 CONCLUSIONS

Major efforts are being made to establish new ways of creating products in a more environmentally friendly manner to ensure a smooth transition to a sustainable society. With regard to energy storage, it is rapidly becoming an integral area of interest in energy management, since it is a way to address renewable energy's shortcomings by ensuring a consistent power supply while conserving energy as well. In the field of bio-based and even biodegradable devices, considerable progress has already been made and biomass plays a significant role in these applications due to their diverse chemical and structural compositions. The overall efficiency of SCs is determined by the surface chemistry and intrinsic pores of the electrodes, ionic sizes of the electrolytes, and their interaction. With the growing need for renewable energy products, biowaste-derived supercapacitors have the potential to address climate change and the greenhouse effect. It is therefore anticipated that HFSCs made from bio-derived carbon could be realized as next-generation devices if more architectures, materials, and processes are developed.

REFERENCES

1. Bull SR (2001) Renewable Energy Today and Tomorrow. *Proc IEEE* 89:1216–1226.
2. Lund H (2007) Renewable Energy Strategies for Sustainable Development. *Energy* 32:912–919.
3. Simon P, Gogotsi Y (2010) Charge Storage Mechanism in Nanoporous Carbons and Its Consequence For Electrical Double-Layer Capacitors. *Philos Trans R Soc A Math Phys Eng Sci* 368:3457–3467.
4. Ranaweera CK, Kahol PK, Ghimire M, Mishra SR, Gupta RK (2017) Orange-Peel-Derived Carbon – Designing Sustainable and High-Performance Supercapacitor Electrodes. *J Carbon Res* 3:25.
5. Zequine C, Ranaweera CK, Wang Z, Dvornic PR, Kahol PK, Singh S, Tripathi P, Srivastava ON, Singh S, Gupta BK, Gupta G, Gupta RK (2017) High-Performance Flexible Supercapacitors Obtained via Recycled Jute – Bio-Waste to Energy Storage Approach. *Sci Rep* 7:1–12.
6. Using HS, Kahol P, Gupta R (2019) Waste Coffee Management – Deriving Nitrogen-Doped Coffee-Derived Carbon. C 44.
7. Bhoyate S, Ranaweera CK, Zhang C, Morey T, Hyatt M, Kahol PK, Ghimire M, Mishra SR, Gupta RK (2017) Eco-Friendly and High Performance Supercapacitors for Elevated Temperature Applications Using Recycled Tea Leaves. *Glob Challenges* 1:1700063.
8. Yu D, Chen C, Zhao G, Sun L, Du B, Zhang H, Li Z, Sun Y, Besenbacher F, Yu M (2018) Biowaste-Derived Hierarchical Porous Carbon Nanosheets for Ultrahigh Power Density Supercapacitors. *ChemSusChem* 11:1678–1685.
9. US EPA (2018) National Overview – Facts and Figures on Materials, Wastes and Recycling | Facts and Figures about Materials, Waste and Recycling.
10. Wang H, Li Z, Tak JK, Holt CMB, Tan X, Xu Z, Amirkhiz BS, Harfield D, Anyia A, Stephenson T, Mitlin D (2013) Supercapacitors Based on Carbons with Tuned Porosity Derived from Paper Pulp Mill Sludge Biowaste. *Carbon N Y* 57:317–328.
11. Salanne M, Rotenberg B, Naoi K, Kaneko K, Taberna PL, Grey CP, Dunn B, Simon P (2016) Efficient Storage Mechanisms for Building Better Supercapacitors. *Nat Energy* 1:1–10.
12. Bhat MY, Yadav N, Hashmi SA (2019) Pinecone-Derived Porous Activated Carbon for High Performance All-Solid-State Electrical Double-Layer Capacitors Fabricated with Flexible Gel Polymer Electrolytes. *Electrochim Acta* 304:94–108.
13. Goswami S, Dillip GR, Nandy S, Banerjee AN, Pimentel A, Joo SW, Martins R, Fortunato E (2019) Biowaste-Derived Carbon Black Applied to Polyaniline-Based High-Performance Supercapacitor Microelectrodes –Sustainable Materials for Renewable Energy Applications. *Electrochim Acta* 316:202–218.
14. Singu DC, Joseph B, Velmurugan V, Ravuri S, Grace AN (2018) Combustion Synthesis of Graphene from Waste Paper for High Performance Supercapacitor Electrodes. *Int J Nanosci* 17:1–5.
15. Bai X, Wang Z, Luo J, Wu W, Liang Y, Tong X, Zhao Z (2020) Hierarchical Porous Carbon with Interconnected Ordered Pores from Biowaste for High-Performance Supercapacitor Electrodes. *Nanoscale Res Lett* 15:88.
16. Adan-mas A, Alcaraz L, Arévalo-cid P, López-gómez FA, Montemor F (2021) Coffee-Derived Activated Carbon from Second Biowaste for Supercapacitor Applications. *Waste Manag* 120:280–289.
17. Ji Y, Deng Y, Chen F, Wang Z, Lin Y, Guan Z (2020) Ultrathin Co_3O_4 Nanosheets Anchored on Multi-Heteroatom Doped Porous Carbon Derived from Biowaste for High Performance Solid-State Supercapacitor. *Carbon N Y* 156:359–369.
18. Li Z, Gao S, Mi H, Lei C, Ji C, Xie Z, Yu C, Qiu J (2019) High-Energy Quasi-Solid-State Supercapacitors Enabled by Carbon Nanofoam from Biowaste and High-Voltage Inorganic Gel Electrolyte. *Carbon N Y* 149:273–280.

19. Li Z, Bai Z, Mi H, Ji C, Gao S, Pang H (2019) Biowaste-Derived Porous Carbon with Tuned Microstructure for High-Energy Quasi-Solid-State Supercapacitors. *ACS Sustain Chem Eng* 7:13127–13135.
20. Yang P, Xie J, Zhong C (2018) Biowaste-Derived Three-Dimensional Porous Network Carbon and Bioseparator for High-Performance Asymmetric Supercapacitor. *ACS Appl Energy Mater* 1:616–622.
21. Wang B, Wang Y, Peng Y, Wang X, Wang J, Zhao J (2018) 3-Dimensional Interconnected Framework of N-Doped Porous Carbon Based on Sugarcane Bagasse for Application in Supercapacitors and Lithium Ion Batteries. *J Power Sources* 390:186–196.
22. Xiong S, Zhang X, Chu J, Wang X, Zhang R, Gong M, Wu B (2018) Hydrothermal Synthesis of Porous Sugarcane Bagasse Carbon/MnO_2 Nanocomposite for Supercapacitor Application. *J Electron Mater* 47:6575–6582.
23. Chen J, Qiu J, Wang B, Feng H, Yu Y, Sakai E (2017) Polyaniline/Sugarcane Bagasse Derived Biocarbon Composites with Superior Performance in Supercapacitors. *J Electroanal Chem* 801:360–367.
24. Vighnesha KM, Shruthi, Sandhya, Sangeetha DN, Selvakumar M (2018) Synthesis and Characterization of Activated Carbon/Conducting Polymer Composite Electrode for Supercapacitor Applications. *J Mater Sci Mater Electron* 29:914–921.
25. Vijayan BL, Mohd Zain NK, Misnon II, Reddy MV, Adams S, Yang CC, Anilkumar GM, Jose R (2020) Void Space Control in Porous Carbon for High-Density Supercapacitive Charge Storage. *Energy and Fuels* 34:5072–5083.
26. Nagamuthu S, Ryu K-S (2019) MOF-Derived Microstructural Interconnected Network Porous Mn_2O_3/C as Negative Electrode Material for Asymmetric Supercapacitor Device. *CrystEngComm* 21:1442–1451.
27. Zhu K, Wang Y, Tang JA, Qiu H, Meng X, Gao Z, Chen G, Wei Y, Gao Y (2016) In Situ Growth of MnO_2 Nanosheets on Activated Carbon Fibers –A Low-Cost Electrode for High Performance Supercapacitors. *RSC Adv* 6:14819–14825.
28. Vijayan BL, Misnon II, Anil Kumar GM, Miyajima K, Reddy MV, Zaghib K, Karuppiah C, Yang CC, Jose R (2020) Facile Fabrication of Thin Metal Oxide Films on Porous Carbon for High Density Charge Storage. *J Colloid Interface Sci* 562:567–577.
29. Zhang Y, Gao Z, Song N, Li X (2016) High-Performance Supercapacitors and Batteries Derived from Activated Banana-Peel with Porous Structures. *Electrochim Acta* 222:1257–1266.
30. Wei M, Wang D (2020) A Novel Utilized Method of Tar Derived from Biomass Gasification for Fabricating Binder-Free All-Solid-State Hybrid Supercapacitors. *Int J Hydrogen Energy* 45:4793–4803.
31. Li Z, Chen D, An Y, Chen C, Wu L, Chen Z, Sun Y, Zhang X (2020) Flexible and Anti-Freezing Quasi-Solid-State Zinc Ion Hybrid Supercapacitors Based on Pencil Shavings Derived Porous Carbon. *Energy Storage Mater* 28:307–314.
32. Shan B, Cui Y, Liu W, Zhang Y, Liu S, Wang H, Sun L, Wang Z, Wu R (2018) Fibrous Bio-Carbon Foams – A New Material for Lithium-Ion Hybrid Supercapacitors with Ultrahigh Integrated Energy/Power Density and Ultralong Cycle Life. *ACS Sustain Chem Eng* 6:14989–15000.
33. Li H, Wang M, Qi G, Xia Y, Li C, Wang P, Sheves M, Jin Y (2020) Oriented Bacteriorhodopsin/Polyaniline Hybrid Bio-Nanofilms as Photo-Assisted Electrodes for High Performance Supercapacitors. *J Mater Chem A* 8:8268–8272.
34. Ramkumar B, Yuvaraj S, Surendran S, Pandi K, Ramasamy HV, Lee YS, Selvan RK (2018) Synthesis and Characterization of Carbon Coated $LiCo1/3Ni1/3Mn1/3O_2$ and Bio-Mass Derived Graphene Like Porous Carbon Electrodes for Aqueous Li-Ion Hybrid Supercapacitor. *J Phys Chem Solids* 112:270–279.
35. Wang X, Lu X, Liu B, Chen D, Tong Y, Shen G (2014) Flexible Energy-Storage Devices – Design Consideration and Recent Progress. *Adv Mater* 26:4763–4782.

36. Senthilkumar ST, Selvan RK (2015) Flexible Fiber Supercapacitor Using Biowaste-Derived Porous Carbon. *ChemElectroChem* 2:1111–1116.
37. Senthilkumar ST, Fu N, Liu Y, Wang Y, Zhou L, Huang H (2016) Flexible Fiber Hybrid Supercapacitor with $NiCo_2O_4$ Nanograss@Carbon Fiber and Bio-Waste Derived High Surface Area Porous Carbon. *Electrochim Acta* 211:411–419.

25 Hybrid Biowaste Materials for Supercapacitors

Prashant Dubey
CSIR-National Physical Laboratory (CSIR-NPL)

Ashwinder Kaur
CSIR-Central Scientific Instruments Organization (CSIR-CSIO)
Punjabi University

Vishal Shrivastav
CSIR-Central Scientific Instruments Organization (CSIR-CSIO)

Isha Mudahar
Punjabi University

Sunita Mishra
CSIR-Central Scientific Instruments Organization (CSIR-CSIO)

Shashank Sundriyal
CSIR-National Physical Laboratory (CSIR-NPL)

CONTENTS

DOI: 10.1201/9781003178354-29

25.1 INTRODUCTION

Energy plays a pivotal role in everybody's life and the development of science and technology for the future. Till date, energy storage devices such as batteries and supercapacitors have gained immense popularity because of their wide applications. Supercapacitors (SCs) are empowered with high power density, rapid charge/discharge rate, moderate energy density, and extraordinary cycle life [1]. Owing to these features, SCs find vast applications in a wide range of devices including portable electronic devices and hybrid electric vehicles (HEVs). The term supercapacitor came into existence in 1978 when NEC used it for the first time, but it originated in 1957 when a group of General Electric engineers conducted device experimentations utilizing porous carbon electrode and spotted the electric double-layer capacitor effect for the very first time [2].

SCs are generally grouped into two according to the charge storage mechanisms: (i) electric double-layer capacitors (EDLCs) and (ii) pseudocapacitors. EDLCs, which involve a non-faradaic process where capacitance is achieved via pure electrostatic interaction, are generated due to the adsorption of electrolytes ion over the electrode–electrolyte interface, while psuedocapacitors involve fast and reversible faradic processes due to electroactive species that are responsible for charge storage [3]. Among the various challenges faced by supercapacitors, the problem of low energy density requires the utmost concern of the researchers. Therefore, designing novel high-performance electrodes exhibiting enhanced specific capacitance is a prolific way to achieve high energy density without compromising the power density. The most commonly used electrode materials for supercapacitors are carbon materials. However, all these materials originate from non-renewable sources of energy, especially fossil fuels, which are being used up at a rapid rate. Therefore, an alternative green electrode material is the need of an hour.

Hence, in recent times, researchers have emphasized more on the renewable biowaste-derived porous activated carbon materials. Biowaste materials mainly consist of organic matter related to plants and animals, popularly known as biodegradable material. Biowaste materials involve plant waste, forest waste, human waste, animal waste, and municipal waste. Biowaste materials generally demonstrate varied structures such as hierarchical, three-dimensional, periodic patterns, and some other intriguing nanostructures. Biowaste precursors can be directly pyrolyzed and activated via physical or chemical means to produce highly porous nanocarbon materials [4]. BWCs are blessed with intriguing features such as high conductivity, high surface area, and hierarchical pore size distribution. However, they exhibit reduced properties as compared to the newly explored materials such as MXenes and MOFs. This is because biowaste-derived activated carbon materials only impart EDLC characteristics, which limits its supercapacitor performance.

For this, biowaste-derived activated carbon is coupled with different pseudocapacitive materials such as metal oxides and conducting polymers to develop biowaste hybrid materials to cherish the synergistic effects of two different materials. In addition, biowaste hybrid materials can also be developed by doping heteroatoms into BWC. These heteroatoms could provide some pseudocapacitive properties to the biowaste-derived activated carbon material and hence could lead to enhanced electrochemical performance. Hence, significant attention should be devoted to developing new hybrid biowaste materials for their application in supercapacitor devices for enormous electrochemical performance.

25.2 CLASSIFICATION OF HYBRID BIOWASTE MATERIALS

Based on the properties of coupled materials, hybrid biowaste materials can be subcategorized into different kinds of hybrid biowaste materials such as biowaste/conducting polymers hybrid, biowaste/metal oxides hybrid, heteroatoms-doped biowaste hybrid, and other biowaste hybrid materials.

25.2.1 Conductive Polymers/Biowaste Hybrid

Various conducting polymers such as polyaniline (PANI), polypyrrole (PPy), and poly(3,4-ethylenedioxythiophene) (PEDOT) have significantly been regarded as the ideal electrode materials for pseudocapacitors. However, conducting polymers displayed irreversible degradation due to continuous volume expansion and compression and lead to poor cyclic stability. To resolve this issue, introduction of carbon-based nanomaterials with conducting polymers is an effective tool [5,6]. Hence, conducting polymers have been deposited over pristine BWC to develop hybrid biowaste materials, thereby combining EDLC-based biowaste-derived carbon and pseudocapacitor-based conducting polymers. The enhanced electrochemical performance of conducting polymers/biowaste hybrid materials is derived from the strong Π–Π stacking between the conjugated backbone of conducting polymers and the biowaste-derived carbon. Conducting polymers/biowaste hybrid materials are predominantly prepared using different procedures such as chemical oxidative polymerization, interfacial polymerization, and electrochemical polymerization. Broadly, the synthesis methods can be categorized into two groups: (i) chemical and (ii) electrochemical deposition. The chemical methods for the synthesis of conducting polymers/biowaste hybrid materials usually involve the typical in situ chemical oxidative polymerization by employing an oxidizing agent. This synthesis approach is simple and cost-effective and can be easily scaled up for large-scale commercial production.

The electrochemical synthesis approach involves electrochemical deposition of conducting polymers on hybrid biowaste material. Potentiostatic or galvanostatic deposition of conducting polymers from electrolyte on the BWC's surface that has already been mounted on a substrate such as graphene sheets in an electrochemical cell is a common electrochemical deposition technique [7,8]. During the electrodeposition process, deposition time, electrode potential, and current density can be varied to monitor the amount of conducting polymers deposited on BWC as well as the morphology and porosity of the resulting conducting polymers/biowaste hybrid materials. However, this method cannot be scaled up for large-scale commercial production.

25.2.2 Metal Oxides/Biowaste Hybrid

Transition metal oxides (TMOs) are more promising metal oxides in energy storage applications due to their multiple transition states, good electrical conductivity, and reversible redox reactions compared to other metal oxides. Thus, introducing metal oxides into the matrix of porous carbon allows charge storage by both faradaic and non-faradaic mechanisms while enhancing the cycle life of SCs. The choice of raw biowaste, metal oxide, reaction time, and temperature significantly influence the structure, surface chemistry, and interface of the composite.

The present combinations of composites are numerous, but the most used ones are MnO_2, NiO, V_2O_5/biowaste hybrid materials used for supercapacitors. The chemical and hydrothermal methods are the most followed synthesis procedures for hybrid composites. In a simple in situ chemical deposition, the desired amount of BWC is mixed with a metal precursor ($MnCl_2$) solution followed by ultrasonic treatment. The addition of a strong oxidizing agent ($KMnO_4$) initiates the chemical reaction in the supply of continuous heat for the required time [9]. In another approach, the growth of iron oxide particles on bagasse-derived AC provides uniform distribution of Fe_2O_3/Fe_3C particles and the synergistic effect of the graphitic layers of the carbon with Fe_2O_3/Fe_3C particles leads to a high specific capacitance [10]. The hydrothermal method of synthesis involves the occurrence of chemical reaction at high temperature and pressure. Biowaste hybrid structures have successfully been synthesized with hydrothermal treatment with different morphologies of the metal oxides such as nanoparticles, nanosheets, nanorods, nanoflowers, and nanowires. Figure 25.1 shows the synthesis process of 3D MnO_2/HC from the precursors, hemp stems, and metal precursors [11].

25.2.3 Heteroatoms-Doped Biowaste Hybrid

Activated carbons are synthesized from various biowaste precursors such as pomelo peels, soybean residue, banana peels, coir piths, coconut shell, neem leaves, and willow catkins. Nowadays, researchers explore various effective modification methods to obtain hybrid biowaste-derived carbon and heteroatoms doping is one of the most

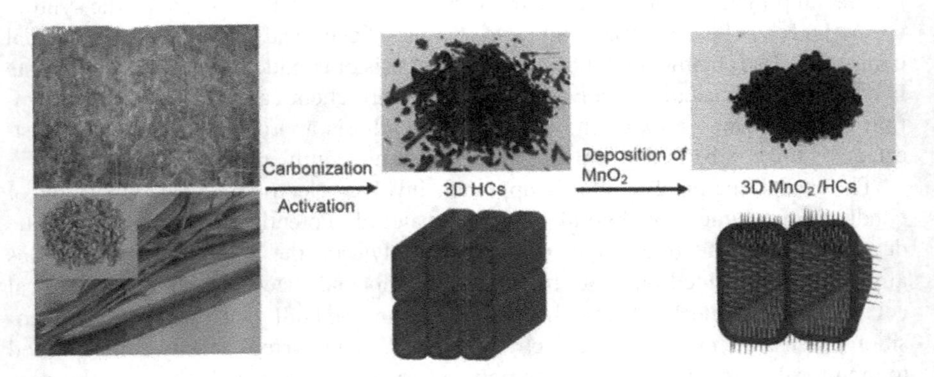

FIGURE 25.1 Schematic diagram of the preparation of 3D MnO_2/HCs. (Reproduced with permission from Ref. [11]. Copyright (2017) American Chemical Society.)

effective and cost-effective methods. Introducing heteroatoms into BWC leads to heteroatoms-doped biowaste hybrid materials. In particular, many biowaste materials are naturally rich in heteroatoms such as nitrogen and oxygen, and after undergoing pyrolysis and activation, they can be converted into stable heteroatom/biowaste hybrid materials without any additional modification [12].

However, in some biowaste precursors, the inherent heteroatoms content is relatively small or absent and hence some modifications need to be done. In general, the heteroatoms/biowaste hybrid materials are believed to impart pseudocapacitance characteristics that are contributed from the redox reactions of these electrochemically active species containing heteroatoms [13]. A variety of methods for introducing heteroatoms to the BWC's surfaces have been investigated, such as chemical treatment, plasma, and oxidation. Phosphoric acid as a chemical activation agent can be considered helpful to inject heteroatoms into biowaste products. Doping of nitrogen atoms can be achieved by treating carbon frameworks with air and ammonia gas; however, it results in lower nitrogen content. This shortcoming can be rectified by utilizing nitrogen-containing materials as precursors, such as urea. In addition, for nitrogen and sulfur doping, BWC frameworks are treated with thiourea followed by suitable activation route [14]. In addition, air plasma treatment which introduces oxygen- and nitrogen-containing groups on the biowaste carbon surface is an alternative method for the surface modification of biowaste-derived carbon materials [15,16].

25.2.4 Other Biowaste Hybrid Materials

In addition to the aforementioned materials, carbon nanotubes (CNTs), sulfides, metal–organic frameworks (MOFs), and metal nitrides can also be coupled with biowaste materials to get high-performance supercapacitors. For instance, biowaste-derived AC and carbon nanotube (CNT) electrodes are used because CNTs have a high electrical conductivity and unique porous structure. A unique method to produce the hybrid of AC and CNT was described, in which the preparation of green monoliths from biowaste and CNTs is performed by carbonization and activation routes [17]. Biowaste–metal chalcogenides composites are also gaining interest as a potential electrode in energy storage devices. It is because of the tunable stoichiometric compositions, rich redox sites, unique crystal structures, and high electronic conductivities that could participate in meeting the requirements for supercapacitor applications. Biowaste-derived activated carbon and graphene composites are also interesting materials for energy storage applications. Graphene serves as an ideal framework to load active materials and have a high specific surface area as well as outstanding conductivity. The biowaste-derived carbons and graphene oxide (GO) combine by π–π interactions and hydrogen bonding, due to the presence of functional groups on the surface of GO as well as carbon [18].

The metal–organic frameworks (MOFs) are 3D mesoporous structures that are prepared by the coordination of the metal centers and connecting ligands. The use of MOFs with activated carbons is a new concept to enhance the characteristics of carbon. The MOF/biowaste hybrid materials are normally synthesized by a solvothermal method, in which (i) BWC is dispersed into well-dissolved MOF precursors and

(ii) mixtures are sonicated and subjected to solvothermal reactions. The hydrothermal treatment product formed with uniform growth of MOF on the porous surface of activated carbon with mesopores, which further enhance the electron transmissions.

25.3 ADVANTAGES AND LIMITATIONS OF HYBRID BIOWASTE MATERIALS

Hybrid biowaste materials are so different when compared to the pristine biowaste-derived carbonaceous materials, because the former imparts both EDLC and pseudocapacitance behavior, while in the latter, only EDLC characteristics prevail. Besides modifying the pore structure, doping with heteroatoms can improve the wettability of carbon materials and lead to partial pseudocapacitance characteristics of the material. In addition, conducting polymers/biowaste hybrid materials are proved to be more advantageous over pristine biowaste materials because of their diverse synthesis methods, and high electrical conductivity and pseudocapacitive nature of the conducting polymers [19]. Besides these intriguing advantages of conducting polymers/biowaste hybrid materials, few issues need the immediate attention of the researchers. Conducting polymers are notorious for their low cyclability, which is largely due to the high-volume transitions that occur during the charge–discharge cycle. The incorporation of BWC into the conducting polymer matrix prevents extensive expansion and shrinkage of the conducting polymer network during cycles. Due to the weaker biowaste-derived carbon surface interconnection, in situ chemical oxidative polymerization of conducting polymers/biowaste hybrid from aqueous solution can suffer from poor carbon material dispersion, conducting polymer aggregation, and increased resistance. BWCs with functionalization, which is mostly accomplished by concentrated acid treatment and heteroatom doping, could solve this problem.

The most commonly used are biowaste-derived activated carbons and metal oxides, conducting polymers, and heteroatoms-doped hybrid materials. Activated carbon is blessed with tunable porosity, but unfortunately, the micropores are normally wetted by the electrolytes and left with an exposed area that remains no longer useful to store charge [20]. Metal oxides/mixed metal oxides–biowaste hybrid materials in aqueous electrolytes exhibit mesoporous structures to allow effective charge storage. The multiple oxidation states of the transition metal allow fast and reversible redox activity and permit fast electron transport due to the improvement of the structure and conductivity of the hybrid materials [21].

To some extent, biowaste hybrid materials are found with some discrepancies based on their structures and properties accordingly. The excessive growth of the nanostructures in/on a matrix of porous AC caused the disappearance of mesopores and formation of micropores, thereby resisting the flow of ions [20]. Carbon variant-based hybrid materials such as CNTs, graphene oxide/biowaste hybrid possess very low equivalent series resistance. CNTs and graphene oxide (GO) act as conductive veins to transport ions quickly and hinders the aggregation of the carbon nanosheets. The incorporation of CNTs or GO in the carbon matrix improves the electrical conductivity and buffers the mechanical stress of hybrid over cycles [22]. In the MOF/biowaste hybrid materials synthesized by a solvothermal technique, the problem lies with the lattice mismatch between the ACs and MOF structures and control over the growth of the MOFs on the surface of AC. The excessive growth of the MOFs on the surface of AC may lead to a decrease in specific capacitance. To cope with this issue, if the amount of carbon is increased, it further leads to some unexpected coordination reactions and destroys the structure of hybrid material and even decreases the porosity.

25.4 APPLICATIONS OF HYBRID BIOWASTE MATERIALS FOR SUPERCAPACITORS

Hybrid biowaste materials have great potential in supercapacitor applications due to the existence of both the EDLC and pseudocapacitance nature of the biowaste/hybrid materials. Hybrid biowaste materials due to their enhanced electrical conductivity, extra electroactive space for redox species, shorter transportation path, and wettability are deemed as the next-generation electrode material for supercapacitor applications.

25.4.1 Conducting Polymers/Biowaste Hybrid as an Electrode for Supercapacitors

Conducting polymers/biowaste hybrid materials are rapidly growing hybrid biowaste materials that show great caliber for supercapacitor applications. Incorporation of different conducting polymers into BWC not only provides pseudocapacitive characteristics, but also enhances the electrical conductivity of the as-prepared biowaste hybrid materials. Conducting polymers/biowaste hybrid materials render superior supercapacitor performance than pristine biowaste-derived carbon materials. This is stated based on the array of the previous studies. For instance, the in situ chemical oxidation method was used to synthesize the hybrid of Eichhornia crassipes-derived nanocarbon material and polyaniline [23]. In this method, the Eichhornia crassipes-derived carbon and aniline were ultrasonicated to form a homogeneous mixture and then oxidants such as ammonium persulfate ($(NH_4)_2S_2O_8$) were used to initiate the polymerization process and then the solution was left under ice bath for the completion of polymerization. The resultant PANI/biowaste hybrid material exhibited an excellent cyclic stability and a highest specific capacitance of 1,542 Fg^{-1}, which is much higher when compared to the pristine biowaste material (293 Fg^{-1}) at a constant current density of 0.78 Ag^{-1}. This is due to the existence of conducting PANI and the Π–Π interaction between the carbon framework and PANI, which reduces the length of the conducting path and also favors charge transfer action through the diffusion-controlled mechanism. Figure 25.2a and b depicts the schematic diagram of the synthesis of PANI/biowaste hybrid and illustrates the Π–Π interaction present between the biowaste-derived carbon and PANI, respectively.

In another report, PEDOT was used as a conducting polymer to synthesize brewer's spent grain-derived carbon/PEDOT hybrid via in situ chemical oxidation method [24]. In this approach, in situ chemical oxidation of EDOT/biowaste (brewer's spent grain) was done using $Fe(Tos)_3$ as an oxidizing agent. Electrode fabrication to achieve the maximum compatibility between electrode–electrolyte systems is shown in Figure 25.3. The hybrid rendered a high specific capacitance of 25 mF/cm^2, which is much higher than their parental counterparts.

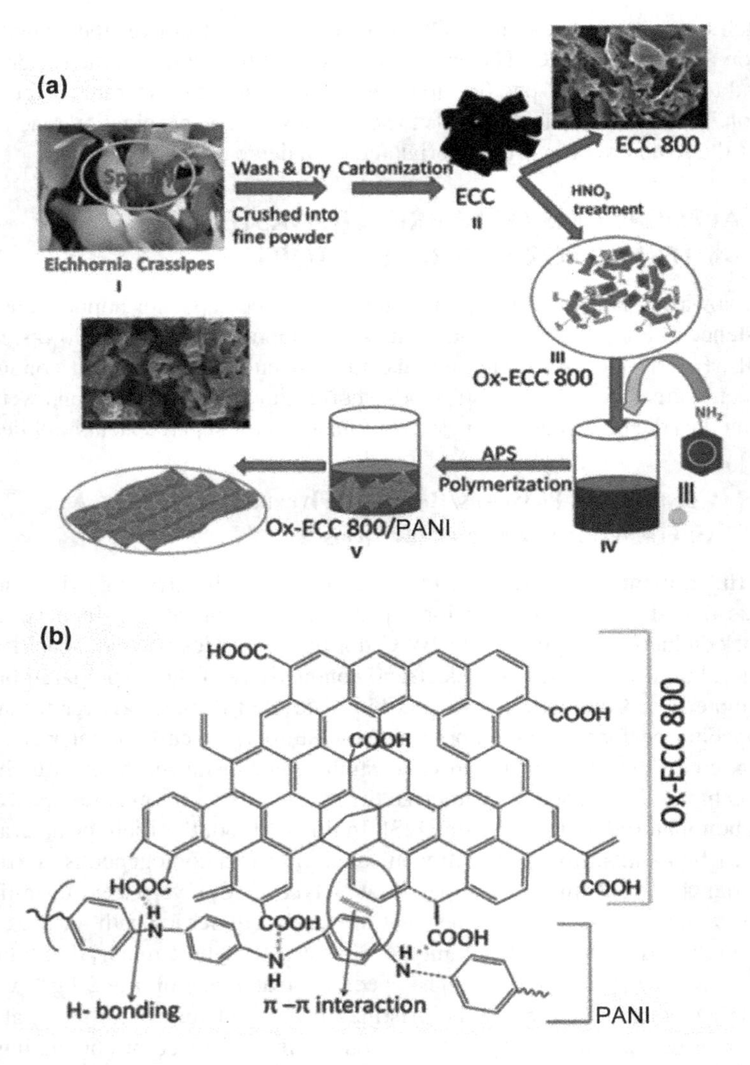

FIGURE 25.2 (a) Schematic route to the synthesis of Eichhornia crassipes/PANI hybrid material and (b) Π–Π interaction between the carbon framework and PANI. (Reproduced with permission from Ref. [23]. Copyright (2020) Elsevier.)

Apart from the in situ chemical oxidation method, conducting polymers/biowaste hybrid materials can also be synthesized via electrochemical deposition. This approach involves different factors such as current density and potential, and these factors can also be employed to synthesize electrode material with desirable thickness and porous network. Several carbonaceous materials such as activated carbon, CNT, and graphene oxide were used earlier to fabricate conducting polymers/carbon hybrid, which showed a great supercapacitor performance [25–27]. BWC has not been utilized yet to design conducting polymers/biowaste hybrid materials via electrodeposition method.

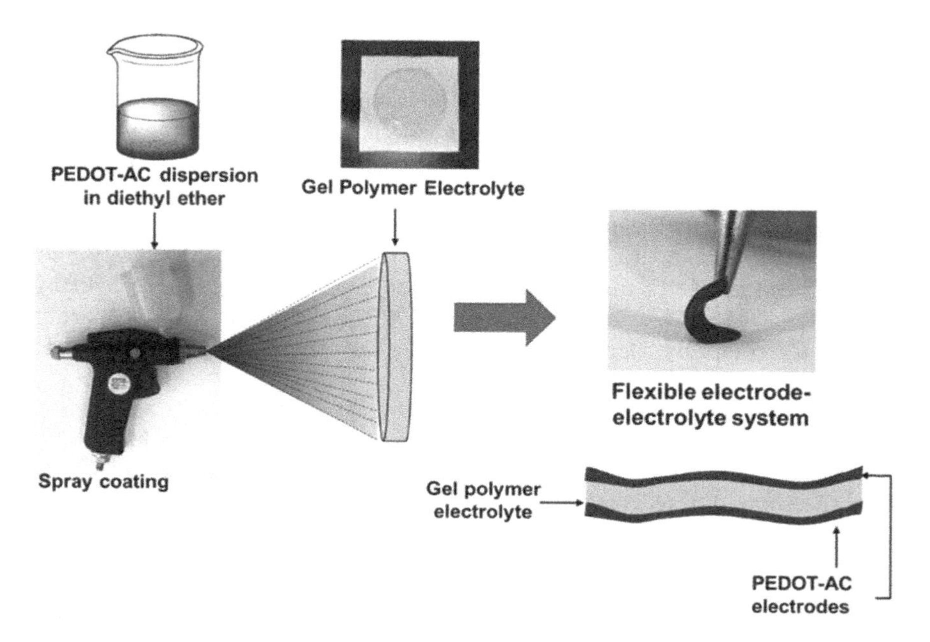

FIGURE 25.3 Schematic of the fabrication of a flexible electrode–electrolyte system. (Reproduced with permission from Ref. [24]. Copyright (2020) Multidisciplinary Digital Publishing Institute.)

25.4.2 Metal Oxides/Biowaste Hybrid as an Electrode for Supercapacitors

Metal oxides undergo fast and reversible redox reactions at the interface of metal oxide surface and electrolyte; however, their high cost and toxic nature limit their application. The poor characteristics of oxides are resolved by developing hybrid materials of metal oxides with highly porous, economical, nature-friendly biowaste-derived activated carbon. The choice of metal oxides depends particularly on the pore dimensions and electrochemical characteristics of BWCs. In this view, a hybrid material is prepared using MnO_2 and 3D honeycomb-like activated carbon. The hybrid material is prepared using hydrothermal technique via in situ growth of MnO_2 nanowires in a matrix of hemp stem-derived 3D activated carbon. An asymmetric SC was fabricated with 3D MnO_2/HC and HC electrodes to store about 33.3 Wh/kg of energy with a power delivery of 14.8 kW/kg [11]. The direct growth of cubic Co_3O_4 on mollusk shell-derived activated carbon shows the uniform alignment of cubic Co_3O_4. This enhances the electron/ion transport efficiency, which in turn results in a high specific capacitance of 1,307 F/g at a current density of 1 A/g with capacitance loss of only 16% after 3,000 cycles [21]. Ternary metal oxides being another branch of metal oxide find remarkable space to be used with activated carbons in SC applications. For example, a mollusk shell-derived honeycomb activated carbon combined with spinel $NiCo_2O_4$ nanowires delivered a specific capacitance of 1,696 F/g. The solid-state symmetric SC prepared based on this composite had an energy density of 8.47 Wh/kg at a current density of 1 A/g with high cycle stability over 10,000 cycles [28]. In an iron oxide/carbide biowaste hybrid material, the oxygen-containing functional groups are methoxy, carboxyl, and phenolic, as sugarcane bagasse allows homogeneous embedding of Fe^{3+} ions during the carbonization of carbon matrix. An increasing iron content resulted in the formation of mesopores, but simultaneously, a decrease in specific surface area was observed, which might be due to the excessive loading of iron oxide/carbide particles on void spaces. Fe_3C formation arises because, at high temperature of ~1,000°C, Fe^{3+} ions get converted to Fe_3O_4 or simple Fe by C reduction, in which the reduced C diffuses into Fe_3O_4 particles. The synergistic effects of AC layers with Fe_3O_4/Fe_3C particles facilitate the composite electrode of SCs to display a specific capacitance of 211.6 F/g at a current density of 0.4 Ag^{-1} [10].

An orthorhombic V_2O_5@C composite was successfully prepared from the Hibiscus sabdariffa family by a solvothermal process to control crystal size, shape, and structure at a moderate temperature. The asymmetric cell based on V_2O_5@C composite showed a cycle stability of 98.9% up to 25,000 cycles with a capacity retention of 88%. These studies ensured that the metal oxides/carbon composites can be used as prospective electrodes for supercapacitive applications [29].

25.4.3 Heteroatoms/Biowaste Hybrid as an Electrode Material for Supercapacitors

BWC materials render excellent supercapacitor performance owing to their high surface area, hierarchical pore size network, and moderate electrical conductivity. Still, to meet the future requirements, continuous efforts have been put forward to develop novel electrode materials for high-performance supercapacitors. Heteroatom doping is the most efficient way to augment the electrochemical performance of the electrode material. Heteroatoms/biowaste hybrid materials are the new-generation electrode materials for high-performance supercapacitors. This is due to the enhanced wettability, extra surface redox reactions, and high electrical conductivity offered by the hybrid biowaste materials. Some biowastes are blessed with inherent heteroatoms, and hence after pyrolysis at a high temperature, they act as heteroatoms-doped biowaste-derived carbon materials. For example, a bamboo shell was utilized to develop activated carbon by the KOH activation process [30]. Bamboo shells are rich in proteins, amino acids, and carbohydrates, thereby introducing heteroatoms into the final activated carbon product. The resultant bamboo shell-derived AC/heteroatoms hybrid material supercapacitor delivered a high energy density of 13.15 Wh/kg at a decent power density of 546.6 W/kg. Similarly, coffee waste ground is comprised of proteins that have inherent O and N atoms and, by subsequent activation process, results in an O, N-doped biowaste hybrid material, which can achieve a high specific capacitance of 190 F/g at 5 mV/s with a remarkable capacitance retention of 92% after 2,000 cycles [31].

Furthermore, some biowaste precursors have very small or no inherent heteroatoms. Hence, extra modification is done to introduce heteroatoms into biowaste via various methods such as chemical methods and air plasma methods to enhance the electrochemical performance of the supercapacitor. For example, heteroatoms were introduced in coir pith biowaste via a chemical method (by treating coir pith with thiourea and subsequent activation at high temperature) [32]. The as-synthesized biowaste hybrid material displayed a maximum specific capacitance of 247.1 F/g at a current density of 0.2 A/g. Furthermore, the assembled supercapacitor device rendered a high cyclic stability of 82% capacitance retention and 100% of columbic efficiency after 10,000 charge–discharge cycles. Furthermore, O, N-doped hybrid biowaste materials can be synthesized by using various dopants such as $(NH_4)_2CO_3$ and urea that contain both nitrogen and oxygen heteroatoms.

In another approach of synthesizing heteroatoms/biowaste hybrid materials, the activating agent is itself is used as the doping agent to introduce heteroatoms into the biowaste-derived carbon matrix. An activating agent such as NH_4Cl, H_3PO_4, and $KMnO_4$ was utilized to develop heteroatoms/biowaste hybrid materials. In this line, $KMnO_4$ was utilized as an activating agent and as a template precursor with hydroxyl and amidogen functional groups to synthesize a heteroatoms-doped chitin biowaste hybrid material [33]. The as-prepared heteroatoms/biowaste hybrid displayed an ultra-high specific capacitance of 412.5 F/g at 0.5 A/g, along with good rate capability. Additionally, camellia pollen as a biowaste precursor was activated and doped with heteroatoms at the same time by using different agents such as NH_4Cl, urea, and $(NH_4)_2CO_3$, which act as both dopant and activating agents [34]. Among them, NH_4Cl is proved to be the best porogen (pore generator), while urea is found to be the most effective dopant and delivered superior electrochemical performance. Furthermore, the as-prepared urea-doped biowaste hybrid electrode delivered a maximum specific capacitance of 300 F/g at 1 A/g, while the urea-doped electrode-based supercapacitor rendered a remarkable energy density of 14.3 Wh/kg. Similarly, H_3PO_4 and $(NH_4)_2HPO_4$ can also be used to perform both as an activating agent and a heteroatom dopant [35,36]. $(NH_4)_2HPO_4$ is used as both an activating agent and dopant to synthesize an O, N, and P-doped durian shell hybrid biowaste material [36]. The schematic route to prepare the O, N, and P-doped biowaste hybrid material is shown in Figure 25.4. The resultant product demonstrated a high specific capacitance of 560 F/g at 2 A/g with a high energy density of 12 Wh/kg at a high-power density of 312 W/kg.

25.4.4 Other Biowaste Hybrid Materials as an Electrode for Supercapacitors

Supercapacitors based on carbon nanotubes and biowaste hybrid materials are in the race of becoming efficient energy storage materials. The electrochemical performance of CNTs and coconut shell-derived AC composite was examined in non-aqueous electrolytes, and at 15 and 50 wt.% of CNTs, it achieved a capacitance of 88 and 50 F/g, respectively [37]. Green monoliths (GMs) were synthesized from self-adhesive carbon grains derived from empty fruit bunch fibers of an oil palm tree (90%), CNTs (5%), and potassium hydroxide (5%). The GMs were carbonized under flow of N_2 gas at various temperatures of 600 °C, 700 °C, and 800 °C and were activated at 800 °C for 1 h to form activated carbon monoliths (ACM6, ACM7, and ACM8). Among them, ACM8 showed superior results, which might be due to the higher number of electrons and ions participating in the charge/discharge process. However, with the addition

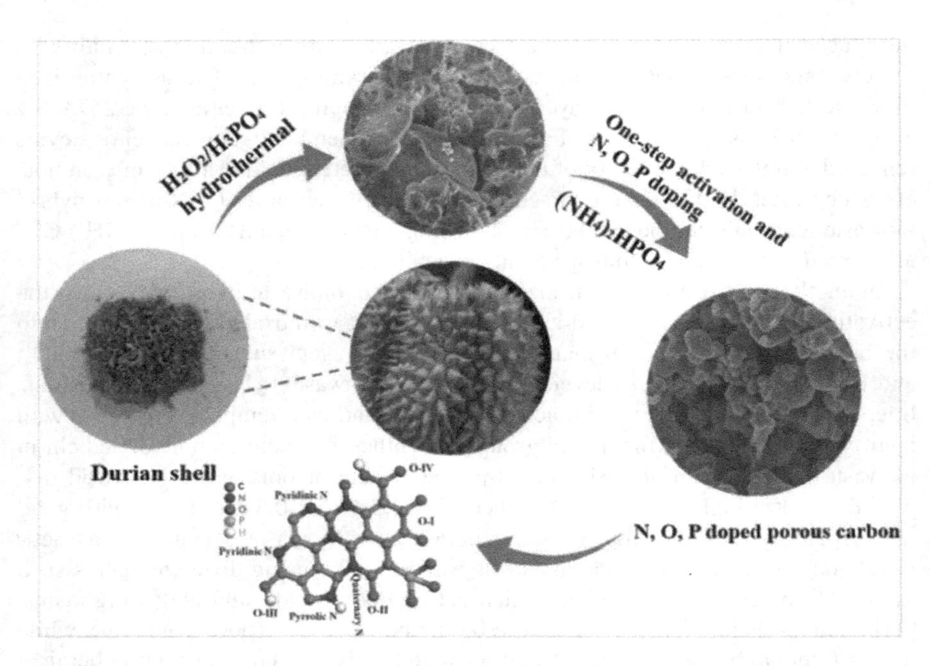

FIGURE 25.4 Schematic route to the preparation of N, O, and P heteroatoms-doped durian shell-derived hybrid biowaste material. (Reproduced with permission from Ref. [36]. Copyright (2020) Springer Nature.)

of CNTs, the specific capacitance decreased due to a decrease in surface area, but a decrease in equivalent series resistance (ESR) elevated the power density [17].

Biowaste–metal dichalcogenide composites were synthesized by hydrothermal treatment to develop facile biowaste hybrid materials. For instance, the washed and treated dry waste (human hair or kapok fibers) was added to a solution of ammonium molybdate and thiourea and the resultant solutions were treated hydrothermally at 150 °C for 24 h and named H-MoS_2/carbon and K-MoS_2/carbon, which displayed a specific capacitance of 143 F/g and 254 F/g, respectively. The incorporation of the MoS_2 nanoflowers on the carbon surface improved the electronic conductivity of the hybrid material. Therefore, this study shows an effective way to use multifunctional biowaste–MoS_2 hybrid materials from waste human hair and kapok fiber as a potential electrode material for SCs [38]. A biowaste-derived carbon dots and reduced graphene oxide composite was synthesized by one-step hydrothermal method. Pre-synthesized cauliflower leaves-derived carbon dots were dispersed in GO solution and treated hydrothermally at 200 °C for 10 h to form RGO/carbon dot. The carbon dots served as effective spacers to avoid restacking of graphene nanosheets, resulting in a greater surface area as well as pore volume. The RGO/carbon dot composite with GO:carbon dot mass ratio of 2:1 exhibited a specific capacitance of 278 F/g at 0.2 A/g [39].

25.5 CONCLUSIONS AND FUTURE OUTLOOK

The performance of supercapacitors highly depends on the fabrication method and the electrochemical properties of the electrode material. Over the last decade, significant advances have been made in exploring electrochemical properties of biowaste hybrid electrode materials. It is desirable to provide a simple and effective way of synthesizing various hybrid biowaste electrodes with low resistance, high energy density, and high-power density. Hybrid biowaste electrodes have the advantages of both biowaste-derived activated carbon and amalgamated materials (metal oxides, conducting polymers, heteroatoms, and other materials such as MOFs and nitride). Biowaste hybrid materials have emerged as one of the key electrode materials for developing high-performance SCs because of their large surface area and high electrical conductivity. This leads to improved rapid ion diffusion into or out of internal electrodes using short paths, and enhanced ion accessibility to the surface. The efficiency of supercapacitors can be vastly improved as research into the production of biowaste hybrid-based supercapacitors progresses. In reality, the research and development of hybrid biowaste-based supercapacitors is still in its early stages, with several problems to be resolved. At the theoretical forefront, the charging mechanisms of hybrid biowaste-based supercapacitors must be investigated and explained. Several researchers have attempted to investigate the charging mechanism of hybrid biowaste supercapacitors in different electrolyte conditions, but the exact charging mechanism of hybrid biowaste supercapacitors is still unknown.

Different charging pathways such as faradaic processes, for example redox reactions, electrosorption, and intercalation, or non-faradaic charge aggregation on the surface of hybrid biowaste can be involved in a reversible series because hybrid biowaste allows an asymmetric supercapacitor (ASC). An asymmetric supercapacitor must be manufactured with the utmost compatibility of electrode materials to increase the specific capacitance. The optimization study of the potential window between the two distinctive asymmetric electrode materials is still a challenge. To completely comprehend and harness the capacitive properties of hybrid biowaste, it is essential to recognize the involved charging processes and their equilibrium. Also, exploring the enhancement of specific capacitance considering both the electrode and electrolyte and their interaction, depending on device-level configuration, remains a major challenge. On practical and research forefronts, new approaches and improvements to existing routes of fabricating hybrid biowaste materials are required. Future studies will focus on improving the interfacial compatibility of biowaste materials and amalgamated materials to take advantage of high theoretical specific capacitance and high electrical conductivity of amalgamated materials (metal oxides, conducting polymers, heteroatoms, etc.) and mechanical stability and high surface area of biowaste-derived carbon. In retrospect, we can confidently conclude that hybrid biowaste produces high-performing electrochemical energy storage units, but further work needs to be done to satisfy industrial supercapacitor specifications.

REFERENCES

1. Conway, B.E., Transition from “supercapacitor” to “battery” behavior in electrochemical energy storage. *Journal of the Electrochemical Society*, 1991. **138**(6): pp. 1539–1548.
2. Winter, M. and R.J. Brodd, What are batteries, fuel cells, and supercapacitors? *Chemical Reviews*, 2004. **104**(10): pp. 4245–4270.
3. Raza, W., et al., Recent advancements in supercapacitor technology. *Nano Energy*, 2018. **52**: pp. 441–473.
4. Sundriyal, S., et al., Advances in bio-waste derived activated carbon for supercapacitors: Trends, challenges and prospective. *Resources, Conservation and Recycling*, 2021. **169**: p. 105548.
5. Frackowiak, E., et al., Supercapacitors based on conducting polymers/nanotubes composites. *Journal of Power Sources*, 2006. **153**(2): pp. 413–418.
6. Magu, T.O., et al., A review on conducting polymers-based composites for energy storage application. *Journal of Chemical Reviews*, 2019. **1**(1): pp. 19–34.
7. Chen, J.H., et al., Electrochemical synthesis of polypyrrole/carbon nanotube nanoscale composites using well-aligned carbon nanotube arrays. *Applied Physics A*, 2001. **73**(2): pp. 129–131.
8. Muthulakshmi, B., et al., Electrochemical deposition of polypyrrole for symmetric supercapacitors. *Journal of Power Sources*, 2006. **158**(2): pp. 1533–1537.
9. Li, D., et al., MnO_2 nanosheets grown on N-doped agaric-derived three-dimensional porous carbon for asymmetric supercapacitors. *Journal of Alloys and Compounds*, 2020. **815**: p. 152344.
10. Manippady, S.R., et al., Partially graphitized iron–carbon hybrid composite as an electrochemical supercapacitor material. *ChemElectroChem*, 2020. **7**(8): pp. 1928–1934.
11. Yang, M., et al., MnO_2 nanowire/biomass-derived carbon from hemp stem for high-performance supercapacitors. *Langmuir*, 2017. **33**(21): pp. 5140–5147.
12. Chen, D., et al., Effect of self-doped heteroatoms in biomass-derived activated carbon for supercapacitor applications. *ChemistrySelect*, 2019. **4**(5): pp. 1586–1595.
13. Qian, L., et al., Recent development in the synthesis of agricultural and forestry biomass-derived porous carbons for supercapacitor applications: A review. *Ionics*, 2020. **26**(8): pp. 3705–3723.
14. Wang, B., et al., A simple and universal method for preparing N, S co-doped biomass derived carbon with superior performance in supercapacitors. *Electrochimica Acta*, 2019. **309**: pp. 34–43.
15. Adusei, P.K., et al., Fabrication and study of supercapacitor electrodes based on oxygen plasma functionalized carbon nanotube fibers. *Journal of Energy Chemistry*, 2020. **40**: pp. 120–131.
16. Hussain, S., et al., Nitrogen plasma functionalization of carbon nanotubes for supercapacitor applications. *Journal of Materials Science*, 2013. **48**(21): pp. 7620–7628.
17. Basri, N. and B. Dolah, Physical and electrochemical properties of supercapacitor electrodes derived from carbon nanotube and biomass carbon. *International Journal of Electrochemical Science*, 2013. **8**: pp. 257–273.
18. Huang, J., et al., Hierarchical porous graphene carbon-based supercapacitors. *Chemistry of Materials*, 2015. **27**(6): pp. 2107–2113.
19. Kandasamy, S.K. and K. Kandasamy, Recent advances in electrochemical performances of graphene composite (graphene-polyaniline/polypyrrole/activated carbon/carbon nanotube) electrode materials for supercapacitor: A review. *Journal of Inorganic and Organometallic Polymers and Materials*, 2018. **28**(3): pp. 559–584.
20. Minakshi, M., et al., A hybrid electrochemical energy storage device using sustainable electrode materials. *ChemistrySelect*, 2020. **5**(4): pp. 1597–1606.
21. Liu, Y., et al., Co_3O_4@ highly ordered macroporous carbon derived from a mollusc shell for supercapacitors. *RSC Advances*, 2015. **5**(92): pp. 75105–75110.

22. Lyu, L., et al., Recent development of biomass-derived carbons and composites as electrode materials for supercapacitors. *Materials Chemistry Frontiers*, 2019. **3**(12): pp. 2543–2570.
23. Verma, C.J., et al., Polyaniline stabilized activated carbon from Eichhornia Crassipes: Potential charge storage material from bio-waste. *Renewable Energy*, 2020. **162**: pp. 2285–2296.
24. González, F.J., et al., 'In-situ' preparation of carbonaceous conductive composite materials based on PEDOT and biowaste for flexible pseudocapacitor application. *Journal of Composites Science*, 2020. **4**(3): p. 87.
25. Bai, M.-H., et al., Electrodeposition of vanadium oxide–polyaniline composite nanowire electrodes for high energy density supercapacitors. *Journal of Materials Chemistry A*, 2014. **2**(28): pp. 10882–10888.
26. Gao, S., et al., Electrodeposition of polyaniline on three-dimensional graphene hydrogel as a binder-free supercapacitor electrode with high power and energy densities. *RSC Advances*, 2016. **6**(64): pp. 58854–58861.
27. Hu, L., et al., In situ electrochemical polymerization of a nanorod-PANI–Graphene composite in a reverse micelle electrolyte and its application in a supercapacitor. *Physical Chemistry Chemical Physics*, 2012. **14**(45): pp. 15652–15656.
28. Xiong, W., et al., Composite of macroporous carbon with honeycomb-like structure from mollusc shell and $NiCo_2O_4$ nanowires for high-performance supercapacitor. *ACS Applied Materials & Interfaces*, 2014. **6**(21): pp. 19416–19423.
29. Ngom, B., et al., Sustainable development of vanadium pentoxide carbon composites derived from Hibiscus sabdariffa family for application in supercapacitors. *Sustainable Energy & Fuels*, 2020. **4**(9): pp. 4814–4830.
30. Han, J., et al., Heteroatoms (O, N)-doped porous carbon derived from bamboo shoots shells for high performance supercapacitors. *Journal of Materials Science: Materials in Electronics*, 2018. **29**(24): pp. 20991–21001.
31. Hossain, R., et al., In-situ O/N-heteroatom enriched activated carbon by sustainable thermal transformation of waste coffee grounds for supercapacitor material. *Journal of Energy Storage*, 2021. **33**: p. 102113.
32. Karuppannan, M., et al., Nitrogen and sulfur co-doped graphene-like carbon sheets derived from coir pith bio-waste for symmetric supercapacitor applications. *Journal of Applied Electrochemistry*, 2019. **49**(1): pp. 57–66.
33. Wang, Y., et al., Heteroatoms-doped hierarchical porous carbon derived from chitin for flexible all-solid-state symmetric supercapacitors. *Chemical Engineering Journal*, 2020. **384**: p. 123263.
34. Cao, L., et al., Comparison of the heteroatoms-doped biomass-derived carbon prepared by one-step nitrogen-containing activator for high performance supercapacitor. *Diamond and Related Materials*, 2021. **114**: p. 108316.
35. Nirosha, B., et al., Elaeocarpus tectorius derived phosphorus-doped carbon as an electrode material for an asymmetric supercapacitor. *New Journal of Chemistry*, 2020. **44**(1): pp. 181–193.
36. Wang, K., et al., Durian shell-derived N, O, P-doped activated porous carbon materials and their electrochemical performance in supercapacitor. *Journal of Materials Science*, 2020. **55**(23): pp. 10142–10154.
37. Taberna, P.-L., et al., Activated carbon–carbon nanotube composite porous film for supercapacitor applications. *Materials Research Bulletin*, 2006. **41**(3): pp. 478–484.
38. Durairaj, A., et al., Facile synthesis of waste-derived carbon/MoS_2 composite for energy storage and water purification applications. *Biomass Conversion and Biorefinery*, 2021: pp. 1–12.
39. Nguyen, L.H. and V.G. Gomes, High efficiency supercapacitor derived from biomass based carbon dots and reduced graphene oxide composite. *Journal of Electroanalytical Chemistry*, 2019. **832**: pp. 87–96.

26 Polymeric Wastes for Supercapacitors

Fabeena Jahan, Deepthi Panoth, Sindhu Thalappan Manikkoth, Kunnambeth M. Thulasi, Anjali Paravannoor, and Baiju Kizhakkekilikoodayil Vijayan
Kannur University

CONTENTS

26.1 INTRODUCTION

The exponential growth in the production and use of various polymers for numerous purposes ultimately generates enormous amount of solid wastes in the form of discarded tires, masks, compact CDs, and plastic wastes such as bottles and pipes, and the landfills generated pose environmental hazards and also health issues [1]. Currently, due to the increasing population, there is a decrease in the abundance of fossil fuels and an increase in white pollution-associated environmental problems that highly demand the research and development sector to progress in the development of renewable and sustainable energy storage devices. Upcycling of hazardous polymeric wastes to produce value-added carbon materials for innovative applications such as rechargeable energy storage could be considered as a Holy Grail for green engineering of electrode materials. Synthetic polymers such as polyolefin, polypropylene (PP), polyethylene terephthalate (PET), polyethylene (PE), polyurethane(PU), polyvinyl chloride (PVC), and polystyrene (PS) due to their high chemical stability

DOI: 10.1201/9781003178354-30

cannot be degraded easily and are considered as widely produced wastes that are hazardous to humankind and environment [2]. As these polymers are abundant in carbon elements, the most promising and eco-friendly recycling way for these materials will be their carbonization and the resultant carbon material could possess a morphology that could be used for various applications such as supercapacitor/battery electrodes, catalysts, and sorbents. These carbonized products do not require any complex retreatment, and carbonization is an eco-friendly recycling approach to keeping plastic wastes out of landfills and the oceans. So far, many polymers have effectively been carbonized to porous carbon nanostructure, carbon nanotubes (CNTs), graphite, carbon hollow spheres, etc. [3].

Supercapacitors are energy storage devices possessing a much outstanding performance with high power density, long life cycle, quick charging and discharging, superior safety, improved cyclic stability, longer robustness, low maintenance cost, a wide range of working temperatures, and better efficiency over rechargeable batteries. Compared to lithium-ion batteries, supercapacitors do not have much energy density, so scientists were greatly encouraged to improve the supercapacitor's energy density by proposing novel active electrode materials of different shapes possessing high energy density and power density and electrolytes with a wide operating voltage range. Various materials based on carbon are used as electrodes including activated carbon (AC), CNTs, mesoporous and microporous carbon, conductive polymers, and graphene for supercapacitor manufacturing. Many polymer materials composed of carbon as the main component have been proved to yield carbon-based materials via the carbonization process. For example, PS waste is modified by carbonization to electrode materials for supercapacitor application. Thus, the environmental concern due to the plastic wastes can be overcome by implementing polymer waste as the material for electrodes for the manufacturing of high-voltage supercapacitors having increased energy density and cycle stability (Figure 26.1) [4].

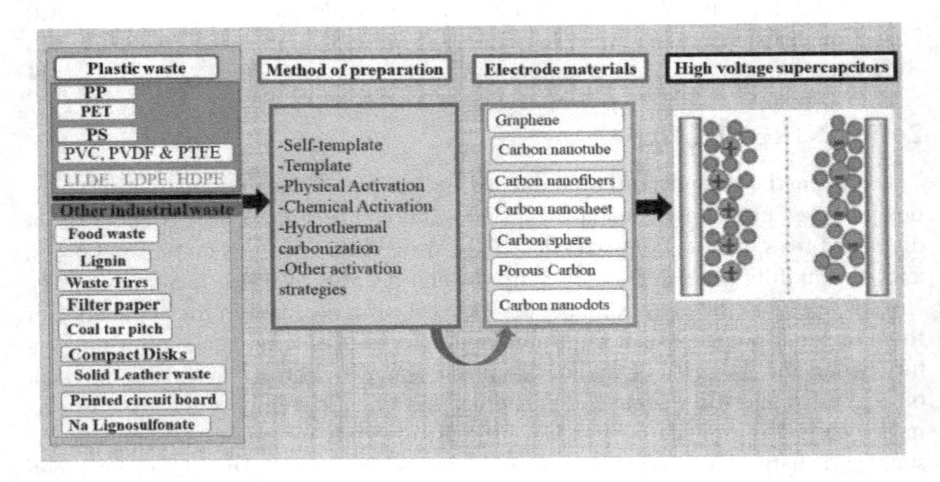

FIGURE 26.1 Schematic representation for the transformation of various polymeric wastes to carbonaceous electrode materials for supercapacitor application. (Adapted with permission from Ref. [5]. Copyright (2020) Elsevier.)

The most effective technique to get porous carbon-based supercapacitors having higher energy and power densities is to develop them with an optimal size of micropores and with a suitable proportion of mesopores. The fundamental factor for attaining high capacitance is obtained by using electrodes that are electrically conducting and having a high specific surface area. Graphene has all these conditions as required, with a high conductivity, electrochemical compatibility, stability, and porosity. CNTs and graphene although widely studied for their supercapacitor applications have several demerits including their expense and tiresome processes involved in their production. The electronic and ionic conductivities are also essential parameters for the evaluation of how good a supercapacitor can be [5]. Conducting polymers, such as polyaniline, offer high capacitance and are cheaply available and environmentally friendly, but their great expansion and shrinkage of volume during the doping process cause breakdown of their structure and a subsequent shorter life span. Moreover, almost all organic conducting polymer-based electrodes are found to have a drop in their capacitance to half the initial value after 1,000 cycles [6]. ACs are the most beneficial among other materials due to their ease of availability, high surface area, affordable price, effective electrical conductivity, long life cycle, anti-corrosivity, high electrochemical stability, improved porosity, and environment-friendly nature. Precursors of AC can be from different sources such as carbonaceous materials (coal, petroleum, and coke), agricultural wastes (coconut shell, banana peels, seaweeds, etc.), and industrial materials, and activation is done by the chemical or physical activation process. Activation temperature and time significantly affect the properties such as pore size distribution and surface area of carbon precursor. The ACs produced from the carbonaceous material face the problem of scarcity of fossil fuels and also are costly and quickly exhaustible due to the huge consumption of energy. Hence, agro-based AC precursors such as coconut shell draw more attraction than the costlier ACs manufactured from coke/petroleum products [7]. So, for an innovative category of carbon to replace coconut shell-derived AC, it must have a high working potential range of 3.5–4 V with a noticeable increase in the energy density over coconut shell-based AC or otherwise it should be comparatively cheap and easy to fabricate. Thus, for the scientific researchers, the main challenge is to identify new precursors alternative to agro-based wastes that are inexpensive, are easily available, and have a high carbon content that will offer a potential chance to synthesize AC-based electrodes for supercapacitors [5]. Polymeric wastes including plastic bags, rubber tires, PS, PP masks, and PET bottles were used as an alternative to agricultural waste-based carbon precursors to produce AC for application in electrode fabrication for supercapacitors. However, all these approaches result in only limited production of carbon content, so in order to cope with problems such as decreased carbon yield, low porosity, and low specific capacitance, it is recommended to dope it with various chemicals and minerals such as boron, sulfur, nitrogen, and phosphorus before carbonization to tune the final material to have improved electrical properties. Doping with heteroatoms not only improves the carbon yield, but also improves the physical as well as the chemical characteristics of the porous carbon materials [3,8]. The present chapter briefly discusses the various synthetic methods of carbon-based electrode materials from various polymeric wastes such as PE, PET, PS, and PVC for supercapacitor applications.

26.2 CARBON-BASED ELECTRODE FROM POLYMER WASTE FOR SUPERCAPACITOR APPLICATIONS

The use of polymer waste for the production of value-added carbon materials is the most praising and innovative method in the collection of polymer waste recycling techniques. As the main component of plastic waste is polyolefin, the polymer with a carbon content of 86% undoubtedly serves as the major source for the industrial-scale synthesis of carbon materials for high-end applications [9]. A vast number of waste polymer materials have been employed for the production of carbon electrode-based supercapacitors, including PE, PET, PS, and PVC. Making of these "raw materials" into activated carbon materials having diverse pore distribution, surface character, and morphologies is truly challenging [5]. Several carbonization techniques including pyrolysis and activation methods have been practiced for the synthesis of activated carbon materials from polymers. A detailed discussion on the preparation methods is given below.

26.2.1 Synthetic Methods of Carbon Electrode Materials

Techniques that have been proposed for revitalizing industrial waste can be either physical or chemical techniques. Conjugation of both these techniques also leads to the formation of highly activated carbon materials with excellent properties by carefully observing time, temperature, and chemical agents. The preparation of carbon electrode from several carbon precursors is mainly carried out via three methods, namely activation, template-based, and carbonization methods.

26.2.1.1 Activation Method

Activation involves the process of converting industrial waste materials into superior carbonaceous products. Pyrolysis is the most common activation method, where the decomposition of materials takes place at elevated temperatures in an inert or minimal oxygen atmosphere. As already described, the activation method can be executed in two different routes, viz. physical activation and chemical activation.

26.2.1.1.1 Physical Activation Method

Physical activation is generally performed just after pyrolysis by a direct route under steam, dry air, CO_2, or their combination as activating agents. In this method, pores in the carbon are opened in the gaseous atmosphere, leading to efficient microporous and macroporous activated carbon materials [10]. For instance, CO_2 gas serves as an effective activating agent that reacts with carbon in the polymer chain. It helps to increase the surface area due to pore opening and prevents oxygen from entering the pores. However, for the production of activated carbon with desirable features, it is very essential to pay attention to gas flow rate and activation time.

In a particular physical activation method, the source is initially heated in the temperature between 400 °C and 900 °C to remove massive volatile compounds followed by an incomplete gasification process at a temperature range of 350 °C–1,000 °C in the presence of an oxidizing gas. The active oxygen primarily burns out the remaining pyrolysis side products trapped inside the openings of the carbon framework, thereby opening the clogged pores. Moreover, as the oxidant burns away, volatile areas induce the formation of microporous morphology and result in CO and CO_2. The time required for the complete process depends on the gas flow rate and the activation temperature. From both economic and quality considerations, steam is the most favorable activating agent owing to its abundance, low cost, and lack of side products during activation. The steam activation can be carried out in one pot as the steam can be mixed with pyrolysis products. The process paves the way for the synthesis of advanced carbon materials with rich oxygen functionalities, which enhances the capacitance and wettability of carbonaceous products. For example, Qiao et al. synthesized microporous carbon fibers from PVC with a very high surface area using steam activation in an earlier work. The plastic wastes were kept at 260 °C for 2 h and at 410 °C for 1–2 h in a two-stage heat treatment. A 58% weight loss is achieved in the initial step with the successive removal of HCl, and a 23.2% weight loss is achieved in the second step, leaving 18.4% carbon for activated carbon synthesis. The sample is carbonated and steam-activated at 900 °C for 30–90 min. A final yield of 4%–8% of the initial mass of PVC was obtained with the virgin resin material [11].

26.2.1.1.2 Chemical Activation Method

Compared to physical activation, chemical activation is better as it yields activated carbon with a high quality and wider specific surface area. The activation process is somewhat more efficient because it takes a shorter time and doesn't require a higher temperature for activation. These advantages suppress the concerns of high cost, complexity in the agent removal process, and corrosive nature of activating agents.

In the chemical activation process, chemical activating agents such as NaOH, LiOH, Na_2CO_3, H_3BO_3, $CaCl_2$, Na_2S, KOH, KCl, K_2CO_3, $CaCl_2$, P_2O_5, CsOH, $MgCl_2$, $AlCl_3$, $FeCl_3$, $ZnCl_2$, $FeSO_4$, RbOH, HCl, H_2SO_4, HNO_3, H_3PO_4, and H_2O_2 were impregnated on a carbon precursor and heated at a temperature range of 400°C–1,200°C in an inert atmosphere. The pyrolysis and the activation can be done either in a single step or in multiple steps under various gases. Chemical activation is usually executed at the laboratory level, and it is highly beneficial in the context of pore size distribution. However, the method is inconvenient in the sense that the washing procedures require more amount of water to eliminate impurities formed as by-products during the activation process.

Among the different activating agents, alkalis such as KOH are the best, effective, and broadly employed agents for turning industrial waste materials into high-grade carbonaceous materials for supercapacitor applications. The chemical activation of different precursors of carbon using KOH provides benefits in terms of higher yield, low activation temperature and time, high specific surface area, and improved microporous geometry. The mechanism of activation by KOH is illustrated using the following equations [12]:

$$6KOH + 2C = 2K + 3H(6)_2 + 2K_2CO_3 \tag{26.1}$$

$$K(7)_2\,CO_3 + C = K_2O + 2CO \tag{26.2}$$

$$K(8)_2\,CO_3 = K_2O + CO_2 \tag{26.3}$$

$$2K + CO(9)_2 = K_2O + CO \tag{26.4}$$

KOH melts at 360 °C and results in a solid–liquid interface where subsequent reactions will take place. KOH oxidizes carbon at 400°C and results in the formation of metallic K, H_2, and K_2CO_3. At about 600°C, all the KOH is used up and further reaction occurs between C and K_2CO_3, which were significantly converted to CO_2 and K_2O at above 700 °C. Simultaneously, K_2O and K_2CO_3 can be further reduced by carbon to form metallic K, which can intercalate into the carbon lattice that results in the formation of a micropore. At around 800 °C, K_2CO_3 gets completely vanished and the released CO_2 is further reduced to form CO at elevated temperatures. Besides, the released H_2, CO_2, and CO also aid in pore formation in the carbon framework.

H_3PO_4 and $ZnCl_2$ are efficient activating agents, and the process can be accomplished in a single-step pyrolysis–activation process as they react easily with natural polymers such as cellulose and lignin [13]. For instance, at low temperatures, H_3PO_4 catalyzes the bond breaking of aryl ester bond and glycosidic linkage in lignin and cellulose, respectively. At higher temperatures, the cyclization and condensation of volatile substances occur and swelling of biomass structure facilitates the separation of wood microfibres. The gasification of microfibres and the voids formed due to micropore separation create porous structures, a most desirable property for supercapacitor applications. Similarly, $ZnCl_2$ activation can impede the formation of large molecular bio-oil and enables the formation of ultra-porous geometry. However, the mechanism of activation using H_3PO_4 and $ZnCl_2$ suggests that they are not appropriate for the post-activation of carbonaceous materials. Potassium ferrate (K_2FeO_4), K_2CO_3, and $KMnO_4$ were also demonstrated as activating agents in the literature [14]. K_2FeO_4 can act as both an activating agent and a catalyst in the activation process. The activation mechanism of K_2CO_3 is the same as a fragment of KOH activation. $KMnO_4$ is a strong oxidizing agent that reacts with carbon atoms at ambient temperature. The moderate activation condition and MnO_2 deposition on hierarchically porous carbon manifest tremendous performance in supercapacitor applications.

26.2.1.2 Template Method

Template methods are very powerful for the synthesis of hierarchical porous carbon with a well-defined and controlled porosity from non-structured materials. In this technique, a template with a precise geometry is embedded into the carbon precursors followed by high-temperature carbonization and chemical etching, leading to the formation of activated carbon with a replicated structure of the template. Generally, the replicated pores are meso- and macropores. However, micropores can also be achieved due to the shrinkage of carbon sources or the release of volatile materials. Numerous templates are employed for the synthesis of activated carbon, including hard template, soft template, or its combination, and zeolites are used for the conversion of industrial raw materials into carbonaceous materials for supercapacitor applications.

26.2.1.2.1 Hard Template Method

In this method, solid particles such as silica, and polymer colloids are used as a sacrificial template for constructing meso- or macropores. The mechanism involves the infiltration of carbon precursors into the vacant spaces in the colloidal particle, and subsequent penetration results in a solid form. Pyrolysis will be conducted after that in an inert atmosphere, which facilitates the carbon formation from the solid fillers, and the elimination of template occurs via chemical etching. The hard carbon structure is designed after eliminating template particles that mount the macropores remaining in the native sites of the particles [15]. Ice crystals, fish scale, yeast cell, wood, cotton fabrics have also been used as hard templates regarding their natural structure. Plastics including PS and PU can also be used as a hard template.

26.2.1.2.2 Soft Template Method

Soft template methods have advantages over hard template methods when considering the formation of carbon materials with diverse morphologies, mild experimental conditions, and exemption from template creation and exclusion [16]. Surfactants are efficient soft templates owing to their bipolar structure and capacity to synthesize micelles [17]. Heteroatom-rich compounds can also be used as a soft template for the synthesis of doped hierarchal porous carbon for supercapacitor electrodes. The mixing of the hard and soft templates has also been employed for preparing porous carbon with the existence of mesopores and macropores together, which is advantageous for potential application in supercapacitor devices [18]. Due to the tiny space of hard templates, the soft templates of comparatively bigger sizes will fill the voids between hard colloids. The macropores-based hierarchically porous carbon helps in the transport of ions and electrolyte access and also improves the dispersion and loading of active materials, thereby enhancing the capacitance of the supercapacitor device.

26.2.1.3 Hydrothermal Carbonization Method

The hydrothermal carbonization method forms hydrochar, an incompletely carbonized product that displays lower condensation degree and large oxygen functionalities. However, poor porosity and lower surface area demand further carbonization/activation. Hierarchical porous carbon processed from hydrochar has a vast number of heteroatoms and a surface containing functional groups which bolster the capacitance of the electrode in both aqueous and organic electrolytes. The electrochemically inactive functional groups on the surface of carbon enhance the electrode wettability, thereby increasing the specific capacitance via complete utilization of ions and surface accessibility [19].

26.3 POLYMERIC WASTE-DERIVED ELECTRODE MATERIALS FOR SUPERCAPACITORS

Porous carbon has grabbed wide attention as an energy storage material owing to its ultra-high specific surface area, controllable pore geometry, and rich surface functional groups together with stable physical and chemical properties, high mechanical strength, and high electrical conductivity. Among the different carbon precursors, plastics are relatively more promising because the texture and morphology can be tuned by optimized reaction conditions. Plastics have a high carbon content and thus are suitable for activated carbon material synthesis. Thermoplastics such as PE, PS, PP, PET, fluorine, and chlorine-based polymers, and thermosetting plastics such as PUs, phenolic resins, and polypyrroles are the most used carbon precursors for value-added products. Few among them are discussed in this chapter.

26.3.1 Polyethylene

The conventional methods for the treatment of plastic waste are mostly chemical degradation, incineration, and landfill. However, most of the plastics are hard to degrade and may form toxic secondary products. Polyethylene, a widely available and low-cost polymer material with an abundancy in the carbon content of about 86%, can be potentially used for energy storage applications. However, the low melting point in the range of 120 °C–180 °C and poor thermal stability during the carbonization process limit the use of PE. Therefore, an efficient and cost-effective synthetic method is urgently needed to produce activated carbon from PE for high-end applications. Additionally, the incorporation of functional groups on the PE surface prior to pyrolysis also requires extra harsh chemical treatments for desired functionality. Among the different carbon morphologies obtained from PE, nanostructured mesoporous carbons are the most capable electrode materials for supercapacitor applications, while PE-derived mesoporous carbons are rarely investigated.

Yang et al. demonstrated the preparation of hierarchical porous carbon from PE waste via ball milling and carbonization techniques with a magnesium carbonate pentahydrate (MCHP) as a flame-retardant agent followed by NH_3 activation. MCHP improves the thermal stability of PE in the carbonization process and provides an in situ MgO template during pyrolysis. The PE-derived carbon exhibits a high surface area and mesoporosity, leading to the fabrication of supercapacitor devices with a very high capacitance (Figure 26.2). The supercapacitor electrode shows 97.1% capacitive retention even after 10,000 cycles with excellent cyclic stability. An energy density of 43 Wh/kg has been achieved for the symmetrical supercapacitor device, which indicates the high purity of the sample [20].

In the same year, Zhang and co-workers pioneered the quick conversion of PE waste into hollow mesoporous carbon cages with a very high specific surface area by direct pyrolysis with Mg powder. The as-prepared magnesium nitride (Mg_3N_2) and magnesium oxide (MgO) mixture was used as a hard template, which catalyzes the decomposition of PE, and the products are deposited on its surface. The final product is formed by etching the template, which results in high-quality carbon materials that exhibit an excellent specific capacity and good rate performance in a three-electrode system and validates its potential as an electrode material for supercapacitors [21].

Lian et al. developed graphene/mesoporous carbon (G@PE40-MC700) electrode materials from waste PE with graphene oxide and flame retardant by carbonization at 700 °C. The sample displays a high surface area (1,175 m^2/g) and a substantial quantity of mesopores (2.30 cm^3/g) and offers an enhanced electrochemical performance with a wide potential window. The assembled hybrid supercapacitor with

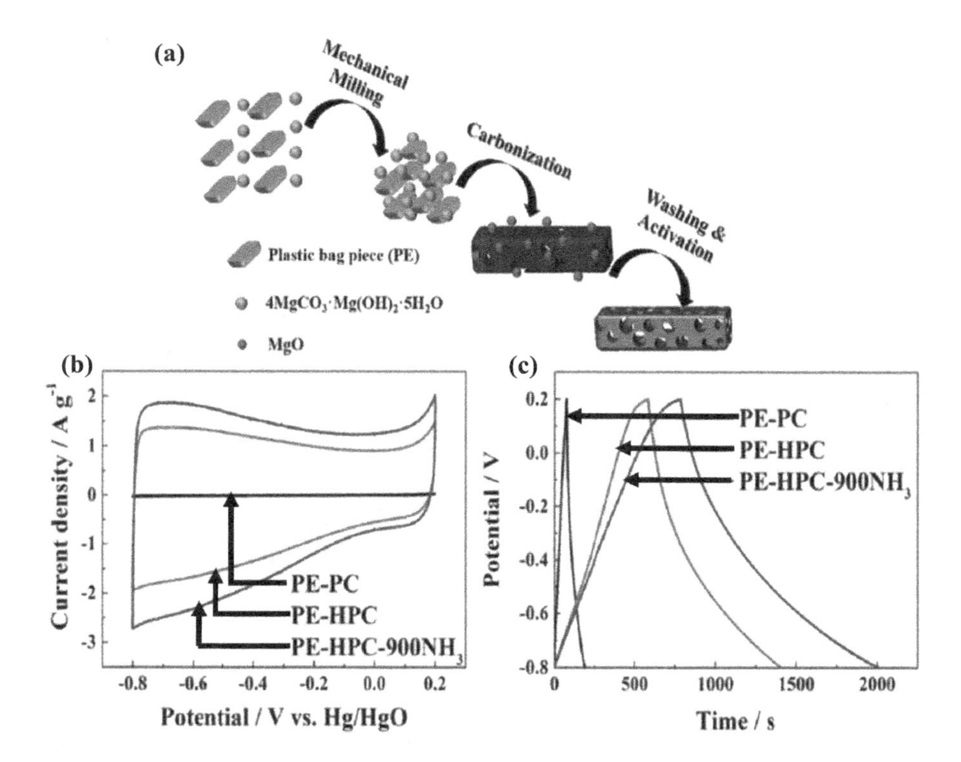

FIGURE 26.2 (a) Schematic diagram for the fabrication of PE-HPC-900NH_3, (b) CV curves of the as-prepared samples at a scan rate of 10 mV/s, (c) galvanostatic charge/discharge curves of the as-prepared samples at 0.2 A/g. (Adapted with permission from Ref. [20]. Copyright (2019) Elsevier.)

G@PE40-MC700 as anode and $LiMn_2O_4$ as cathode exhibits an energy density of 47.8 Wh/kg at a power density of 250 W/kg and displays a high cycling stability of 83.8% after 5,000 cycles [22].

26.3.2 Polystyrene

Another most commonly used non-biodegradable plastic that consists of plastic waste in high ratio and causes severe ecological harm is PS. However, since there is no satisfactory method for effective recycling, the recycle rate of PS wastes is extremely low in comparison with PE and PET with a recycling rate of approximately 90%. Thus, an economical way to reutilize PS wastes is to convert them into functional supercapacitor carbon materials as PS has a high carbon content.

The manufacturing of carbon material from PS wastes for supercapacitors is very challenging as PS cannot be readily transformed into carbon materials in the absence of templates or catalysts. Only a few synthetic procedures were known to be published for the carbonization of PS waste, and the reported methods needed organic solvents or extremely sophisticated synthetic procedures. Moreover, the as-prepared carbon materials exhibited uncontrollable structural characteristics. Some authors reported the use of PS-derived carbon electrodes for supercapacitors, and their performance seemed to be very poor [23,24]. Sun et al. synthesized carbon spheres from PS-based macroreticular resin spheres by carbonization and activation [23]. A specific capacitance of 182 F/g in 6 mol/L of KOH electrolyte was exhibited by the PS-derived carbon sphere activation after an activation time of 2 h. Xu et al. reported the supercapacitive properties of PS-based hierarchical porous carbon (PS-HPC), which exhibited a capacitance retaining proportion and a capacitance of 84% and 28.7 mF/cm^2, respectively [24].

The control of morphology and structure of pores of PS-derived carbon materials is very difficult and essential to regulate the electrochemical performance of carbon-based supercapacitors. According to recent reports, the template carbonization method can be used effectively for the conversion of PS wastes to carbon materials with potential applications in supercapacitors [25]. The parameters of template-derived materials, such as porosity, morphology, and conductivity, can be easily tuned by the template method. These special parameters of PS-derived carbon materials are highly dependent on the template used. To this date, MgO has been regarded as the most favorable template and catalyst for the synthesis of highly supercapacitive carbon materials because of its low price, easy purification, controllable pore size, and morphologies. Moreover, MgO templates can be easily removed by a non-corrosive acid [2]. Ma et al. reported the template-based carbonization of PS wastes into hierarchical porous carbon nanosheets (HPCNs) with coexisting micropores/mesopores [25]. In the synthesis process, the mixture of PS waste and MgO template is carbonized at 700°C in the presence of an inert argon atmosphere for 1 h. The porous MgO template introduces mesopores in the carbon nanosheets. The pores are then further controlled by KOH activation and resulted in HPCNs with a specific surface area of 2,650 m^2/g and a pore volume of 2.43 cm^3/g. The HPCNs exhibited a specific capacitance of 323 F/g at 0.5 A/g and 222 F/g at 20 A/g in 6 mol/L of KOH electrolyte in a three-electrode system. The PCNs were also found to exhibit a 92.6% capacitance retaining power even after 10,000 cycles. In an organic electrolyte, an energy density and a power density of 44.1 Wh/kg and 757.1 W/kg, respectively, were found to be displayed.

Min et al. reported the synthetic procedure for porous carbon flakes (PCFs) via direct pyrolysis of PS waste using flake-MgO as the template in a stainless steel autoclave (Figure 26.3) [2]. MnO_2 nanosheets were then deposited selectively on the surface of resultant PCFs successively to form hybrid structures (PCF-MnO_2). Due to the high specific capacity of MnO_2 combined with the increased conductivity and specific surface area of PCFs, a positive synergistic interaction between MnO_2 and PCF happens and hence the resulting hybrid PCF-MnO_2 exhibited an ultra-high capacitance of 308 F/g at 1 mV/s and 247 F/g at 1 A/g in LiCl electrolyte. PCF-MnO_2 also exhibited an excellent durability of 93.4% capacitance retaining power over 10,000 cycles at 10 A/g in a symmetric supercapacitor device. This work demonstrated an easy technique for the production of high-performance low-cost electrode material for supercapacitors.

Wang et al. demonstrated a new method to prepare nitrogen-doped porous carbon nanosheets (NPCNs) from PS wastes [26]. Here, in this synthesis method, Zn and

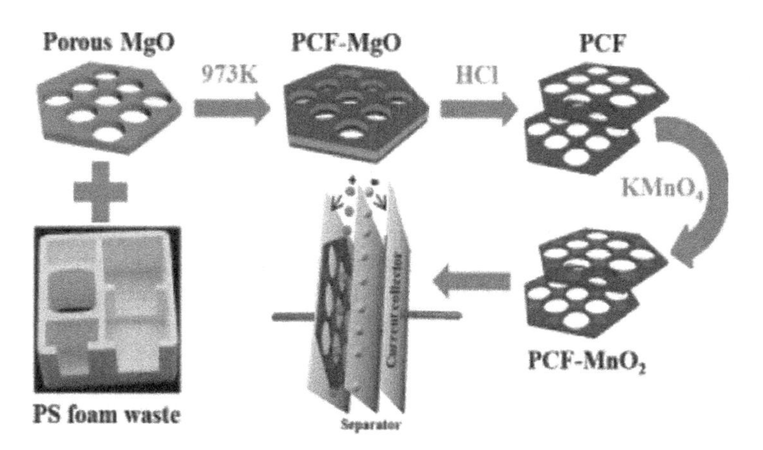

FIGURE 26.3 A synthesis scheme of PCF-MnO_2 hybrid. (Adapted with permission from Ref. [2]. Copyright (2019) Elsevier.)

Co bimetallic zeolitic imidazolate framework (CoZn-ZIF) nanocrystals were developed on magnesium hydroxide sheets [$Mg(OH)_2$] [called $Mg(OH)_2$@CoZn-ZIFs] and were used as a template. NPCNs are synthesized by carbonizing the mixture of $Mg(OH)_2$@CoZn-ZIFs and PS directly. $Mg(OH)_2$ is converted to magnesia (MgO) upon high-temperature treatment, which helps in the conversion of carbon precursors in PS to decompose into a carbon skeleton. The degree of graphitization of NPCNs is also promoted by the presence of Co species. The pyrolysis of imidazole ligands aided the introduction of nitrogen into the carbon framework in situ. The evaporation of Zn from the template resulted in the generation of more micro- and mesopores in the carbon material. After the synthesis process, the templates and Co species were removed and NPCNs with very large specific surface areas were obtained. The NPCN electrodes exhibited a specific capacitance of 149 F/g at a current density of 0.5 A/g in 6 M KOH electrolyte and an excellent cycling stability of 97.6% even after 5,000 cycles.

26.3.3 Polyethylene Terephthalate

Polyethylene terephthalate is extensively used in food and consumer products as packaging materials. PET causes severe environmental problems, as it needs around 500–700 years to be biodegraded. Terephthalic acid and ethylene glycol are the degradation products of PET, and these chemicals are difficult to be carbonized in comparison with the degradation products of PP, PE, PS, etc. Thus, controllable carbonization of PET is very challenging. Although Przepiorski et al. prepared a porous carbon by pyrolysis of PET, it was very hard to control the microstructure of the carbon product [27]. The process consisted of heating the mixture of PET and magnesium carbonate under argon gas flow, followed by acid washing. The thermal decomposition of magnesium carbonate resulted in the formation of MgO, which acted as a catalyst for the carbonization of PET into porous carbon materials. Essawy et al. reported the synthesis of graphene from waste PET bottles at 800 °C in an autoclave [28]. Wei et al. synthesized carbon spheres by pyrolyzing PET wastes in supercritical CO_2 at 500 °C–650 °C [29]. However, the high pressure inside the autoclave hinders the large production of carbon materials from PET waste.

Recently, in 2019, Kamali et al. produced amorphous carbon materials from waste PET bottles using the molten salt method [30]. Unfortunately, the as-produced carbon materials exhibited a poor supercapacitor performance. Chen et al. reported the conversion of PET wastes into highly valuable electrode materials for supercapacitor applications [31]. Here, they introduced an easy method to efficiently transform waste PET bottles into porous carbon nanosheets (PCNSs) through catalytic carbonization followed by KOH activation. PCNSs exhibited a hierarchical porous architecture, an ultra-high specific surface area (2,236 m^2/g), and a large pore volume (3 cm^3/g), which contribute to the outstanding supercapacitive performance of 169 F/g (6 M KOH). Furthermore, PCNSs exhibited a high capacitance of 121 F/g and an energy density of 30.6 Wh/kg at 0.2 A/g in a novel organic electrolyte.

26.3.4 Polymer Waste Based on Fluorine and Chlorine

Plastic consumption has drastically increased with industrialization. Plastic products containing halogens such as chlorine and fluorine, mainly PVC, polytetrafluoroethylene (PTFE), and polyvinylidene fluoride (PVDF), are considered to be highly hazardous for human life and wildlife as they are capable of releasing toxic substances for long term into the environment. The emergence of disposable plastics is growing serious day by day due to the application of plastics at a large scale in various industries and domestic life. Recycling is still thought to be the best way to tackle the issue caused by plastic accumulation in the environment even though the recycling rate is very low. The proper disposal of halogen plastics is still highly challenging. These fluorine- and chlorine-based plastic wastes are potential and promising candidates for synthesizing high-performance carbon electrodes for supercapacitor applications. Waste plastic products of PVC, PTFE, and polyvinylidene fluoride have attracted the attention of researchers owing to their high carbon content. Turning plastic waste products into high-end application carbon materials is one of the efficient utilization of plastic waste products. So far, using waste plastics for carbon materials with diverse morphology such as porous carbon nanosheets, carbon spheres, carbon nanotubes, and graphene have been developed [26].

PVC is a universal polymer and accounts for the second largest production volume of thermoplastics with numerous applications in various areas, including flooring, pipes, insulation, packaging, roofing, bottling, and healthcare materials. The flexibility of the compound with various processing techniques, such as calendaring, extrusion of plastic, injection molding, and the relatively low cost make it a versatile product, which causes post-consumer waste disposal problems due to their increased use and growing demand in the industry [32]. PVC is highly toxic as it contains dioxins, heavy metals, phthalates, and bisphenol A (BPA). However, PVC remains very popular in the production of consumer goods. Every year, a large amount of PVC wastes are scattered and piled up in the environment, which can cause severe environmental and public health risks. The incineration of PVC wastes or using them as a landfill is an inappropriate technology as they release other by-products such as hydrogen chloride, chlorine, and organochlorine, which are dangerous for both humans and the environment [33,34]. Chang and co-workers proposed an efficient and green synthesis technique to develop highly porous carbonaceous materials from PVC plastic wastes such as table cloth, plastic wrap, document bag, and PVC tube via potassium hydroxide-assisted dehalogenation at room temperature, along with clean by-products of potassium chloride and water. PVC plastic wrap-transformed porous carbon electrode displayed a high performance for aqueous symmetric supercapacitor with a specific capacitance of 399 F/g in 6.0 mol/L KOH and 363 F/g in 1.0 mol/L H_2SO_4 (Figure 26.4) [35]. Chen and his group developed a facile and effective

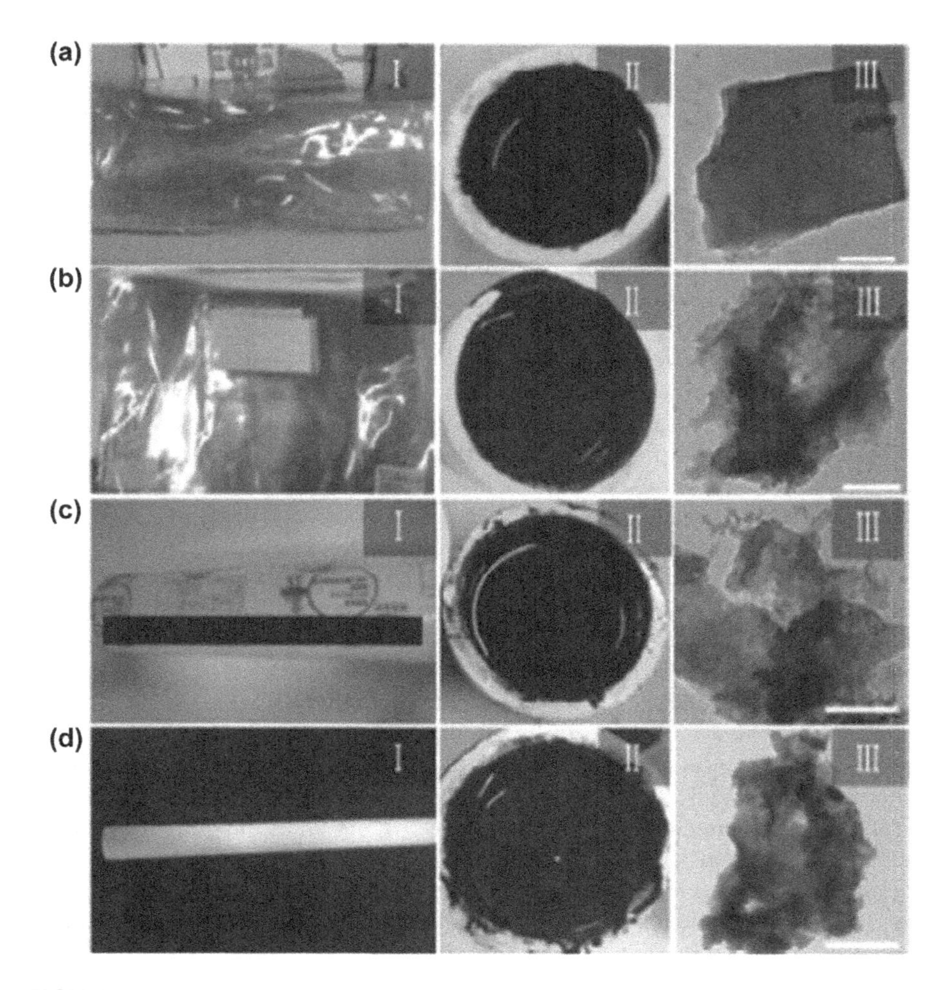

FIGURE 26.4 Column I. Optical images of different PVC plastic products: (a) table cloth (TC), (b) document bag (DB), (c) plastic wrap (PW), and (d) PVC tube; Column II. PVC-dehalogenated carbon slurry; Column III. Corresponding TEM micrographs of the PVC plastic products. (Adapted with permission from Ref. [35]. Copyright (2018) American Chemical Society.)

template carbonization technique to convert plastic PVC waste to MnO_x-doped nanoporous carbon obtained by the redox reaction of carbon with potassium permanganate ($KMnO_4$) solution. The carbon-blank sample prepared using PVC waste and $Mg(OH)_2$ with the mass ratio of 1:2 at a carbonization temperature of 700 °C were mixed with $KMnO_4$ in the ratio 1:1, and the resultant sample exhibited a specific capacitance of 751.5 F/g at 1.0 A/g [36].

In PTFE, the fluorine atoms shield the carbon atoms, which results in their high chemical resistance to acids and alkalis. Moreover, the high electronegativity of fluorine repels everything; thus, it is most commonly used as coatings for non-stick metal cooking pans. The manufacturing process commonly employed for the production of PTFE products is hot sintering or ram extrusions. PTFE wastes can neither be incinerated nor be dumped. During the incineration process, highly corrosive vapors are released, which can destroy the incineration plant itself [37]. Ying et al. transformed PTFE plastic waste into nanoporous carbon via template carbonization method, with zinc powder as a potential hard template. Carbonizing PTFE and zinc powder in the mass ratio 1:3 at 700 °C exhibited a specific capacitance of 313.7 F/g at 0.5 A/g along with a high capacitance retention of 93.10% after 5,000 cycles. Moreover, this same method can be followed to obtain nanoporous carbon from PVC and polyvinylidene fluoride [38]. Jiang et al. successfully converted PTFE waste to nanoporous carbon spheres via a simple template carbonization technique, where calcium carbonate ($CaCO_3$) served as a hard template. The electrochemical performance of the resultant nanoporous carbon spheres was improved by incorporating some amount of urea ($CO(NH_2)_2$ into the carbon matrix. When PTFE, $CaCO_3$, and $CO(NH_2)_2$ were in the mass ratio 2:1:2, they exhibited a very high surface area of 1,048.2 m^2/g and pore volume of nearly 1.03 cm^3/g and this PTFE composite exhibited a specific capacitance of 237.8 F/g with a current density of 1 A/g, using 6 mol/L KOH as electrolyte (Figure 26.5) [39].

Polyvinylidene fluoride is one of the most significant fluorinated polymers broadly used in engineering plastics. It possesses high chemical stability, strength to bases and acids, and inertness to organic solvents and does not undergo hydrolysis. In recent years, the disposal of PVDF waste materials has become a serious concern as their production has increased drastically [40]. One of the efficient ways to tackle this issue is to utilize this polymer waste as precursors for the production of porous carbon structures, which are commonly used as a potential supercapacitor electrode material. Cheng and co-workers using waste PVDF as carbon precursor and nickel(II) nitrate ($Ni(NO_3)_2.6H_2O$) as the graphitic catalyst prepared nanoporous graphitic carbon materials (NGCMs). Carbonization temperature plays a vital role in enhancing their electrochemical performance as the porosity of the sample decreased with the increase in carbonization temperature from 800 °C to 1,200 °C. They also introduced 4-(4-nitrophenylazo)-1-naphthol (NPN) as redox additive in 2 mol/L KOH electrolyte for the sample prepared at 800 °C to enhance their electrochemical performance. The specific capacitance of the pristine sample increased 2.98 times with the addition of 4 mmol/L of NPN [41].

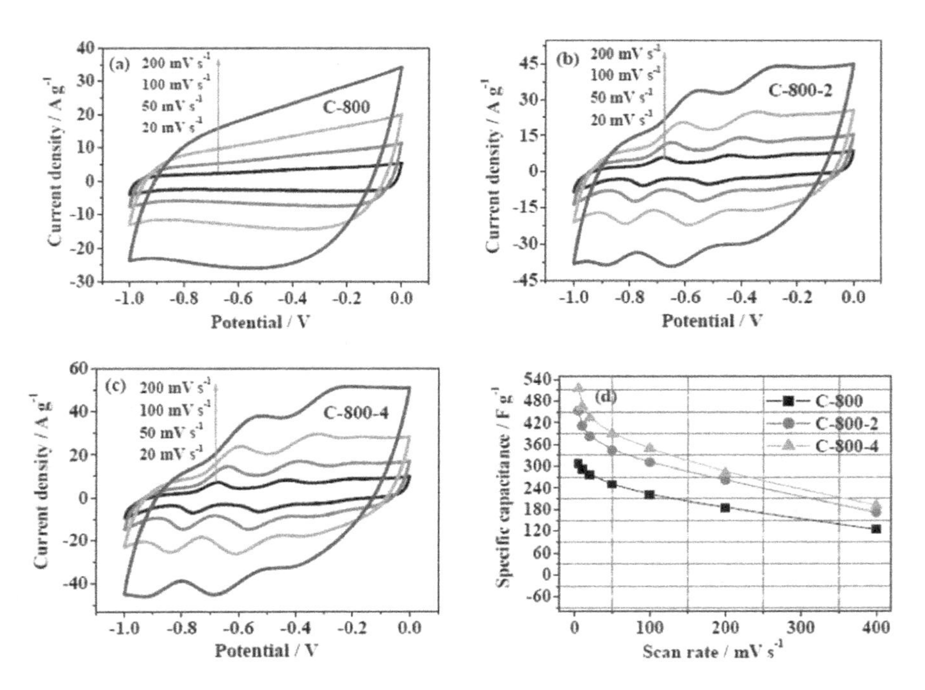

FIGURE 26.5 Cyclic voltammetry curves of samples at different scan rates ranging from 20 to 200 mV/s. (a) Sample C-800 (with carbonization temperature 800 °C), (b) Sample C-800-2 (with 2 mmol/L of NPN), (c) Sample C-800-4 (with 4 mmol/L of NPN), and (d) specific capacitances from CV curves. (Adapted with permission from Ref. [41]. Copyright (2015) American Chemical Society.)

26.4 CONCLUSIONS

The plastic waste and industrial waste materials are dumped in the environment and thus are cheaply available in bulk form, so for the science and research development community, converting this waste to promising electrode materials for energy storage devices has become a milestone to be achieved. In summary, the quickly increasing and environmentally harmful toxic polymeric waste is used as a raw material to produce high value-added carbon electrodes by various activation processes for the manufacturing of supercapacitor with better performance of the increased surface area, porosity, life span, capacitance, etc. Many polymeric wastes such as PS, PE, PET, and PP could be carbonized to active materials through a facile method, which could be employed as an electrode for supercapacitors. For the upcoming energy storage systems, supercapacitors should be capable of hoarding and producing energy at increased rates and could be able to have many cycles of charging and discharging; thus, attention is focused on the carbonization of plastic wastes with heteroatoms such as halogen, nitrogen, and oxygen. And the uses of the resulting carbon materials for different streams other than for supercapacitors should also be studied. At the same time, appropriate and eco-friendly carbonization methods are needed to convert many waste plastics into carbon materials that are usually difficult to process. The application possibilities of polymeric materials that are processed through carbonization could also be extended to other energy storage systems such as lithium-ion batteries or to wastewater purification, where materials possessing high surface area and better wettability are required.

REFERENCES

1. Boota, M.; Paranthaman, M. P.; Naskar, A. K.; Li, Y.; Akato, K.; Gogotsi, Y. Waste Tire Derived Carbon-Polymer Composite Paper as Pseudocapacitive Electrode with Long Cycle Life. *ChemSusChem*, **2015**, *8* (21), 3576–3581. https://doi.org/10.1002/cssc.201500866.
2. Min, J.; Zhang, S.; Li, J.; Klingeler, R.; Wen, X.; Chen, X.; Zhao, X.; Tang, T.; Mijowska, E. From Polystyrene Waste to Porous Carbon Flake and Potential Application in Supercapacitor. *Waste Manag.*, **2019**, *85*, 333–340. https://doi.org/10.1016/j.wasman.2019.01.002.
3. Hu, X.; Lin, Z. Transforming Waste Polypropylene Face Masks into S-Doped Porous Carbon as the Cathode Electrode for Supercapacitors. *Ionics*, **2021**, *27* (5), 2169–2179. https://doi.org/10.1007/s11581-021-03949-7.
4. Farzana, R.; Rajarao, R.; Bhat, B. R.; Sahajwalla, V. Performance of an Activated Carbon Supercapacitor Electrode Synthesised from Waste Compact Discs (CDs). *J. Ind. Eng. Chem.*, **2018**, *65*, 387–396. https://doi.org/10.1016/j.jiec.2018.05.011.
5. Utetiwabo, W.; Yang, L.; Tufail, M. K.; Zhou, L.; Chen, R.; Lian, Y.; Yang, W. Electrode Materials Derived from Plastic Wastes and Other Industrial Wastes for Supercapacitors. *Chin. Chem. Lett.*, **2020**, *31* (6), 1474–1489. https://doi.org/10.1016/j.cclet.2020.01.003.
6. Goswami, S.; Dillip, G. R.; Nandy, S.; Banerjee, A. N.; Pimentel, A.; Joo, S. W.; Martins, R.; Fortunato, E. Biowaste-Derived Carbon Black Applied to Polyaniline-Based High-Performance Supercapacitor Microelectrodes: Sustainable Materials for Renewable Energy Applications. *Electrochim. Acta*, **2019**, *316*, 202–218. https://doi.org/10.1016/j.electacta.2019.05.133.
7. Rajagopal, R. R.; Aravinda, L. S.; Rajarao, R.; Bhat, B. R.; Sahajwalla, V. Activated Carbon Derived from Non-Metallic Printed Circuit Board Waste for Supercapacitor Application. *Electrochim. Acta*, **2016**, *211*, 488–498. https://doi.org/10.1016/j.electacta.2016.06.077.
8. Mu, X.; Li, Y.; Liu, X.; Ma, C.; Jiang, H.; Zhu, J.; Chen, X.; Tang, T.; Mijowska, E. Controllable Carbonization of Plastic Waste into Three-Dimensional Porous Carbon Nanosheets by Combined Catalyst for High Performance Capacitor. *Nanomaterials*, **2020**, *10* (6), 1097. https://doi.org/10.3390/nano10061097.
9. Wu, C.; Wang, Z.; Wang, L.; Williams, P. T.; Huang, J. Sustainable Processing of Waste Plastics to Produce High Yield Hydrogen-Rich Synthesis Gas and High Quality Carbon Nanotubes. *RSC Adv.*, **2012**, *2* (10), 4045. https://doi.org/10.1039/c2ra20261a.
10. González, J. F.; Román, S.; González-García, C. M.; Nabais, J. M. V.; Ortiz, A. L. Porosity Development in Activated Carbons Prepared from Walnut Shells by Carbon Dioxide or Steam Activation. *Ind. Eng. Chem. Res.*, **2009**, *48* (16), 7474–7481. https://doi.org/10.1021/ie801848x.
11. Qiao, W. M.; Yoon, S. H.; Korai, Y.; Mochida, I.; Inoue, S.; Sakurai, T.; Shimohara, T. Preparation of Activated Carbon Fibers from Polyvinyl Chloride. *Carbon*, **2004**, *42* (7), 1327–1331. https://doi.org/10.1016/j.carbon.2004.01.035.
12. Wang, J.; Kaskel, S. KOH Activation of Carbon-Based Materials for Energy Storage. *J. Mater. Chem.*, **2012**, *22* (45), 23710. https://doi.org/10.1039/c2jm34066f.
13. Jagtoyen, M.; Derbyshire, F. Activated Carbons from Yellow Poplar and White Oak by H_3PO_4 Activation. *Carbon*, **1998**, *36* (7–8), 1085–1097. https://doi.org/10.1016/S0008–6223(98)00082–7.

14. Gong, Y.; Li, D.; Luo, C.; Fu, Q.; Pan, C. Highly Porous Graphitic Biomass Carbon as Advanced Electrode Materials for Supercapacitors. *Green Chem.*, **2017**, *19* (17), 4132–4140. https://doi.org/10.1039/C7GC01681F.
15. Abudu, P.; Wang, L.; Xu, M.; Jia, D.; Wang, X.; Jia, L. Hierarchical Porous Carbon Materials Derived from Petroleum Pitch for High-Performance Supercapacitors. *Chem. Phys. Lett.*, **2018**, *702*, 1–7. https://doi.org/10.1016/j.cplett.2018.04.055.
16. Chen, C.; Wang, H.; Han, C.; Deng, J.; Wang, J.; Li, M.; Tang, M.; Jin, H.; Wang, Y. Asymmetric Flasklike Hollow Carbonaceous Nanoparticles Fabricated by the Synergistic Interaction between Soft Template and Biomass. *J. Am. Chem. Soc.*, **2017**, *139* (7), 2657–2663. https://doi.org/10.1021/jacs.6b10841.
17. Sun, L.; Zhou, Y.; Li, L.; Zhou, H.; Liu, X.; Zhang, Q.; Gao, B.; Meng, Z.; Zhou, D.; Ma, Y. Facile and Green Synthesis of 3D Honeycomb-like N/S-Codoped Hierarchically Porous Carbon Materials from Bio-Protic Salt for Flexible, Temperature-Resistant Supercapacitors. *Appl. Surf. Sci.*, **2019**, *467–468*, 382–390. https://doi.org/10.1016/j.apsusc.2018.10.192.
18. Wang, Q.; Cao, Q.; Wang, X.; Jing, B.; Kuang, H.; Zhou, L. Dual Template Method to Prepare Hierarchical Porous Carbon Nanofibers for High-Power Supercapacitors. *J. Solid State Electrochem.*, **2013**, *17* (10), 2731–2739. https://doi.org/10.1007/s10008-013-2166–4.
19. Gu, W.; Sevilla, M.; Magasinski, A.; Fuertes, A. B.; Yushin, G. Sulfur-Containing Activated Carbons with Greatly Reduced Content of Bottle Neck Pores for Double-Layer Capacitors: A Case Study for Pseudocapacitance Detection. *Energy Environ. Sci.*, **2013**, *6* (8), 2465. https://doi.org/10.1039/c3ee41182f.
20. Lian, Y.; Ni, M.; Huang, Z.; Chen, R.; Zhou, L.; Utetiwabo, W.; Yang, W. Polyethylene Waste Carbons with a Mesoporous Network towards Highly Efficient Supercapacitors. *Chem. Eng. J.*, **2019**, *366*, 313–320. https://doi.org/10.1016/j.cej.2019.02.063.
21. Zhang, Y.; Yu, Y.; Liang, K.; Liu, L.; Shen, Z.; Chen, A. Hollow Mesoporous Carbon Cages by Pyrolysis of Waste Polyethylene for Supercapacitors. *New J. Chem.*, **2019**, *43* (27), 10899–10905. https://doi.org/10.1039/C9NJ01534E.
22. Lian, Y.-M.; Utetiwabo, W.; Zhou, Y.; Huang, Z.-H.; Zhou, L.; Faheem, M.; Chen, R.-J.; Yang, W. From Upcycled Waste Polyethylene Plastic to Graphene/Mesoporous Carbon for High-Voltage Supercapacitors. *J. Colloid Interface Sci.*, **2019**, *557*, 55–64. https://doi.org/10.1016/j.jcis.2019.09.003.
23. Sun, G.; Wang, J.; Li, K.; Li, Y.; Xie, L. Polystyrene-Based Carbon Spheres as Electrode for Electrochemical Capacitors. *Electrochim. Acta*, **2012**, *59*, 424–428. https://doi.org/10.1016/j.electacta.2011.10.067.
24. Xu, F.; Cai, R.; Zeng, Q.; Zou, C.; Wu, D.; Li, F.; Lu, X.; Liang, Y.; Fu, R. Fast Ion Transport and High Capacitance of Polystyrene-Based Hierarchical Porous Carbon Electrode Material for Supercapacitors. *J. Mater Chem.*, **2011**, *21* (6), 1970–1976. https://doi.org/10.1039/C0JM02044C.
25. Ma, C.; Liu, X.; Min, J.; Li, J.; Gong, J.; Wen, X.; Chen, X.; Tang, T.; Mijowska, E. Sustainable Recycling of Waste Polystyrene into Hierarchical Porous Carbon Nanosheets with Potential Applications in Supercapacitors. *Nanotechnology*, **2020**, *31* (3), 035402. https://doi.org/10.1088/1361-6528/ab475f.
26. Wang, G.; Liu, L.; Zhang, L.; Fu, X.; Liu, M.; Zhang, Y.; Yu, Y.; Chen, A. Porous Carbon Nanosheets Prepared from Plastic Wastes for Supercapacitors. *J. Electron. Mater.*, **2018**, *47* (10), 5816–5824. https://doi.org/10.1007/s11664-018-6497–x.
27. Przepiórski, J.; Karolczyk, J.; Takeda, K.; Tsumura, T.; Toyoda, M.; Morawski, A. W. Porous Carbon Obtained by Carbonization of PET Mixed with Basic Magnesium Carbonate: Pore Structure and Pore Creation Mechanism. *Ind. Eng. Chem. Res.*, **2009**, *48* (15), 7110–7116. https://doi.org/10.1021/ie801694t.
28. El Essawy, N. A.; Ali, S. M.; Farag, H. A.; Konsowa, A. H.; Elnouby, M.; Hamad, H. A. Green Synthesis of Graphene from Recycled PET Bottle Wastes for Use in the Adsorption of Dyes in Aqueous Solution. *Ecotoxicol. Environ. Saf.*, **2017**, *145*, 57–68. https://doi.org/10.1016/j.ecoenv.2017.07.014.

29. Wei, L.; Yan, N.; Chen, Q. Converting Poly(Ethylene Terephthalate) Waste into Carbon Microspheres in a Supercritical CO_2 System. *Environ. Sci. Technol.*, **2011**, *45* (2), 534–539. https://doi.org/10.1021/es102431e.
30. Kamali, A. R.; Yang, J.; Sun, Q. Molten Salt Conversion of Polyethylene Terephthalate Waste into Graphene Nanostructures with High Surface Area and Ultra-High Electrical Conductivity. *Appl. Surf. Sci.*, **2019**, *476*, 539–551. https://doi.org/10.1016/j.apsusc.2019.01.119.
31. Wen, Y.; Kierzek, K.; Min, J.; Chen, X.; Gong, J.; Niu, R.; Wen, X.; Azadmanjiri, J.; Mijowska, E.; Tang, T. Porous Carbon Nanosheet with High Surface Area Derived from Waste Poly(Ethylene Terephthalate) for Supercapacitor Applications. *J. Appl. Polym. Sci.*, **2020**, *137* (5), 48338. https://doi.org/10.1002/app.48338.
32. Janajreh, I.; Alshrah, M.; Zamzam, S. Mechanical Recycling of PVC Plastic Waste Streams from Cable Industry: A Case Study. *Sustain. Cities Soc.*, **2015**, *18*, 13–20. https://doi.org/10.1016/j.scs.2015.05.003.
33. Glas, D.; Hulsbosch, J.; Dubois, P.; Binnemans, K.; De Vos, D. E. End-of-Life Treatment of Poly(Vinyl Chloride) and Chlorinated Polyethylene by Dehydrochlorination in Ionic Liquids. *ChemSusChem*, **2014**, *7* (2), 610–617. https://doi.org/10.1002/cssc.201300970.
34. Rahimi, A.; García, J. M. Chemical Recycling of Waste Plastics for New Materials Production. *Nat. Rev. Chem.*, **2017**, *1* (6), 0046. https://doi.org/10.1038/s41570-017–0046.
35. Chang, Y.; Pang, Y.; Dang, Q.; Kumar, A.; Zhang, G.; Chang, Z.; Sun, X. Converting Polyvinyl Chloride Plastic Wastes to Carbonaceous Materials via Room-Temperature Dehalogenation for High-Performance Supercapacitor. *ACS Appl. Energy Mater.*, **2018**. https://doi.org/10.1021/acsaem.8b01252.
36. Cheng, L. X.; Zhang, L.; Chen, X. Y.; Zhang, Z. J. Efficient Conversion of Waste Polyvinyl Chloride into Nanoporous Carbon Incorporated with MnO_x Exhibiting Superior Electrochemical Performance for Supercapacitor Application. *Electrochimica Acta*, **2015**, *176*, 197–206. https://doi.org/10.1016/j.electacta.2015.07.007.
37. Lakshmanan, A.; Chakraborty, S. K. Recycling of Polytetrafluoroethylene (PTFE) Scrap Materials. In *Sintering Techniques of Materials*; Lakshmanan, A., Ed.; InTech, **2015**. https://doi.org/10.5772/59599.
38. Chen, X. Y.; Cheng, L. X.; Deng, X.; Zhang, L.; Zhang, Z. J. Generalized Conversion of Halogen-Containing Plastic Waste into Nanoporous Carbon by a Template Carbonization Method. *Ind. Eng. Chem. Res.*, **2014**, *53* (17), 6990–6997. https://doi.org/10.1021/ie500685s.
39. Jiang, W.; Jia, X.; Luo, Z.; Wu, X. Supercapacitor Performance of Spherical Nanoporous Carbon Obtained by a $CaCO_3$-Assisted Template Carbonization Method from Polytetrafluoroethene Waste and the Electrochemical Enhancement by the Nitridation of $CO(NH_2)_2$. *Electrochimica Acta*, **2014**, *147*, 183–191. https://doi.org/10.1016/j.electacta.2014.09.050.
40. Castagnet, S.; Gacougnolle, J.-L.; Dang, P. Correlation between Macroscopical Viscoelastic Behaviour and Micromechanisms in Strained α Polyvinylidene Fluoride (PVDF). *Mater. Sci. Eng. A*, **2000**, *276* (1–2), 152–159. https://doi.org/10.1016/S0921–5093(99)00320–2.
41. Cheng, L. X.; Zhu, Y. Q.; Chen, X. Y.; Zhang, Z. J. Polyvinylidene Fluoride-Based Carbon Supercapacitors: Notable Capacitive Improvement of Nanoporous Carbon by the Redox Additive Electrolyte of 4-(4-Nitrophenylazo)-1-Naphthol. *Ind. Eng. Chem. Res.*, **2015**, *54* (41), 9948–9955. https://doi.org/10.1021/acs.iecr.5b02490.

27 Carbon Nanostructures Derived from Polymeric Wastes for Supercapacitors

Vanessa Hafemann Fragal and
Elisangela Pacheco da Silva
UEM – State University of Maringa

Elizângela Hafemann Fragal
University of Claude Bernard Lyon 1

Michelly Cristina Galdioli Pellá
UEM – State University of Maringa

Thiago Sequinel
Federal University of Grande Dourados

Rafael Silva
UEM – State University of Maringa

Cristian Tessmer Radmann and
Vanessa Bongalhardo Mortola
Federal University of Rio Grande

Luiz Fernando Gorup
University of São Carlos
Federal University of Rio Grande
Federal University of Pelotas

DOI: 10.1201/9781003178354-31

CONTENTS

27.1 INTRODUCTION

The pollution of the environment is one of the biggest challenges faced in the 21st century. It may come from the gas emitted by industry or by a car, plastics disposals, sewage discharge, harmful substances used for agricultural purposes, or so many other sources. These harmful substances pollute water bodies and the soil and also affect the global temperature. The increasing consequences of human activities over the environment raised awareness regarding the importance of preserving planet Earth. Since ending pollution is still not possible, researchers have been studying alternatives to minimize the effects of pollution on the environment. Alternative fuel sources and recycling are examples of already available and used methods to decrease pollution. Driven by this global concern, this chapter will focus on plastic wastes recycling for energy application. Plastics are buoyant and durable polymeric-based materials capable of adsorbing and transporting toxic substances while traveling in water bodies [1,2].

Recycling represents one of the alternatives to minimize the pollution caused by them. It allows the use of plastic for energy production (incineration), the extraction of specific atoms (such as carbon), and the conversion of plastic waste into value-added materials such as electrodes for electric supercapacitors (ESCs). Carbon-based electrodes derived from polymer waste are an excellent alternative to replace other expensive materials since they are low-cost raw materials and can be prepared by sustainable routes. ESCs are energy storage devices known for their relatively high energy storage capacity, high rates of charge and discharge, extended battery life, high power pulse, and little degradation over hundreds of high-efficiency cycles [3,4]. Their properties are in between those of batteries and traditional capacitors. However, ESCs have been being improved more rapidly than batteries and traditional capacitors, allowing their use in applications that go from renewable energy to microscopic electronics [5,6]. For example, your mobile phone may have better sound and flash that works at distances ten times longer than before because the previous conventional capacitor was replaced by a supercapacitor. Thus, this alternative approach of recycling polymer wastes and converting them into functional carbon materials is an intelligent strategy to mitigate pollution and to contribute with new technologies for renewable energy. From the technological point of view, the development of ESCs, from a sustainable route, suitable for high-power applications has raised much excitement and speculation throughout the manufacturing industry.

Inspired by this innovative field, this chapter starts resuming the importance of supercapacitors in society, their added value, and basic notions regarding their configurations and properties. Next, we resume the different types of polymeric wastes and the recycling processes of plastics. The subsequent section deals with several synthetic methods used to convert polymeric waste into carbon nanostructures for supercapacitors. Finally, the future directions and challenges related to the polymeric recycling technology to convert the trash into carbon materials for supercapacitors are presented.

27.2 MARKET VALUE

The global market for electric supercapacitors of the type electric double-layer capacitor (EDLC) was valued at USD 980 million in 2019 and USD 834 million in 2020. There was around 17% market reduction in 2020 caused by the pandemic of COVID-19. The pandemic situation has been ruining several countries around the world and causes severe challenges to energy storage companies [7]. The EDLC manufacturers must also face fluctuations in demand for raw materials, resulting in disruptive effects on production, revenues, and supply. However, the industry can have a fast development due to mass vaccination and the heating of the economy. The market is anticipated to grow with a healthy growth rate of more than 21.8% over the forecast period 2021–2026 [8]. The manufacture of supercapacitors is focused on East Asia, representing 55% of the world's production. North America is the second-largest producer (28%), followed by Europe (7%) [9]. These EDLCs find extensive applications in industries such as the transportation and consumer electronics sectors. The key market trends are the surging demand for EDLCs in the automotive industry [10]. The world is seeing a major shift in the automobile market from conventional oil-powered engines to electric and hybrid automotive [11]. It generates the demand for electrical equipment in automotive production. Manufacturers of fuel cells are keen on using the inherent advantages of supercapacitors to create better products. In this way, the increasing demand for energy storage devices with high-end features, technological advancements, low cost, and safety devices are urgent and strategic in the industry of electric supercapacitors [12,13]. Thus, the development of ESCs suitable for high-power applications has raised much excitement and speculation throughout the manufacturing industry.

Companies worldwide can produce and commercialize supercapacitor devices based on several technologies, governmental standards, and regulations in different countries concerning gas emission control. In the future, it tends to be unified, focusing on reducing waste generation, encouraging recycling, and using clean energy. The transformation of industrial wastes in technological products with commercial value is one of the factors expected to drive the growth of the market, opening new opportunities to use domestic and industrial wastes to produce carbon-based materials for supercapacitor application. As widely known, the transformation of polymeric wastes into high-value materials is a big incentive for polymeric recycling. The reuse even saves energy compared to the manufacturing of virgin materials. Also, evolving green energy applications, advancements in technologies for ESCs production, improving price/performance ratios, and the growth of new applications across several industries are driving the market for EDLCs [14].

27.3 CLASSIFICATION OF ENERGY STORAGE DEVICES

Efficient materials are increasingly needed for energy storage, meeting the growing demand for energy consumption. ESCs have become the most promising candidates among other energy storage systems because they can provide not only a higher energy density than conventional dielectric capacitors, but also a higher power density compared to batteries (Ragone plot graph, Figure 27.1a). ESCs are a type of efficient electricity storage system that stores electrical energy at the interface between a solid electrode and an electrolyte (Figure 27.1b). The mechanisms in which supercapacitors store and release charges are completely reversible, making them able to withstand a large number of charge/discharge cycles without any visible degradation (Figure 27.1c–e). It is worth mentioning that ESCs can also store or release energy very quickly and can operate over a wide range of temperatures [15].

Despite their relatively low storage capacity, ESCs have a high potential to be used in commercial applications such as consumer electronics, backup memory systems, and mobile electrical systems. Thus, an ESC with a bigger loading storage capacity is attractive for a wide range of applications. ESCs are generally assembled using

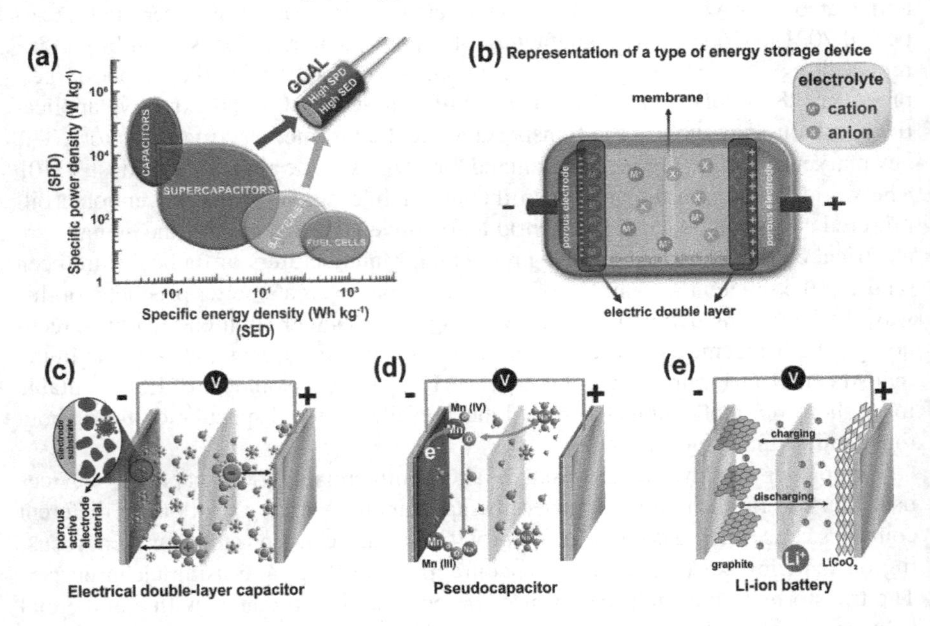

FIGURE 27.1 Schematic representation of (a) Ragone plot for energy storage and conversion devices, (b) ESC storage devices; (c) electric double-layer capacitor, (d) pseudocapacitor, and (e) LIBs.

electrodes of specific materials, which are capable of (i) forming an electric double layer or the non-faradaic process on their surfaces, which leads to capacitance, and (ii) being subjected to processes of surface oxidation/reduction, generating the separation of the charge at the electrolyte interface (faradaic electrode process) or the so-called pseudocapacitance. Thus, the energy stored in ESCs can be of capacitive or pseudocapacitive nature (Figure 27.1c and d). It is possible to create hybrid supercapacitors by combining the properties of EDLCs (e.g., excellent stability and power performance) with the high specific capacitance of the pseudocapacitors. Moreover, it is possible to increase the cell voltage by combining the correct electrodes, leading to an improvement in power and energy density, obtaining a supercapacitor with excellent performance. The development of hybrid electrodes appears as an alternative to EDLC-type electrodes and faradaic processes [16]. The ESCs shown in Figure 27.1b are an intermediate energy storage devices system between capacitors and batteries, which have the properties of large capacitance, characteristics of higher energy density and greater storage capacity at a lower cost. These particular properties make ESCs suitable for several applications such as power electronics, spatial, and military purposes. ESCs also have a huge demand for hybrid electric vehicles (HEVs), helping in the starting and stopping functions by providing peak power for improved acceleration for energy recovery.

As known, the electrochemical properties of electrode materials mainly depend, among others, on the following factors: (i) appropriate size; (ii) morphology of the material; (iii) stable chemical structure; (iv) adequate surface area, and (v) high conductivity [17–19]. Although researchers have already studied a considerable number of materials, they have not yet found an electrode material that meets all the conditions above with excellence. Therefore, researchers have put a lot of effort into developing carbon materials with properties that are as close to those five desired ones as possible.

Currently, the electrochemical efficacy of supercapacitors is determined by active substances in the electrode, mainly composed of carbon derivatives. Also, it is worth mentioning that the continuous development of electronic equipment requires safer, more stable, and high-capacity materials to be used in the electrodes. The reduction in costs and the limitations in the manufacturing methods (pollutants and toxic by-products from the processes) are other fertile fields for advances, discovering processes that are not only suitable for execution in the laboratory, but scalable for industrial production.

27.4 TYPES AND RECYCLING METHODS OF POLYMER WASTES

Polymer waste has been a suitable source to produce -value-added carbons for supercapacitors since it is a low-cost raw material rich in carbon elements. Converting polymer waste into value-added carbon is a fascinating alternative to reuse the already used polymeric materials and to decrease the environmental pollution caused by their accumulation.

Waste derived from polymer can be classified into two categories: natural and synthetic (Figure 27.2). Natural polymers (e.g., silk, cellulose, wood, starch, and chitin) occur in nature, and they derive from many sources (e.g., plants, animals, and microorganisms). They are often water based and less harmful to the environment due to their biodegradability. Synthetic polymers, on the other hand, are human-made polymers, derived, in most cases, from petroleum oil (non-renewable sources). Nylon, polytetrafluoroethylene (PTFE), epoxy, poly(ethylene terephthalate) (PET), polyethylene (PE), and polypropylene (PP) are examples of synthetic polymers. They can be subdivided into two classes: elastomers and plastics. The second class is responsible for a major environmental impact on the world, so it will be the focus of this chapter.

Plastics can be subdivided into thermoset and thermoplastic (Figure 27.2). Polyurethane (PU), silicone, vinyl esters, and epoxy resins are examples of thermosets. After heating, these polymers soften and undergo a curing process, forming cross-links between the polymer chains. Consequently, the thermosets cannot be remolded. In this case, further heating will only degrade the polymer because the inter- and intramolecular bonds are covalent, making this initial processing an irreversible chemical transformation. Thermoplastics are polymers capable of softening and flowing when subjected to increased temperature and pressure. Without the influence of these factors, the material will solidify according to the given shape. If temperature and pressure are reapplied, the material will soften and flow again because, in this case, the curing process is completely reversible due to the absence of a chemical bond between the chains. This characteristic allows thermoplastic materials to be reshaped and recycled without negatively affecting the physical properties of the polymer [20]. For instance, polyamides (PAs), PE, PP, polycarbonates (PCs), expanded polystyrene (EPS), polystyrene (PS), and polyvinyl chloride (PVC) are included in this class of polymer.

The recycling processes for plastic polymers are classified into four types, as shown in Figure 27.3. The primary recycling involves the mechanical reprocessing of scrap material with controlled history for the production of materials with equivalent properties. The secondary recycling is also a mechanical process for materials used in products that require inferior properties. In tertiary recycling, waste is chemically processed and leads to the recovery of valuable chemical constituents such as monomers, carbons, or additives. In quaternary recycling, the process requires the energy use of the material to be recycled, which means that recycling refers to the recovery of energy [21]. These four recycling processes can

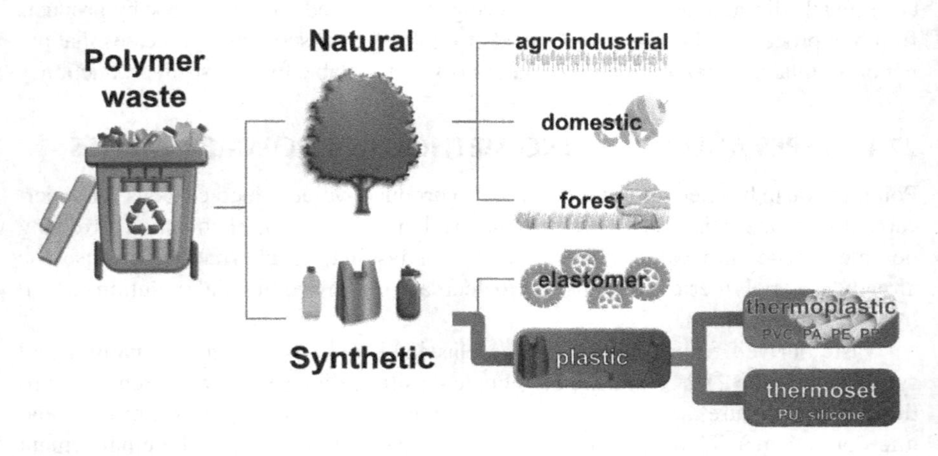

FIGURE 27.2 Types of polymer wastes.

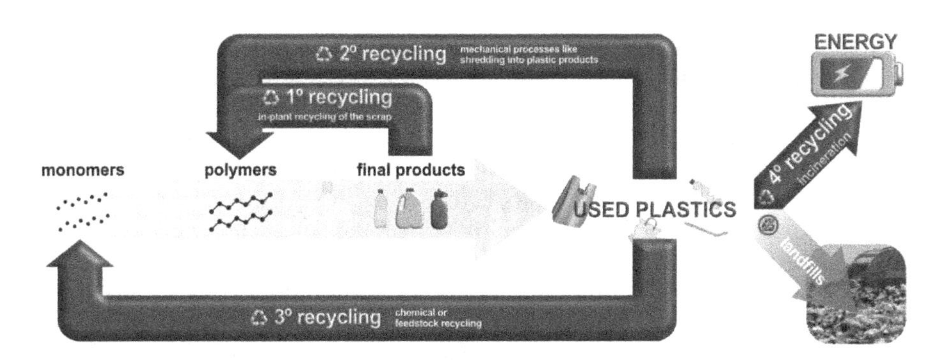

FIGURE 27.3 Four types of polymeric waste recycling.

be used in thermoplastic polymers, but thermosets are only suitable for reusing purposes (zero-order recycling) due to their type of bond that causes the degradation of the material before reaching the melting temperature. Meanwhile, thermoplastic polymers form physical bonds between the chains, which makes it possible to melt the material several times.

Chemical recycling is an advanced technology that transforms plastic polymers into smaller molecules, mostly liquids or gases, owing to the total or partial depolymerization of the polymer. The monomers can be polymerized one more time, recomposing the original polymer, or redirected into other applications such as fuels [22]. The tertiary recycling process yields products consisting of high carbon contents with minor or no heteroatom contribution (depending on the polymer waste source), which is valuable for ESCs as electrodes. Several methods can be used in chemical recycling: hydrogenation, gasification, chemical depolymerization, pyrolysis, catalytic fracture, photodegradation, and degradation in a microwave reactor. Furthermore, it represents an effective alternative to overcome environmental issues because it reduces the production of new wastes [21].

27.5 POLYMER WASTES MANAGEMENT FOR SUPERCAPACITORS

The criteria for producing a high-performance supercapacitor electrode include high specific capacitance, large rate capability, and high cycle stability. Moreover, the cost of the production of the active material is still the major challenge that needs to be overcome. Only then this material will become commercially viable. Thus, developing ideal devices with excellent electrochemical properties is a challenge for researchers. In this way, carbon materials have been the most plentiful resource for EDCL use. Their excellent properties such as high surface area (greater than 1,500 m^2/g), high porosity (micro- and mesoporous), high conductivity (greater than 1.0 S/cm), high purity, and cost–performance make these materials valuable for supercapacitor application [1]. In general terms, the first three of these properties are critical to the development of supercapacitor electrodes and they can be manipulated and optimized during the synthesis process.

The properties of carbon-based electrodes for supercapacitor applications are strongly dependent on the raw material and the synthesis process. The chemical composition of the polymer waste sources and their structural behavior may limit some features, for example the graphitization degree. For instance, the literature reports that biomass (wood, nutshells, plant, etc.), some coals, and polymers (polyvinylidene chloride, PVDC) cannot be converted to graphite under high temperatures [23,24]. In another way, the presence of trace amounts of inorganic minerals in some wastes can act as the porosity activator agent [25]. Also, some sources of polymer waste contain heteroatoms. In this case, the heteroatom acts as the doping agent in the final carbon and can improve the capacitance and the total power density of the electrode due to the possible faradaic process that occurs on the particle surface.

The special electronic structure of carbon allows this element to form a wide range of structures, known as "allotropy". Diamond, amorphous carbon, and graphite are three typical known examples of natural allotropes of carbon. However, various new allotropic forms have been derived from synthetic processes, including graphene (Gr), carbon nanotubes (CNTs), fullerenes, and carbon-based quantum dots (CQDs) [26–28]. They can be classified into zero (0D), one (1D), two (2D), and three (3D) dimensions, and they are usually produced using processes based on pyrolysis (Figure 27.4). Each one of these spatial dimensions possesses unique advantages and disadvantages that contribute to the electrochemical performance.

0D carbons are materials in which all the dimensions are measured within the nanoscale. It can be represented by carbon dots, carbon nanoparticles, and so forth. In 1D carbon nanomaterials, one dimension is outside the nanoscale. It allows several types of morphology, including nanotubes and nanofibers. The structure of such materials has high length-to-diameter aspect ratios and oriented growth direction, which provides an efficient transport pathway for both electrons and ions, besides similar benefits of 0D nanostructures [29,30]. 2D carbon nanomaterials are better described by graphene and many other layered materials. They have two dimensions outside the nanoscale. Layered nanomaterials possess the ability to intercalate ions in their interlayer, providing an effective pathway to ionic transfers, being highly

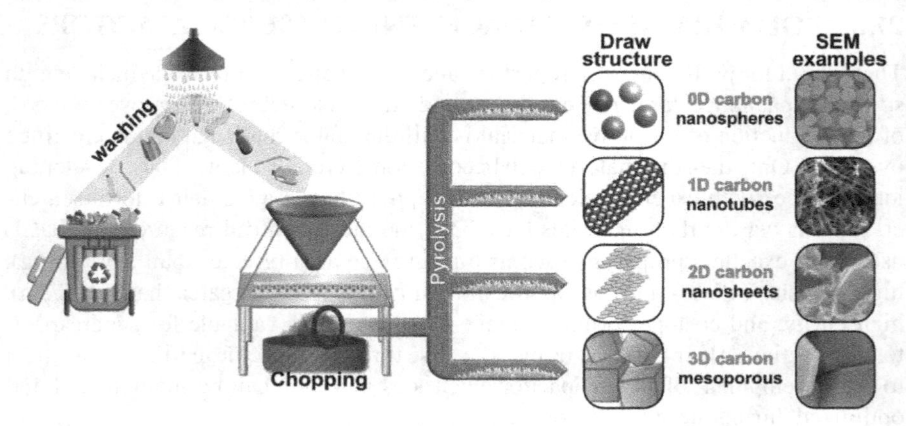

FIGURE 27.4 Schematic diagram of the main method (pyrolysis) used to convert polymer wastes into 0D–3D carbon materials.

desirable as electrodes for supercapacitors. Finally, mesoporous carbon, hydrogel, and aerogel are classified as 3D carbon nanostructures (no dimension is confined to the nanoscale). Among the dimensions aforementioned, 3D is the most studied in the supercapacitor field. The great interest in materials in three dimensions is because these structures can overcome the disadvantages inherent in 0D, 1D, and 2D materials, such as easy aggregation and non-adjustable specific surface area [30,31]. These materials are unquestionable choices for supercapacitor application due to their intrinsic characteristics, including high specific surface area, high porosity, and high chemical and thermal stability. As electrodes for supercapacitors, nanoporous carbon offers abundant electrochemical active sites to accommodate numerous charges besides providing a fast charge/discharge route, improving the performance of the device [32].

Several methodologies are used to convert plastic waste into carbon-based nanostructures, including chemical vapor deposition (CVD), pyrolysis–activation methods, template methods, hydrothermal carbonization (HTC), spray pyrolysis, and activation (physical and chemical) [33]. Some preparation processes can be of high cost, use dangerous chemicals, and have low production yield.

27.5.1 Chemical Vapor Deposition

Chemical vapor deposition is a synthesis process in which the chemical constituents react in the vapor phase near or on a heated substrate to form a solid deposit. It is a versatile technology to develop new carbon nanostructures such as graphene, nanotubes, nanofibers, and fullerenes. CVD has proven to be the preferable method for the mass production of carbon nanotubes from plastic waste. In the process, hydrocarbon precursors are deposited onto a substrate in the presence of a catalyst. The main advantages of this technique are the high yield of nanotubes and a lower temperature requirement. However, the disadvantages are related to the high consumption of electrical power and the injection of H_2, which is a highly inflammable gas becomes a key challenge to enable the use of this method in industries [34,35].

27.5.2 Hydrothermal Carbonization

Hydrothermal carbonization is a process where carbonaceous precursors are usually dispersed in an aqueous solution and heated at temperatures between 120 °C and 260 °C. The solution containing the precursor is placed in an autoclave with autogenerated pressure in the absence of a chemical catalyst. Nevertheless, during the process, the dielectric constant and polarity of the water change, which acts as a catalyst. Few works reported the hydrothermal carbonization of plastics [36–38]. Plastic wastes have high thermal stability. Therefore, a small amount of decomposition is obtained by the HTC method. Furthermore, most of the carbon materials obtained by HTC derive from biomass. The advantage of this method is that it can take place at low reaction conditions compared to the pyrolysis method. However, HTC is mostly influenced by the feedstock type because precursors with complex structures require excessively high temperature and pressure, making the process unfeasible for large-scale uses [39].

27.5.3 Pyrolysis

Pyrolysis is a commonly used method to convert polymer waste into carbon-based nanomaterials. There are two types of pyrolysis: at low temperatures and at high temperatures. Both are performed in an inert atmosphere. Carbonaceous waste can be converted into carbon materials using the method of pyrolysis at high temperatures (usually higher than 600 °C). Under high temperatures, volatile organic compounds are decomposed and high yields of carbon are reached. The pyrolysis process might occur with or without a catalyst. The most commonly used catalysts used to enhance the efficiency of the process was Cu-Al_2O_3, zeolites, $Al(OH)_3$, $Ca(OH)_2$, and Fe_2O_3, and others. The catalysts have a critical role in reducing the temperature and time involved in the process, improving the pyrolysis processing of plastic waste [40]. From economic and commercial points of view, pyrolysis has been considered a sustainable method to convert polymer waste feedstocks into several carbon nanostructures (0D, 1D, 2D, and 3D). The possibility of conducting the process as a batch, the use of low pressure, and minimal feedstock preprocessing are some advantages of this technique. However, the process has high operational costs and requires high investments. Nevertheless, it is still considered the main technique to convert waste into valuable products by the industries. Some properties such as porosity and specific surface area can be controlled by the adjustment of the carbonization temperature. This simple method can produce a large number of carbon networks composed of interconnecting porous channels that depend on the polymer waste source. However, for some waste sources, the high temperature can limit the specific surface area and the porosity, with a further activation step being necessary [41].

27.5.4 Chemical and Physical Activations

Chemical and physical activations are two processes used in most researches to produce nanoporous carbon materials from carbon-rich precursors. In the physical activation method, the carbonization of the precursor is followed by the activation in a carbon dioxide or steam atmosphere. Although CO_2 is preferred in physical activation, steam has shown to be of low cost and pollution-free. During the process, the temperature can reach 1,100 °C. It has been considered a good method to obtain carbon since it is free of chemical additives. On the other hand, chemical activation is more common and involves a one-step heat treatment in the presence of activating agents (acids, bases, or salts). Generally, these chemical agents are used after the pyrolysis method to introduce hierarchical nano- and/or mesopores into the carbonaceous materials. But they can also be used before. The temperature used in chemical activation depends on the precursor and the catalyst. However, the common range is between 450 °C and 900 °C [40,42]. Several chemical agents such as zinc chloride ($ZnCl_2$), potassium permanganate ($KMnO_4$), potassium hydroxide (KOH), and calcium carbonate ($CaCO_3$) are used as activating agents. In particular, KOH is the mostly employed activating agent for deriving carbons from wastes for supercapacitor applications. The high specific surface area in association with high porosity enhances the electrochemical performance of carbon nanomaterials.

There are excellent reviews that suggest the mechanism of KOH activation to produce nanopores [42,43]. However, it is still not completely understood due to the complexity of experimental variables.

27.6 GENERAL CONCLUSIONS AND FUTURE PERSPECTIVES

Our work highlights the importance of the technological development of ESCs. They have properties that lie in between batteries and traditional capacitors, but ESCs are being improved more rapidly than the other two. ESCs have a huge diversity of applications due to their versatile storage energy capacity. However, since the last decade, ESCs have been of demand mainly in the automotive industry. The world is seeing a major shift in the automobile market from conventional oil-powered engines to electric and hybrid automotive. Manufacturers of fuel cells are keen on using the inherent advantages of ESCs to create better products. Accordingly, polymer waste recycling is an emergent alternative to overcome environmental issues, pollution, and energy crisis. Many methodologies such as chemical vapor deposition, pyrolysis–activation methods, template methods, hydrothermal carbonization, and activation (physical and chemical) can be applied in the preparation of a wide variety of carbon-based electrodes. Pyrolysis is the most commonly used method to convert polymer waste into carbon-based nanomaterials such as nanofibers, nanotubes, graphene, and mesoporous carbon. However, the process still involves high operational and investment costs. Thus, there is still much research to be carried out for advancements in the development of ESC materials, mainly on topics such as safe energy, reduced costs, and reduced pollutants and toxic substances. The production of materials using the principles of green chemistry will reduce the impacts on the environment and human beings. Therefore, ESCs became a fertile field for advances regarding cost reduction and manufacturing limitations. The increasing demand for energy storage devices, high-end features, and technological advancements, low cost, safety devices, and reuse of domestic and industrial wastes are the primary factors expected to drive the growth of the market. However, governmental standards and regulations in different countries concerning gas emission control with a focus on reducing waste generation, encouraging recycling, and using clean energy are secondary factors expected to drive the growth of the market.

Therefore, the future works should concentrate on the methodologies of producing of advanced materials from domestic and industrial wastes, with a focus on increased efficiency and low cost.

ACKNOWLEDGMENTS

The authors acknowledge the Brazilian agencies Sao Paulo Research Foundation (FAPESP) (grant numbers: 2012/07067-0, 2013/23572-0, 2016/019405, and 2013/07296) for financial support and the concession of a scholarship. Special thanks are due to CEPID (2013/07296-2), INCTMN (2008/57872-1), and CNPq (573636/2008-7, 435975/2018-8, 309711/2019-3, and 421648/2018-0). This study was financed in part by the Coordination for the Improvement of Higher Education

Personnel (CAPES – Brazil) – Finance Code 001, CAPES:CAPES-EPIDEMIAS (Programa Estratégico Emergencial de Prevenção e Combate a Surtos, Endemias, Epidemias e Pandemias Número do Processo: 88887.513223/2020-00).

REFERENCES

1. W. Utetiwabo, L. Yang, M.K. Tufail, L. Zhou, R. Chen, Y. Lian, W. Yang, Electrode materials derived from plastic wastes and other industrial wastes for supercapacitors, *Chinese Chem. Lett.* 31 (2020) 1474–1489. https://doi.org/https://doi.org/10.1016/j.cclet.2020.01.003.
2. N.V. Challagulla, M. Vijayakumar, D. Sri Rohita, G. Elsa, A. Bharathi Sankar, T. Narasinga Rao, M. Karthik, Hierarchical activated carbon fibers as a sustainable electrode and natural seawater as a sustainable electrolyte for high-performance supercapacitor, *Energy Technol.* 8 (2020) 2000417. https://doi.org/https://doi.org/10.1002/ente.202000417.
3. L. Chang, Y.H. Hu, Breakthroughs in designing commercial-level mass-loading graphene electrodes for electrochemical double-layer capacitors, *Matter.* 1 (2019) 596–620. https://doi.org/https://doi.org/10.1016/j.matt.2019.06.016.
4. L. Wei, G. Yushin, Nanostructured activated carbons from natural precursors for electrical double layer capacitors, *Nano Energy.* 1 (2012) 552–565. https://doi.org/https://doi.org/10.1016/j.nanoen.2012.05.002.
5. S. Manopriya, K. Hareesh, The prospects and challenges of solar electrochemical capacitors, *J. Energy Storage.* 35 (2021) 102294. https://doi.org/https://doi.org/10.1016/j.est.2021.102294.
6. P. Sharma, T.S. Bhatti, A review on electrochemical double-layer capacitors, *Energy Convers. Manag.* 51 (2010) 2901–2912. https://doi.org/https://doi.org/10.1016/j.enconman.2010.06.031.
7. EDLC (2021) Rep. Electr. Double-Layer Capacit. Mark. Is Segmented by End User (Consumer Electron. Ind. Automotive), Geogr. (n.d.). https://www.mordorintelligence.com/industry-reports/electric-double-layer-capacitor-market#:~:text=Market Overview, the forecast period 2021–2026 (accessed April 30, 2021).
8. 360 Market Updates (2021) Glob. Electr. Double-Layer Capacit. Mark. Size 2021 Res. Reports Collect Inf. Useful Extensive, Tech. Mark. Commer. Study Futur. Growth by 2026. (n.d.). https://www.thecowboychannel.com/story/43184727/global-electric-double-layer-capacitor-edlc-market-size-2021-research-reports-collect-information-useful-for-the-extensive-technical-market-oriented (accessed April 30, 2021).
9. D.P. Harrop (2021) Rep. Electrochem. Double Layer Capacit. Supercapacitors 2015–2025 Ultracapacitors, EDLC, Electrochem. Capacit. Supercabatteries, AEDLC Electron. Electr. Eng. (n.d.). https://www.idtechex.com/en/research-report/electrochemical-double-layer-capacitors-supercapacitors-2015-2025/378 (accessed April 30, 2021).
10. M. Hedlund, J. Lundin, J. De Santiago, J. Abrahamsson, H. Bernhoff, Flywheel energy storage for automotive applications, *Energies.* 8 (2015). https://doi.org/10.3390/en81010636.
11. O. Bethoux, Hydrogen fuel cell road vehicles: State of the art and perspectives, *Energies.* 13 (2020). https://doi.org/10.3390/en13215843.
12. Q. Zhang, J.-Q. Huang, W.-Z. Qian, Y.-Y. Zhang, F. Wei, The road for nanomaterials industry: A review of carbon nanotube production, post-treatment, and bulk applications for composites and energy storage, *Small.* 9 (2013) 1237–1265. https://doi.org/https://doi.org/10.1002/smll.201203252.

13. H. Peng, X. Sun, W. Weng, X. Fang, 6 – Energy Storage Devices Based on Polymers, in: *Book Polymer Materials for Energy and Electronic Applications*, H. Peng, X. Sun, W. Weng, X.B.T.-P.M. for E. and E.A. Fang (Eds.), Academic Press, 2017: pp. 197–242. https://doi.org/https://doi.org/10.1016/B978–0–12–811091–1.00006–9.
14. D. Majumdar, M. Mandal, S.K. Bhattacharya, Journey from supercapacitors to supercapatteries: Recent advancements in electrochemical energy storage systems, *Emergent Mater.* 3 (2020) 347–367. https://doi.org/10.1007/s42247-020-00090–5.
15. P. Staiti, F. Lufrano, A study of the electrochemical behaviour of electrodes in operating solid-state supercapacitors, *Electrochim. Acta.* 53 (2007) 710–719. https://doi.org/https://doi.org/10.1016/j.electacta.2007.07.039.
16. Z.S. Iro, C. Subramani, S.S. Dash, A brief review on electrode materials for supercapacitor, *Int. J. Electrochem. Sci.* 11 (2016) 10628–10643. https://doi.org/10.20964/2016.12.50.
17. J. Luo, J. Wang, S. Liu, W. Wu, T. Jia, Z. Yang, S. Mu, Y. Huang, Graphene quantum dots encapsulated tremella-like $NiCo_2O_4$ for advanced asymmetric supercapacitors, *Carbon N. Y.* 146 (2019) 1–8. https://doi.org/https://doi.org/10.1016/j.carbon.2019.01.078.
18. A. Ray, A. Roy, S. Saha, M. Ghosh, S. Roy Chowdhury, T. Maiyalagan, S.K. Bhattacharya, S. Das, Electrochemical energy storage properties of Ni-Mn-oxide electrodes for advance asymmetric supercapacitor application, *Langmuir.* 35 (2019) 8257–8267. https://doi.org/10.1021/acs.langmuir.9b00955.
19. B. Li, F. Dai, Q. Xiao, L. Yang, J. Shen, C. Zhang, M. Cai, Nitrogen-doped activated carbon for a high energy hybrid supercapacitor, *Energy Environ. Sci.* 9 (2016) 102–106. https://doi.org/10.1039/C5EE03149D.
20. W.D. Callister, *Materials Science and Engineering : An Introduction*, John Wiley & Sons, New York, 2007.
21. M. Okan, H.M. Aydin, M. Barsbay, Current approaches to waste polymer utilization and minimization: A review, *J. Chem. Technol. Biotechnol.* 94 (2019) 8–21. https://doi.org/https://doi.org/10.1002/jctb.5778.
22. V. Sinha, M.R. Patel, J.V. Patel, Pet waste management by chemical recycling: A review, *J. Polym. Environ.* 18 (2010) 8–25. https://doi.org/10.1007/s10924-008-0106–7.
23. K. László, A. Bóta, L.G. Nagy, I. Cabasso, Porous carbon from polymer waste materials, *Colloids Surfaces A Physicochem. Eng. Asp.* 151 (1999) 311–320. https://doi.org/https://doi.org/10.1016/S0927–7757(98)00390–2.
24. A.G. Pandolfo, A.F. Hollenkamp, Carbon properties and their role in supercapacitors, *J. Power Sources.* 157 (2006) 11–27. https://doi.org/https://doi.org/10.1016/j.jpowsour.2006.02.065.
25. Q. Ma, Y. Yu, M. Sindoro, A.G. Fane, R. Wang, H. Zhang, Carbon-based functional materials derived from waste for water remediation and energy storage, *Adv. Mater.* 29 (2017) 1605361. https://doi.org/https://doi.org/10.1002/adma.201605361.
26. K.D. Patel, R.K. Singh, H.-W. Kim, Carbon-based nanomaterials as an emerging platform for theranostics, *Mater. Horizons.* 6 (2019) 434–469. https://doi.org/10.1039/C8MH00966J.
27. J.R. Siqueira, O.N. Oliveira, 9 – Carbon-Based Nanomaterials, in: *Book Nanostructures*, A.L. Da Róz, M. Ferreira, F. de Lima Leite, O.N.B.T.-N. Oliveira (Eds.), William Andrew Publishing, 2017: pp. 233–249. https://doi.org/https://doi.org/10.1016/B978–0–323–49782–4.00009–7.
28. R. Rauti, M. Musto, S. Bosi, M. Prato, L. Ballerini, Properties and behavior of carbon nanomaterials when interfacing neuronal cells: How far have we come? *Carbon N. Y.* 143 (2019) 430–446. https://doi.org/https://doi.org/10.1016/j.carbon.2018.11.026.
29. P.M. Visakh, M.J.M. Morlanes, *Nanomaterials and Nanocomposites: Zero-to Three-Dimensional Materials and Their Composites*, John Wiley & Sons, United States, 2016.

30. Z. Yu, L. Tetard, L. Zhai, J. Thomas, Supercapacitor electrode materials: Nanostructures from 0 to 3 dimensions, *Energy Environ. Sci.* 8 (2015) 702–730. https://doi.org/10.1039/C4EE03229B.
31. S. Kumar, G. Saeed, L. Zhu, K.N. Hui, N.H. Kim, J.H. Lee, 0D to 3D carbon-based networks combined with pseudocapacitive electrode material for high energy density supercapacitor: A review, *Chem. Eng. J.* 403 (2021) 126352. https://doi.org/https://doi.org/10.1016/j.cej.2020.126352.
32. V.H. Fragal, E.H. Fragal, T. Zhang, X. Huang, T.S.P. Cellet, G.M. Pereira, A. Jitianu, A.F. Rubira, R. Silva, T. Asefa, Deriving efficient porous heteroatom-doped carbon electrocatalysts for hydrazine oxidation from transition metal ions-coordinated casein, *Adv. Funct. Mater.* 29 (2019) 1808486. https://doi.org/https://doi.org/10.1002/adfm.201808486.
33. X.-L. Zhou, H. Zhang, L.-M. Shao, F. Lü, P.-J. He, Preparation and application of hierarchical porous carbon materials from waste and biomass: A review, *Waste Biomass Valoriz.* 12 (2021) 1699–1724. https://doi.org/10.1007/s12649-020-01109–y.
34. Y.M. Manawi, Ihsanullah, A. Samara, T. Al-Ansari, M.A. Atieh, A review of carbon nanomaterials' synthesis via the chemical vapor deposition (CVD) method, *Mater. (Basel, Switzerland).* 11 (2018) 822. https://doi.org/10.3390/ma11050822.
35. N. Mishra, S. Shinde, R. Vishwakarma, S. Kadam, M. Sharon, M. Sharon, MWCNTs synthesized from waste polypropylene plastics and its application in super-capacitors, *AIP Conf. Proc.* 1538 (2013) 228–236. https://doi.org/10.1063/1.4810063.
36. X. Zhao, L. Zhan, B. Xie, B. Gao, Products derived from waste plastics (PC, HIPS, ABS, PP and PA6) via hydrothermal treatment: Characterization and potential applications, *Chemosphere.* 207 (2018) 742–752. https://doi.org/https://doi.org/10.1016/j.chemosphere.2018.05.156.
37. J. Poerschmann, B. Weiner, S. Woszidlo, R. Koehler, F.-D. Kopinke, Hydrothermal carbonization of poly(vinyl chloride), *Chemosphere.* 119 (2015) 682–689. https://doi.org/https://doi.org/10.1016/j.chemosphere.2014.07.058.
38. M.E. Iñiguez, J.A. Conesa, A. Fullana, Hydrothermal carbonization (HTC) of marine plastic debris, *Fuel.* 257 (2019) 116033. https://doi.org/https://doi.org/10.1016/j.fuel.2019.116033.
39. Y. Shen, A review on hydrothermal carbonization of biomass and plastic wastes to energy products, *Biomass and Bioenergy.* 134 (2020) 105479. https://doi.org/https://doi.org/10.1016/j.biombioe.2020.105479.
40. S. Joseph, G. Saianand, M.R. Benzigar, K. Ramadass, G. Singh, A. Gopalan, J.H. Yang, T. Mori, A.H. Al-Muhtaseb, J. Yi, Recent advances in functionalized nanoporous carbons derived from waste resources and their applications in energy and environment, *Adv. Sustain. Syst.* 5 (2021) 2000169.
41. A. Dutta, J. Mahanta, T. Banerjee, Supercapacitors in the light of solid waste and energy management: A review, *Adv. Sustain. Syst.* 4 (2020) 2000182. https://doi.org/https://doi.org/10.1002/adsu.202000182.
42. Z. Heidarinejad, M.H. Dehghani, M. Heidari, G. Javedan, I. Ali, M. Sillanpää, Methods for preparation and activation of activated carbon: A review, *Environ. Chem. Lett.* 18 (2020) 393–415. https://doi.org/10.1007/s10311-019-00955–0.
43. J. Wang, S. Kaskel, KOH activation of carbon-based materials for energy storage, *J. Mater. Chem.* 22 (2012) 23710–23725. https://doi.org/10.1039/C2JM34066F.

28 Supercapacitors Based on Waste Generated in Automobiles

Souhardya Bera and Subhasis Roy
University of Calcutta

CONTENTS

28.1 INTRODUCTION

Personal prosperity and success are considerable contributors to increasing material consumption, and the automotive industry belongs to one of the most lucrative consumer markets. There are various factors for why one buys an automobile [1], but an increase in disposable income and easy finance has been seen to promote automobile sales lately [2]. But as the vehicle population increases, the number of cars and subsequent parts coming in for scraping, recycling, and reusing is expected to increase consequently. But then, automobiles are considered as one of the most recyclable material consumer products. Being such a big contributor to customer consumption, automobiles require an organized, systematic, and modern end-of-life vehicle recycling industry adjacent to the thriving automotive industry. Hereby, the industry is highly regulated in terms of waste management [3], because 95% of automobiles go through the recycling process at the end of their usable life [4].

DOI: 10.1201/9781003178354-32

As a part of constructing automobiles, more emphasis needs to be provided on creating more materials to recycle and fewer materials waste to dispose of, consequently increasing the viability of the automotive industry. While recycling is already conceived as a less viable option by many, technologies developed to increase automobiles' fuel efficiency, if found adversely affecting the industry, may be deemed risky on implementation. Hence, continuous support and emphasis on the automobile recycling industry is of paramount importance [5]. Presently, different governments implementing various regulations to reduce the negative environmental impact of the vehicles and the associated disposable batteries, rubber, and oil are propelling the growth of the recycling industry. Manufacturers have already started reusing waste materials to resolve the shortage in the supply of resources during the manufacturing process, allowing them to address environmental degradation and resource depletion. In this context, storing energy is of great interest and researchers are searching for energy storage systems that are renewable, sustainable, and efficient. An indispensable part of this clean energy-storage portfolio is electrochemical energy storage. It benefits from the fact that electrical and chemical energy has the same charge carrier, the electron. Batteries, capacitors, and fuel cells are usually taken as the common forms of electrochemical energy storage and conversion.

Supercapacitors (SCs) are considerably recent developments in the field of electrochemical energy storage. Comparatively higher capacitance values, longer cycle life, higher power density, and nearly no maintenance cost make SCs [6–7] a definite choice for clean energy [8–9]. In cases where SCs are used complimentarily to batteries, results have proven to enhance the operating efficiency and overall performance of the energy storage system [10]. Application in diverse consumer electronics requires thinner, lighter, flexible, and transparent SCs [11]. Very recently, novelties have been assembled to utilize various solid waste materials from a wide range of sources, especially industries, agriculture, and electronics. The scenario of waste management remains to revolve around electrochemical devices such as batteries, and works on SCs have still been few and relatively new [12]. Primarily, SCs derived from automobile wastes are still a novelty and have a long way to make a sustainable presence in the energy storage market.

28.2 AUTOMOBILE WASTE – RECYCLING VS DUMPING

There remains a widespread misunderstanding regarding the extent to which automobiles can be recycled and that the motor vehicles are already part of the most recyclable product portfolio, with more than 75% by weight being recycled [13]. As prices of raw materials increase and countries become climate-conscious, focusing on waste treatment, disposal, and sustainability in operation becomes increasingly essential for all companies, ranging from large-scale conglomerates to small-scale enterprises. Reducing the resources needed to produce new materials and keeping wastes out of landfills are among the many advantages by which recycling benefits the environment, health, and safety. More than just their realization, environmental policies have forced institutions to condone their corporate responsibility seriously, which is just not developing product information from cradle-to-grave to "take

TABLE 28.1
Advantages and Disadvantages of Automobile Waste Recycling

Advantages	Disadvantages
1. Reduces pollution	1. Increased capital costs
2. Environment-friendliness	2. Unhygienic
3. Minimizes global warming	3. Doubts around material quality
4. Reduces landfill wastes and sites	4. Absence of widespread options
5. Conserves natural resources	5. Concerns around durability
6. Creates employment	
7. Reduces energy consumption	
8. Increases sustainability	

back and recycle" to develop new technologies. This might also need structural adjustments in the organization of the industry. Since these decisions are closely intertwined, most future strategies might incorporate waste management at an early stage to have a competitive advantage both in the manufacturing sector and in the consumer market (Table 28.1) [14].

28.3 SUPERCAPACITORS DERIVED FROM DIFFERENT AUTOMOBILE WASTES

28.3.1 From Waste Engine Oil (WEO)

Engine oil is an important part of an automobile, which can be regarded as the "blood" of an automobile, which plays a crucial role in lubrication and cleaning in the engine. In contrast, engine oil as waste is already known as toxic and hazardous. A number of degraded lubrication additives and undesired substances are present in the WEO, such as soot, complex mixture of aliphatic hydrocarbons, polycyclic aromatic hydrocarbons (PAHs), di-aromatic hydrocarbons, inorganic compounds, metallic particles, and organometallic species such as Cr, Pb, Ni, Mg, Zn, and Cu [15–19]. PAHs have also proven to be powerful carcinogens for humans [20]. Given the dreadful impact of WEO on the environment, improper disposal can have serious environmental consequences. Recycling, regenerating, pyrolysis [21], or combustion [22] for energy production have presently been attempted, but they come with associated environmental problems and high costs.

WEO contains 15–50 "C" atoms-based hydrocarbons, which can be used for their cost-efficiency and ready access in large quantities as a carbon precursor. Carbon precursors for manufacturing carbon-based products such as graphene [23], carbon nanotubes [24], carbon spheres [25,26], porous carbons [27], and carbon single crystals [28] are usually saturated and aromatic hydrocarbons. Therefore, the reuse of WEO has the potential for being an essential raw material for carbon materials, which would provide the most essential sustainable route and an environment-friendly approach to WEO utilization.

28.3.1.1 Hierarchical Porous Carbon Nanosheets (HPCNs)

Until adequately tuned, the usual electrode material consists of dominating micropore characteristics and tortuous channel structure, leading to a poor electrochemical performance. This pore size distribution requires enhanced structuring to improve the performance of SCs, and properties such as conductivity and surface wettability become important criteria to prepare porous carbon with a well-designed hierarchical pore structure.

Electrochemical storage applications have seen few works on HPCNs incorporating WEO as raw materials. Li et al. devised HPCNs from WEO and proved them to be an ideal electrode material for SC application [29].

The WEO was carbonized in the Ar atmosphere at 600° C–800 °C for 2h, and then porous carbon nanosheets with various KOH chemical activation ratios were produced. The manufactured HPCN displayed a high specific surface area of 2,276 m^2/g, which can be deemed relatively high. Although the value remains potentially lower than that of some biomass-derived porous carbon [30], WEO's massive supply at low cost can trigger its use. Similarly, this fabricated WEO-based HPCN can also act as a substitute to commercial activated carbon produced from wood and coal as raw materials [31,32]. The fabricated HPCN exhibited a distinctive capacitance of 352 F/g, while the current density increased from 0.5 to 20 A/g by maintaining more than 87.7% of the original capacities, signifying the enhanced wettability to the excellent rate capability achieved in the samples. High energy storage capability, good rate capability, and perfect style stability can be attributed to the superior performance. There was no apparent decrease in capacitance, even after 5,000 charge and discharge cycles, with capacitance retention reaching 99.6%, which can be credited to the excellent pore stability, hierarchical porous structure, adequate conductivity, and enhanced wettability.

28.3.1.2 Porous Carbon/ZnS Nanocomposite

Activated nanoporous carbon is an impressive material for SC applications [33–35] and has been obtained using various impressive, low-cost, and earth-abundant materials. Derived from various biomass sources, the activated porous carbon material proved to be an efficient electrode material, owing to its unique structure, high surface area, pore size, high electrocatalytic activity, and stability [36–38]. Moreover, to be a helpful electrode material, the raw material needs to be abundantly present, and the increasing abundance of vehicles has led to the growing availability of WEO.

WEO collected was exposed to carbonization at 700 °C for 4h under Ar atmosphere [39]. The sample incorporating carbon and ZnS as the primary component is expressed as porous carbon/ZnS nanocomposite (CZS). The carbonization, followed by acid wash using 10% piranha solution for 10min, helps eliminate the inorganic impurities, and the resultant sample is then distilled with water to bring the pH of the sample to 6.5–7. The sample is then identified as activated porous carbon/ZnS nanocomposite material (ACZS). The ACZS-based SC device displayed a high specific energy of 13.8 Wh/kg and specific power of 4.5 KW/kg. Even after 10,000 continuous charge–discharge cycles, the capacitance retention reached 95%, higher than various biomass raw materials such as grapefruit peels [40], wheat straw [41], protein [42], willow catkin fiber [43],

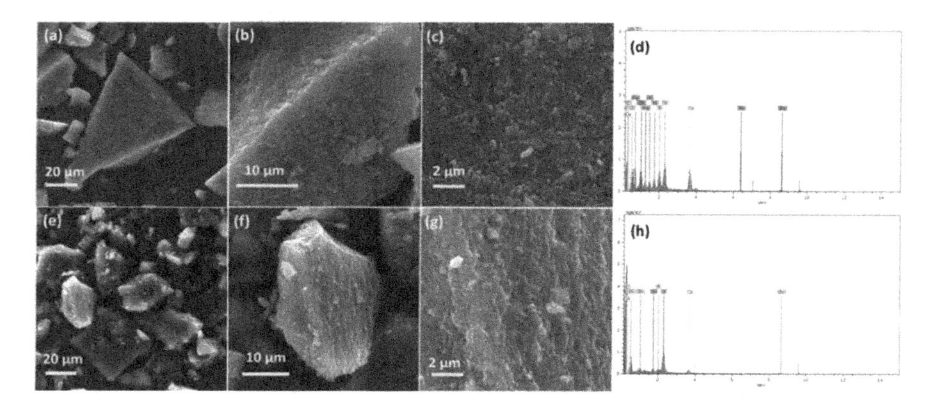

FIGURE 28.1 (a–c) SEM images of CZS at different magnifications, (d) SEM-EDS spectra of CZS material, (e–g) SEM images of ACZS at different magnifications, and (h) SEM-EDS spectra of ACZS material [39]. (Reprinted with permission. Copyright (2020) Elsevier.)

and swim bladders [44]. The unusually less bulk resistance in addition to low interfacial resistance indicates the good electrochemical charge storage properties and excellent electrochemical stability of ACZS-based SCs (Figure 28.1).

28.3.2 From Scrap Waste Tires

Given the precarious condition of environmental pollution and degradation due to disposing and dumping of presently non-recyclable items, a prominent suitor of this "Waste-to-Use" category is scrap waste tires from automobiles, which can reduce the enormous load on Earth's traditionally used resources. Being non-degradable [45], they stay in nature for a sufficiently long time, leading to contagious and unknown diseases. Utilizing waste tires for subtle value-added products can realize sustainable environmental mitigation policies. Waste tires were reused for new rubber products such as a playground, rubber-modified asphalts, doormats, and gaskets, and were recycled by grinding, crumbling, pyrolysis, and combustion [46]. And among them, the pyrolysis-based recycle strategy has received widespread attention due to its ability to affect the environment minimally [47].

Pyrolysis is a thermal decomposition process of waste tires in an oxygen-free environment. Thermally decomposed at 400 °C, waste tires produce by-products of 40–60 wt.% tire pyrolysis oil (TPO), 5–20 wt.% pyro-gas, and 30–40 wt.% pyro-char [48]. Pyro-char contains a very high level of carbon that can serve as an alternative raw material precursor for the production of porous activated carbon for potentially use in energy storage applications such as batteries and SCs [49]. Waste tires, therefore, have a low expenditure and moderate carbon yield, coupled with high economic feasibility (Figure 28.2).

28.3.2.1 Activated Carbons

The potential application of pyro-char has not been much discussed or reported, but the production of activated carbon (AC) from pyro-char by activation at high temperature with steam or CO_2 has been studied [50–52]. The char needs pre-treatment to remove

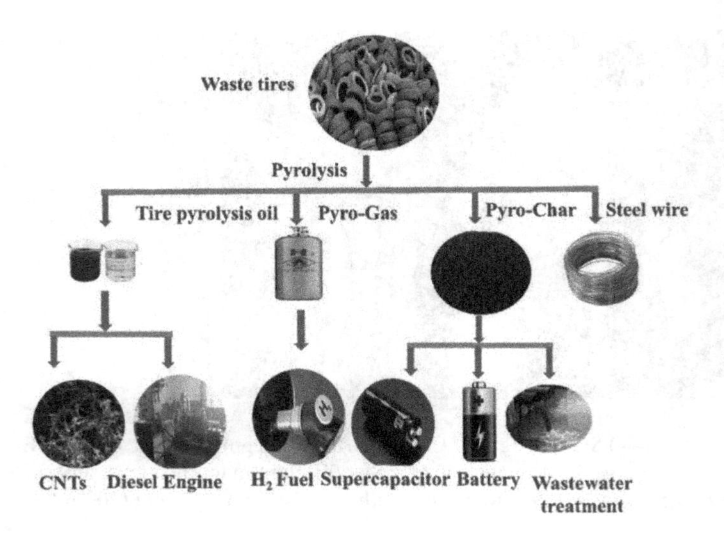

FIGURE 28.2 The schematic diagram for the waste tire pyrolysis and its application [48]. (Reprinted with permission. Copyright (2019) Elsevier.)

rudimental pyrolysis oil and inorganic ashes, if any. A chemical treatment increases the surface area and helps to decrease the concentration of contaminants. Activation temperature also plays an essential role in the yield of ACs. Functional groups on carbon also enhance the wettability of the electrode and improve the capacitive performance [53]. AC plays an indispensable role in supercapacitors due to its low cost and probable large-scale production. They have a high surface area and, therefore, have common usage as an electrode material for SCs. Thus, using scrap tires as a precursor to producing this material would be beneficial and sustainable for future energy storage systems.

SCs developed from scrap tires need to consider the effect of pore structures created by the removal of inorganic compounds such as ZnO and ZnS deposits. The pore structure – micropore or mesopore, porosity, and some other physical properties significantly affect the specific capacitance and rate capability of AC-based SCs. Notably, both mesopores and micropores have paramount importance in predicting the electrochemical behavior of SCs, and optimization requires tuning in specific surface areas with the micropores and mesopores [54]. The activation parameters prove essential, especially in calibrating the specific surface area, total pore volume, mesopore volume, and micropore volume. Experiments conclude that the value of specific capacitance depends only on the AC electrode's micropore volume. The rate capability was dependent on the mesopore/micropore volume ratio. Temperature can prove to be a critical parameter, as high temperatures may destroy the AC structure by reducing surface area and pore volume [48]. Moreover, with a rapid increase in surface area and pore volume, the micropore volume decreases, suggesting a continuous development of mesopores and macropores at the expense of micropores [55].

This electrochemical capacitance can be modified and tuned by physical activation and chemical activation treatments, which now depends on pore structure. Activation conditions allow the manufacture of ACs with surface having acidic characteristics by electrochemical treatment, cold plasma treatment, chemical oxidation, and gaseous oxidation [56–59]. Such treatments would enhance wettability and capacitance through additional faradic reactions (the pseudocapacitance effect) [60]. Han et al. studied AC produced out of pyro-char and consequently treated it with nitric acid for comparison. AC-HNO_3 displayed a higher specific capacitance, enhanced cycling performance, superior stability, and lower equivalent series resistance. Nitric acid treatment was found to be directly responsible for the increase in the surface oxygen-containing functional groups and a remarkable augmentation in the SC performance [61].

The rate capability of electric double-layer capacitance (EDLC)-based SC electrodes depends on four different processes: (i) ion diffusion in the liquid electrolyte, (ii) charge transfer at the electrolyte/electrode interface, (iii) ion adsorption on the surface of the solid electrode, and (iv) electron transport in the solid electrode and the complete circuit. The absolute values of rate capability and specific capacitance depend on the rate-limiting factor of the samples taken for study in fabricating the AC electrode (Table 28.2).

TABLE 28.2
Different Parts and Properties of the SCs Achieved from AC and Their Conclusions

Fabrication Methodology	Results Obtained	Remarks on Obtained Results	Reference
AC obtained by steam activation of industrial pyro-char	With activation energy increasing from 700 °C to 900 °C, the yield activated for 2 h decreased from 87.2% to 3.1%.	This decrease was caused due to the increase in C–H_2–O reaction rate with the temperature increase.	[55]
	The best electrode devised had 89.3% of capacitance maintained when the current density was increased from 0.5–4 A/g.	The higher capacitance retention can be attributed to a smaller value of equivalent series resistance – 0.34 Ω.	
	The electrode delivered a maximum energy density of 13.9 Wh/kg at a power density of 500 W/kg and retained 13.3 Wh/kg at 4,000 W/kg.	These values indicate efficient and ultrafast electrolyte ion transport.	
AC tailored by H_3PO_4 activation process	The best electrode had a surface area of 563.2 m^2/g and the highest pore volume of 0.201 cm^3/g.	The activation process helps to tailor this surface area and pore volume.	[54]
	The best electrode showed a specific capacitance of 106.4 F/g with good capacity retention even after 1,000 cycles.	Such good absolute values state that these carbon electrodes can be operated in practice.	
AC treated with conc. nitric acid	The surface area was found to be 915 m^2/g, and the total pore volume was 0.95 cm^3/g.	This can be ascribed to the increase in surface oxygen-containing functional groups.	[61]
	The specific capacitance was 72% as compared to 56% before acid treatment (for 1,000 cycles).	Acid treatment enhanced the utilization efficiency of the surface area.	

28.3.3 From $PM_{2.5}$ Pollutant

Particulate matter 2.5 ($PM_{2.5}$) (diameter of 2.5 μm or less) is a serious contender that causes most of the airborne diseases in humans, whose long-term exposure to which can cause premature mortality due to ischemic heart diseases (IHDs), stroke, chronic respiratory diseases (e.g., COPD), and lung cancer [62]. The usual morbidities related to $PM_{2.5}$ are respiratory, cardiovascular, cerebrovascular, and diabetes-related morbidities [63]. The World Health Organization guidelines maintain that the concentration of $PM_{2.5}$ should not exceed 10 μg/m^3 annual mean, or 25 μg/m^3 24-hour mean [64]. $PM_{2.5}$ consists mainly of elemental carbon, organic carbon, and inorganic salt, making it even more interesting for potential applications, especially in renewable energy and energy storage [65]. Not only filtering out the particulate, but also reusing or recycling in any method possible is a matter of concern.

28.3.3.1 TPF-Derived SCs

Diesel Particulate Filter (DPF) and Triboelectric Particular Filter (TPF) reduce particulate matter exhaust. Due to their physical filtering process, DPF is a very inefficient process, and TPF depends on the coupling effect of the triboelectric and electrostatic absorption [66]. The coupling effect implies that the emission remains much low and, to a much extent, is controlled and collected.

Wang et al. fabricated a SC out of the carbon collected from a TPF filter, providing great insights into how this carbon can be recycled with a much enhanced efficiency for energy storage and devised in a self-powered system of triboelectric nanogenerator (TENG) [67]. This solves two significant problems: (i) meeting the increasing power demand of electronics and (ii) converting environmental polluters to electricity and energy storage devices. Micro-/nanostructured carbon particles are essential for SC electrode materials due to their high specific capacitance and high chemical stability [68]. Carbon-based SCs are usually the best choice for TENG since they match the pulse output characteristics.

Carbon captured from TPF possesses a natural nanostructure and a slight degree of graphitization. The captured carbon further shows regular nanoscale granular particles, with a size close to 100 nm. The nanostructure plays an intriguing role in increasing the specific surface area and capacitance of the double-layered capacitor. In contrast, the graphitization might increase the electric conductivity and decrease the electrode's resistance, which can significantly benefit fabricating an electrode material for a SC. Such a fabricated carbon is found to have a good electrochemical performance, thus operating at a high charging–discharging rate and, consequently, matching TENG's pulse output characteristics. It also displays an excellent cycling ability with quite a stable capacitance of the initial value even after 100,000 continuous cycles, suggesting a good electrochemical reversibility (Figure 28.3).

28.3.3.2 Diesel Vehicle-Derived $PM_{2.5}$ Carbon Nanoparticles (PM-CNPs)

Diesel engines are primary exhausts and significant sources of global black carbon emissions, dominated by carbon soot particulate emissions (with diameter less than 100 nm) and agglomerate chain structures of carbon soot particles forming larger sizes [69,70]. Those porous carbon-derived electrode materials are the first-class

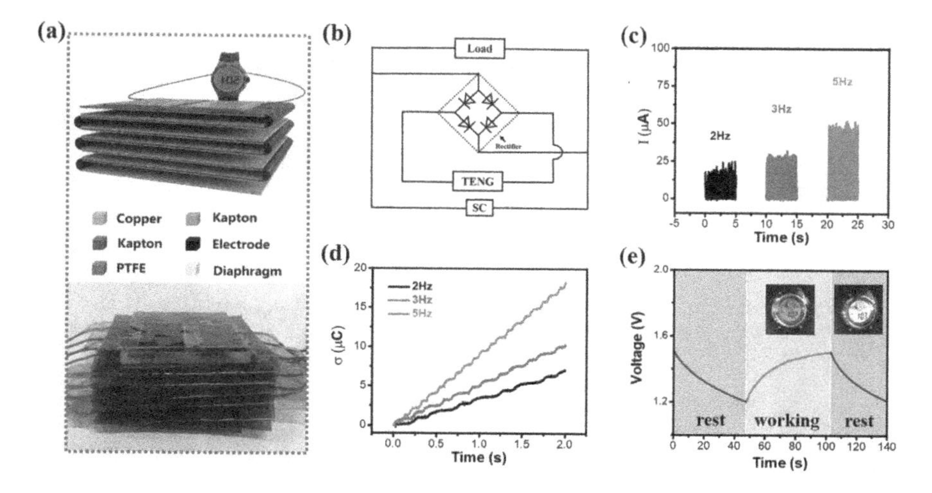

FIGURE 28.3 Performance of the self-powered system based on the SCs and the folding multilayer TENG. (a) Diagrammatic model and image of the self-powered system. (b) Circuit diagram of the self-powered system. (c) Rectified short-circuit current output of the self-powered system at various frequencies. (d) Triboelectric transferred charge of the self-powered system at various frequencies. (e) An electronic watch is powered by the TENG-based self-powered system [67]. (Reprinted with permission. Copyright (2019) Elsevier.)

choices for attractive electrode materials, and studies on nitrogen-doped carbon materials have been by far low, but with great potential [71].

Zhu et al. fabricated $PM_{2.5}$-derived carbon nanoparticles (PM-CNPs), obtained from diesel engines and further doped by nitrogen element [72]. The waste and harmful environmental polluter can be transformed into a useful electrode material for use in the energy conversion and storage field by the facile "collection–annealing–purification" approach. The nitrogen-doped PM-CNPs exhibit a large specific surface area, porous structure, and good conductivity. Notably, the PM-CNP-based SCs exhibit an excellent specific capacity and good stability over 5,000 cycles, which can be further enhanced with proper tuning if experimented with. The carbonaceous waste recycled in such a manner was a novelty. It can provide high aspiration for varied carbon precursors and new inspiration for air pollution control and managing automobile waste (Figure 28.4).

28.4 CONCLUSIVE REMARKS

Technically, much of the fundamental research, studies, and experiment revolve around the fabrication of carbon electrodes for SCs, as the excellent performance comes from ion adsorption on the porous carbon electrode. SCs are technologies of tomorrow and require much more energy storage capacity or excellent capacity resistance even after some thousands of charging–discharging cycles. And this methodology and mechanism open up opportunities to model carbon electrodes from diverse carbon precursors. Studies need to clarify whether it's only due to specific surface area, pore structure, geometry, or a combination of all three factors. Unutilized waste

FIGURE 28.4 (a) Optical images of $PM_{2.5}$ sample collected by the PTFE filter membrane in air sampler and the product of PM-CNPs after thermal annealing and purification, (b) SEM image, (c) TEM image, and (d) HRTEM image of the as-obtained PM-CNPs [72]. (Reprinted with permission. Copyright (2017) Elsevier.)

materials produced in various systems can effectively provide the carbon required for these energy storage devices. Moreover, this approach would resolve worry and consternation around environmental pollution and sustainable policies to tackle them and would be an excellent example of applying creativity to studying the sustainable, green methodology in the principles of "waste-to-wealth".

Automobiles being the primary consumers of hydrocarbon-based fuels such as gasoline and diesel, their waste would be expected to contain a considerable amount of elemental carbon, which can be extracted in forms useful for further use. Most of the wastes either land up in a landfill, causing extravagant environmental pollution. Studies have been conducted to treat those wastes, but merely processing those is not enough. Their use in other sectors needs to be studied, for umpteen probabilities and possibilities can be found in various sectors. Energy storage devices are simply one of them. SCs require carbon precursors for electrode materials, and automobile waste has much to offer. Structuring such structures would be helpful to idealize and realize what automobile wastes have to offer.

Overall, automobile waste provides an excellent specific surface area and has good stability. The porosity also plays an important role, and the electrochemical performance was noteworthy. Studies suggested that the SCs devised had good capacity retention after quite a good number of cycles, which is necessary for a SC. Altogether, in this market movement of the 21st century, automobile waste can help realize the "Waste-to-Use" mechanism, but needs further study, scrutiny, and contemplation.

ACKNOWLEDGMENTS

This work was supported by the Department of Science and Technology (DST), Ministry of Science and Technology, Government of India, funded project under the MI IC#5 "Conversion of Sunlight to Storable fuels" issued by DBT-DST Joint Funding Opportunity, Central Government of India, through Mission Innovation Programme DST (DST/TMD(EWO)/IC5–2018/06) (Subhasis Roy).

REFERENCES

1. Janosi PED (1959) Factors influencing the demand for new automobile, *J. Mark.*, 23:412–418.
2. Joshi D, Bhatt V (2018) A study on factors influencing consumer's preference while making purchase decision of first own car in Ahmedabad city, *Roots Int. J. Int. Res.*, 2:67–76.
3. Sharma P, Sharma A, Sharma A, Srivastava P (2016) Automobile waste and its management, *Res. J. Chem. Environ. Sci.*, 4:01–07.
4. Sivakumar GD, Godwin S, Anatharam S (2014) Indian automobile material recycling management, *Int. J. Innov. Res. Sci. Eng. Technol.*, 3:2754–2758.
5. Das S, Curlee TR, Rizy CG, Schexnayder SM (1995) Automobile recycling in the United States: Energy impacts and waste generation, *Resour. Conserv. Recycl.*, 14:265–284.
6. Zhang LL, Zhao XS (2009) Carbon-based materials as supercapacitor electrodes, *Chem. Soc. Rev.*, 38, 2520–2531.
7. Balducci A, Dugas R, Taberna P, Simon P, Plee D, Mastragostino M, Passerini S (2007) High temperature carbon-carbon supercapacitor using ionic liquid as electrolyte, *J. Power Sources*, 165:922–927.
8. Largeot C, Portet C, Chmiola J, Taberna P, Gogotsi Y, Simon P (2008) Relation between the ion size and pore size for an electric double-layer capacitor, *J. Am. Chem. Soc.* 130:2730–2731.
9. Kandalkar S, Dhawale D, Kim C, Lokhande C (2010) Chemical synthesis of cobalt oxide thin film electrode for supercapacitor application, *Synth. Met.*, 160:1299–1302.
10. Gonzalez A, Goikolea E, Barrena JA, Mysyk R (2016) Review on supercapacitors: Technologies and materials, *Renew. Sust. Energ. Rev.*, 58:1189–1206.
11. Poonam, Sharma K, Arora A, Tripathi SK (2019) Review of supercapacitors: Materials and devices, *J. Energy Storage*, 21:801–825.
12. Dutta A, Mahanta J, Banerjee T (2020) Supercapacitors in the light of solid waste and energy management: A review, *Adv. Sustain. Sys.*, 4:2000182.
13. Klimisch RL (1994) The greening of industrial ecosystems: Designing the modern automobile for recycling, NA, pp. 165–170.
14. Groenewegen P, Hond FD (1993) Product waste in the automotive industry: Technology and environmental management, *Bus. Strategy Environ.*, 2:1–12.
15. Mikeska LA (1936), Chemical structure of lubricating oil, *Ind. Eng. Chem.*, 28:970–984.
16. Dominguez-Rosado E, Pitchel J (2003) Chemical characterization of fresh, used and weathered motor oil via GC/MS, NMR and FITR Techniques, *Proc. Indiana Acad. Sci.*, 112:109–116.
17. Suriani AB, Alfarisa S, Mohamed A, Isa IM, Kamari A, Hashim N, Mamat MH, Mohamed AR, Rusop M (2015) Quasi-aligned carbon nanotubes synthesised from waste engine oil, *Mater. Lett.*, 139:220–223.
18. Maceiras R, Alfonsin V, Morales FJ (2017) Recycling of waste engine oil for diesel production, *Waste Manage.*, 60:351–356.

19. He YM, Zhao FF, Zhou Y, Ahmad F, Ling ZX (2015) Extraction induced by emulsion breaking as a tool for simultaneous multi-element determination in used lubricating oils by ICP-MS, *Anal. Method*, 7:4493–4501.
20. Dreij K, Mattsson A, Jzarvis IWH, Lim H, Hurkmans J, Gustafsson J, Bergvall C, Westerholm R, Johansson C, Stenius U (2017) Cancer risk assessment of airborne PAHs based on in vitro mixture potency factors, *Environ. Sci. Technol.*, 51:8805–8814.
21. Lam SS, Liew RK, Jusoh A, Chong CT, Ani FN, Chase HA (2016) Progress in waste oil to sustainable energy, with emphasis on pyrolysis techniques, *Renew. Sust. Energy Rev.*, 53:741–753.
22. Zimmermann T, Jepsen D (2018) A framework for calculating waste oil flows in the EU and beyond – the cases of Germany and Belgium 2015, *Resour. Conserv. Recycl.*, 134:315–328.
23. Lu Y, Yang X (2015) Molecular simulation of graphene growth by chemical deposition on nickel using polycyclic aromatic hydrocarbons, *Carbon*, 81:564–573.
24. Segawa Y, Ito H, Itami K (2016) Structurally uniform and atomically precise carbon nanostructures, *Nat. Rev. Mater.*, 1:15002.
25. Zhang P, Qiao ZA, Dai S (2015) Recent advances in carbon nanospheres: Synthetic routes and applications, *ChemComm*, 51:9246–9256.
26. Yu Q, Guan D, Zhuang Z, Li J, Shi C, Luo W, Zhou L, Zhao D, Mai L (2017) Mass production of monodisperse carbon microspheres with size-dependent supercapacitor performance via aqueous self-catalyzed polymerization, *ChemPlusChem*, 82:872–878.
27. Zhu C, Li H, Fu S, Du D, Lin Y (2016) Highly efficient nonprecious metal catalysts towards oxygen reduction reaction based on three-dimensional porous carbon nanostructures, *Chem. Soc. Rev.*, 45:517–531.
28. Shen G, Sun X, Zhang H, Liu Y, Zhang J, Meka A, Zhou L, Yu C (2015) Nitrogen-doped ordered mesoporous carbon single crystals: Aqueous organic–organic self-assembly and superior supercapacitor performance, *J. Mater. Chem. A.*, 3:24041–24048.
29. Li Y, Zhang D, He J, Wang Y, Zhang X, Zhang Y, Liu X, Wang K, Wang Y (2019) Hierarchical porous carbon nanosheet derived from waste engine oil for high-performance supercapacitor application, *Sustain. Energy Fuels*, 3:499–507.
30. Kumar MA, Katja K, Yufei Z, Hao L, Chengyin W, Bing S, Guoxiu W (2017) Nitrogen-doped porous carbon nanosheets from eco-friendly eucalyptus leaves as high-performance electrode materials for supercapacitors and lithium-ion batteries, *Chem. Eur. J.*, 23:3683–3690.
31. Ardekani PS, Karimi H, Ghaedi M, Asfaram A, Purkait MK (2017) Ultrasonic assisted removal of methylene blue on ultrasonically synthesized zinc hydroxide nanoparticles on activated carbon prepared from wood of cherry tree: Experimental design methodology and artificial neural network, *J. Mol. Liq.*, 229:114–124.
32. Gao S, Ge L, Rufford TE, Zhu Z (2017) The preparation of activated carbon discs from tar pitch and coal powder for adsorption of CO_2, CH_4 and N_2, *Microporous Mesoporous Mater.*, 238:19–26.
33. Kaipannan S, Govindarajan K, Sundaramoorthy S, Marappan S (2019) Waste toner-derived carbon Fe_3O_4 nanocomposite for high performance supercapacitor, *ACS Omega*, 4:15798–15805.
34. Roy S, Kargupta K, Chakraborty S, Ganguly, S (2008) Preparation of polyaniline nanofibers and nanoparticles via simultaneous doping and electro-deposition. *Mater. Lett.*, 62:2535–2538.
35. Karnan M, Subramani K, Sudhan N, Ilayaraja N, Sathish M (2016) Aloe vera derived activated high-surface area carbon for flexible and high-energy supercapacitors, *ACS Appl. Mater. Interfaces*, 8:35191–35202.

36. Sudhan N, Subramani K, Karnan M, Ilayaraja N, Sathish M (2017) Biomass-derived activated porous carbon from rice straw for a high-energy symmetric supercapacitor in aqueous and non-aqueous electrolytes, *Energy Fuels*, 31:977–985.
37. Karnan M, Subramani K, Srividhya PK, Sathish M (2017) Electrochemical studies on corncob derived activated porous carbon for supercapacitors application in aqueous and non-aqueous electrolytes, *Electrochim. Acta*, 228:586–596.
38. Veeramani V, Sivakumar M, Chen SM, Madhu R, Alamri HR, Alothman ZA, Hossain MSA, Chen CK, Yamauchi Y, Miyamoto N, Wu KCW (2017) Lignocellulosic biomass derived, graphene sheet-like porous activated carbon for electrochemical supercapacitor and catechin sensing, *RSC Adv.*, 7:45668–45675.
39. Kaipannan S, Ganesh PA, Manickavasakam K, Sundaramoorthy S, Govindarajan K, Mayavan S, Marappan S (2020) Waste engine oil derived porous carbon/ZnS nanocomposite as Bi-functional electrocatalyst for supercapacitor and oxygen reduction, *J. Energy Storage*, 32:101774.
40. Wang YY, Hou BH, Lu HY, Lu CL, Wu XL (2016) Hierarchically porous N-doped carbon nanosheets derived from grapefruit peels for high performance supercapacitors, *ChemistrySelect*, 1:1441–1447.
41. Zhang S, Tian K, Cheng BH, Jiang H (2017) Preparation of N-doped supercapacitor materials by integrated salt templating and silicon hard templating by pyrolysis of biomass wastes, *ACS Sustain. Chem. Eng.*, 5:6682–6691.
42. Li Z, Xu Z, Tan X, Wang H, Holt CMB. Stephenson T, Olsen BC, Mitlin D (2013) Mesoporous nitrogen-rich carbons derived from protein for ultra-high-capacity battery anodes and supercapacitors, *Energy Environ. Sci.*, 6:871–878.
43. Tan H, Wang X, Jia D, Hao P, Sang Y, Liu H (2017) Structural dependent electrode properties of hollow carbon micro-fibers derived from platanus fruit and willow catkins for high-performance supercapacitors, *J. Mater. Chem. A*, 5:2580–2591.
44. Hu L, Hou J, Ma Y, Li H, Zhai T (2016) Multi-heteroatom self-doped porous carbon derived from swim bladders for large capacitance supercapacitors, *J. Mater. Chem. A*, 4:15006–15014.
45. Ramarad S, Khalid M, Ratnam CT, Chuah AL, Rashmi W (2015) Waste tire rubber in polymer blends: A review on the evolution, properties and future, *Prog. Mater. Sci.*, 72:100–140.
46. Rowhani A, Rainey TJ (2016) Scrap tyre management pathways and their use as a fuel-a review, *Energies*, 9:888.
47. Williams PT, Bottrill RP (1995) Sulphur-polycyclic aromatic hydrocarbons in tire pyrolysis oil, *Fuel*, 74:736–742.
48. Sathiskumar C, Karthikeyan S (2019) Recycling of waste tires and its energy storage application of by-products: A review, *SM&T*, 22:e00125.
49. Gnanaraj J, Lee R, Levine A, Wistrom J, Wistrom S, Li Y, Li J, Akato K, Naskar A, Paranthaman MP (2018) Sustainable waste tire derived carbon material as a potential anode for lithium-ion batteries, *Sustainability*, 10:2840.
50. Aranda A, Murillo R, Garcia T, Callen MS, Mastral AM (2007) Steam activation of tyre pyrolytic carbon black: Kinetic study in a thermo-balance, *Chem. Eng. J.*, 126:79–85.
51. Murillo R, Navarro MV, Lopez JM, Garcia T, Callen MS, Aylon E, Mastral AM (2004) Activation of pyrolytic tire char with CO_2: Kinetic study, *J. Anal. Appl. Pyrolysis*, 71:945–957.
52. Minguel GS, Fowler GD, Sollars CJ (2003) A study of the characteristics of activated carbons produced by steam and carbon dioxide activation of waste tyre rubber, *Carbon*, 41:1009–1016.
53. Wang Y, Shi ZH, Huang Y, Ma YF, Wang CY, Chen MM, Chen YS (2009) Supercapacitor devices based on graphene materials, *J. Phys. Chem. C*, 113:13103–13107.

54. Dey A, Dhar A, Roy S, Das BC (2017) Combined Organic-Perovskite Solar Cell Fabrication as conventional Energy substitute. *Materials Today: Proceedings*, 4(14):12651–12656.
55. Zhao P, Han Y, Dong X, Zhang C, Liu S (2015) Application of activated carbons derived from scrap tires as electrode materials for supercapacitors, *ECS J. Solid State Sci. Technol.*, 4:M35.
56. Momma T, et al. (1996) Electrochemical modification of active carbon fiber electrode and its application to double-layer capacitor, *J. Power Sources*, 60:249–253.
57. Ishikawa M, et al. (1996) Effect of treatment of activated carbon fiber cloth electrodes with cold plasma upon performance of electric double-layer capacitors, *J. Power Sources*, 60:233–238.
58. Khomenko V, Raymundo-Pinero E, Beguin F (2010) A new type of high energy asymmetric capacitor with nano porous carbon electrodes in aqueous electrolyte, *J. Power Sources*, 195:4234–4241.
59. Hsieh CT, Teng H (2002) Influence of oxygen treatment on electric double-layer capacitance of activated carbon fabrics, *Carbon*, 40:667–674.
60. Lufrano F, Staiti P (2010) Influence of the surface-chemistry of modified mesoporous carbon on the electrochemical behavior of solid-state supercapacitors, *Energy Fuels*, 24:3313–3320.
61. Han Y, Zhao PP, Dong XT, Zhang C, Liu SX (2014) Improvement in electrochemical capacitance of activated carbon from scrap tires by nitric acid treatment, *Front. Mater. Sci.*, 8:391–398.
62. Burnett R, Chen H, Szyszkowicz M, Fann N, Hubbell B, Pope CA (2018) Global estimates of mortality associated with long-term exposure to outdoor fine particulate matter, *PNAS USA*, 115:9592–9597.
63. Sharma S, Chandra M, Kota SH (2020) Health effects associated with $PM_{2.5}$: A systematic review, *Curr. Pollut. Rep.*, 6:345–367.
64. WHO air quality guidelines for particulate matter, ozone, nitrogen dioxide and sulphur dioxide, Global Update 2005, Summary of Risk Assessment.
65. Han XL, Naeher LP (2006) A review of traffic-related air pollution exposure assessment studies in the developing world, *Environ. Int.*, 32:106–120.
66. Han CB, Jiang T, Zhang C, Li XH, Zhang CY, Cao X, Wang ZL (2015) Removal of particulate matter emissions from a vehicle using a self-powered triboelectric filter, *ACS Nano*, 9:12552–12561.
67. Li X, Yin X, Wang W, Huabo Z, Liu D, Zhou L, Zhang C, Wang J (2019) Carbon captured from vehicle exhaust by triboelectric particular filter as materials for energy storage, *Nano Energy*, 56:792–798.
68. Wang G, Zhang L, Zhang J (2012) A review of electrode materials for electrochemical supercapacitors, *Chem. Soc. Rev.*, 41:797–828.
69. Burtscher H (2005) Physical characterization of particulate emissions from diesel engines: A review, *J. Aerosol Sci.*, 36:896–932.
70. Ning Z, Chan KL, Wong KC, Westerdahl D, Mocnik G, Zhoue JH, Cheunge CS (2013) Black carbon mass size distributions of diesel exhaust and urban aerosols measured using differential mobility analyzer in tandem with Aethalometer, *Atmos. Environ.*, 80:31–40.
71. Lin T, Chen IW, Liu F, Yang C, Bi H, Xu F, Huang F (2015) Nitrogen-doped mesoporous carbon of extraordinary capacitance for electrochemical energy storage, *Science*, 350:1508–1515.
72. Zhu G, et al. (2017) Recycling $PM_{2.5}$ carbon nanoparticles generated by diesel vehicles for supercapacitors and oxygen reduction reaction, *Nano Energy*, 33:229–237.

29 Halogenated Polymeric Wastes for Green Functional Carbon Materials

Yingna Chang, Zongge Li, and Guoxin Zhang
Shandong University of Science and Technology

CONTENTS

29.1 INTRODUCTION

Plastic materials are no stranger to us, and their appearances are widely distributed in many scenes of human production and life, but discarded plastic products (also called plastic wastes) also bring serious problems, namely "white pollution". The enormous applications possessed by plastics are mostly because plastics are economical (compared with metal-based counterparts), are chemically stable, and have excellent mechanical properties. While, in return, their high chemical stability greatly hinders their recycling and reuse, they may take hundreds of years to decay and decompose simply by burying them. Due to the inability to degrade naturally, plastic wastes have become one of the top enemies of human beings; therefore, it is of great significance to develop efficient treatment and utilization technologies for plastic wastes. At present, the main treatment method is the heating treatment, which achieves a certain degree of recycling. However, most plastics are prone to produce toxic and harmful gases when heated, such as toluene when polystyrene is burned; hydrogen chloride and dioxins (a well-known carcinogenic substance) when polyvinyl chloride is burned. Therefore, simple heat treatments that are easily causing secondary pollution can no longer meet the human demands for environmental protection.

DOI: 10.1201/9781003178354-33

In particular, halogenated polymer plastics, represented by PVC, have low preparation/processing costs and their products are self-extinguishing, so they have widely been used in the construction field, especially sewer pipes, plastic windows and doors, panels, wires and cables, and artificial leather, which account for about 10% of all plastic products (take the European plastics market as an example) [1]. PVC has a softening point of 80 °C and starts to decompose at 130 °C, releasing hydrogen chloride, dioxins, and other toxic gases [2], and the products are mainly olefins (at low temperatures), carbon (at high temperatures) [3], etc. It can hardly return to its monomeric state (vinyl chloride). As can be seen in Table 29.1, most halogenated polymer plastics do not support recycling and their recovery rate is close to 0; meanwhile, harmful substances can be easily produced in the process of treatment or degradation. Therefore, an efficient, clean, and environmentally friendly method is urgent for the treatment and utilization of halogenated polymer wastes.

As told in organic chemistry textbooks, the breakage and removal of carbon–halogen bonds can be achieved under mild conditions because most carbon–halogen bonds have lower bond energies and longer bond lengths compared to conventional functional groups such as carbon–oxygen and carbon–nitrogen groups. According to our previous research works, we found typical halogenated polymers such as polyvinylidene chloride (PVDC) can be decomposed by KOH at room temperature by simple hand grinding, resulting in the removal of C–Cl functionalities and the generation of value-added carbon materials; the by-products are clean water and KCl. Such treatment is completely different from the traditional treatments of polymer plastics; that is, in order to achieve the reuse of polymer plastics, the latter treatments need high energy and complex treatment processes [4]. At the same time, compared to the use of other carbon sources, the use of halogenated polymers as a carbon source has additional merits: Clean, environmentally friendly by-products (for example, in the above-mentioned use of PVDC and KOH, the main by-products are water and KCl) can be directly discharged to the environment. Other commonly used carbon sources, such as an oxygenated biomass glucose, need to experience higher temperatures (200 °C–800 °C) and produce a lot of CO, smoke, and other pollution to the environment. In the next sections, we will talk about the research progress of using halogenated polymer materials, especially PVC as carbon sources, to prepare functional carbon materials in efficient and green ways and discuss their applications in electrochemical energy storage and conversion.

29.2 BRIEF INTRODUCTION TO DEHALOGENATION STRATEGY

The use of halogenated organics as carbon sources for the preparation of carbon nanomaterials (CNMs) dates back to 1998. Yadong Li et al. prepared micron-sized nanodiamonds by using small-molecule CCl_4 as a carbon source [5] and reacting it with metal Na under the promotion of Ni–Co catalyst and high temperature of 700 °C (as shown in Equation 29.1). The boom of producing CNMs from halogenated organics was thereafter initiated. However, compared with other functionalized organics, relatively fewer works have been done to study the mechanism and processes of carbonization of halogenated organics, probably because most of them are much more expensive. Meanwhile, at the beginning stage, the synthesis of

TABLE 29.1
Comparison of Different Types of Polymeric Plastics and Their Recovery Rates from Wastes

Full Name	Chemical Structure	Uses	Currently Recyclable?	Recovery Rate (%)	Harmful Toxicity
Polyethylene terephthalate	$[-O-CO-C_6H_4-CO-O-CH_2-CH_2-]_n$	Disposable bottles for drinks, medicines, and many other consumer products	Yes	19.5	No
High-density polyethylene	$[-CH_2-CH_2-]_n$	More durable containers, such as those for detergent, bleach, shampoo, or motor oil	Yes	10	Yes
Polyvinyl chloride	$[-CH_2-CHCl-]_n$	Piping cables, garden furniture, fencing, and carpet backing	No	0	Yes
Polyvinylidene chloride	$[-CH_2-CCl_2-]_n$	Used for packaging of food, medicine, and military products	No	0	Yes
Polyvinylidene difluoride	$[-CH_2-CF_2-]_n$	Petrochemical, electrical and electronic and fluorocarbon coatings	No	0	Yes

(Continued)

TABLE 29.1 (*Continued*)
Comparison of Different Types of Polymeric Plastics and Their Recovery Rates from Wastes

Full Name	Chemical Structure	Uses	Currently Recyclable?	Recovery Rate (%)	Harmful Toxicity
Polytetrafluoroethylene	$-[CF_2-CF_2]_n-$	Industrial and marine operations such as chemical, petroleum, textile, food, paper, medicine, electronics, and machinery	No	0	Yes
F-graphite polymer		Lubricant; active materials for batteries and electrodes	No	0	Yes
Chlorinated polyethylene	$-[CH_2-CH(Cl)-CH_2-CH_2]_n-$	Cables, wires, hoses, tapes, rubber, plastic products, sealing materials, flame-retardant conveyor belts, waterproof membranes, films, and various profiles	No	0	No

CNMs using halogenated organics is still resembling the treatment of other polymeric organics, mainly focusing on the use of high-temperature treatment (typically 600 °C–1,100 °C) to drive the removal of halogenated functional groups and the coupling carbonization of the remaining carbon chains, which were not fully reflecting the advantages of halogenated organics as carbon sources.

$$CCl_4 + 4Na \xrightarrow{700°C, Ni-Co} C(diamond) + 4NaCl \qquad (29.1)$$

In recent years, an increasing amount of work has targeted the low-temperature reactivity of halogenated functional groups to develop a series of functional carbon materials (FCMs) under mild conditions. Related works mainly focus on the following three aspects: (i) bottom-up synthesis of fine graphene structures, such as graphene nanoribbons (width less than 2 nm) [6] and a series of aromatic carbon skeletal materials [7,8], using halogenated (especially brominated and chlorinated) aromatic molecules by low-temperature interfacial induction. (ii) Carbon alkyne materials were prepared under mild conditions using chain halogenated polymers such as polyvinylidene chloride (PVDC); furthermore, carbon alkyne can be extended to prepare organosulfur carbon composites, which can be directly applied as lithium–sulfur battery cathodes [9–11]. (iii) Under energy injection (e.g., grinding or stirring), halogenated functional group can react with alkalis to remove the functional groups and enable carbonization to obtain high-quality CNMs. For example, Jong-Beom Baek et al. used CCl_4 as a carbon source to react with potassium at room temperature under high-energy ball milling, resulting in high-quality nanographene materials [12]. In our group, we used PVDC as a typical carbon source and manually ground it at room temperature, allowing it to react with potassium hydroxide. The carbonization of PVDC can be accomplished in a processing time as short as 1 min, obtaining CNMs with a carbon content of more than 75%, with clean water and KCl as by-products [13]. Furthermore, based on the polymer dehalogenation strategy developed above, we have extensively investigated the carbonization of halogenated polymers such as PVDC, polyvinylidene fluoride (PVDF), and polyvinyl chloride (PVC) wastes into FCMs, and their electrochemical application potentials. In the next sections, we will describe relevant advancements regarding the efficient dehalogenation and functionalization of halogenated polymers-derived CNMs.

29.3 DEHALOGENATION FOR TUNABLE COMPOSITIONS IN CARBON

Clean and efficient accesses to CNMs with adjustable heteroatom doping are of great appeal to the sustainable utilization of energy in technologies such as fuel cells, supercapacitors, and batteries [14]. The commonly used organic carbon sources for the fabrication of doped carbon materials (DCMs), usually functionalized with oxygen, nitrogen, or sulfur, though inexpensive and abundant, demand rigorous synthesis conditions for the defunctionalization and carbonization. As mentioned above, halogen functional groups are very good leaving groups in terms of defunctionalization from carbon chains of polymers due to their longer bond lengths. Based on that, we have found that PVDC can be dechlorinated at room temperature to form carbonaceous materials. Furthermore, taking advantage of highly reactive *in situ* dehalogenated carbon sites, functionalization of dehalogenated carbon sites can be realized by simply adding heteroatom dopant in the reaction system. A variety of DCMs have successfully been pioneered, including single-element doping (nitrogen, sulfur, phosphorus, and boron) [15–17], binary doping (nitrogen/sulfur [18,19] and nitrogen/phosphorus [20]), and ternary doping (boron/nitrogen/phosphorus [21]). What is more attractive is that the above efficient and green heterodoping can be directly transferred into PVC carbonization system [22–24]. By using a similar preparation method, we can achieve the efficient doping of PVC-converted FCMs for value-added and electrochemical application development.

Further investigations revealed that this room-temperature polymer dehalogenation strategy can also be extended to the efficient and green preparation of functionalized graphene materials. The preparation of water-soluble functionalized graphene was achieved by grinding fluorinated graphite (F–G) with an alkali at room temperature [25], which was similar to that of the preparation of CNMs using PVDC as carbon sources. By using special alkalis that contain hetero-elements, such as $NaNH_2$ and Na_2S, it is feasible to obtain water-soluble doped graphene materials. This method requires simpler operation and higher safety compared with the conventional method of obtaining functional graphene materials involving strong acid and strong oxidative. Similarly, superdoping to graphitic carbon materials can be realized by introducing a simple pretreatment of fluorination, followed by annealing in a dopant source. The superdoping of N, S, and B can reach up to 29.82, 17.55, and 10.79 at% for graphene, respectively, which strongly supports that DCMs can be efficiently fabricated by using dehalogenation strategy.

29.4 DEHALOGENATION STRATEGY FOR MATERIALS STRUCTURING AND PORE MANAGEMENT

Beneficial porosity and structures of CNMs hold great importance due to their ability to mediate mass communication between bulk medium and accessible interfaces of CNMs, such as gases and liquids. The mass exchange issues are widely considered as key factors for the improvement of the performance of clean energy utilization devices such as fuel cells, rechargeable batteries, and supercapacitors [26]. In particular, the so-called open pores, mainly including mesopores and macropores, are more attractive in the above-mentioned fields. In order to generate meso-/macropores in CNMs, solid or liquid templates with narrow size distributions are usually demanded, which increases the costs of material synthesis and makes the material fabrication more tedious.

Different from other organic carbon sources, halogenated organics can generate by-products of different material states during the low-temperature dehalogenation process; the in situ generated by-products are manageable to serve as gaseous or solid templates to create pores and alter the morphology of the resultant CNMs. For instance, reacting PVDC with sodium ethoxide results in solid NaCl nanocrystals and ethanol liquid/gas (Equation 29.2); the former by-product can serve as solid template to create macropores, and the latter can bubble the polymeric carbonaceous substrate into 2D forms when heated [13]. If assisted by KOH activation, the resultant CNMs in 2D layered form can achieve a very high specific surface area (SSA) of 1,735.5 m^2/g. Also, taking ZnO as an example, reacting PVDC with ZnO at medium temperature obtains solid $ZnCl_2$ and gaseous H_2O (Equation 29.3), $ZnCl_2$ is a commonly used agent to activate carbon for micropore generation, and the gaseous H_2O can bubble the polymeric carbon into sheet morphology [15]. The resultant CNMs were characterized to have a very large SSA of 1,499 m^2/g and large content of mesopores.

$$(CH_2 - CCl_2)_n + 2nC_2H_5ONa = C_{2n} + 2nNaCl + 2nC_2H_5OH \qquad (29.2)$$

$$(CH_2 - CCl_2)_n + nZnO = C_{2n} + nZnCl_2 + 2nH_2O \qquad (29.3)$$

In addition, hierarchical porous carbon (HPC) materials with well-defined meso-/macroporosity can be obtained by the dehalogenation of PVDF by $NaNH_2$. The mechanism of carbon formation is described by Equation 29.4, and accordingly, two important by-products, NaF and NH_3, were formed, which are suitable for in situ templating and bubbling of intermediate carbon structures with meso-/macropores [27]. Meanwhile, the intentionally left $NaNH_2$ can further activate the carbon to obtain abundant microporosity. Electron microscopy studies and Brunauer–Emmett–Teller (BET) measurements confirmed that the structure of the HPC contains multiscale pores assembled in a hierarchical pattern with most of its volume contributed by mesopores, as shown in Figure 29.1. HPC-M7 with larger macropores was obtained

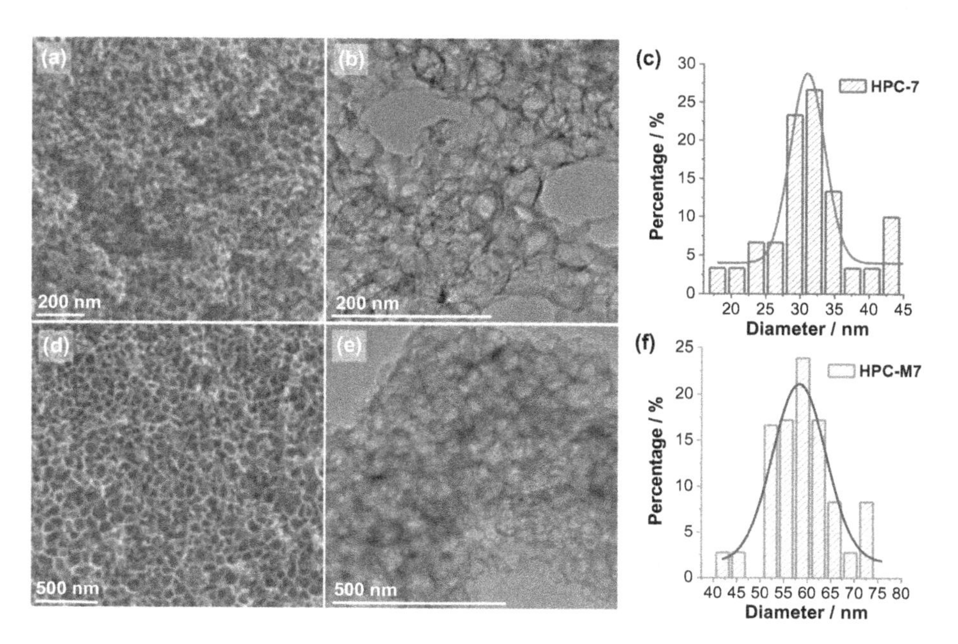

FIGURE 29.1 (a) SEM and (b) TEM images of HPC-7 and (d) SEM and (e) TEM images of HPC-M7. Distribution of pore diameters of (c) HPC-7 and (f) HPC-M7. (Adapted with permission from Ref. [27]. Copyright (2018) Elsevier.)

by adding decomposable and evaporable components – melamine – before the start of the dehalogenation reaction. This specific finding related to the secondary evaporable components may serve as a powerful tool to tune the pore types and their ratios in porous CNMs.

$$(CH_2 - CF_2)_n + NaNH_2 + C_3N_3(NH_2)_3 \xrightarrow{\Delta} N - C + NaF + NH_3 \quad (29.4)$$

As illustrated above, alkalis can efficiently remove the functional group of halogenated organics, and under the circumstances of no use of dopant source, the as-formed surface is pure carbon without functionality, as shown in Equations 29.2 and 29.3. This feature of halogenated organics is completely different from other carbon sources, take oxygenated organic carbon sources as an example, most of which require ultra-high temperature annealing to completely remove the oxygenated functional groups to form a pure carbon surface. The extent of the functionality on the carbon surface determines whether the carbon layer can subsequently occur on it. In other words, when the carbon surface was much less functionalized, it will be difficult to load the subsequently formed carbon. This specialty allows this dehalogenation method to be an efficient method for the self-terminating carbon formation on the substrate of different morphologies. For example, the occurrence of dehalogenation on the surface of 2D substrates such as graphene oxide (GO) enables thin-layer coating of carbon on GO, and prevents further coating of dehalogenated carbon due to lack of functional group of the as-formed carbon layer; therefore, an ultra-thin carbon coating layer can be realized on the GO surface (thickness <2 nm) [28]. Furthermore, the alkalinity of 2D layered double hydroxide (LDH) promotes the defunctionalization of PVDC, leading to the formation of thin dehalogenated carbon layer on LDH and termination of the contact between the PVDC carbon source and the protected LDH [29]. Similarly, we can use other substrate materials with basic surface and desirable morphology and structure and perform dehalogenation reactions on their surface to enable thin-layer carbon coating and obtain different carbon structures.

29.5 ELECTROCHEMICAL APPLICATIONS OF DEHALOGENATED CARBON

Carbon materials are now widely used in electrochemical applications such as energy storage and conversion. The surface properties of electrode materials determine the behavior of electrons and ions on electrode interfaces, while the structure and pores of electrode materials possess a significant impact on the interaction behavior of mass (ions, electrolyte, and gas) between inner accessible active surface of the electrode and the bulk solution/gas tank. It is generally accepted that electrochemical reactions occur at the active interfaces; therefore, increasing the activity of sites and expanding the flux of current per unit time are both effective ways to boost the electrochemical performance. The former can be realized by altering the composition of electrode materials, and the latter can be reached by optimizing the morphology, structure, and porosity of electrode materials. The dehalogenated carbon materials, widely obtainable from PVC and PVDC, can be expected with superior electrochemical performance using the above-mentioned material synthesis and optimization strategies.

To address the requirements of capacitive energy storage for electrode materials, carbon materials with element doping and multiscale pore structures need to be realized to promote: (i) the efficient and dense ion adsorption on the electrode interface, (ii) the occurrence of pseudocapacitance, and (iii) the rapid interaction of charge carriers between the inner accessible surface of the electrode material and the bulk solution. Typically, based on the organic dehalogenation strategy, Guoxin Zhang et al. selected two types of alkalis with different properties: potassium hydroxide (KOH) and sodium ethanol (NaOEt), for the dehalogenation reaction of PVDC to fabricate hierarchically porous structures. KOH served as the main dehalogenation agent and micropore formation promoter, and NaOEt was mainly used to fabricate macropores, and the *in situ* generated nanoscale by-products KCl and NaCl can be utilized as templates to fabricate abundant mesopores. The resultant dehalogenated carbon was found to have a high specific surface area (SSA) of 1,735.5 m^2/g, abundant mesopores and macropores, and heavy nitrogen doping (~10 at.%). Applied as a supercapacitor electrode material, the dehalogenated carbon material HPC-3 showed a high specific capacitance (~328.0 F/g at a current density of 0.5 A/g), excellent rate capability (~62.8% at a current density of 100.0 A/g), and ultra-high cycling stability (99.3% capacity retention after 5,000 cycles under charge/discharge rate at 50.0 A/g).

In general, different types of halogenated polymers including PVDC, PVDF, PVC, and F-graphite (also called the two-dimensional F-polymer) can be made into functional carbon materials with various morphologies, dopants, and rich porosities. As summarized in Table 29.2, the SSA of dehalogenated carbon materials are typically around 1,000 m^2/g and, with the assistance of KOH activation, a SSA of 1,500 m^2/g or above can be achieved. The specific capacitances (Cs) of dehalogenated carbon materials measured with aqueous electrolytes fell in the range of 300–500 F/g, achieving specific energy of 10–20 Wh/kg, which are all among the upper class of carbonaceous supercapacitor electrode materials. In particular, PVC plastic tube-derived porous N-doped carbon materials exhibited a typical sheet-like morphology and a SSA of 641 m^2/g, obtaining a remarkable Cs of 430 F/g. If applied with a higher current density of 20 A/g, the as-made porous N-doped carbon still rendered a very large Cs of 280 F/g. The above discussion indicates that halogenated polymers can be converted into high-quality functional carbon materials with satisfactory supercapacitor performance.

The commonly used tactics such as functionalization of metallic components and heterodoping are very effective to improve the Cs of carbon electrode materials. However, one drawback of doing such variations is most metallic components and heterodopants are active for water splitting, rendering narrowed working potential range. For instance, the working potential window of common alkaline supercapacitors of heteroatom-doped carbon is ~0.8–1.0 V and can be extended to ~1.5–1.8 V by using neutral electrolytes. To further push the working potential window to the limit, two efficient routes have been developed based on the polymer dehalogenation strategy. One is done by achieving high completeness of defunctionalization of PVDF, which guarantees the formation of non-doped carbon interface to degrade the

TABLE 29.2
Supercapacitor Performance of Dehalogenated Carbon (DH-C) Derived from Halogenated Polymers

Source	Name of DH-C	General Morphology	Dopant	SSA (m^2/g)	Electrolyte	C_s^a (F/g) at 1 A/g	C_s^a (F/g) at 100 A/g	E_s^b (Wh/kg)	Reference
PVDC	HPC-3	Sheet	N	1,736	6 M KOH	322	205	10.4	[13]
PVDC	600-NS-DCM	3D porous structure	N, S	669.0	1 M H_2SO_4	427	251	-	[19]
PVDC	GO@NdC-3K	Thin layer	N	1,508	6 M KOH	354 @0.5 A/g	-	11.8	[28]
PVDC	HPDC	Porous structure	N	1,540	1 M Li_2SO_4	188.9[b]	-	21.5	[30]
PVDC	BNP-C	Porous structure	B, N, P	1,119	1 M H_2SO_4	442	260	12.0	[21]
PVDC	NP-C700	Irregular sheet-like structure	N, P	943	6 M KOH	347.6	285 @50 A/g	-	[20]
PVDF	C-800	Porous structure	N	1,049	2 M KOH	180	109 @20 A/g	-	[31]
PVDF	O-PC-8	Porous structure	N	1,772	1 M Li_2SO_4	120	50 @10 A/g	18.1	[32]
PVDF	HPC6	Porous structure	F	3,003	1 M Na_2SO_4	240 @0.5 A/g	37.3 @20 A/g	10.2	[33]
PVDF	HPC-M7	Porous structure	N, F	1,498	1 M Li_2SO_4	33.8[b]	30.0 @10 A/g	18.8	[27]
PVC	NPM-4	Porous network	N	500	1 M H_2SO_4	251	186 @30 A/g	-	[34]
PVC	PT-C	Typical sheet-like morphology	N, S	641	1 M H_2SO_4	430	280 @20 A/g	-	[35]
PVC	NS-C	Porous structure	N, S	1,230	1 M Na_2SO_4	290.2	204.1	-	[23]
F-graphite	ODA-G	Graphene sheets	N, F	1,985	-	328.5 @0.5 A/g	-	19.5 (Wh/L)	[36]
F-graphite	G–pPDA–PANI	Wrinkled, fluffy, and spongy structure	N, F	636	PVA-H_2SO_4 film	638	580 @20 A/g	18.0 (Wh/L)	[37]

[a] Values obtained in a three-electrode measurement.
[b] Values obtained with symmetric layout.

water-splitting activity [32]. The resultant carbon materials exhibited a stable potential window as wide as ~2.5 V in a three-electrode setup, and its assembled symmetric supercapacitor in the neutral electrolyte can stably work with 2.0 V at 5.0 A/g for 5,000 cycles. The other route is to dope carbon materials with high-loading water-splitting inactive boron [21], as shown in Figure 29.2; heavy boron doping at the carbon interface can effectively expand the workable voltage window in alkaline and acid electrolyte to at least 1.5 V while maintaining large Cs, obtaining Cs of 518 F/g at 0.5 A/g and 260 F/g at 100 A/g in 1.0 M H_2SO_4.

According to previous studies, pore generation and heteroatom doping in carbon materials enable their capability of entrapping lithium polysulfides through physical confinement and chemical adsorption, respectively. Also, by employing the polymer dehalogenation strategy, porous N, S-co-doped carbon materials, obtained from PVC, functionalized with heavy N, S-loading (6.5 at.% N and 7.2 at.% S), can be used as an efficient S host for high-performance Li–S battery. The NS-C/S cathode with a sulfur loading of ~1.6 mg/cm^2 can achieve a high reversible capacity of ~1,230 mAh/g (at 0.1 C), high rate performance (capacity retention of ~71% at 1 C), and good cycling performance (capacity loss of ~0.058% per cycle in 500 cycles) [24]. At a high mass loading of ~5 mg/cm^2, the NS-C/S composite still maintains very high capacity retention of about 80% after 200 cycles (at 1 C), confirming the importance of heterodoping of carbon surface for the immobilization and confinement of S and lithium polysulfides. The heavy N- and S-moieties can be also utilized to chem-absorb metal cations; for instance, the NS-C obtained from PVDC exhibited a desalination capacity of 141 mg/g for Na^+ [19], and the NS-C obtained from PVC showed about 50 mg/g capacities for various heavy metal cations including Cd^{2+}, Ni^{2+}, Co^{2+}, Pd^{2+}, Fe^{2+}, and Cu^{2+}, serving as a reliable "use waste to treat waste" strategy for the treatment of PVC waste plastics and heavy metal pollutions [23].

The dechlorination of PVDC by ZnO, as previously illustrated in Section 29.4, can generate by-products of water and $ZnCl_2$, which were used for altering the material morphology and creating hierarchical pores consisting of micropores, mesopores, and macropores. In the presence of melamine, the resultant dechlorinated carbon can be managed into hierarchically porous N-doped carbon (PDC) for the exploration of electrocatalytic applications. In particular, the proper layout of micropores, mesopores, and macropores enabled by multiple pore formation routes greatly shortened the diffusion length of solvated ions and enlarged the electrochemically accessible surface area during electrocatalysis. Consequently, when submitted to catalyze oxygen reduction reaction (ORR), the PDC-900 (PDC samples annealed at 900 °C) exhibited a higher half-wave potential (0.84 V vs. RHE) and larger limiting current density than the commercial 20 wt.% Pt/C [15], as shown in Figure 29.3. Heavy N-doping into dehalogenated carbon is also the key to reserve more Co-N_x moieties for the construction of highly active metal–nitrogen sites. Heavily N-doped carbon materials can be obtained via the dehalogenation of PVDC or PVC in the presence of melamine, as previously illustrated in Section 29.3. Electrochemical measurements showed that Co-N-C materials from PVC exhibited a very good ORR performance comparable to commercial 20 wt.% Pt/C, obtaining an onset potential of ~0.86 V (compared to ~0.93 V for Pt/C) and a

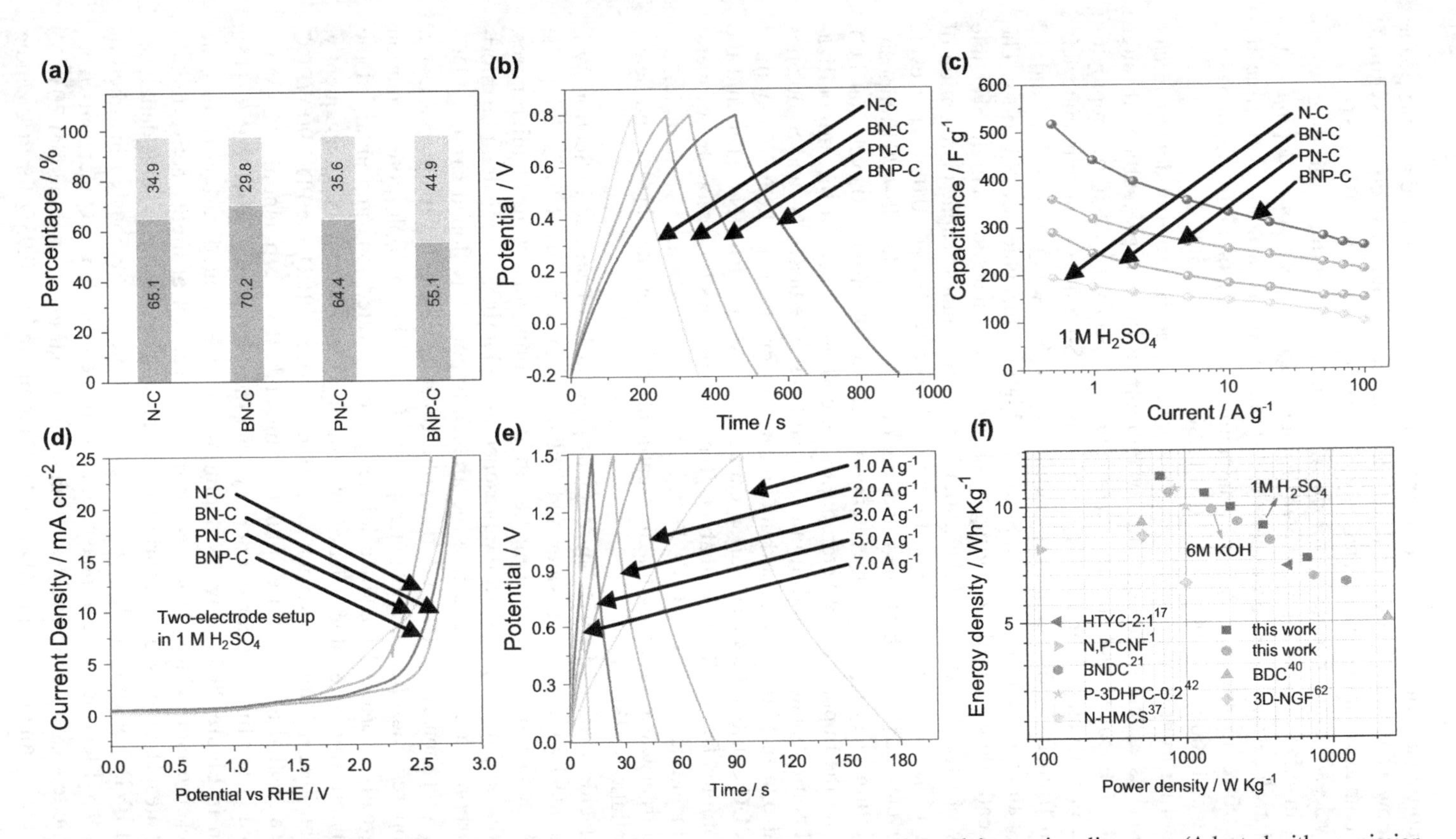

FIGURE 29.2 Electrochemical measurements of BNP-C with the comparison of control samples and the previous literature. (Adapted with permission from Ref. [21]. Copyright (2020) Elsevier.)

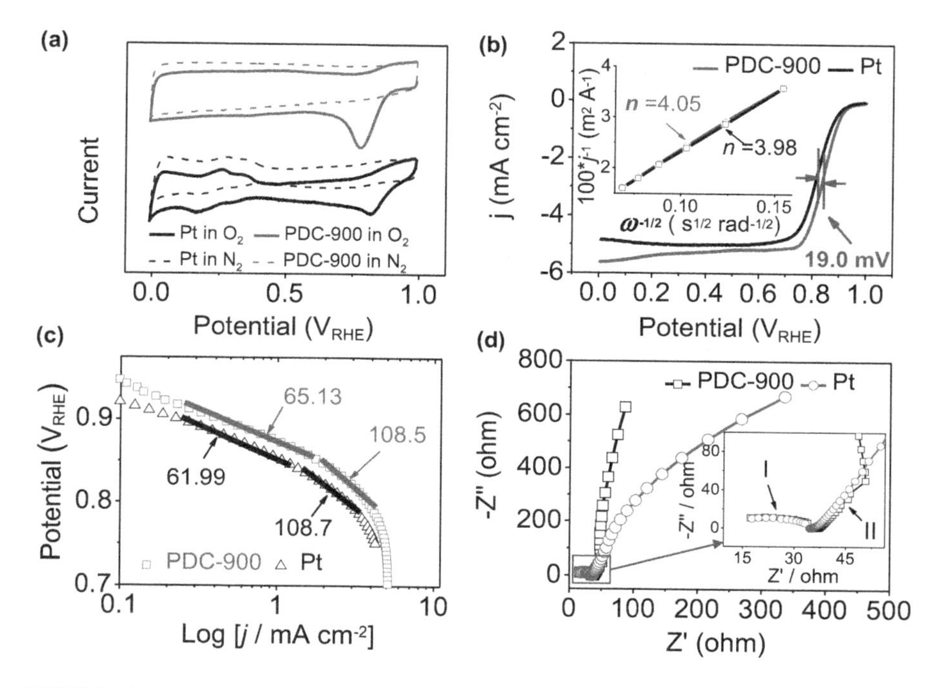

FIGURE 29.3 ORR performance of PDC-900 in 0.1 mol/L KOH. (Reproduced with permission from Ref. [15]. Copyright (2016) Elsevier.)

half-wave potential of 0.79 V (compared to ~0.81 V for Pt/C). The success of using PVC or even PVC plastic scrap as a carbon source to fabricate Co-N-C materials with a high ORR performance has the potential to open the production of scalable, low-cost metal–nitrogen decorated carbon materials for a wide range of electrocatalytic applications.

29.6 CONCLUSIONS AND PERSPECTIVES

Carbon materials are widely sourced, are inexpensive, have abundant conversion paths and high tunability and controllability, and have now shown very important applications in the fields of electrochemical energy storage and conversion. At the same time, halogenated polymer plastics are widely used, due to their excellent mechanical properties and resistance to acid and alkaline corrosion; however, their wastes are extremely difficult to decompose and treatments such as natural piling, burial, or heating can easily to produce toxic gases such as HCl and dioxins, thereby causing enormous damage to our environment. The newly established strategy based on polymer dehalogenation can manage the room-temperature, clean, and efficient treatment of halogenated organic polymer wastes (typically PVC) and enables the conversion to functional carbon materials. In particular, the derived highly reactive carbon intermediates produced during the dehalogenation process can also achieve multiple heterodoping of the resultant carbon materials under mild conditions and impart functionalization to the products. By using different types of dehalogenation agents, the modulation of pore structures and morphology of the dehalogenated carbon materials can be reached. By optimizing the doping sources, pore structure, and morphology of the dehalogenated carbon materials, a series of functional carbon materials that possess excellent performance for electrocatalysis and energy storage have been designed and successfully applied. The above researches regarding polymer dehalogenation have enriched the methods for the controlled synthesis of functional carbon materials and provided a variety of high-performance and inexpensive electrode materials for electrochemical catalysis and energy storage applications. More importantly, the dehalogenation strategy also provides feasible solutions to reducing the hazards caused by the huge amount of halogenated polymer wastes present in the environment. More researchers are expected to contribute to the relevant researches and applications of high-quality functional carbon materials obtained from halogenated polymers.

REFERENCES

1. J. Yu, L. Sun, C. Ma, Y. Qiao, H. Yao, Thermal degradation of PVC: A review, *Waste Manag.* 48 (2016) 300–314.
2. N. Lingaiah, M.A. Uddin, K. Morikawa, A. Muto, Y. Sakata, K. Murata, Catalytic dehydrochlorination of chloro-organic compounds from PVC containing waste plastics derived fuel oil over $FeCl_2/SiO_2$ catalyst, *Green Chem.* 3(2) (2001) 74–75.
3. Q. Cao, G. Yuan, L. Yin, D. Chen, P. He, H. Wang, Morphological characteristics of polyvinyl chloride (PVC) dechlorination during pyrolysis process: Influence of PVC content and heating rate, *Waste Manag.* 58 (2016) 241–249.
4. Z. Sun, Z. Yan, J. Yao, E. Beitler, Y. Zhu, J.M. Tour, Growth of graphene from solid carbon sources, *Nature* 468(7323) (2010) 549–552.
5. Y. Li, Y. Qian, H. Liao, Y. Ding, L. Yang, C. Xu, F. Li, G. Zhou, A reduction-pyrolysis-catalysis synthesis of diamond, *Science* 281(5374) (1998) 246–247.
6. P. Ruffieux, S. Wang, B. Yang, C. Sanchez-Sanchez, J. Liu, T. Dienel, L. Talirz, P. Shinde, C.A. Pignedoli, D. Passerone, T. Dumslaff, X. Feng, K. Mullen, R. Fasel, On-surface synthesis of graphene nanoribbons with zigzag edge topology, *Nature* 531(7595) (2016) 489–492.
7. H. Kong, S. Yang, H. Gao, A. Timmer, J.P. Hill, O. Diaz Arado, H. Monig, X. Huang, Q. Tang, Q. Ji, W. Liu, H. Fuchs, Substrate-mediated C-C and C-H coupling after dehalogenation, *J. Am. Chem. Soc.* 139(10) (2017) 3669–3675.
8. C. Morchutt, J. Björk, S. Krotzky, R. Gutzler, K. Kern, Covalent coupling via dehalogenation on Ni(111) supported boron nitride and graphene, *Chem. Commun.* 51(12) (2015) 2440–2443.
9. Y. Liu, W. Wang, A. Wang, Z. Jin, H. Zhao, Y. Yang, N-doped carbyne polysulfide as cathode material for lithium/sulfur batteries, *Electrochim. Acta* 232 (2017) 142–149.
10. Y. Zhang, Y. Peng, Y. Wang, J. Li, H. Li, J. Zeng, J. Wang, B.J. Hwang, J. Zhao, High sulfur-containing carbon polysulfide polymer as a novel cathode material for lithium-sulfur battery, *Sci. Rep.* 7(1) (2017) 11386.
11. B. Duan, W. Wang, A. Wang, K. Yuan, Z. Yu, H. Zhao, J. Qiu, Y. Yang, Carbyne polysulfide as a novel cathode material for lithium/sulfur batteries, *J. Mater. Chem. A* 1(42) (2013) 13261–13267.
12. S.M. Jung, E.K. Lee, M. Choi, D. Shin, I.Y. Jeon, J.M. Seo, H.Y. Jeong, N. Park, J.H. Oh, J.B. Baek, Direct solvothermal synthesis of B/N-doped graphene, *Angew. Chem. Int. Ed.* 53(9) (2014) 2398–2401.

13. G. Zhang, L. Wang, Y. Hao, X. Jin, Y. Xu, Y. Kuang, L. Dai, X. Sun, Unconventional carbon: Alkaline dehalogenation of polymers yields N-doped carbon electrode for high-performance capacitive energy storage, *Adv. Funct. Mater.* 26(19) (2016) 3340–3348.
14. X. Liu, L Dai, Carbon-based metal-free catalysts, *Nat. Rev. Mater.* 1(11) (2016) 16064.
15. G. Zhang, H. Luo, H. Li, L. Wang, B. Han, H. Zhang, Y. Li, Z. Chang, Y. Kuang, X. Sun, ZnO-promoted dechlorination for hierarchically nanoporous carbon as superior oxygen reduction electrocatalyst, *Nano Energy* 26 (2016) 241–247.
16. G. Zhang, X. Jin, H. Li, L. Wang, C. Hu, X. Sun, N-doped crumpled graphene: Bottom-up synthesis and its superior oxygen reduction performance, *Sci. China Mater.* 59(5) (2016) 337–347.
17. G. Zhang, J. Wang, B. Qin, X. Jin, L. Wang, Y. Li, X. Sun, Room-temperature rapid synthesis of metal-free doped carbon materials, *Carbon* 115 (2017) 28–33.
18. Q. Dang, X. Zhang, W. Liu, Y. Wang, G. Dou, H. Zhang, G. Zhang, Hierarchical porous N, S-codoped carbon material derived from halogenated polymer for battery applications, *Nano Select* 2(3) (2020) 581–590.
19. Y. Chang, G. Zhang, B. Han, H. Li, C. Hu, Y. Pang, Z. Chang, X. Sun, Polymer dehalogenation-enabled fast fabrication of N, S-codoped carbon materials for superior supercapacitor and deionization applications, *ACS Appl. Mater. Interfaces* 9(35) (2017) 29753–29759.
20. H. Cheng, X.· Chen, H. Yu, M. Guo, Y. Chang, G. Zhang, Hierarchically porous N, P-codoped carbon materials for high-performance supercapacitors, *ACS Appl. Energy Mater.* 3(10) (2020) 10080–10088.
21. Y. Chang, H. Shi, X. Yan, G. Zhang, L. Chen, A ternary B, N, P-Doped carbon material with suppressed water splitting activity for high-energy aqueous supercapacitors, *Carbon* 170 (2020) 127–136.
22. G. Dou, K. Du, Q. Dang, X. Chen, X. Zhang, M. Guo, Y. Wang, G. Zhang, Dehalogenated carbon-hosted cobalt-nitrogen complexes for high-performance electrochemical reduction of oxygen, *Carbon* 139 (2018) 725–731.
23. Y. Chang, Q. Dang, I. Samo, Y. Li, X. Li, G. Zhang, Z. Chang, Electrochemical heavy metal removal from water using PVC waste-derived N, S co-doped carbon materials, *RSC Adv.* 10(7) (2020) 4064–4070.
24. C. Hu, Y. Chang, R. Chen, J. Yang, T. Xie, Z. Chang, G. Zhang, W. Liu, X. Sun, Polyvinylchloride-derived N, S co-doped carbon as an efficient sulfur host for high-performance Li–S batteries, *RSC Adv.* 8(66) (2018) 37811–37816.
25. G. Zhang, K. Zhou, R. Xu, H. Chen, X. Ma, B. Zhang, Z. Chang, X. Sun, An alternative pathway to water soluble functionalized graphene from the defluorination of graphite fluoride, *Carbon* 96 (2016) 1022–1027.
26. W. Xu, Z. Lu, X. Sun, L. Jiang, X. Duan, Superwetting electrodes for gas-involving electrocatalysis, *Acc. Chem. Res.* 51(7) (2018) 1590–1598.
27. M. Guo, Y. Li, K. Du, C. Qiu, G. Dou, G. Zhang, Fabricating hierarchically porous carbon with well-defined open pores via polymer dehalogenation for high-performance supercapacitor, *Appl. Surf. Sci.* 440 (2018) 606–613.
28. C. Hu, G. Zhang, H. Li, C. Zhang, Y. Chang, Z. Chang, X. Sun, Thin sandwich graphene oxide@N-doped carbon composites for high-performance supercapacitors, *RSC Adv.* 7(36) (2017) 22071–22078.
29. C. Zhang, G. Zhang, H. Li, Y. Chang, Z. Chang, J. Liu, X. Sun, Interfacial dehalogenation-enabled hollow N-doped carbon network as bifunctional catalysts for rechargeable Zn-air battery, *Electrochim. Acta* 247 (2017) 1044–1051.
30. H. Li, G. Zhang, R. Zhang, H. Luo, L. Wang, C. Hu, I. Samo, Y. Pang, Z. Chang, X. Sun, Scalable fabrication of hierarchically porous N-doped carbon electrode materials for high-performance aqueous symmetric supercapacitor, *J. Mater. Sci.* 53(7) (2017) 5194–5203.

31. L.X. Cheng, Y.Q. Zhu, X.Y. Chen, Z.J. Zhang, Polyvinylidene fluoride-based carbon supercapacitors: Notable capacitive improvement of nanoporous carbon by the redox additive electrolyte of 4-(4-nitrophenylazo)-1-naphthol, *Ind. Eng. Chem. Res.* 54(41) (2015) 9948–9955.
32. L. Wang, G. Zhang, B. Han, Y. Chang, H. Li, J. Wang, C. Hu, Z. Chang, Z. Huo, X. Sun, A two-volt aqueous supercapacitor from porous dehalogenated carbon, *J. Mater. Chem. A* 5(14) (2017) 6734–6739.
33. H. Zhang, L. Zhang, J. Chen, H. Su, F. Liu, W. Yang, One-step synthesis of hierarchically porous carbons for high-performance electric double layer supercapacitors, *J. Power Sources* 315 (2016) 120–126.
34. K. Liu, M. Qian, L. Fan, S. Zhang, Y. Zeng, F. Huang, Dehalogenation on the surface of nano-templates: A rational route to tailor halogenated polymer-derived soft carbon, *Carbon* 159 (2020) 221–228.
35. Y. Chang, Y. Pang, Q. Dang, A. Kumar, G. Zhang, Z. Chang, X. Sun, Converting polyvinyl chloride plastic wastes to carbonaceous materials via room-temperature dehalogenation for high-performance supercapacitor, *ACS Appl. Energy Mater.* 1 (2018) 5685–5693.
36. F.-G. Zhao, Y.-T. Kong, B. Pan, C.-M. Hu, B. Zuo, X. Dong, B. Li, W.-S. Li, In situ tunable pillaring of compact and high-density graphite fluoride with pseudocapacitive diamines for supercapacitors with combined predominance in gravimetric and volumetric performances, *J. Mater. Chem. A* 7(7) (2019) 3353–3365.
37. Y. Sang, L. Bai, B. Zuo, L. Dong, X. Wang, W.-S. Li, F.-G. Zhao, Transfunctionalization of graphite fluoride engineered polyaniline grafting to graphene for high–performance flexible supercapacitors, *J. Colloid Interface Sci.* 597 (2021) 289–296.

30 Waste Mechanical Energy Harvesting from Vehicles by Smart Materials

Ömer Faruk Ünsal and Ayşe Çelik Bedeloğlu
Bursa Technical University

CONTENTS

30.1 INTRODUCTION

The industrial revolution, which started in the 1700s and evolved step by step until today, has forever paved the way for the use of more efficient fossil energy resources (coal, oil, and natural gas), which allows production to pass from body power to machine power and to technological, economic, and social progress. Fossil fuels, which still have a dominant role in meeting the energy needs throughout the world, also harm the health of living things with the air pollution they create and become the biggest trigger of the global temperature increase, due to the harmful greenhouse gases, especially the CO_2 they generate when they are burned. According to the report prepared by BP [1], oil consumption, which is the raw material of vehicle fuel, has increased by an average of 0.9% per day (0.9 million barrels (b/day)) and natural gas consumption has also increased by 2% (78 billion cubic meters (bcm)).

Due to the gradual depletion of fossil fuels and their irreversible damage to the environment, researchers have turned to renewable energy sources and developing production technologies. Hydropower, solar, wind, geothermal, bioenergy, waves, and tides are the main renewable energy sources. According to BP's report [1], the share of renewable energy in electricity generation increased from 9.3% to 10.4% in 2019, surpassing nuclear for the first time. The production shares of coal decreased by 1.5 points to 36.4%. Wind was the largest contributor to renewable energy growth

DOI: 10.1201/9781003178354-34

(1.4 EJ), followed closely by solar (1.2 EJ). Despite this, carbon emissions from energy use increased by 0.5% worldwide in 2019, resulting in an average increase of 1.1% over the last 10 years. While one-third of oil consumption is used by industry and more than half is used in the transport sector (such as rail, shipping, air, and road transport), the remaining one-fifth is currently used by 900 million cars worldwide (about 19 Mb/d). On the other hand, it is assumed that every tonne of oil consumed produces three tonnes of CO_2 [2]. CO_2 emissions from vehicles can be reduced by making vehicles more efficient or by changing the fuel used. Therefore, electric vehicles have begun to attract more and more consumers in recent years, thanks to their increased range, battery life, efficiency, and affordability. For example, the total number of electric cars is projected to reach around 70 m in 2035, accounting for just less than one-tenth of the total increase in global car numbers [2]. Although an electric car is less environmentally friendly than a car with an internal combustion engine when considering the production and disposal of it, the level of emissions from electric vehicles varies depending on how electricity is produced and can become more environmentally friendly with the of renewable or hybrid power generation technologies [3]. Therefore, the use of waste mechanical energy by integrating electricity production systems into vehicles can be a promising way for environmentally friendly energy generation.

In the last 20 years, an improvement on consumer electronics and electrical vehicles caused the intention of the researches on energy conversion systems in mechanical, electrical, and materials sciences. Due to that, many innovative energy conversion systems, which were made by smart materials, have been presented in materials science researches. Smart materials are the materials that sense, respond, and/or adapt themselves to an external stimulus. If the material changes the geometrical or material properties by an external stimulus, it is classified as an active smart material [4]. The external stimulus can be light, electrical field, or mechanical stress. In an active smart mechanism, there is an energy transducing occurrence. If the material cannot transduce the energy, by the way, if the material only can sense the stimulus, the material can be classified as a passive material [5]. Shape memory materials, piezoelectric materials, magnetostrictive materials, thermochromics, and self-healing materials are the most known examples for active smart materials due to their shape or material property changing ability under magnetic, electrical, thermal, or mechanical stimulus [5]. On the other hand, fiber optic cables are defined as passive–active materials, because they can just sense and conduct the photonic energy without any reaction. Smart materials also have relaxation capability as soon as excitation ability.

Waste mechanical energy conversion is one of the most growing areas for materials science since the discovery of the nanogenerator concept. The energy sources required for the generation of electrical energy from waste mechanical energy can be a vibrating or moving structure, device, object, or air, or water. Previously, too many studies were conducted by researchers on waste energy conversion mechanisms based on Faraday's electromagnetic induction law [6]. In this mechanism, the magnetic field of a magnet is captured by a conductive cage or coil when the magnet approaches the cage or coil [7]. However, the electromagnetic energy conversion mechanism is a heavyweight and mechanical system; on the contrary, others use clever materials that promise cheapness, lightness, and comfort. With the advent of piezoelectric and triboelectric nanogenerators, waste mechanical energy conversion studies have focused on the concept of nanogenerators.

Nanogenerators are devices that consist of an active layer and electrodes mounted on the active layer. The material type of the active layer determines the working principle of the nanogenerator. The use of piezoelectric material or dielectric material in the active layer is the most common mechanical energy conversion mechanism for nanogenerators; they are also called piezoelectric nanogenerators or triboelectric nanogenerators, respectively. With smart materials, energy harvesting from vehicles can be realized with piezoelectric, triboelectric, photovoltaic, electromagnetic, and pyroelectric materials [8–10]. While some of these materials (piezoelectric and triboelectric) harvest electrical energy from waste mechanical energy, others (pyroelectric, thermoelectric, and photovoltaic) convert radiative energy (light, heat, etc.) to electricity. Vibration energy, tire pressure on the ground, and speed-dependent wind energy can be potential sources of waste mechanical energy from a vehicle [11–13]. On the other hand, waste thermal energy in the engine, exhaust system, oil pan, or brake disks could be used for energy harvesting [8,14–16]. Moreover, these smart systems have the potential to be used as sensors for safety and condition monitoring systems [17]. However, this section is based on the studies of waste mechanical energy from vehicles.

Although the issue of generating electrical energy from waste mechanical energy is very promising, it has not been among the other major renewable energy generation technologies due to its very few applications and not being scalable. In addition to large energy generation resources, due to the advancement of today's technology, there is a need for efficient and durable electrical energy generating systems that require minimal or no maintenance, especially for small-sized portable and wearable electronics and wireless communication networks. For this reason, technologies that produce electrical energy from waste mechanical energy, especially by vibration, are very suitable to be used in devices or systems that do not require much power to operate; they communicate with each other, thanks to various communication protocols, and have created a smart network by connecting and sharing information. Due to the increasing demand for global vibration energy harvesting systems, it is expected to reach approximately $ 253 million in 2024 [18]. The development of devices to power small-scale devices such as sensors, wireless communication networks, and portable and wearable electronics is of great interest. However, there are still some challenges to overcome, such as complex structure, flexibility, lightness, portability, autonomous operation, and low cost in order for these power generation devices to be used widely and for a long time.

30.2 PIEZOELECTRIC AND TRIBOELECTRIC EFFECTS

Generally, the most abundant type of waste energy that can be used in the environment or cities is mechanical and thermal energies [19]. For this reason, much research on the recovery of waste energy is published by scientists every year. The most common sources of waste mechanical energy are human movements, vehicles, and natural sources (wind, avalanche, etc.). Particularly, vehicles can also produce waste thermal energy, too, with thermoelectrics and pyroelectrics [17]. Piezoelectric and triboelectric effects are able to convert waste energy to a usable energy form; these devices are named nanogenerators. These two types of energy harvesting principles are recovering waste energy according to Maxwell's displacement current principle [19]. The displacement current can be basically defined as the time-dependent charge density in a dielectric material [20]. Increasing charge density causes a current to occur; if the material is connected to a circuit, the current is observed [21]. Different mechanisms reveal charge density in different regions of active layers. In the piezoelectric effect, the absence of a symmetric central atom causes electron density difference when mechanical force is applied. On the other hand, in triboelectric nanogenerators, the driving force to form electrically polarized regions comes from the dissimilar dielectric or electron affinity character of the active layer materials.

30.2.1 Piezoelectric Effect

Piezoelectric effect can be defined as the occurrence of electrical potential under mechanical stress (Figure 30.1) such as bending, compression, and vibration. When a piezoelectric material is exposed to mechanical stress, electronegative and electropositive atoms create the partial electrical polarization on different surfaces of the material and this polarization causes the electrical energy generation [22]. A piezoelectric material can also change its geometry under an electric field of sufficient magnitude. The absence of an atomic center of symmetry in piezoelectric materials reveals atomic/molecular polarization [23]. The atoms that make up the material, the degree of polarization, the applied force, the nature of the electrode integration, and the surface area of the piezoelectric material are the determining factors for the piezoelectric energy harvesting efficiency. Piezoelectric energy conversion systems can be designed with different configurations: d_{31} and d_{33} modes. In d_{31} mode, electrodes and the direction of applied force are parallel to the polarization direction. On the contrary, in d_{33} design, the directions of applied force and polarization are perpendicular [24]. This design preference is determined by using the area of nanogenerators such as human body motion, vehicles, and wind energy conversion.

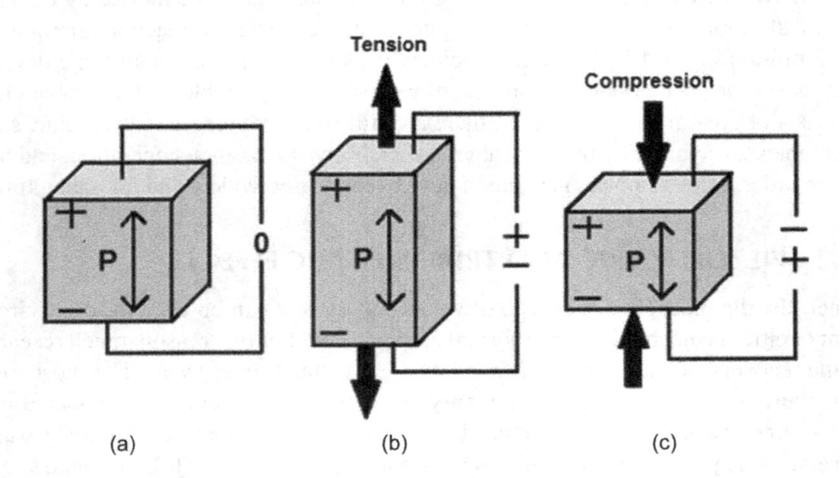

FIGURE 30.1 Schematic illustration of piezoelectric effect. (Adapted with permission from Ref. [25]. Copyright (2020) Elsevier.)

30.2.2 Triboelectric Effect

The triboelectric effect is based on static electricity. Electric current is observed when two different dielectric materials come into contact [24]. The electrons of the surface atoms of dielectric materials can leave their orbits when the materials touch each other and move to the surface atom orbit of the other material. Since this movement will disrupt the electronic balance, electric current is observed from the electron acceptor material to the electron donor material when the materials in contact are separated by completing the circuit [26]. In addition, the dielectric layers of triboelectric systems that respond to generating static electricity can come into contact with four different types (Figure 30.2). Even these four types of contact mechanisms for triboelectric nanogenerators involve sliding or impacting motions; basically, all these modes are based on the contact and separation of dielectric materials. However, especially during sliding movements, apart from electrical energy, frictional heat energy is produced.

Contact–separation and single-electrode (Figure 30.2a–c) designs are based on basic contact and separation movements. The contact–separation mode operates according to the charge separation by the contact of the two dielectric layers. In the single-electrode mode, only one electrode material and one dielectric material are used, and the electrode material is used as the tribo-positive layer. The sliding and independent modes (Figure 30.2b and d) operate with frictional action. The dielectric materials in triboelectric nanogenerators are key to the performance of the nanogenerator. The charge difference between dielectric materials determines the magnitude of the charge separation in contact. Therefore, the materials used in triboelectric nanogenerators are classified as tribo-positive and tribo-negative materials. Materials containing halogen, oxygen, or nitrogen atoms are classified as tribo-negative materials. It can also be said that with increasing electrical conductivity, materials acquire tribo-positive character.

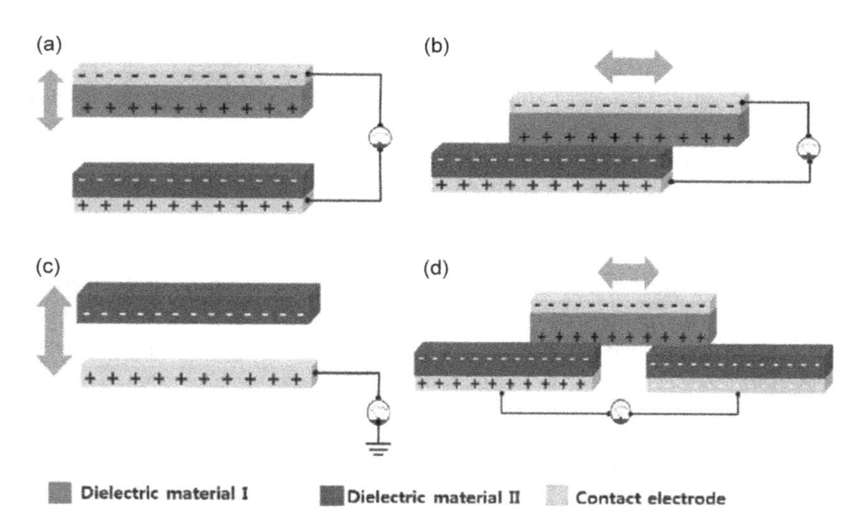

FIGURE 30.2 Contact modes of triboelectric systems. (Adapted with permission from Ref. [27]. Copyright (2019) Elsevier.)

30.3 PIEZOELECTRIC APPLICATIONS

Piezoelectric energy harvesting is a method that works with the pressure-based deformation of the material. Meanwhile, in order to generate electrical energy from piezoelectric material, a mechanical stimulus must be present. The main motivation for waste mechanical energy is the presence of many sources of waste mechanical energy on a vehicle. Engine vibrations, airflow on exterior parts, pressure on seats based on occupant weight, and pressure on tires due to total mass are the most common examples of a vehicle's waste energy sources [28].

Since the piezoelectric effect is a surface activity principle, material size and surface area are the most important parameters for energy harvesting studies [29]. The other critical point is the molecular structure that provides hermetic piezoelectricity. Piezoelectric materials can be classified according to their origin: organic and inorganic piezoelectric materials. Würtzite crystals (ZnO, ZnS, GaN, etc.) and perovskites (PZT, $BaTiO_3$, etc.) are the most known piezoelectric inorganics [25]. However, due to the fragility of inorganic piezoelectric materials and the difficulty of applying them, the need for alternative piezoelectric materials has emerged [30]. Polyvinylidene fluoride and its copolymers have become widely used in energy applications of piezoelectric materials due to their both relatively flexible and piezoelectric characteristics [31]. The lack of centrosymmetry for würtzite and perovskite crystals produces atomic-electrical polarization under pressure. The pressure must be of a size that allows the crystal structure to deform [24]. The deformability of piezoelectric crystals provides them with a low stimulus threshold and high energy conversion efficiency and makes them suitable for sensor applications [32].

Its flexibility and ability to be processed with easy polymer processing methods make PVDF polymer a great candidate for piezoelectric energy conversion to be applied to the automotive industry. Different copolymers of vinylidene fluoride monomer (for example, trifluoroethylene and hexafluoropropylene) are also used for piezoelectric energy applications. The sensing and energy recovery feature of PVDF and its copolymers can be used in vehicles [33]. In the literature, using the vibration energy of the engine system of an automobile, the speed value of the engine could be characterized [34]. In automobiles, tires are preferred for energy harvesting and sensing applications. The friction that occurs in the rotating car wheels while traveling is a great loss of energy under normal conditions, and rolling resistance uses 5%–7% of the total energy [35]. While the vehicle is in motion, the abundant vibration and pressure energy density in a tire and the high frequency of rotational motion can be collected from the tires as waste energy [36]. In particular, flexible piezoelectric materials (inorganic piezoelectric-doped polymers or directly organic piezoelectric materials) can be used for energy scavenging or wireless sensor applications from tires (Figure 30.3) [11].

The mechanism of energy harvesting of piezoelectric materials from vehicles also depends on moving properties. For example, car tires are subject to high-frequency pressure force, unlike bicycles. Here, the mechanical excitation of piezoelectric materials is divided into two mechanisms: to hit and to vibrate (or press) [37]. The vibration mechanism is suitable for high-frequency applications, while the hitting mechanism is more efficient for low-frequency vehicle movements [37]. Waste energy

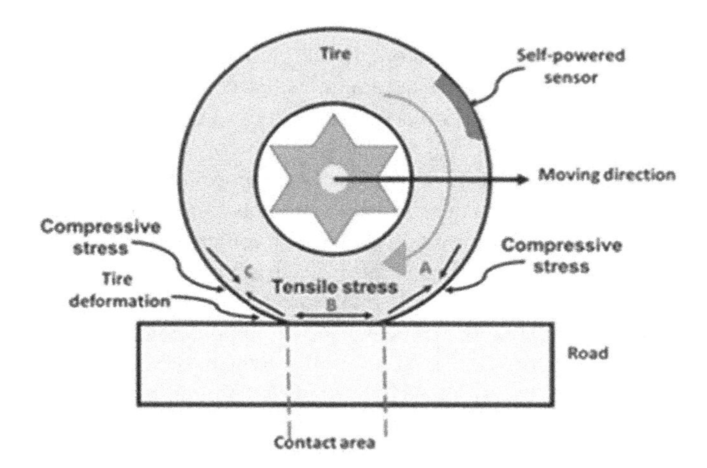

FIGURE 30.3 Self-powered piezoelectric tire sensor. (Adapted with permission from Ref. [11]. Copyright (2018) Elsevier.)

collection from vehicles also includes road applications. Highways provide piezoelectric energy generators with high frequency–high mechanical power density. In terms of highway applications, piezoelectric materials must meet the requirements of "high frequency–high mechanical strength" effects [28]. For this purpose, inorganic piezoelectric materials and/or rigid composites of organic piezoelectric materials are used for highway waste energy collection applications [38].

30.4 TRIBOELECTRIC APPLICATIONS

Triboelectric nanogenerators, as mentioned above, are devices that convert waste mechanical energy into electrical energy with the static electrification mechanism. Professor Zhong Lin Wang demonstrated in 2017 that triboelectric nanogenerators work in accordance with Maxwell's displacement current mechanism [19]. When the surfaces of two different materials touch each other, there will be a transfer of electrons from the partially positively charged one to the other. Polytetrafluoroethylene and silicone polymers are commonly used as negatively charged triboelectric layers, while polyamide metals and polyamides are negatively charged layers [39]. Also, materials with high dielectric properties (such as high dielectric constant and low tangential loss) show better triboelectric energy conversion for nanogenerators [40].

The most important feature in performance of triboelectric nanogenerators, after the atomic configuration, which determines the electrostatic charge properties, is the triboelectrification area [41]. The triboelectrification area can be determined as the contact surface area of the triboelectric layers for nanogenerators [19]. Different fabrication methods such as stereolithography, nanomodeling, or sanding can be used to improve the contact surface area. Today, many sensors are used for different purposes such as driving safety in cars, conditional monitoring, and comfortable driving. This large number of sensors consumes a significant amount of energy from the car battery for hybrid and electric cars [42]. Conversion of sensors to self-powered systems will enable cars to use batteries more efficiently, and one of the alternatives for this conversion is triboelectric nanogenerators. In addition, nanogenerator systems can be used as self-charging power plants in automobiles. By combining different triboelectric mechanisms, devices that reach very small values in the "charge time/discharge time" ratio can be produced.

As a reason for the design of most triboelectric nanogenerators, inefficient energy conversion and handicaps due to the gap between the two triboelectric layers indicate that triboelectric nanogenerators are not suitable for vehicle applications [43]. Despite these integrated triboelectric nanogenerators as traditional automotive parts such as cord fabric and seat sponges, it will make the triboelectric nanogenerator more compatible with vehicles [13]. In triboelectric tire applications, besides the charge density, mechanical properties are also very important in the selection of the triboelectric layer. It requires the use of flexible materials due to excessive bending and compression forces [43]. Brake disks of a car can be cited as one of the most suitable components for integrating energy generators with tires. Frictional forces from tight contact are the main source of waste energy for nanogenerators in an automobile [44]. However, excessive frictional forces will shorten the life of the nanogenerator due to the heat and mechanical stress produced. Therefore, the lateral shift mode triboelectric nanogenerator is not suitable for this application. Brake-integrated triboelectric nanogenerators can also be used as sensors and energy harvesters [44,45]. However, in both applications, the electrical energy produced must be regulated by electronics in order to use and store it. This requirement is due to the instability of the highway, which affects the amount of energy production [45]. Triboelectric nanogenerators can also use inertial force (Figure 30.4) when placed into a matter which has a velocity

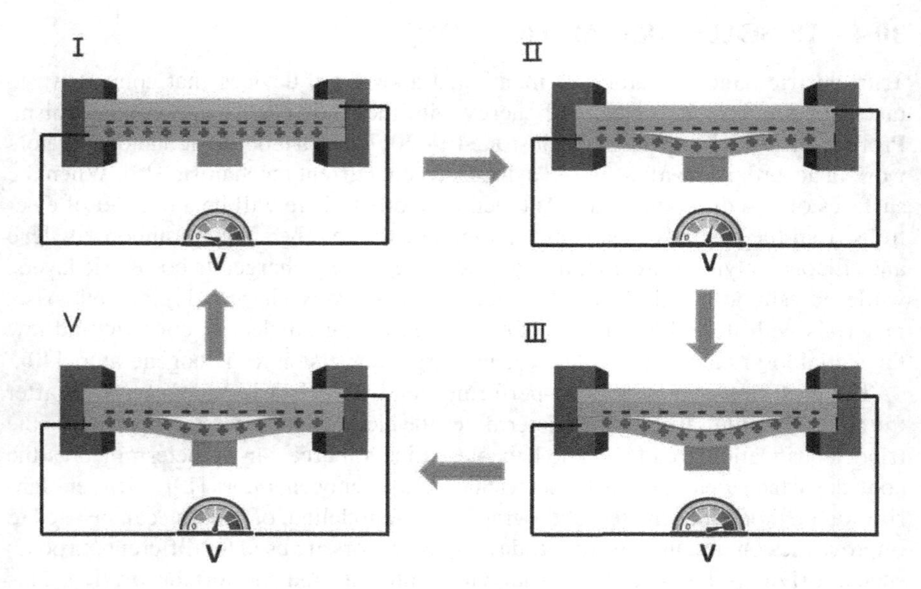

FIGURE 30.4 Triboelectric nanogenerator-based acceleration sensor for automobiles. (Adapted with permission from Ref. [12]. Copyright (2018) Elsevier.)

[46]. The contact–separation movements of the two triboelectric layers with different magnitudes of inertia forces will cause the output energy difference to occur. Meanwhile, the acceleration and current velocity can be detected by this type of triboelectric nanogenerators. Acceleration sensors can operate on piezoelectric or piezoresistive principles. However, this type of acceleration sensors requires an external power supply [12]. A triboelectric nanogenerator using the g-force of a car can clearly detect acceleration and velocity with proper material selection and processing [46]. Moreover, the combination of equal triboelectric nanogenerator-based acceleration sensors placed on different axes can create a multi-axis acceleration sensor system.

30.5 HYBRID APPLICATIONS

Energy harvester systems also involve more discipline to achieve higher efficiency from combined devices. Hybrid energy harvesters combine piezoelectric, pyroelectric, photovoltaic, etc., materials. These are energy collectors created by combining different types of combines such as harvester's principles all have different disadvantages, such as applicability or efficiency. By combining different types of generators, they eliminate the disadvantages of single-mechanism generators [25]. For example, the thermal energy produced in friction-based triboelectric nanogenerators mentioned in the "triboelectric effect" section can be recovered with thermoelectric or pyroelectric components [47].

Piezoelectric and triboelectric effects can also complement each other in terms of disadvantages or advantages. While triboelectric nanogenerators have a high voltage output with low current output, piezoelectric nanogenerators show the opposite of triboelectric nanogenerators [48]. The hybridization of piezoelectric and triboelectric nanogenerators significantly increases the output current and voltage of the device compared to single piezoelectric or single triboelectric types. In addition, piezoelectric materials have a relatively low excitation threshold to generate electrical signals [32]. Meanwhile, the use of piezoelectric materials in triboelectric systems allows them to be used as sensor systems.

Hybrid mechanical energy harvesters for automobiles can come in various combinations other than piezoelectric–triboelectric hybridization. The hybrid generator can be combined with a mechanical energy harvester and a solar or thermal harvester. Vehicles are exposed to various conditions such as sunlight and extreme cold or hot weather. For example, a triboelectric generator manufactured for the purpose of collecting blade energy in the vehicle may be combined with a thermoelectric or pyroelectric generator. Alternatively, a thermoelectric or pyroelectric material can be used as the tribo-positive or tribo-negative material for the hybrid nanogenerator. As mentioned above, the use of triboelectric or piezoelectric nanogenerators is a trend for energy harvesting tire applications. Tires are also exposed to a lot of heat energy due to the heating of the rubber material while the vehicle is in motion, and the heat energy on the asphalt can be absorbed. Therefore, triboelectric or piezoelectric tires are well suited to be combined with smart materials that harvest thermal energy. Solar cells also have the potential to be used in vehicles with mechanical energy harvesting smart materials. Also, multi-link hybrid generators for energy harvesting are possible. It is possible to use triboelectric and piezoelectric energy harvesters in order to catch waste mechanical energy as mentioned. Furthermore, thermoelectric or pyroelectric materials can be used to convert the produced heat energy, with tribo-piezo hybrid energy harvesters because of the high heat energy production on tires during vehicle movement [47]. Photovoltaics also can be used with these materials, according to the region of use of the energy harvester on a vehicle, as a third smart material component of the hybridized energy harvester. Additionally, electromagnetic induction systems can maintain the mechanical energy conversion efficiency of hybrid systems [48].

On the basis of energy conversion, nanogenerators also can be used as sensors as mentioned. Furthermore, a hybrid harvester system can include different sensors with different energy conversion mechanisms. For example, one component of a hybrid nanogenerator can work as a condition sensor (pH, humidity, etc.), while the other works as a movement or acceleration sensor [49]. In addition to this, a hybrid nanogenerator system can consist of a power source and a nanogenerator-based sensor.

Other approaches used to convert vehicle-related mechanical energy into electrical energy include energy harvesting in vehicle traffic over speed bumps [49,50] and with road coatings and pavements [51–53]. In recent years, the efficiency, cost, and system performance of various hybrid energy-collecting prototypes produced on the roads, especially the energy used by vehicles to slow down in areas where speed is limited, and the energy collected by the pavements or coatings on the road, have been evaluated. However, their power conversion efficiency and economic applicability remain limited.

30.6 CONCLUSIONS AND FUTURE PROSPECTS

There is a great need for renewable and alternative energy technologies due to the rapidly increasing use of fossil fuels, which brings global warming to the point of no return, and the ever-increasing global energy demand of the increasing population. In recent years, there has been a significant increase in the number of studies carried out to produce electrical energy from mechanical energy in a renewable and environmentally friendly way. The collection of waste mechanical energy, which is abundant and readily available in our daily life, is one of the effective and promising ways of energy production. The automotive industry is one of the largest industries in the world. The developing technology offers more comfortable, safer, and environmentally friendly vehicles through the automotive sector. The emergence of electric vehicles and various types of sensors used in all kinds of land vehicles has increased the demand for electric energy in automobiles. Therefore, triboelectric and piezoelectric energy harvesters are the two biggest candidates for harvesting automotive-derived waste mechanical energy. Auxiliary sectors such as textile, rubber, semiconductor, or polymer industries are focusing on these technologies in parallel with the increasing energy demand.

Besides, for electric vehicles, it could be a promising solution to harvest the vehicle's wasted energy and use that wasted energy to increase the electric vehicle's battery capacity and range. Further, in recent years, the widely available, low-cost, and flexible polymers have been used to collect the energy produced by mechanical action with scalable low-cost production techniques. In addition, nanotechnology is used in material design, production, functionalization, and analysis to increase the efficiency of systems supported by modeling and simulations.

REFERENCES

1. BP, *Statistical Review of World Energy* 66 (2020).
2. S. Dale and T. D. Smith *Back to the future: electric vehicles and oil demand* (2016).
3. F. Knobloch, S. V. Hanssen, A. Lam, H. Pollitt, P. Salas, U. Chewpreecha, M. A. J. Huijbregts, and J. F. Mercure, *Nature Sustainability* **3**, 437 (2020).
4. C. Dry, *Journal of Intelligent Material Systems and Structures* **4**, 420 (1993).
5. S. Kamila, *American Journal of Applied Sciences* **10**, 876 (2013).
6. A. Kumar, S. S. Balpande, and S. C. Anjankar, in *Procedia Computer Science* (Elsevier, Netherlands, 2016), pp. 785–792.
7. B. E. Lewandowski, D. Advisor, and K. J. Gustafson, An implantable, stimulated muscle powered piezoelectric generator (2009).
8. Y. Tabbai, A. Alaoui-Belghiti, R. El Moznine, F. Belhora, A. Hajjaji, and A. El Ballouti, *International Journal of Precision Engineering and Manufacturing - Green Technology* **8**, 487 (2021).
9. L. Eldada, *Journal of Nanophotonics* **5**, 051704 (2011).
10. J. Pei, F. Guo, J. Zhang, B. Zhou, Y. Bi, and R. Li, *Journal of Cleaner Production* **288**, 125338 (2021).
11. D. Maurya, P. Kumar, S. Khaleghian, R. Sriramdas, M. G. Kang, R. A. Kishore, V. Kumar, H. C. Song, J. M. (Jerry) Park, S. Taheri, and S. Priya, *Applied Energy* **232**, 312 (2018).
12. K. Dai, X. Wang, F. Yi, C. Jiang, R. Li, and Z. You, *Nano Energy* **45**, 84 (2018).
13. W. Seung, H. J. Yoon, T. Y. Kim, M. Kang, J. Kim, H. Kim, S. M. Kim, and S. W. Kim, *Advanced Functional Materials* **30**, 1 (2020).
14. P. Fernández-Yáñez, A. Gómez, R. García-Contreras, and O. Armas, *Journal of Cleaner Production* **182**, 1070 (2018).
15. Y. Choi, A. Negash, and T. Y. Kim, *Energy Conversion and Management* **197**, 111902 (2019).
16. M. Aljaghtham and E. Celik, *Energy* **200**, 117547 (2020).
17. H. Askari, E. Hashemi, A. Khajepour, M. B. Khamesee, and Z. L. Wang, *Nano Energy* **53**, 1003 (2018).
18. Global vibration energy harvesting systems market by product (non-linear systems, rotational systems, and linear systems), by application (transportation, power generation, industrial, building and home automation and others), competition, forecast and opportunities by region (2024). https://www.researchandmarkets.com/reports/4833522/global-vibration-energy-harvesting-systems-market.
19. Z. L. Wang, *Materials Today* **20**, 74 (2017).
20. X. Cao, M. Zhang, J. Huang, T. Jiang, J. Zou, N. Wang, and Z. L. Wang, *Advanced Materials* **30**, 1704077 (2018).
21. W. Wu, T. Yang, Y. Zhang, F. Wang, Q. Nie, Y. Ma, X. Cao, Z. L. Wang, N. Wang, and L. Zhang, *ACS Nano* **13**, 8202 (2019).
22. Z. L. Wang and J. Song, *Science* **312**, 242 (2006).
23. Ö. F. Ünsal, Y. Altın, and A. Çelik Bedeloğlu, *Journal of Applied Polymer Science* **137**, 48517 (2020).
24. Ö. Faruk Ünsal and A. Çelik Bedeloğlu, *Material Science Research India* **15**, 114–130 (2018).
25. Ö. F. Ünsal, A. Sezer Hiçyilmaz, A. N. Yüksel Yilmaz, Y. Altin, İ. Borazan, and A. Çelik Bedeloğlu, Energy-Generating Textiles (2020).
26. F. R. Fan, Z. Q. Tian, and Z. Lin Wang, *Nano Energy* **1**, 328 (2012).
27. S. Chandrasekaran, C. Bowen, J. Roscow, Y. Zhang, D. K. Dang, E. J. Kim, R. D. K. Misra, L. Deng, J. S. Chung, and S. H. Hur, *Physics Reports* **792**, 1 (2019).
28. Y. Liu, S. Du, C. Micallef, Y. Jia, Y. Shi, and D. J. Hughes, *Energies* **13**, 1 (2020).

29. J. Li, C. Zhao, K. Xia, X. Liu, D. Li, and J. Han, *Applied Surface Science* **463**, 626 (2019).
30. D. B. Deutz, J. A. Pascoe, B. Schelen, S. Van Der Zwaag, D. M. De Leeuw, and P. Groen, *Materials Horizons* **5**, 444 (2018).
31. W. Zhai, Q. Lai, L. Chen, L. Zhu, and Z. L. Wang, *ACS Applied Electronic Materials* **2**, 2369 (2020).
32. L. Kong, T. Li, H. Hng, F. Boey, T. Zhang, S. Li. *Waste energy harvesting: Mechanical and thermal energies*, Springer Science & Business Media (2014).
33. L. Jin, S. Ma, W. Deng, C. Yan, T. Yang, X. Chu, G. Tian, D. Xiong, J. Lu, and W. Yang, *Nano Energy* **50**, 632 (2018).
34. A. A. Khan, M. M. Rana, G. Huang, N. Mei, R. Saritas, B. Wen, S. Zhang, P. Voss, E. A. Rahman, Z. Leonenko, S. Islam, and D. Ban, *Journal of Materials Chemistry A* **8**, 13619 (2020).
35. S. Bhamre, S. Mali, and C. Mane *E3S Web of Conferences* **170**, 01027 (2020).
36. M. A. A. Abdelkareem, L. Xu, M. K. A. Ali, A. Elagouz, J. Mi, S. Guo, Y. Liu, and L. Zuo, *Applied Energy* **229**, 672 (2018).
37. C. Jettanasen, P. Songsukthawan, and A. Ngaopitakkul, *Sustainability (Switzerland)* **12**, 2933-2949 (2020).
38. P. Cahill, N. A. N. Nuallain, N. Jackson, A. Mathewson, R. Karoumi, and V. Pakrashi, *Journal of Bridge Engineering* **19**, 04014034 (2014).
39. C. Wu, A. C. Wang, W. Ding, H. Guo, and Z. L. Wang, *Advanced Energy Materials* **9**, 1 (2019).
40. Z. Fang, K. H. Chan, X. Lu, C. F. Tan, and W. Ho, *Journal of Materials Chemistry A Communication* **6**, 52–57 (2018).
41. H. J. Yoon, D. H. Kim, W. Seung, U. Khan, T. Y. Kim, T. Kim, and S. W. Kim, *Nano Energy* **63**, 103857 (2019).
42. P. Maharjan, T. Bhatta, H. Cho, X. Hui, C. Park, S. Yoon, M. Salauddin, M. T. Rahman, S. S. Rana, and J. Y. Park, *Advanced Energy Materials* **10**, 1 (2020).
43. H. Askari, Z. Saadatnia, A. Khajepour, M. B. Khamesee, and J. Zu, *Advanced Engineering Materials* **19**, 1 (2017).
44. C. Bao Han, W. Du, C. Zhang, W. Tang, L. Zhang, and Z. Lin Wang, *Nano Energy* **6**, 59 (2014).
45. T. Guo, J. Zhao, W. Liu, G. Liu, Y. Pang, T. Bu, F. Xi, C. Zhang, and X. Li, *Advanced Materials Technologies* **3**, 1 (2018).
46. Y. K. Pang, X. H. Li, M. X. Chen, C. B. Han, C. Zhang, and Z. L. Wang, *ACS Applied Materials and Interfaces* **7**, 19076 (2015).
47. Y. Wu, S. Kuang, H. Li, H. Wang, R. Yang, Y. Zhai, G. Zhu, and Z. L. Wang, *Advanced Materials Technologies* **3**, 1800166 (2018).
48. L. Bu, Z. Chen, Z. Chen, L. Qin, F. Yang, K. Xu, J. Han, and X. Wang, *Nano Energy* **70**, 104500 (2020).
49. G. Del Castillo-García, E. Blanco-Fernandez, P. Pascual-Muñoz, and D. Castro-Fresno, *Proceedings of Institution of Civil Engineers: Energy* **171**, 58 (2018).
50. P. Todaria, L. Wang, A. Pandey, J. O'connor, D. Mcavoy, T. Harrigan, B. Chernow, and L. Zuo, Design, Modeling and Test of a Novel Speed Bump Energy Harvester, in *Sensors and Smart Structures Technologies for Civil, Mechanical, and Aerospace Systems*, SPIE, U.S.A., 9435, (2015).
51. F. Duarte, A. Ferreira, and P. Fael, *Proceedings of Institution of Civil Engineers: Energy* **171**, 70 (2018).
52. F. Duarte and A. Ferreira, *Proceedings of Institution of Civil Engineers: Energy* **169**, 79 (2016).
53. F. Duarte, J. P. Champalimaud, and A. Ferreira, *Proceedings of the Institution of Civil Engineers: Municipal Engineer* **169**, 13 (2016).

Index